T0235044

Practical Design and Production of Optical Thin Films

OPTICAL ENGINEERING

1. Electron and Ion Microscopy and Microanalysis: Principles and Applications, *Lawrence E. Murr*
2. Acousto-Optic Signal Processing: Theory and Implementation, *edited by Norman J. Berg and John N. Lee*
3. Electro-Optic and Acousto-Optic Scanning and Deflection, *Milton Gottlieb, Clive L. M. Ireland, and John Martin Ley*
4. Single-Mode Fiber Optics: Principles and Applications, *Luc B. Jeunhomme*
5. Pulse Code Formats for Fiber Optical Data Communication: Basic Principles and Applications, *David J. Morris*
6. Optical Materials: An Introduction to Selection and Application, *Solomon Musikant*
7. Infrared Methods for Gaseous Measurements: Theory and Practice, *edited by Joda Wormhoudt*
8. Laser Beam Scanning: Opto-Mechanical Devices, Systems, and Data Storage Optics, *edited by Gerald F. Marshall*
9. Opto-Mechanical Systems Design, *Paul R. Yoder, Jr.*
10. Optical Fiber Splices and Connectors: Theory and Methods, *Calvin M. Miller with Stephen C. Mettler and Ian A. White*
11. Laser Spectroscopy and Its Applications, *edited by Leon J. Radziemski, Richard W. Solarz, and Jeffrey A. Paisner*
12. Infrared Optoelectronics: Devices and Applications, *William Nunley and J. Scott Bechtel*
13. Integrated Optical Circuits and Components: Design and Applications, *edited by Lynn D. Hutcheson*
14. Handbook of Molecular Lasers, *edited by Peter K. Cheo*
15. Handbook of Optical Fibers and Cables, *Hiroshi Murata*
16. Acousto-Optics, *Adrian Korpel*
17. Procedures in Applied Optics, *John Strong*
18. Handbook of Solid-State Lasers, *edited by Peter K. Cheo*
19. Optical Computing: Digital and Symbolic, *edited by Raymond Arrathoon*
20. Laser Applications in Physical Chemistry, *edited by D. K. Evans*
21. Laser-Induced Plasmas and Applications, *edited by Leon J. Radziemski and David A. Cremers*
22. Infrared Technology Fundamentals, *Irving J. Spiro and Monroe Schlessinger*
23. Single-Mode Fiber Optics: Principles and Applications, Second Edition, Revised and Expanded, *Luc B. Jeunhomme*
24. Image Analysis Applications, *edited by Rangachar Kasturi and Mohan M. Trivedi*
25. Photoconductivity: Art, Science, and Technology, *N. V. Joshi*
26. Principles of Optical Circuit Engineering, *Mark A. Mentzer*
27. Lens Design, *Milton Laikin*
28. Optical Components, Systems, and Measurement Techniques, *Rajpal S. Sirohi and M. P. Kothiyal*
29. Electron and Ion Microscopy and Microanalysis: Principles and Applications, Second Edition, Revised and Expanded, *Lawrence E. Murr*

30. Handbook of Infrared Optical Materials, *edited by Paul Klocek*
31. Optical Scanning, *edited by Gerald F. Marshall*
32. Polymers for Lightwave and Integrated Optics: Technology and Applications, *edited by Lawrence A. Hornak*
33. Electro-Optical Displays, *edited by Mohammad A. Karim*
34. Mathematical Morphology in Image Processing, *edited by Edward R. Dougherty*
35. Opto-Mechanical Systems Design: Second Edition, Revised and Expanded, *Paul R. Yoder, Jr.*
36. Polarized Light: Fundamentals and Applications, *Edward Collett*
37. Rare Earth Doped Fiber Lasers and Amplifiers, *edited by Michel J. F. Digonnet*
38. Speckle Metrology, *edited by Rajpal S. Sirohi*
39. Organic Photoreceptors for Imaging Systems, *Paul M. Borsenberger and David S. Weiss*
40. Photonic Switching and Interconnects, *edited by Abdellatif Marrakchi*
41. Design and Fabrication of Acousto-Optic Devices, *edited by Akis P. Goutzoulis and Dennis R. Pape*
42. Digital Image Processing Methods, *edited by Edward R. Dougherty*
43. Visual Science and Engineering: Models and Applications, *edited by D. H. Kelly*
44. Handbook of Lens Design, *Daniel Malacara and Zacarias Malacara*
45. Photonic Devices and Systems, *edited by Robert G. Hunsperger*
46. Infrared Technology Fundamentals: Second Edition, Revised and Expanded, *edited by Monroe Schlessinger*
47. Spatial Light Modulator Technology: Materials, Devices, and Applications, *edited by Uzi Efron*
48. Lens Design: Second Edition, Revised and Expanded, *Milton Laikin*
49. Thin Films for Optical Systems, *edited by François R. Flory*
50. Tunable Laser Applications, *edited by F. J. Duarte*
51. Acousto-Optic Signal Processing: Theory and Implementation, Second Edition, *edited by Norman J. Berg and John M. Pellegrino*
52. Handbook of Nonlinear Optics, *Richard L. Sutherland*
53. Handbook of Optical Fibers and Cables: Second Edition, *Hiroshi Murata*
54. Optical Storage and Retrieval: Memory, Neural Networks, and Fractals, *edited by Francis T. S. Yu and Suganda Jutamulia*
55. Devices for Optoelectronics, *Wallace B. Leigh*
56. Practical Design and Production of Optical Thin Films, *Ronald R. Willey*
57. Acousto-Optics: Second Edition, *Adrian Korpel*
58. Diffraction Gratings and Applications, *Erwin G. Loewen and Evgeny Popov*
59. Organic Photoreceptors for Xerography, *Paul M. Borsenberger and David S. Weiss*
60. Characterization Techniques and Tabulations for Organic Nonlinear Optical Materials, *edited by Mark Kuzyk and Carl Dirk*

61. Interferogram Analysis for Optical Testing, *Daniel Malacara, Manuel Servín, and Zacarias Malacara*
62. Computational Modeling of Vision: The Role of Combination, *William R. Uttal, Ramakrishna Kakarala, Sriram Dayanand, Thomas Shepherd, Jagadeesh Kalki, Charles F. Lunskis, Jr., and Ning Liu*
63. Microoptics Technology: Fabrication and Applications of Lens Arrays and Devices, *Nicholas F. Borrelli*
64. Visual Information Representation, Communication, and Image Processing, *Chang Wen Chen and Ya-Qin Zhang*
65. Optical Methods of Measurement: Wholefield Techniques, *Rajpal S. Sirohi and Fook Siong Chau*
66. Integrated Optical Circuits and Components: Design and Applications, *edited by Edmond J. Murphy*
67. Adaptive Optics Engineering Handbook, *edited by Robert K. Tyson*
68. Entropy and Information Optics, *Francis T. S. Yu*
69. Computational Methods for Electromagnetic and Optical Systems, *John M. Jarem and Partha P. Banerjee*
70. Laser Beam Shaping: Theory and Techniques, *edited by Fred M. Dickey and Scott C. Holswade*
71. Rare-Earth-Doped Fiber Lasers and Amplifiers: Second Edition, Revised and Expanded, *edited by Michel J. F. Digonnet*
72. Lens Design: Third Edition, Revised and Expanded, *Milton Laikin*
73. Handbook of Optical Engineering, *edited by Daniel Malacara and Brian J. Thompson*
74. Handbook of Imaging Materials, *edited by Arthur S. Diamond and David S. Weiss*
75. Handbook of Image Quality: Characterization and Prediction, *Brian W. Keelan*
76. Fiber Optic Sensors, *edited by Francis T. S. Yu and Shizhuo Yin*
77. Optical Switching/Networking and Computing for Multimedia Systems, *edited by Mohsen Guizani and Abdella Battou*
78. Image Recognition and Classification: Algorithms, Systems, and Applications, *edited by Bahram Javidi*
79. Practical Design and Production of Optical Thin Films: Second Edition, Revised and Expanded, *Ronald R. Willey*

Additional Volumes in Preparation

Ultrafast Lasers: Technology and Applications, *edited by Martin Fermann, Almantas Galvanauskas, and Gregg Sucha*

Practical Design and Production of Optical Thin Films

Second Edition, Revised and Expanded

Ronald R. Willey
Willey Optical, Consultants
Charlevoix, Michigan

CRC Press
Taylor & Francis Group
Boca Raton London New York

CRC Press is an imprint of the
Taylor & Francis Group, an **informa** business

CRC Press
Taylor & Francis Group
6000 Broken Sound Parkway NW, Suite 300
Boca Raton, FL 33487-2742

First issued in paperback 2019

ISBN-13: 978-0-8247-0849-8 (hbk)
ISBN-13: 978-0-367-39603-9 (pbk)

Visit the Taylor & Francis Web site at
http://www.taylorandfrancis.com

and the CRC Press Web site at
http://www.crcpress.com

Preface to the Second Edition

The field of optical thin films continues to expand at a rapid rate in both commercial application and technology. As individuals and as a technical community, we continue to gain new understanding as new tools, techniques, processes, and experimental results become available. Since the publication of the first edition, enough new material has become available to warrant a new edition with expanded usefulness for the reader.

Some of the areas of expansion and addition include: new design visualization and optimization techniques, coating equipment updates, a review of spectral measuring equipment and techniques, a significantly expanded materials section, further ion-assisted and energetic process experience, more detail on and examples of design of experiments methodology, expanded understanding of optical monitoring sensitivity, the use of constrained optimization to promote more reproducible designs, consideration of the rapidly expanding dense wavelength division multiplexing applications of optical coatings, and new simulations of the error compensation effects available in optical monitoring with associated tolerance estimations.

This book continues to serve as a tutorial text for those new to the field and as a reference for those with long experience.

Ronald R. Willey

Preface to the First Edition

This book deals with the basic understanding, design, and practical production of optical thin films, or interference coatings. It focuses on two main subjects that are critical to meeting the practical challenges of producing optical coatings. The first is the design of coatings, an understanding of which allows the practitioner to know the possibilities and limitations involved in reducing, enhancing, or otherwise controlling the reflection, transmission, and absorption of light (visible or otherwise). The second deals with the practical elements needed to actually produce optical coatings. These include equipment, materials, process development and know-how, and monitoring and control techniques.

Emphasis is placed on gaining insight and understanding through new viewpoints not found in existing books. The approaches are primarily empirical and graphical. Extensive mathematical derivations have been well covered in other texts, but there remains a need for additional guidance for coating engineers who must design and produce coatings. As optical coatings have become more sophisticated and demanding over the past few decades, optical engineers have had to better understand the limitations and possibilities of design and production. The widespread availability of computers now makes the optimization of detailed designs easy for anyone in the field. A broadened insight will lead the designer to better starting designs that will converge reliably to workable final results. One aim of this text is to aid the reader by adding global views that enhance the creative thought process and lead to new and improved design solutions. Another aim is to share results and information from practical experience in the areas of equipment, materials, and processes which may shorten the reader's path to the production of practical optical coatings.

This book offers the reader practical concepts, understanding, and approaches to use modern computer tools to get better design results. Some insight and tools are provided for knowing where certain boundaries or limitations lie in thin film design. Additional understanding is provided in the monitoring and control of some of the more complex and difficult-to-manufacture modern optical thin films.

Among the book's unique features is an empirically discovered family of functions of index of refraction versus thickness that are discussed here which provide a selection of various antireflection coatings that can be used as needed. The field of Fourier analysis and synthesis of thin film designs is explored from a new viewpoint. Optical monitoring sensitivity concepts and error correction techniques that can enhance production results are elaborated upon in ways not found in other texts.

This book should be of benefit to every optical and thin film engineer and scientist, whether experienced or new to the field. It should serve as a tutorial text and also as a reference for even masters in the field.

Ronald R. Willey

Contents

1

Fundamentals of Thin Film Optics and the Use of Graphical Methods in Thin Film Design

1.1. INTRODUCTION

Getting started is often the most difficult part of a new task, and it is always the first thing to do. The goal supported by this book is to produce a practical thin film optical coating which meets whatever particular requirements are presented to the reader within the limitations of the technology. We will give a brief overview of what is needed to do this and how one might proceed.

The first thing needed is a clear statement of the requirements and/or goals of the coating such as spectral reflectance versus wavelength, spectral range of concern, substrate characteristics, environment to be encountered (and survived), etc. Often it is necessary to work with the end user or customer to establish these requirements and desires within the framework of what is possible and practical. Chapter 2 provides some assistance in estimating what can be done at this first stage.

The second thing needed by the coating developer is a basic understanding of the underlying principles of optical thin film performance and design. These are set forth in Chapters 1 and 3 from several viewpoints to give a more global perspective of the optical thin film design task. The possibilities, limitations, and options in various types of thin film coating designs are discussed in Chapter 2. This leads to a choice of the type of design to use to meet the requirements and an estimate of the number of layers and materials needed.

The third stage is to select materials which will have the desired properties over the required spectral range and which will perform in the required environmental conditions. These conditions often include temperature, humidity, abrasion, adhesion, salt fog, cleaning solutions, etc. Chapter 5 discusses the more commonly used materials and the processes by which they may be deposited. The equipment to be used in the production of a coating will affect the design and material choices. Chapter 4 reviews the typical equipment currently used for thin film production.

The fourth step is to perform the detail design of the coating to meet the optical performance requirements. If the index of refraction versus wavelength in the region of concern of the chosen materials is well known and stable, and the control of film thicknesses in the deposition process is adequate, the final optical coating product can be expected to be the same as predicted by the detail design. A computer evaluation and optimization program is used for the detail design process. The preliminary or starting design, derived typically from the choices made from Chapters 2 and 5, is evaluated and then the layer thicknesses are optimized with respect to the spectral requirements of the product. Sometimes a few more layers may need to be added to meet some requirement. It is our practice also to try to remove layers from a near-final design and reoptimize to determine if fewer layers might still meet the requirements. The fewest number of layers is generally desirable as long as there is enough margin in the design to allow for expected process variability and still meet the requirements.

The fifth phase of the process is to test the design in actual deposition. The first part of this should be to verify that the indices of the deposited materials are the same as expected and used in the design. If this is true, this phase is simple and short. If significant differences are found, the material processes may need refinement and greater control. This may require extensive effort. Chapter 6 discusses process development to reduce variability and characterize material properties. This can feed back improved information for the material and process know-how covered in Chapter 5.

The sixth stage which leads to the successful production of the required optical thin film is to gain adequate control of the layer thicknesses. The actual strategy for thickness control should be considered at the fourth or design step and influenced by the equipment to be used (Chapter 4). Chapter 7 deals with the many ways that film thickness might be monitored and controlled. It also discusses monitoring strategies, where the different strategies might best be used, and their strengths and weaknesses. When a monitoring process is established with sufficient repeatability, then the only task is to set the film thicknesses to produce the required results. This is typically a "Kentucky windage" process. That is to say that the actual resulting film thickness is usually somewhat different from that given as a parameter to the monitoring system. As long as the difference is the same from run to run, the difference can be subtracted from what is given as a parameter to the system so that the resulting thickness deposited meets the requirements. Section 7.7.3 gives a detailed example of how this was done in a specific case.

The application of the above six steps and the information in the chapters mentioned should lead to a successful coating in the great majority of optical thin films used in the world today.

We would like to mention a few key things which may be helpful to keep in mind when starting a new coating development. These are discussed in more detail in the chapters. They are usually true, but we will not state that there are no exceptions.

It is best to view all optical thin film coatings as affecting the reflectance of a surface as discussed in Chapters 1, 2, and 3. This may be to reduce the reflectance in a given spectral range or to increase it. Transmittance, optical density, etc., are all derivatives of reflectance.

More layers and thereby a thicker coating in a design give more control over fine details of the spectral reflectance profile as shown in Chapter 2. However, the improvements with increasing thickness seem to be asymptotic. For example, more than eight (8) layers in a broadband antireflection coating, where the optical thickness is about one wavelength in the band, typically gives limited reduction in the reflection over the band. It requires an order of magnitude thicker coating to reduce the reflectance to half that of eight layer.

We also show in Chapter 2 that it is desirable to use high and low index materials in a design where the difference in index is as great as practical. There is more reflectance produced at each interface, and therefore fewer layers are needed to produce a given result.

It has further been shown in Chapter 2 that the lowest practical index should be used for the last layer in an antireflection coating.

The fewest layers practical in a design is most desirable from a production point of view. It also has been shown in broadband antireflection coatings that a minimum number of layers is optimum for given designs and more layers cannot produce as good a result.

The simplest solutions which can meet the requirements are usually the most elegant and the most successfully produced.

Absolute values of deposition parameters such as pressure, temperature, etc., are difficult to obtain and may be unnecessary. The most beneficial characteristic of almost any process is its stability or reproducibility. With stable indices of refraction and material distribution in a process, the parameters of a design and process can be adjusted by "Kentucky windage" to meet the requirements.

The design of experiments methodology discussed in Chapter 6 can often minimize the number of experiments necessary to optimize and stabilize a deposition process.

In recent years, "Concurrent Engineering" has been a topic of some interest in the industrial world. Our interpretation of the topic is that it is nothing new, but is a rediscovery of a truth that had been lost by some cultures. The best results are obtained when the engineers/designers are in close communication with the fabricators/producers of the product that they develop and when they understand the details of the production processes needed to produce their designs. Anyone who has worked in a design and production field will have examples where a design was submitted to be fabricated without any prior communication between the designer and the fabricator. This usually leads to problems. The design might call for things which cannot be achieved or might not take advantage of processes or capabilities which exist. The engineers/designers should be in adequate communication with the producers/fabricators to optimize the results of their combined efforts. It has been our practice for decades as engineers/designers to gain as much hands-on experience as practical with the processes for which we design. This allow us to understand the capabilities and limitations of the processes so that our designs are achievable and properly utilize the existing capabilities. We further attempt to discuss design and fabrication ideas at the formative stages with all of the pertinent parties. This often takes the form of going into the shop and saying, "We need to achieve this result. What choices might we have as to how to fabricate it, and how might the details and tolerances best be arranged?" The producers/fabricators should also be included in all of the formal design reviews from preliminary to final review. The essence of Concurrent Engineering in our opinion is communication and the inclusion of the knowledge and experience of all of the appropriate parties involved in a development at all stages of a project. In general, if the designer is not also the producer, the designer and producer need to coordinate as a unit to attain the best result from the application of their combined resources and experience.

The application of Concurrent Engineering to the optical thin film development process is no different from the development of a machine, a building, or a spacecraft.

With these few suggestions in mind, we will now go into the details behind the preceding overview.

This chapter reviews the fundamentals of thin film optics and elaborates on the use of graphical methods to gain understanding and insight into some of the basic principles of optical thin film design. Reflectance diagrams and admittance plots are described in some detail. The insight gained from the admittance plots of antireflection (AR) coatings with ideal inhomogeneous index profiles connect these results to Fourier thin film design techniques discussed in Chapter 3. The limited choice of low index materials for broadband antireflection coatings necessitates a discussion of the possibilities and limitations of equivalent index approximations. A specific example in section 1.7.1 illustrates the design process to achieve the goal of a very broad band and low reflectance AR coating on crown glass.

The emphasis of this chapter is on graphics, concepts, insight, and reduction to practice in optical thin film design.

1.2. REVIEW OF THIN FILM OPTICS PRINCIPLES

Some years ago, I was privileged to hear a presentation by Prof. Robert Greenler (then president of the Optical Society of America) wherein he aptly demonstrated the interference colors in a soap bubble. A mixture of equal parts of dish soap and glycerine (which can be purchased in a pharmacy) is diluted with water until bubbles can be blown which persist for many seconds without breaking. An ordinary flower pot of about 6 inch diameter is dipped upside down in a pie tin of the solution and withdrawn from the solution with a bubble formed on its top lip. With the pot still upside down, one blows into the hole in the bottom of the pot until the bubble becomes hyperhemispheric. The hole is then sealed by wiping some of the solution across the hole with the thumb. The pot is then gently turned right side up and set on a table. Figure 1.1 illustrates the configuration.

If the solution is homogeneous and the air currents around the bubble are

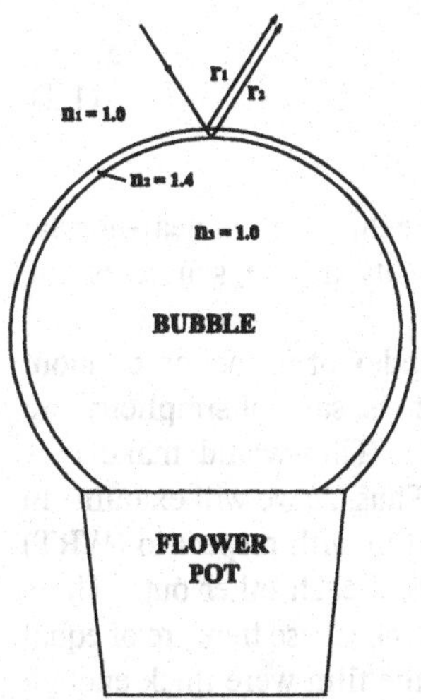

Fig. 1.1 Soap bubble on a flower pot illustrating interference colors.

negligible, the solution will gradually drain from the top of the bubble to the bottom causing the top to become thinner and the bottom sides to become thicker. The colors of the top will change from pale "pastels" to increasingly saturated colors. The top will eventually become white and then black just as the bubble breaks (because it has gotten too thin). The colors will form horizontal bands. The bottom sides will become less saturated colors as they get thicker. These colors illustrate various principles related to our interests, but we will here focus on the top of the bubble. The physics of the situation is explained below.

We will first review some of the underlying principles and the mathematical foundations of thin film design briefly from a fundamental point of view. The more experienced reader might choose to skip ahead to Sect. 1.5.

The Fresnel amplitude reflection coefficient (r) for an interface between two nonabsorbing media at normal incidence can be represented by the following equation:

$$r = \frac{n_1 - n_2}{n_1 + n_2} \tag{1.1}$$

where n_1 and n_2 are the (real) indices of refraction of the two media. For the more general case of absorbing media with complex refractive indices $\mathbf{n}_j = n_j - ik_j$ (j = 1,2,...); the reflection intensity coefficient (R) is the amplitude reflection coefficient (r) times its complex conjugate (r*),

$$R = rr* \tag{1.2}$$

For most of the common applications of absorption-free thin films at near-normal incidence of light it is sufficient to treat the intensity as the square of the amplitude.

The soap bubble is a mixture of water with an index of refraction of about 1.333 plus soap and glycerine of higher indices. For the sake of simplicity, we will assume that the index of the mixture is 1.40. This would make r_1 = (1.0−1.4)/(1.0+1.4) = − 0.1667 while r_2 = +0.1667. Thus, as we will examine in more detail below, if the soap bubble film were very thin with respect to (WRT) a wavelength of light, the two reflections would cancel each other out. There would be no reflection, i.e., it would look black! This is because they are of equal magnitude and opposite sign. If, on the other hand, the film were thick enough to delay r_2 by one half wavelength WRT r1, then the two would be in phase and add to give constructive interference of r = −.3333 or R = .1111 = 11.11%. This would look white in reflection.

In a rigorous sense, the amplitude of the residual reflection resulting from the interaction of light with two interfaces with a phase separation of phi (ϕ) is:

$$r = \frac{r_1 + r_2\, e^{-i\phi}}{1 + r_1\, r_2\, e^{-i\phi}} \tag{1.3}$$

for a more extensive discussion, see Apfel[1].

Let us look at a more general "real life" example. Fig. 1.2 represents a single layer of magnesium fluoride ($n_1 = 1.38$) on crown glass ($n_2 = 1.52$) in a medium of air (or vacuum $n_0 = 1.0$). Equation (1.4) gives the amplitude reflection coefficient at the interface between the air and magnesium fluoride at normal incidence of light:

$$r_1 = \frac{n_0 - n_1}{n_0 + n_1} = \frac{-.38}{2.38} = -.1597 \tag{1.4}$$

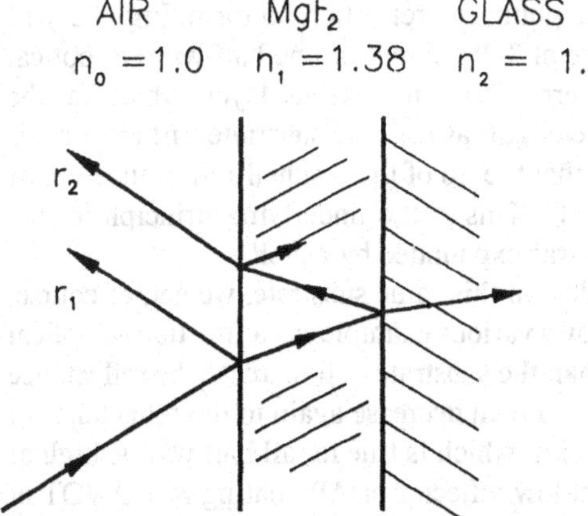

Fig. 1.2 Reflections at a single layer coating of MgF_2 on crown glass.

Equation (1.5) gives the amplitude reflection coefficient at the fluoride-to-glass interface.

$$r_2 = \frac{n_1 - n_2}{n_1 + n_2} = \frac{-.14}{2.9} = -.0483 \qquad (1.5)$$

When these two reflections are in phase, they add to give -0.2080 or an intensity of 0.0433 (4.33% reflectance). However, when we apply Eqn. 1.3 with $e^{-i\phi} = 1$, we find r = .2063 and R = 4.26% which is the same as an uncoated substrate of index 1.52. When the reflections are 180 degrees out of phase ($e^{-i\phi} = -1$), they add destructively (subtract) to give -0.1123 amplitude or 0.0126 intensity. This single beam interference consideration yields approximately the typical residual reflectance of 1.26% for a single layer antireflection coating (SLAR) on crown glass at the design wavelength. Thus, for the exact reflectance value, multiple beam interference was taken into account. This was done using Eqns. 1.3. The derivation of this is also found in Macleod[2], page 51.

1.3. REFLECTANCE DIAGRAMS

Figure 1.3 illustrates the interaction of the reflections as a function of phase angle for a single thin film. When the layer is very thin, the two reflections are nearly in phase. As the layer increases to one quarter wave optical thickness (QWOT) at the reference (or design) wavelength, the phase delay of r_2 with respect to r_1 approaches 180 degrees which gives a minimum residual reflectance. If the layer thickness increases further from this point, the reflection vector in Fig. 1.3 will return to its starting value at a phase of 360 degrees or one half wave of optical thickness. This situation is often referred to as an "absentee layer" which has the same reflectance (at the design wavelength) as the bare substrate surface, i.e., as if it were not there at all. The path that the tip of the residual reflectance vector travels is the circle seen in Fig. 1.3. This is the underlying principle of the reflectance or circle diagram which was expounded by Apfel[1].

For a coating of a different index on this same substrate, we get, of course, different reflectance. Figure 1.4 shows various examples as a function of optical thickness. Layers of higher index than the substrate will increase the reflectance to a maximum at the QWOT point and then decrease again to the reflectance of the bare substrate at the half wave point, which is true for all half waves, high or low index. We see that the simplest low reflector or AR coating is a QWOT of low index on the substrate. The lowest reflectance (R=0) would be obtained by a layer whose index is the geometric mean between the substrate and the medium ($n_1 = (n_0\, n_2)^{0.5}$ as in Fig. 1.2); for this crown glass, that would be 1.233.

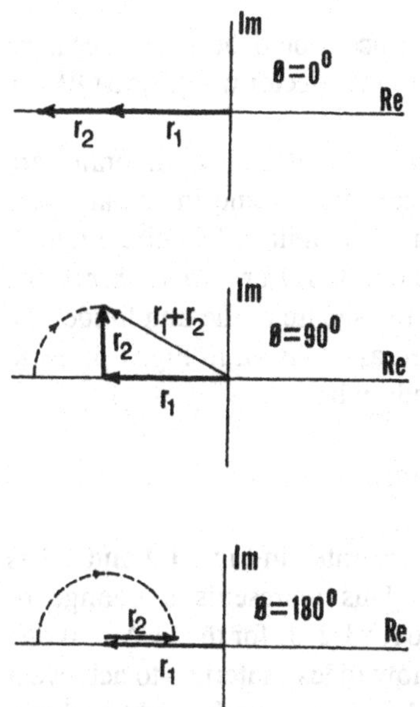

Fig. 1.3 Reflection amplitude vector addition as phase changes with the thickness of the layer on Fig. 1.2.

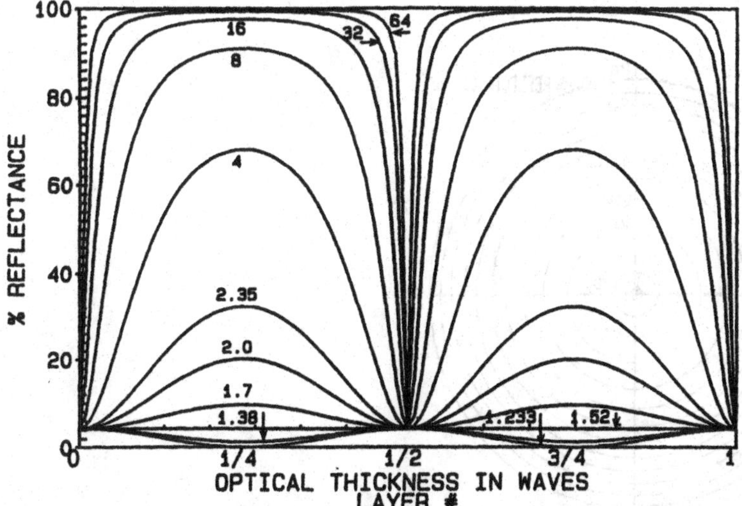

Fig. 1.4 Reflectance intensity as a function of the optical thickness for a single layer of the various indices on a 1.52 index substrate.

The reflectance amplitude from each interface would be equal because: $(n_1 - n_0)/(n_1+n_0) = (n_2 - n_1)/(n_2+n_1)$. The simplest high reflector would be a QWOT layer of the highest available index.

The amplitude reflection coefficients in the reflectance diagram are represented by the distance from the center where the real and imaginary axes intersect. The extreme of this "circle" diagram is the unit radius circle which would represent an amplitude reflection coefficient of 1.0 or 100%. Since the reflection intensity (reflectance R) is effectively the square of the amplitude, the circles of equal reflectance are not equally spaced as is shown in Fig. 1.5. Note how far from the center the 1% reflectance boundary lies.

1.3.1. Low Reflectors, Antireflection Coatings

The single layer antireflection coating (SLAR) illustrated in Figs. 1.2 and 1.3 is shown on a reflectance diagram in Fig. 1.6. This represents a change of reflectance from the bare substrate at R=4.26% to R=1.26% for the SLAR. If one were interested in using two layers of high and low index materials to achieve a zero reflectance coating at one wavelength, Fig. 1.7 shows how it might be done. The last layer (toward the incident medium) would have to have the low index,

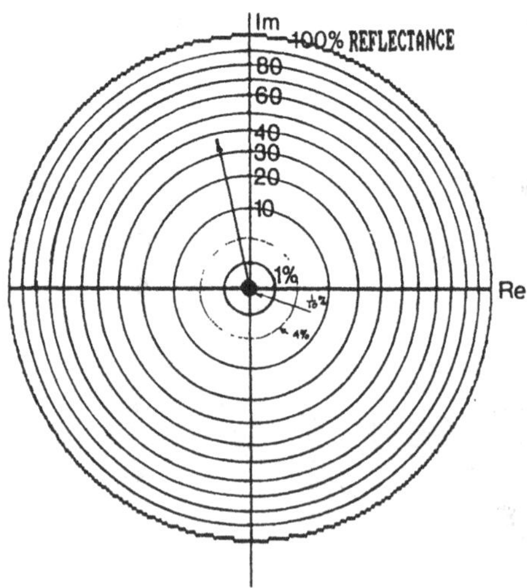

Fig. 1.5 Equal percent reflectance intensity contours on a reflectance circle diagram.

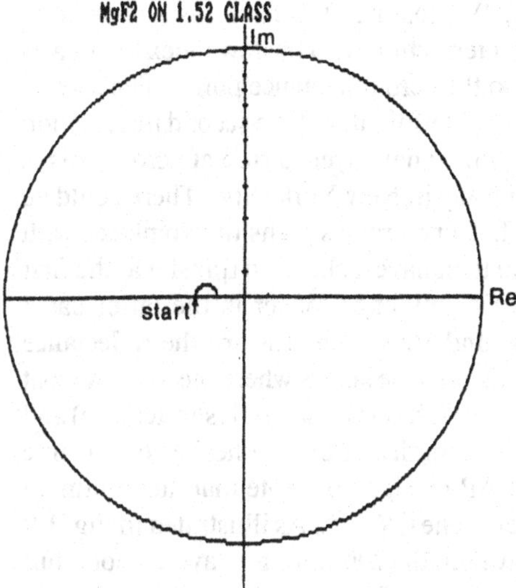

Fig. 1.6 Reflectance locus of the coating in Figs. 1.2 and 1.3 with increasing thickness on a reflectance circle diagram.

and it would have to be represented by a circle passing through the origin. The high index first layer (on the substrate) would have to start at the substrate index and intersect the low index circle (generally at two points).

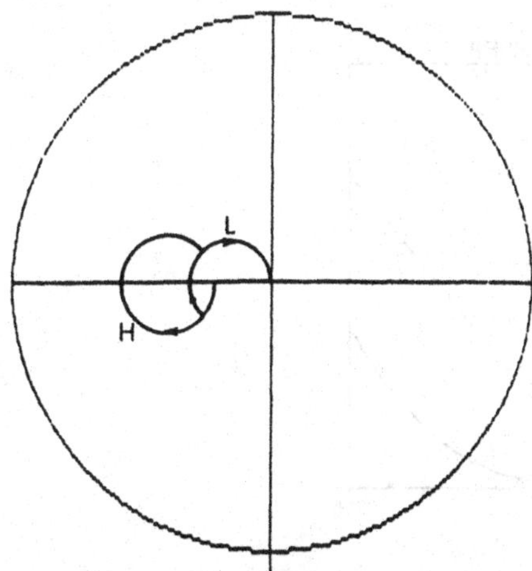

Fig. 1.7 Reflectance diagram of the two solutions to a two layer AR coating (V-coats).

There are two solutions to this classical "V" coating. The first has a short high index layer which terminates at the first intersection with the low index locus, and the low index layer completes the path to the zero reflectance point. The second solution has a longer high index layer which terminates at the second intersection with the low index circle, and a shorter low index layer to stop at zero. This is like traveling between two points on a subway in New York City. There could be an "H" train which leaves from your office and crosses paths in two places with the "L" train that will get you home. You then have a choice to transfer at the first crossing or wait until the second. The ride might be longer in the latter case! Curves V1 (with the shorter first layer) and V2 in Fig. 1.8 are the reflectance versus wavelength of these two V-coats, their shape shows where the name V-coat originates. V1 is generally preferred because it has less material (shorter ride) and is somewhat more tolerant to shifts in wavelength and/or thickness errors. Curve A in Fig. 1.8 is the reflectance of the SLAR of Fig. 1.6. Note that the minimum reflectance wavelength is where the layer is one QWOT. As illustrated in Fig. 1.9 (on an expanded scale), for a longer wavelength (700 nm), the layer is not a full QWOT and therefore the reflectance is higher. For a shorter wavelength (400 nm), the layer is more than a QWOT, and therefore the reflectance is starting to increase again as the optical thickness increases toward a half wave. This same behavior can be seen in Fig. 1.10 to cause the reflectance of either V-coat to rapidly move away from zero as the wavelength changes from the design wavelength.

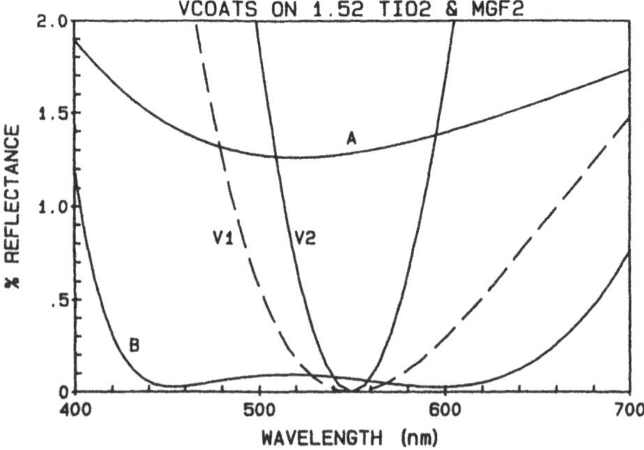

Fig. 1.8 Spectral reflectance: A is a single layer AR coating of MgF_2 on glass, B is a three layer broadband AR from Fig. 1.11, and V1 and V2 are V-coats from Fig. 1.7.

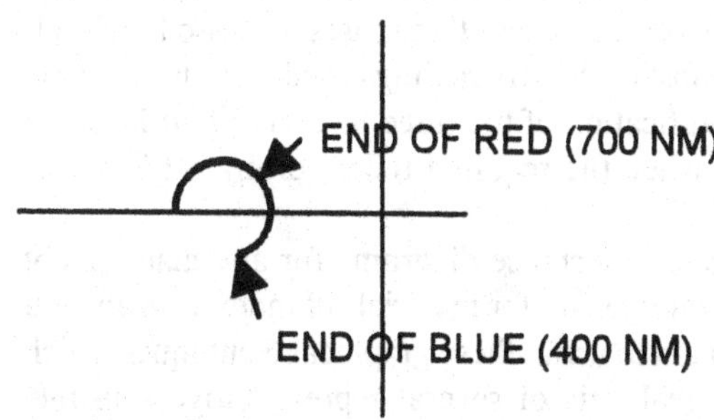

Fig. 1.9 Change in the locus of reflectance curve as in Fig. 1.6 with wavelength.

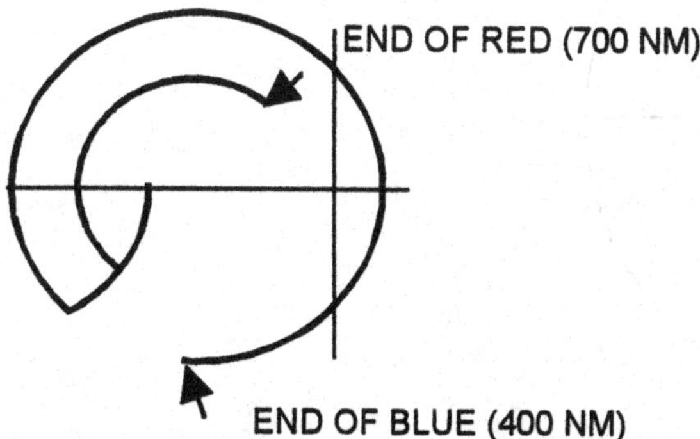

Fig. 1.10 Change in the locus of reflectance curve as in Fig. 1.7 with wavelength.

A special case of the V-coat as seen in Fig. 1.11 can be constructed for a 1.52 substrate with a medium index layer M and low index layer L which are each a QWOT (Fig. 1.11). This happens if n_M = 1.7 and n_L = 1.38. This two layer coating would have similar sensitivity to wavelength or "dispersion" as the other V-coats. However, the addition of a half wave layer of high index (about n_H = 2.25) between the M and L layers will "achromatize" the design or make it insensitive to wavelength changes over a fairly broad range. Curve B in Fig. 1.8 shows the result of this broad band AR of the classical quarter-half-quarter wave (QHQ) design. Figure 1.12 illustrates this compensation effect on the reflectance diagram for the red (650 nm) and Fig. 1.13 shows the blue (450 nm) which comes quite close to zero reflectance at the origin. These three cases discussed so far (1, 2, and 3 layers) form the basis of most of the AR coatings produced in the world today. We will later discuss a modification of the three layer to a four layer for practical reasons of producibility when the required index (of n_M =1.7) is not available.

Apfel[1] showed how to generate reflectance diagrams for any materials of interest. Figs. 1.14 and 1.15, for example, are for materials of index 1.38 and 4.0 in air or vacuum. Figure 1.16 is an aid to using these graphical techniques which shows the starting points on the real axis of several representative substrate materials. The one with the “dumbbell” marks is ordinary glass at n = 1.52. The others range from fused silica at about 1.45 to germanium at about 4.0.

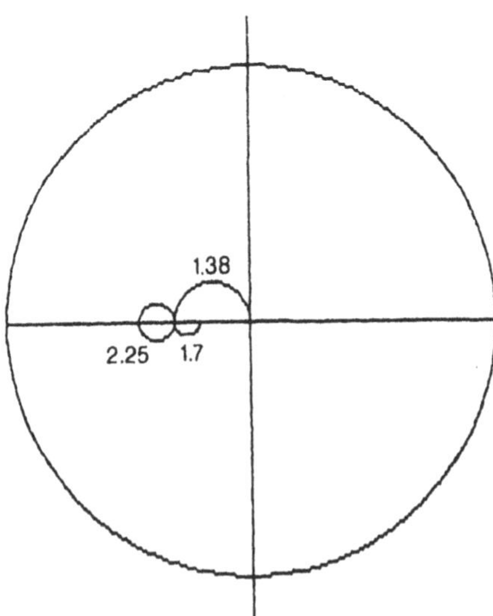

Fig. 1.11 Reflectance diagram of a three layer AR on crown glass.

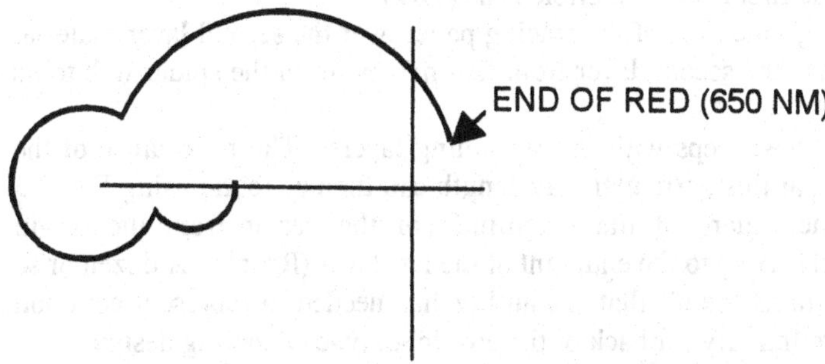

Fig. 1.12 Change in the locus of reflectance curve as in Fig. 1.11 at 650 nm.

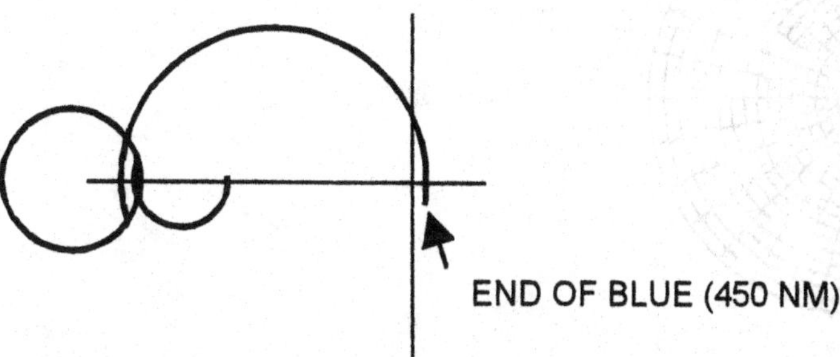

Fig. 1.13 Change in the locus of reflectance curve as in Fig. 1.11 at 450 nm.

These diagrams can be readily used for graphical overlay construction of the reflectance of a thin film design following these steps:

(1) Lay tracing paper over the Fig. 1.16 diagram and mark the substrate starting point and the real and imaginary axes for alignment with the other diagrams.

(2) Now align the axes of the tracing paper over the first layer material "spider web." Using the circle of which starts from that substrate point, trace the path from that point on the reflectance diagram moving clockwise to the termination of the first layer (1/2 circle is a QWOT).

(3) Now align the axes of the tracing paper over the second layer material spider web. Trace the second layer from that new point in the spider web to its termination.

(4) Repeat these steps with the remaining layers. The reflectance of the complete coating at this particular wavelength can then be found using Fig. 1.5 which shows the square of the magnitude of the vector from the origin (intersection of the axes) to the endpoint of the last layer (R=rr*). A dozen or so such spider diagrams are all that the author has needed to represent common materials for preliminary (or back of the envelope) type of coating design.

Note that the "radial" spokes on the spider webs have been plotted to represent 1/20th of a QWOT or 1/80th of a wave of optical (phase) thickness.

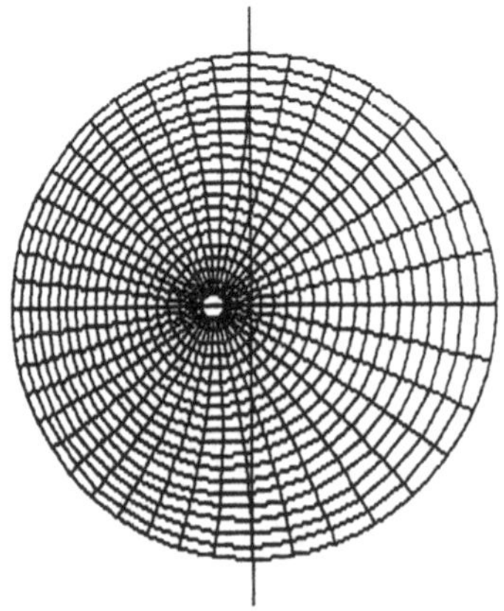

Fig. 1.14 Reflectance loci and equal optical thickness lines for 1.38 index material.

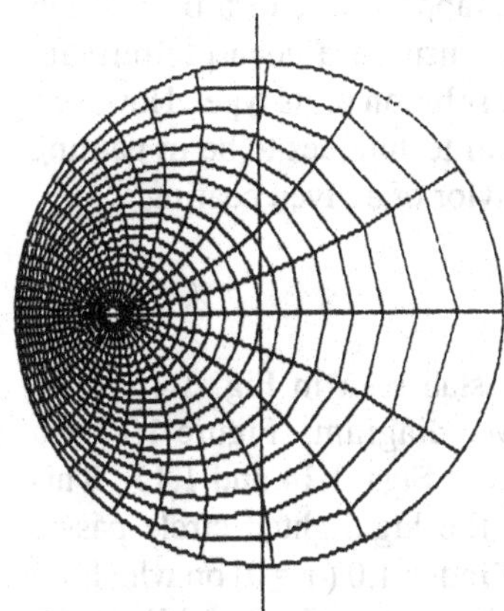

Fig. 1.15 Reflectance loci and equal optical thickness lines for 4.0 index material.

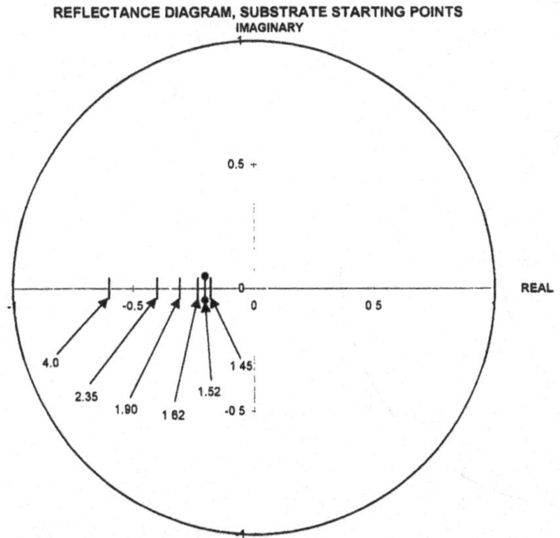

Fig. 1.16 Graphic aid of substrate starting points from indices 1.45 to 4.0.

We can use these to count the "spokes" (and fractions thereof) for each individual layer and thereby obtain an adequate estimate of the thickness of the layer in QWOTs. Such results can then be entered into an appropriate thin film design program as a starting design. The first results are usually a good preliminary design which can be quickly optimized to the best solution of its type. However, we have found the greatest utility of these graphical techniques to be in gaining insight and as an aid to understanding of the behavior of a given coating.

1.3.1.1 Behavior of a High Index Slab

The narrow bandpass behavior of a high index slab seen in Fig. 1.4 can be visualized also on a reflectance circle or spider web diagram. Figure 1.17 is a spider diagram for a hypothetical index of 10.8 as in Figs. 1.14 and 1.15. This index was chosen just for illustration such that the highlighted circle passes through the origin. Here we imagine a substrate of index 1.0 ($r = 0$) on which we start to deposit material of index 10.8. The locus of the total reflection follows the bold circle from the origin in a clockwise direction. At one QWOT, it will have a maximum reflection near $r = -1$ (at the left on the diagram). An additional QWOT, to make a total of one half wave, will bring the reflection back to zero at the origin.

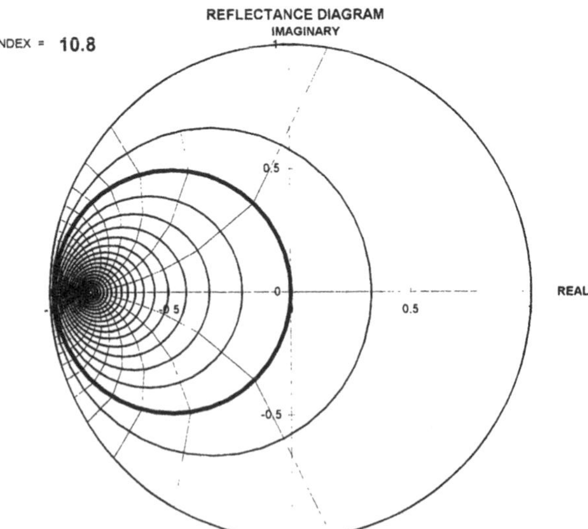

Fig. 1.17 Locus on a reflectance amplitude diagram for a 10.8 index material freestanding in air, illustrating the principle of the NBP filters in Fig. 1.4.

The point to note here is that the lines of equal phase thickness (spokes) are very widely spaced near the origin for high index materials, and they are very closely "packed" at the end of the $-r$ axis on the left. The result is that the reflectance grows very quickly from zero to a high r with only a small addition of phase thickness on the substrate. For a very high index like 99, the first phase thickness line representing 1/20th of a QWOT on such a spider diagram would have an $r = .93$. Almost all of the action in such a case is clustered near the left end of the $-r$ axis. This means that the reflection is high for almost all optical phase thicknesses except those near one half wave, or multiples of a half wave.

With reference to Eqn. 1.3, r_1 and r_2 are equal in magnitude and opposite in sign and have large values close to unity. In such a case, the denominator of Eqn. 1.3 becomes a major factor in the shape of the locus. Without it, the locus of r would start at the origin and circle out of the $r = 1$ boundary as it pivots clockwise about the high index point. However, the denominator tempers this so that as $e^{-i\phi}$ approaches -1, the r approaches $-2/2$. Thus, the behavior seen in Fig. 1.4 can also be visualized on a reflectance/spider diagram such as Fig. 1.17.

1.3.2. High Reflectors

We next address the subject of producing a high reflector and how it would appear on a reflectance diagram. Figure 1.18 shows the reflectance diagram of a stack of QWOT layers of the form (HL)4. This means a pair of high and low index QWOTs is repeated four times for a total of eight layers. The reflectance diagram starts with the first H at the substrate and progresses clockwise until it reaches a maximum reflectance at one QWOT where it intersects the real axis. The next low layer continues clockwise to the real axis again, etc. Figure 1.19 shows the reflectance versus wavenumber (10,000/wavelength in micrometers) after each pair is applied. By the time eight layers have been added to the substrate, the reflectance at the design wavelength has become high. More pairs will make it correspondingly higher and the transmittance correspondingly less. The edges of this "stop" band become steeper as the number of layers increase. On either side of the stop band are pass bands. As illustrated in Fig. 1.20 after Thelen[4], the edges of the stop band can be used to create a long-wave-pass (LWP) filter on one side of the center wavelength of the QWOT stack or a short-wave-pass (SWP) filter on the other. This high reflector stack is the building block of mirrors, LWP and SWP filters, and certain types of bandpass filters. The latter are made by bringing separate edges of a LWP and a SWP filter toward each other in the design until the desired pass band is achieved. The AR class of coatings of Fig. 1.20 have been

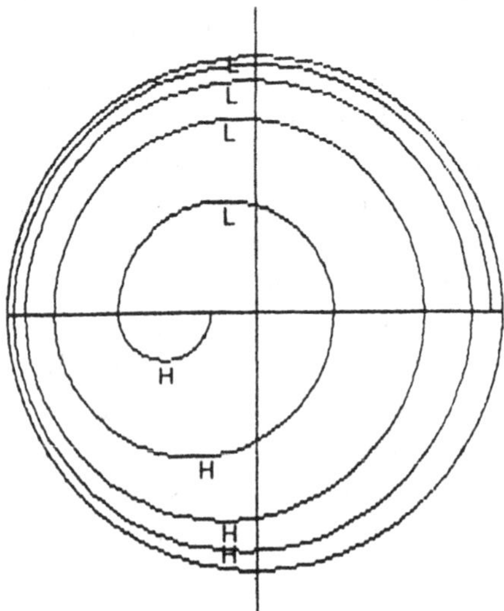

Fig. 1.18 Locus of the reflectance of a high/low QWOT layer stack spiraling outward from the substrate to high reflectance as deposited.

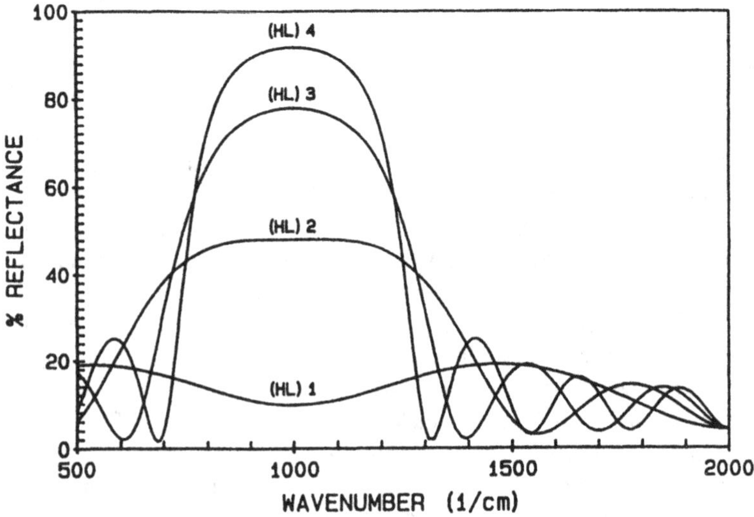

Fig. 1.19 Spectral reflectance as H/L pairs are added to the stack of Fig. 1.18.

mentioned already and the high reflectors in this section. Neutral beamsplitters are related to the high reflectors with only a few layers. The basis of the narrow band pass filter is seen in Fig. 1.4 and will be discussed in more detail below. The minus filter is discussed extensively by Thelen[4] and would be created by a QWOT stack where the difference in indices from H to L was small. The principles of the polarizing beamsplitter will be discussed below in connection with angular effects.

A stack of many periods of (HL) will be the same as the same number of periods of (LH) except at the start and end of the stacks. It is seen in Fig. 1.19 that there are ripples on either side of the stop band. These ripples can cause problems in some applications if they are in a band where high transmittance is required. The cause of the ripples is the lack of a proper "impedance match" or admittance matching layers (AR coatings) between the effective admittance of the stack and the admittances of the substrate and the medium. This subject is treated extensively in Thelen's early paper[3] and more recent book[4]. Basically, the last few layers on both sides of the stack can be adjusted in thickness to simulate a desired admittance and to create an AR coating for the stack over some (usually limited) wavelength band. The region of good "impedance match" can be shifted in wavelength, broadened, and can often be made to reduce the ripple to an acceptable level for the application. We will cover this in later chapters.

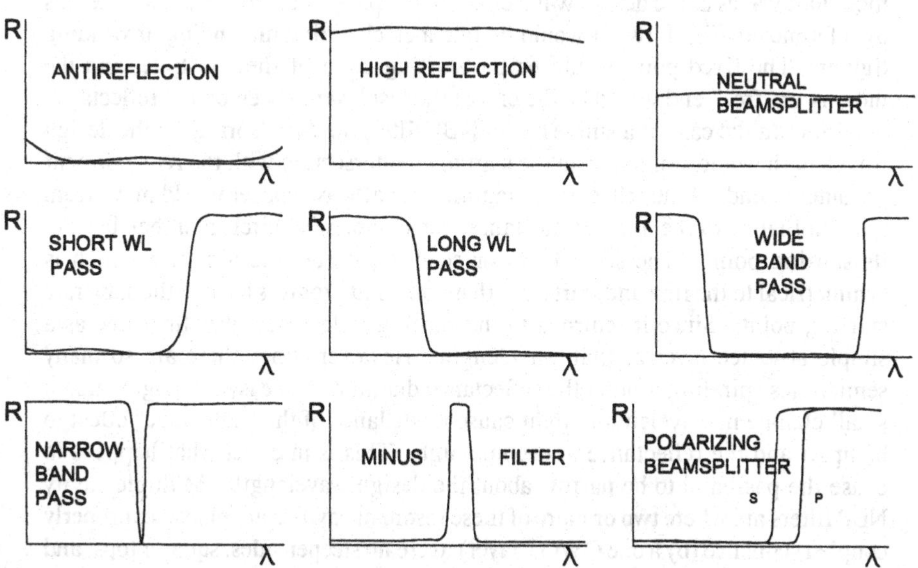

Fig. 1.20 Various types of filters made with only dielectric materials.

1.3.3. Narrow Bandpass Pass Filters

Narrow bandpass (NBP) filters are usually of a somewhat different design than the bandpass constructed of a LWP and SWP filter combination. If we examine Fig. 1.4 and Sect. 1.3.1.1 again, we see that a very high index single layer would create a filter with a narrow passband at 1/2 wave and multiples of 1/2 wave optical thickness. If such very high refractive indices existed, the high Fresnel reflectance at the interfaces with the spacer medium would have high transmission (low reflectance) at all multiples of one halfwave of optical thickness and a high reflectance for all other wavelengths. The passband width gets narrower as the index gets higher. This is in effect a Fabry-Perot interferometer with two mirrors (the interfaces) spaced by a dielectric layer. The early applications of this principle used partially transparent metal coatings such as silver for the mirrors on each side of the spacer layer. Such metal mirrors, however, come with some unavoidable absorption losses that are not inherent in all dielectric designs. Indices higher than 4.2 (germanium) are not commonly available or used, so we have to use some other method to get the same effect. Figure 1.19 shows the way. A high reflector can be made by a stack of high/low layers to give the same reflectance as any desired index interface (at a given wavelength). This mirror stack can then be spaced from a similar symmetric stack by a layer which is a multiple of one or more half waves at the design wavelength of the passband. We can visualize this by reference to Fig. 1.18. It would be like a clock spring unwinding or winding tighter. The fixed point would be the starting point of the substrate, and the moving or "free" end would be the end of the last layer as seen on the reflectance diagram. In the case of a single cavity NBP filter, the first "spring" at the design wavelength would coil out from the start to an intersection with the real axis near the outer bounds of the reflectance diagram. A halfwave spacer would move from the termination of the mirror reflectance locus for one full circle and then back to its starting point. The second mirror stack would be like a spring which is symmetrical to the first and start from that point and progress back to the substrate starting point. Since it returned to the starting reflectance, it is the same as a simple absentee layer at that wavelength. However, since there are so many semicircles spiraling around the reflectance diagram, it is easy to imagine that a small change in wavelength might cause the balance of the "half wave" effect to be upset and the reflectance to become high. This is in effect what happens to cause the passband to be narrow about the design wavelength. Multiple cavity NBP filters are where two or more of these resonant cavities are placed in properly coupled sequence (by a one QWOT layer) to create steeper sides, squarer tops, and greater reflection or rejection outside the passband. Figure 1.21 shows an example of a spectrophotometer trace of an actual three-cavity NBP filter in the infrared

(IR). The three humps at the top of the passband are a clue to the fact that this is a three-cavity filter. Two humps would imply a two-cavity filter. However, with more than three cavities, counting humps is not a reliable clue.

1.3.3.1 NBP Wavelength Effects as Seen on Reflectance Diagrams

In connection with Figs. 1.9-1.10 and 1.12-1.13, we discussed how the reflectance diagram changes with wavelength. Similarly, the "springs" mentioned above will wind and unwind with changes of wavelength from the design wavelength. Figure 1.22 shows the locus or path of the last reflectance point of a coating design as a function of wavelength. The partial circle toward the lower left of Fig. 1.22 is for a single cavity NBP filter centered at 1550 nm. As the wavelength is changed from the design wavelength, the reflectance (r) moves away from the origin of the diagram toward the high reflectance boundary (along this circular locus). The two-cavity design, showing a cardioid-shaped locus in Fig. 1.22, moves more rapidly away from the origin. Its reflectance will therefore increase more rapidly with wavelength than does the one-cavity filter (steeper skirts or sides). Similarly, the three-cavity design moves even more rapidly toward higher reflectance after having gyrated about the origin over a small wavelength range. This gyration is associated with the broader flat top on the three-cavity transmittance spectrum. The points on each locus for ± 0.4 nm are indicated by arrows. The design wavelength of 1550.0 nm is centered on the origin in each case. The reflectance

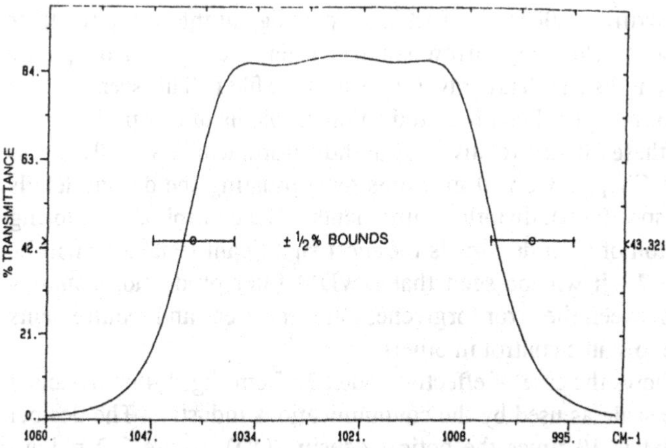

Fig. 1.21 Actual spectral transmittance of a three-cavity NBP filter at about 9.77 μm.

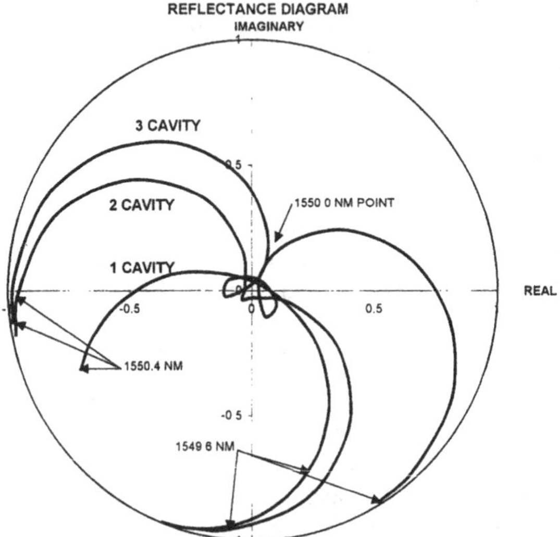

Fig. 1.22 Locus of the final reflectance point of a coating design as a function of wavelength for 1-, 2-, and 3-cavity filters.

is higher still in the blocked regions as the wavelengths move further from the design than we have plotted here. This is another way to view what is happening in NBP filters.

1.3.3.2 Dense Wavelength Division Multiplexing (DWDM) Filters

At the time of this writing (2001), DWDM is a topic of great interest in the fiber optics communications field. Very narrow bandpass filters are required to separate out the many wavelengths which are traveling in a single fiber. This seems to one of the most challenging optical coating production problems of this period.

The design of these filters is relatively straightforward, which we will address in this section. In Chap. 2 we will give aids for estimating the design details needed to achieve specific bandwidth requirements. The control or monitoring during the production of such designs is the most significant issue and will be discussed in Chap. 7. It will be seen that DWDM filter production contains extreme contrasts between the error forgiveness in some areas and requirements for rigid adherence to stable control in others.

Figure 1.23 shows the case of effective index 32 from Fig. 1.4 on a decibel (dB) transmittance scale as used by the communications industry. The decibel scale in this context is 10 times the optical density (OD), where OD = $\log_{10}$ (1/Transmittance).

The communications industry also tends to work with a frequency scale rather than wavelength. The L-band is at a range of frequencies from 185 to 192 teraHertz (THz = 10^{12} Hz) while the C-band is from 192.2 to 196 THz. This covers the contiguous wavelength range from 1528 to 1620 nm. The bandwidth is expressed in gigaHertz (GHz = 10^9 Hz). The bandwidth $\Delta\lambda$ in nanometers can be found from the bandwidth Δf in GHz by: $\Delta\lambda = -c \times \Delta f / f^2$, where 0.2997925 $\times 10^{18}$ nm/second is c (the velocity of light) and f is the frequency. Therefore, 100 GHz represents 0.8 nm bandwidth at 1549 nm and 50 GHz is 0.4 nm.

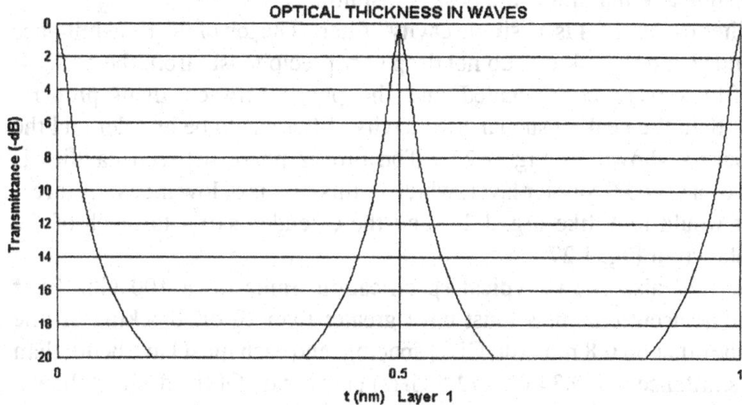

Fig. 1.23 Spectral trace of effective index 32 NBP filter from Fig. 1.4 on a decibel (dB) transmittance scale.

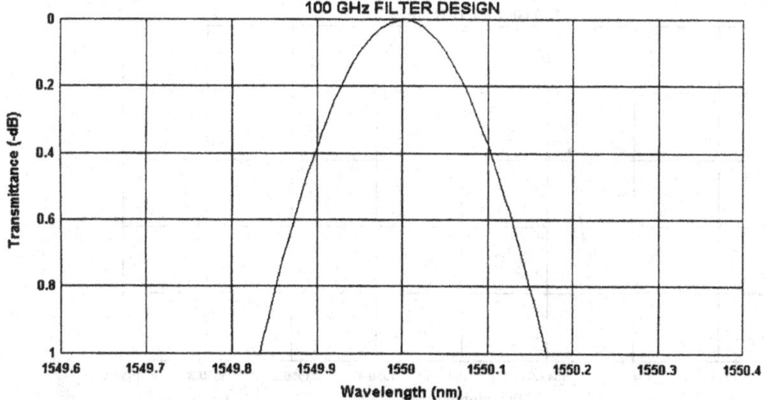

Fig. 1.24 Spectral trace of a "real" one-cavity design to similar Fig. 1.23.

Figure 1.24 shows the peak of a similar "real" design to Fig. 1.23 using materials such as SiO_2 and Ta_2O_5 for the low and high index materials in the mirrors and spacer. This design at 1550 nm is:

(1H 1L)9 4H (1L 1H)8 1L .59917H .76524L

where there are nine (9) pairs of high and low index QWOTs in each mirror and two half waves or four (4) QWOTs in the spacer. The last two layers are an antireflection coating or "impedance matching layers" to the air or vacuum beyond the filter designed to maximize the peak transmittance.

The filter in Fig. 1.24 is a "single cavity" filter. The top of the transmittance peak is rounded and the "skirts" do not drop very precipitously from the peak. If two such filters were concatenated and the phase between them properly maintained, then the peak of such a "two-cavity" filter would be broader and the skirts steeper as shown in Fig. 1.25. The proper phase between cavities is maintained by a QWOT coupler layer, which in this case is of low index. A three-cavity filter would look like Fig. 1.26, and the extended skirts for each of the designs is shown in Fig. 1.27.

Figure 1.27 also shows typical specification limits for a 100 Ghz filter wherein the adjacent channels must have greater than 20 dB blocking of one channel to the other on 0.8 nm (100 GHz) spacing, and each must have better than 0.3 dB transmittance in a 0.34 nm (42.5 GHz) pass band. Of the designs shown, only the three-cavity design meets these requirements.

A simple extension of this concatenation of cavities to more than three will create steeper sides and a broader top. However, another problem appears as seen

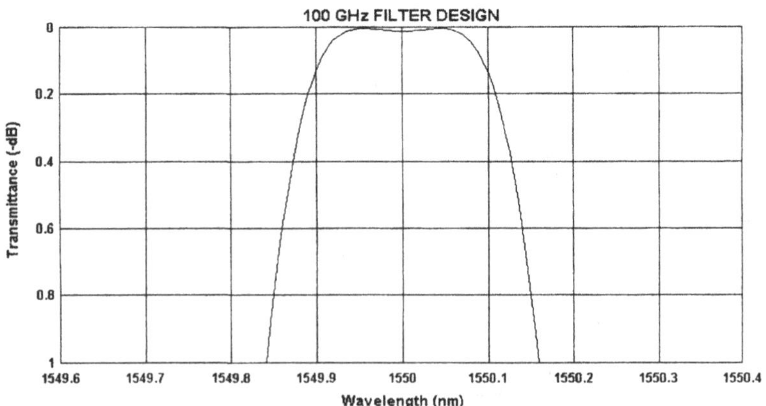

Fig. 1.25 Two-cavity NBP filter for 100 GHz DWDM applications.

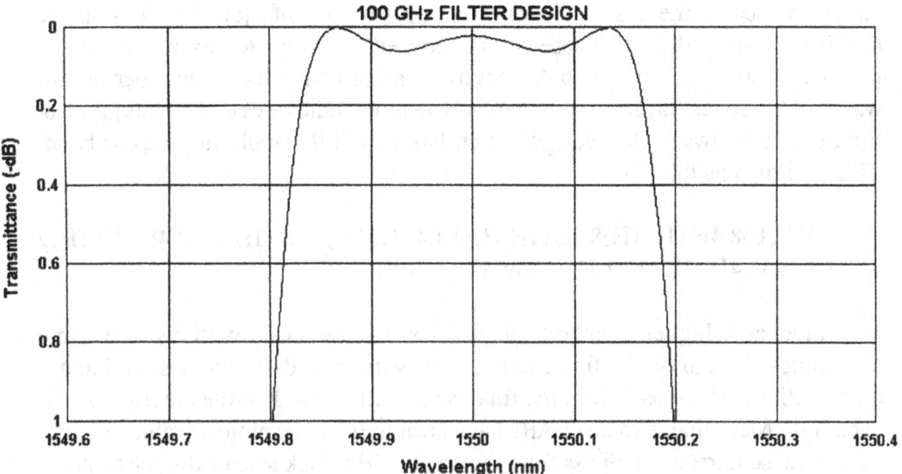

Fig. 1.26 Three-cavity NBP filter for 100 GHz DWDM applications.

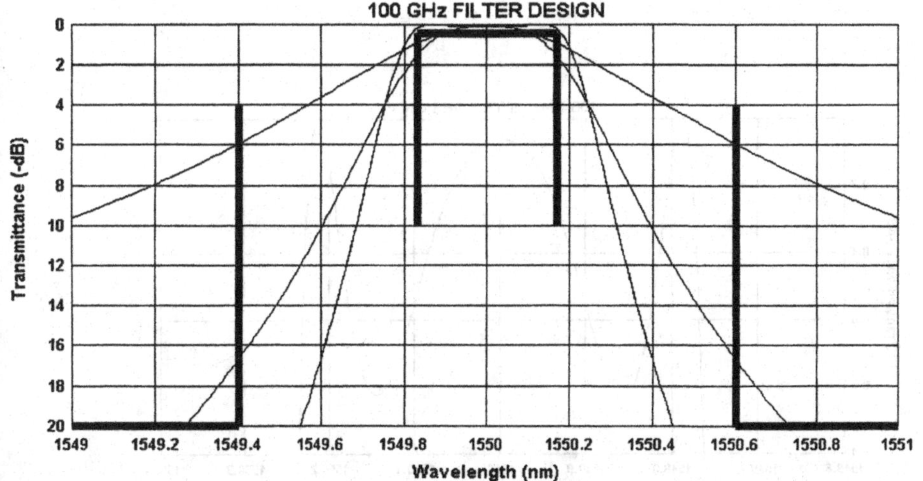

Fig. 1.27 Extended "skirts" of filters and typical specification limits.

in Fig. 1.28. There are "rabbit ears" (as they are called) which hang down, in this case, from the desired 0.0 dB level (or 100% T). Macleod[47] points out that: "The ears are a consequence of a deterioration of the matching towards the edges of the passband because there is no allowance for the variation of equivalent admittance that is very steep there." These can be reduced or eliminated by changing the number of layer pairs in the outer cavities and/or changing the number of half waves in the spacer layers of the cavities. Cushing[50] addresses this technique also. Figure 1.29 shows such a design which has very little ripple in the pass band. This design is as follows:

(1H 1L)8 4H (1L 1H)8 1L (1H 1L)9 4H (1L 1H)9 1L (1H 1L)9 4H (1L 1H)9 1L (1H 1L)8 4H (1L 1H)7 1L .64691H .68219L.

Five and higher numbers of cavities can be dealt with by the same techniques. The antireflection coatings may vary from design to design, but are all essentially a "V-coat" from the final cavity mirror stack to the medium (air or vacuum). More than two layer ARs have been found to be of no advantage in the case of these narrow bandpass filters (except on the back side of the substrates).

Note that, if the back side of a thin substrate is not coated with a proper AR, there can be ripples in the passband which depend on the substrate thickness and constitute an interference effect where the substrate is the spacer layer!

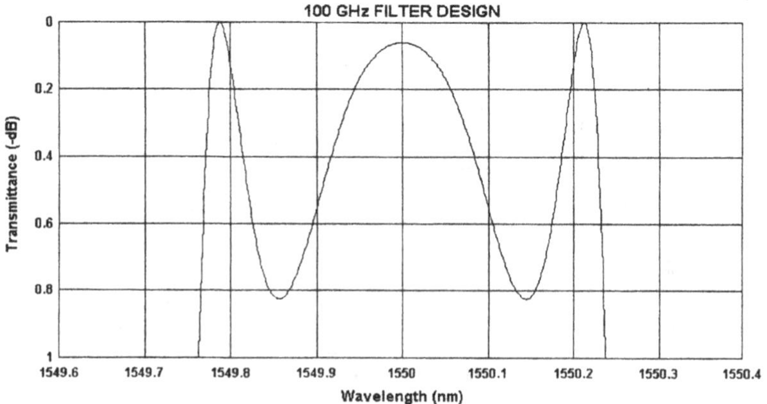

Fig. 1.28 Four-cavity NBP filter for 100 GHz with "rabbit ears".

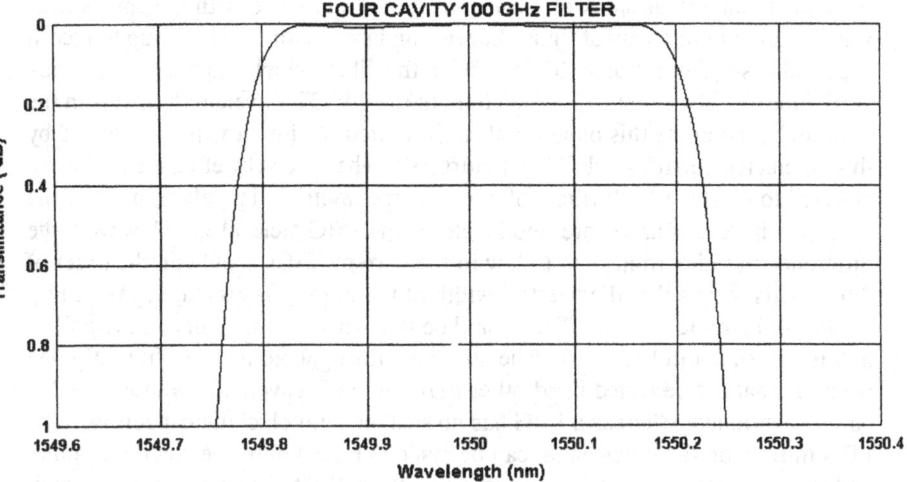

Fig. 1.29 Four-cavity NBP filter for 100 GHz without "rabbit ears".

These filters are generally well suited to DWDM applications as "Add" or "Drop" filters. In the Add case, all other wavelengths are reflected by the filter with low losses, while the Add wavelength passes through the filter from the local station in the direction of the reflected wavelengths and is joined with them. The Drop case allows the selected wavelength to pass through the filter to the local station while all the others are reflected and travel on. The current requirements of 0.3 dB (6.7% R) maximum loss in the transmitted band are acceptable for a drop filter. However, if one were attempting to use such a filter to simultaneously perform an Add ***and*** a Drop, this 0.3 dB from the Dropped signal would combine with the Added signal and could never be eliminated. The requirement for a filter to do both Add and Drop simultaneously and also give 20 dB isolation is that the in-band reflection would be less than 0.0044 dB (1% R). This is far from the current production capabilities at this time.

1.3.3.3 Fiber Bragg Gratings

There has been a great deal of work done in recent years in fiber Bragg gratings (FBG) for DWDM applications such as described in the book by Kashyap.[28] The concept is that a fiber can be made with a core composition which will gain a permanent change in index of refraction when illuminated with the appropriate wavelength and intensity of light. Interfering laser beams can be set up to record a periodic structure along the length of the fiber which resemble the index variations in a QWOT stack. When light of the QWOT wavelength travels in the fiber and encounters this photographically altered section, it will be reflected by this "dielectric mirror stack." Such mirrors can be spaced by effective half wave "layers" to produce NBP filters of one or more cavities. Therefore, most of the ideas we have discussed are applicable to the FBG field also. However, the difference in index from high to low in the current FBGs is only on the order of .001! This means that the spectral width of the high reflecting area (as in Fig. 1.19) will be quite narrow. This would be like what Thelen[4] calls a minus filter and is illustrated in Fig. 1.20. The attractive thing about the thin film DWDM filters is that the reflected band other than the NBP covers the whole C and L bands mentioned, whereas a FBG has no such general blocking capability. The FBG mirror, on the other hand, can be made to reflect only one narrow channel and pass the rest. This is the converse of the DWDM dielectric filters which transmit one channel and reflect the rest. It will be interesting to see how these technologies are mixed for best utility in the future.

1.3.4. Beamsplitters

Optical coatings with reflectance between that of high and low reflectors might be classified as beamsplitters. Combinations of the type of the first few layers shown in Fig. 1.19 could be starting points to make partial reflectors with all dielectric layers. Given some starting design, using any automatic optimizing thin film design program would probably result in a useful beamsplitter design.

1.3.4.1 Effects of Angle of Incidence

As the angle of incidence departs from normal incidence, the *s* and *p* polarizations behave as though they have different indices of refraction. The normal incidence index of refraction for the *s* polarization is multiplied by the cosine of the angle of incidence in the medium ($n_s = n \times \cos(i)$), and the index of refraction for the *p* polarization is divided by the cosine of the angle of incidence in the medium ($n_p = n / \cos(i)$).

It now also becomes necessary to know the angle of the light ray in the medium. This is found using Snell's Law: $n_1 \sin(i) = n_2 \sin(r)$, where i and r (in this case only) are the angles of the incident and refracted rays and n_1 and n_2 are the indices of the media of the incident and refracted rays.

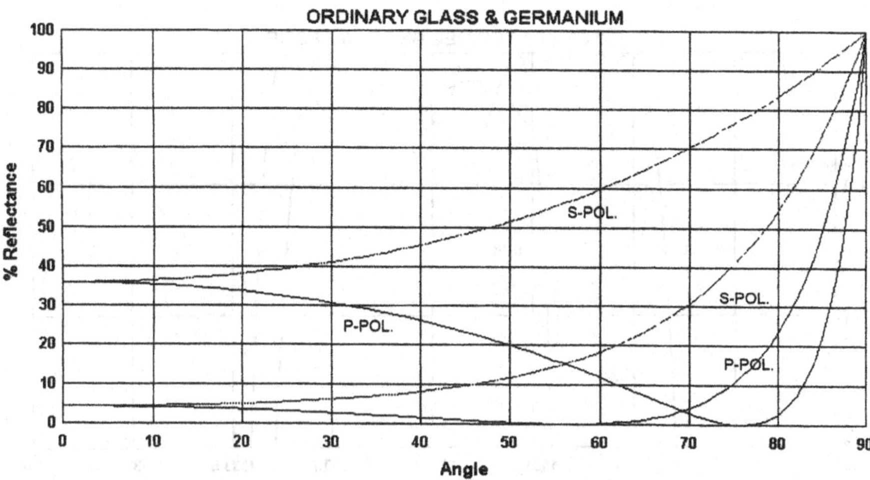

Fig. 1.30 Reflectance versus angle of glass and germanium surfaces.

Figure 1.30 shows these effects on uncoated substrates of ordinary glass (index 1.52) and germanium (index 4.0). Note that the *p* polarization reflectance goes to zero at angles of about 57 and 76° respectively. This is known as the Brewster Angle, and a window tilted at this angle will transmit all of the *p* polarized light while reflecting more *s* polarized light than at normal incidence. Such a window inside a laser beam will cause the laser to be polarized in the *p* orientation since the *s* polarization is much more "lossy".

It can be seen that the polarization effects are small up to 10 or 20 degrees, so we generally can consider this region to be "near normal incidence" and ignore the angular effects. However, some special cases like DWDM filters need more careful attention to polarization. Figure 1.31 shows the effects of a 15° tilt on the 100 Ghz filter shown in Fig. 1.26. The *s* polarization bandwidth gets narrower and has shifted from 1550 to 1533.65 nm. The *p* polarization has gotten wider and shifted further. This is probably not acceptable in the DWDM environment. Figure 1.32 shows the case of the filter tilted 9.9° and thereby shifted 7.2 nm to 1542.8 nm. This is a shift that covers 10 channels of 100 GHz bandwidth. If such distortion of the band were acceptable, the basic filter could be tuned by an appropriate tilt to any of these wavelengths in the range. To shift the band from

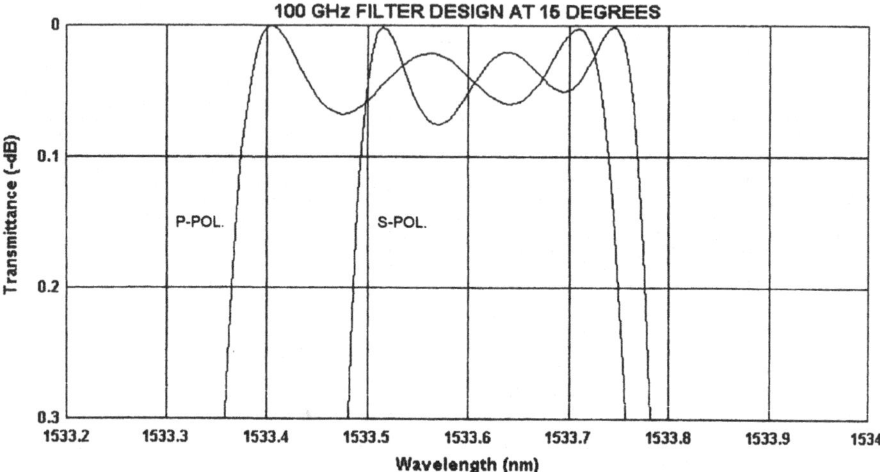

Fig. 1.31 Effects of a 15° tilt on a 3-cavity filter for 100 GHz DWDM applications.

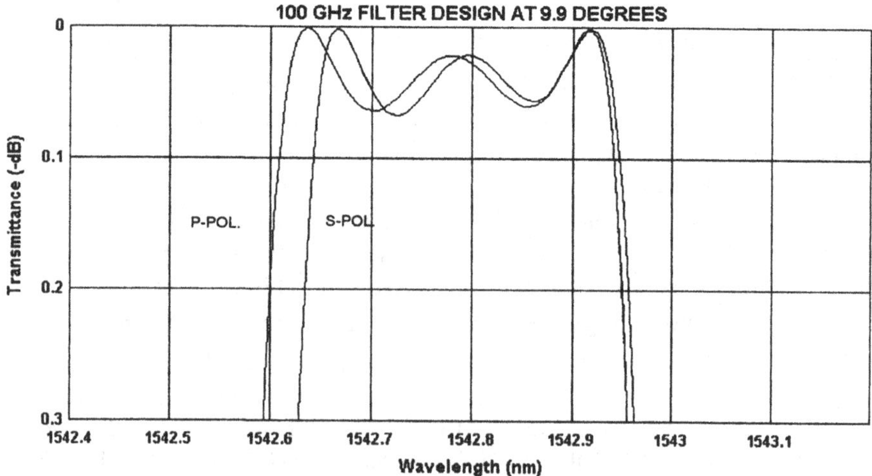

Fig. 1.32 Effects of a 9.9° tilt on a 3-cavity filter for 100 GHz DWDM applications.

1550 to 1549.2 nm or one channel, it takes a tilt of 3.3°. The wavelength shift is a function of the square of the angle of incidence, so the 9.9° (3 × 3.3°) shifts nine channels (3^2). This can give some indication of the alignment tolerances needed for such filters.

1.3.4.2 Polarizing Beamsplitters

The different behavior of the *s* and *p* polarizations at angles of incidence significantly different from normal incidence can be used to make polarizing beamsplitters. The situation is usually most favorable and useful if the beamsplitter coating is immersed in glass at 45° as in a beamsplitter cube. Figure 1.33 shows the results of a simple QWOT stack of 2.1/1.45 (16 layers) cemented in glass of index 1.52. In general, there is a region on both sides of the narrowed *p* polarization reflectance band where the broadened *s* polarization band can have very high reflectance and the *p* is quite low. It is therefore easy to make a polarizing beamsplitter for a narrow band like a laser on either the SWP or LWP side of the reflector. These are narrower if the coating is on a slab rather than immersed, but they still can be useful. This basic starting design can then be adjusted by optimization for the particular needs at hand.

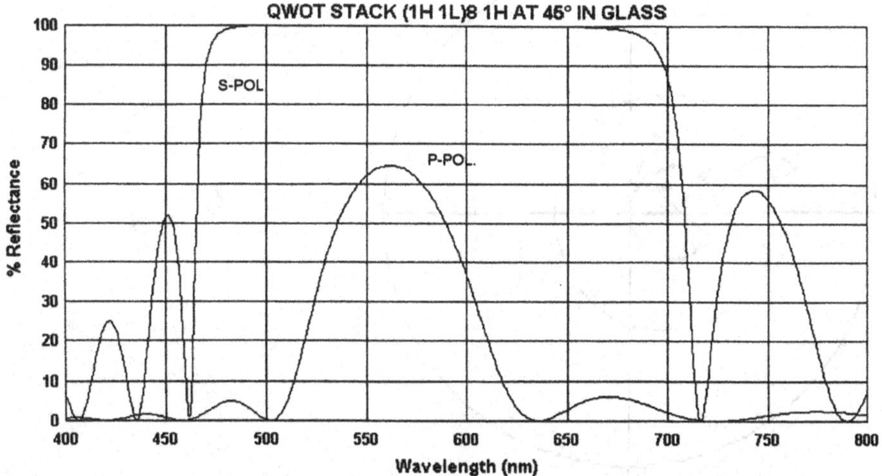

Fig. 1.33 QWOT stack (1H 1L)8 of indices 2.1/1.45 on glass at 45°

The ultimate broad band polarization beamsplitter is where the *p* polarization reflection in the middle of Fig. 1.33 can be reduced to near zero! This is the case of the MacNeille polarizer described in some detail by Macleod[2] on pages 328-332. Basically, conditions are satisfied to make the *p* polarization at the Brewster angle so that none is reflected and all is transmitted. The *s* polarization reflection is then boosted by an appropriate number of layers in the stack.

Cushing[48] described the design of a filter for use at 45° that had extremely low polarization. His approach was related to that of Baumeister[49] in using multiple indices to have less severe changes in index than just the highest to lowest.

Polarizing effects and coatings and any significant angle of incidence are best handled in thin film design software, but we will show a few more effects on admittance diagrams in Sect. 1.6.

1.3.5. Three-Layer AR Coating on Germanium, Example

Let us now apply the principles described above to the preliminary design of a three layer AR coating for a germanium substrate. We will use thorium fluoride at an index of about 1.38 as in Fig. 1.14 and germanium as a coating material at about 4.0 index as in Fig. 1.15.

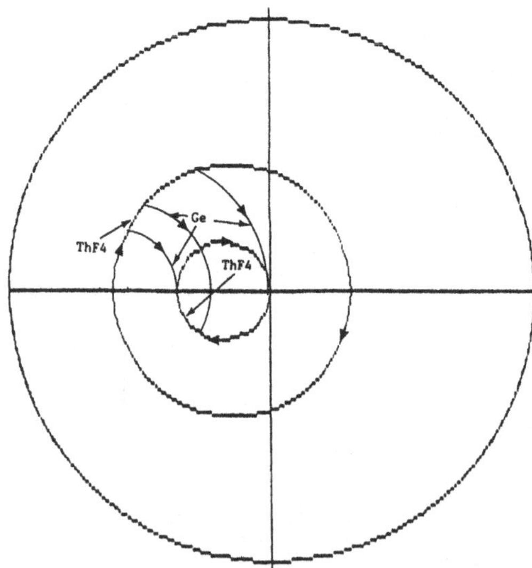

Fig. 1.34 Reflectance diagram of a preliminary design for a three-layer AR on germanium.

We know that the first layer (on the substrate) in this case would have to be thorium fluoride, because a germanium layer would have no effect on the germanium substrate. This means that the second layer would have to be germanium and the third thorium fluoride. We also know that to get zero reflectance the circle of the last thorium fluoride layer must pass through the origin of a reflectance diagram. We further know that the circle of the first layer must pass through the substrate index point of 4.0. We can use tracing paper over graphs like Figs. 1.14, 1.15, and 1.16 to draw these two circles as in Fig. 1.34. We can deduce that the germanium layer must lie in a specific range shown in Fig. 1.34 (or half wave additions thereto). By picking any of the germanium layers from the range and the corresponding thorium fluoride layers, the design can be quickly optimized over a specific (although limited) spectral range with any reasonable thin film design optimization program. It turns out that such a design optimized for a band from 8 to 12 μm will have the shortest first layer that allows the second layer to reach the left end of the circle for the third layer at the horizontal axis. This third (last) layer will be a QWOT. When we get to Fig. 1.80, it will be seen that such a solution is consistent with the principles that will become more apparent at that time.

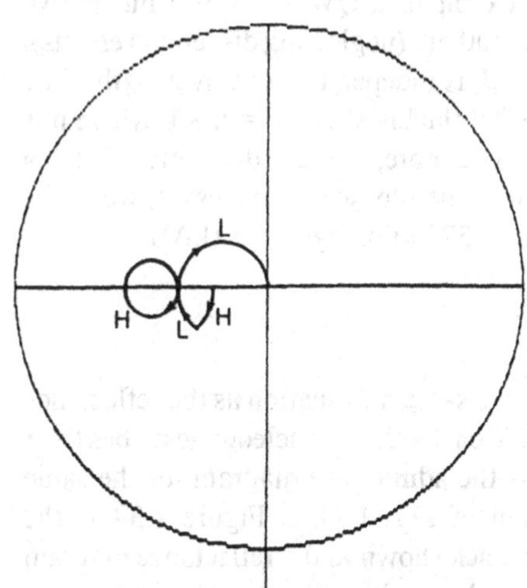

Fig. 1.35 Reflectance diagram of the construction of a two material, four-layer replacement for the three material, three layer AR in Fig. 1.11.

1.3.6. Example Four-Layer Broad Band AR Coating in the Visible

The index of 1.7 needed for the design of Fig. 1.11 mentioned in section 1.3.1 may not be easily available. We could think of using only the 2.25 and 1.38 indices and somehow simulate or approximate the 1.7 index with these materials. We know that the reflectance of the simulated index layer should start and end at the same points as for the QWOT layer with the desired 1.7 index. The only logical solution that comes to mind is a high index layer circle that starts at the substrate and goes clockwise, and a low layer circle that passes through the endpoint of the would-be 1.7 index layer as shown in Fig. 1.35. Where these two circles intersect is the end of the first layer and the beginning of the second layer. The third circle represents the half wave layer of high index, and the last circle the QWOT layer of low index from the three layer design of Fig. 1.11. This approximation of the intermediate layer by a combination of two layers of higher and lower index is attributed to Rock[5]. We discuss further approximations in more detail below.

1.3.7. Physical Thickness versus Optical Thickness

We have referred to the optical thickness of layers and the QWOT. It is sometimes necessary to know the physical thickness of layers. It is also necessary to keep in mind that an optical thickness (OT) is only a given value such as a QWOT at a specific wavelength. For example, a QWOT at 800 nm is two QWOTs or a half wave optical thickness at 400 nm (neglecting dispersion effects). The physical thickness (t), on the other hand, is independent of wavelength. The relationship between the optical and physical thickness is $OT = n \times t$, where n is the index at the wavelength of interest. Therefore, if the index were 2.0 for a QWOT of high index material at 1550 nm, the physical thickness, t, would be 193.75 nm (or 1937.5 Ångstrom units or 1.9375 kiloÅngstroms (kA)).

1.4. ADMITTANCE DIAGRAMS

The admittance diagram gives essentially the same information as the reflectance diagram, as they are conformal mappings of each other. Macleod[2] describes their derivation in some detail. Figure 1.36 is the admittance diagram for the same design shown in the reflectance diagram of Fig. 1.11. Figure 1.38 is the admittance diagram for the high reflector stack shown in the reflectance diagram of Fig. 1.18. The goal for an AR coating to be used in air or vacuum is to bring the admittance at the top of the coating to $1.0-i0.0$ which is the normalized admittance of free space. It can be seen that either representation, reflectance or

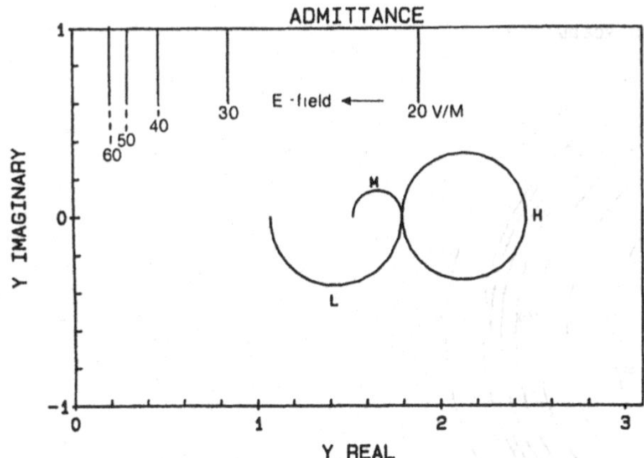

Fig. 1.36 Admittance diagram of the three layer AR in Fig. 1.11 for comparison and also showing the lines of equal electric field.

admittance, works well for designs like AR coatings which stay close to the origin. However, high reflectors quickly move out in admittance "orbits" that go off the page, whereas they remain always inside the unit circle of the reflectance diagram. One advantage of the admittance diagram is that the electric field at any point within the layer system can be determined simply from the distance from the imaginary axis. Macleod[6] gives the derivation of the electric field as a function of the admittance in free space units, i.e., Y = 1.0 in a vacuum. The resulting electric field in volts per meter as a function of the real part of the admittance units is

$$E = 27.46 / [Re(Y)]^{0.5} \tag{1.6}$$

where the incident power density is one watt per square meter. This generates lines of equal field on the admittance diagram which are parallel to the imaginary axis as shown on Fig. 1.36. On a reflectance circle diagram, the lines of equal volts per meter are as shown in Fig. 1.37. Here it is apparent that a design must stay away from the right outer portion of the diagram to avoid high fields. Apfel[7] describes how to design optical coatings for reduced electric field at any particular level in the stack. This is useful in designing laser damage resistant coatings. As we will see in Sect. 1.5, one might want the converse of high fields in absorbing materials to maximize absorption. Smith[8] gave an interesting comparison of admittance diagrams, reflectance diagrams, Smith charts (no relation), and other graphical representations. We use admittance diagrams for the balance of this chapter after the next section.

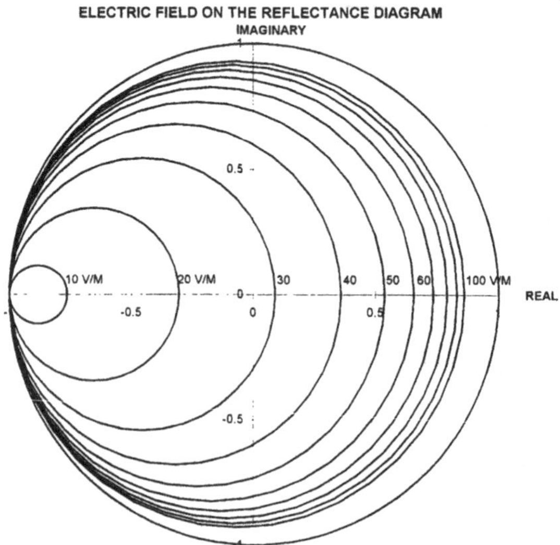

Fig. 1.37 Lines of equal volts per meter on a reflectance circle diagram.

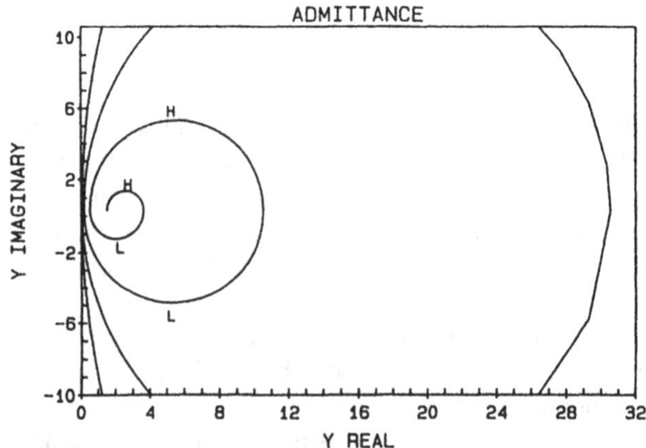

Fig. 1.38 Admittance of the QWOT high reflector stack shown in Fig. 1.18 for comparison.

1.5. TRIANGLE DIAGRAMS

Up to this point, we have dealt only with non-absorbing coating materials. Generally, the full description of index of refraction is a complex quantity, $N = n - ik$. Here the n is the index of refraction for a purely dielectric material that we have been considering thus far, and the k is the extinction coefficient needed to describe absorption in a material. The absorption coefficient, α, is defined as: $\alpha = 4\pi k/\lambda$, where λ is the wavelength. The ratio of the transmitted intensity, I, to the initial intensity, I_0, through an absorbing medium of thickness **x** can be found by: $I/I_0 = e^{-\alpha x}$.

Apfel[9] describes another useful graphical tool, the triangle diagram. These are useful in seeing at a glance what a semi-transparent film will reflect, transmit, and absorb, and what transmittance can be achieved if the layer were coated to reduce its reflectance. We will describe it here and go through some examples. Figure 1.39, taken from Apfel's work, shows the characteristics of twelve semitrans-parent metallic films when used as single layers of varying thickness. The three corners of the triangle diagram represent 100% transmittance (T), reflectance (R), and absorption (A). The sum of the three must equal unity ($A + R + T = 1$).

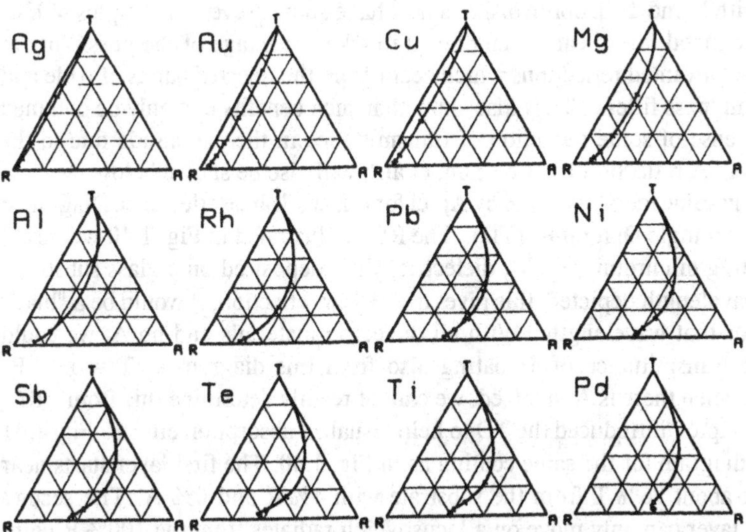

Fig. 1.39 Triangle diagrams plotting R, T, and A for various semitransparent metallic films versus thickness from zero to an opaque film (from Apfel[1.5]).

At the start of deposition with zero thickness (as a freestanding film with no substrate), T equals 100%, and R and A equal 0%. As the thickness increases, the reflectance and absorption increase as a function of the real and imaginary parts of the refractive index. The transmittance moves from the 100% apex to the 0% baseline. Silver, for example, is seen to quickly reach a high reflectance with low absorption losses, while nickel moves quickly to much higher absorption values before reaching a peak R = 60% and A = 40% as the transmittance goes to zero.

Apfel shows that the potential transmittance described by Berning et al.[10] can be found by a straight line from the R= 100% point through the particular point representing the thickness of the metallic film. That is to say that an AR coating on the metallic film could reduce the reflectance to zero and move the T and A to the point where the line of equal potential transmittance intersects the triangle side T to A. The triangle diagram is a relatively new and useful tool for working with metallic (or other absorbing) thin films.

In the balance of the book other than this section, we confine the discussions to non-absorbing materials unless otherwise stated.

1.5.1. Designing Coatings with Absorbing Materials

In much of the literature, A is taken to be zero for most dielectric film studies, or at most a perturbation. However, when metals are used, A becomes an equal partner with R and T. Dobrowolski, et al.[29] have shown several examples of how and where metal layers can be employed to take advantage of the possibility to reduce the unwanted reflections which occur from the rejected bands of wide and narrow bandpass filters. They also show that such benefits can only be obtained at the expense of some reduction of transmittance in the passbands due to the inherent A. A reflection filter by Tan, et al.[30] will also be shown below.

The graphic tools promoted by Apfel for this work are Reflectance Diagrams[1] (RD) and Triangle Diagrams[9] (TD). The RD is illustrated in Fig. 1.40 where a 3 layer coating of chromium (Cr), dielectric, Cr is deposited on a glass substrate. For the wavelength depicted, this gives a very low reflection; it would be a "black mirror" for that wavelength. With all dielectric materials and no A, we could know the transmittance of a coating also from this diagram as T = 1 - R. However, when there is A involved, we cannot readily determine this from a RD. However, Apfel[9] introduced the TD to help visualize absorption effects. Fig. 1.41 is such a diagram for the same coating as in Fig. 1.40. The first layer starts near the top at about 96% T from the substrate with 4% R and 0% A. The second dielectric layer can only move on a locus which radiates from the 100% R point through the point where that layer starts. With the proper combination of dielectric layers, it is possible to move on such a locus to the 0% R line (on the

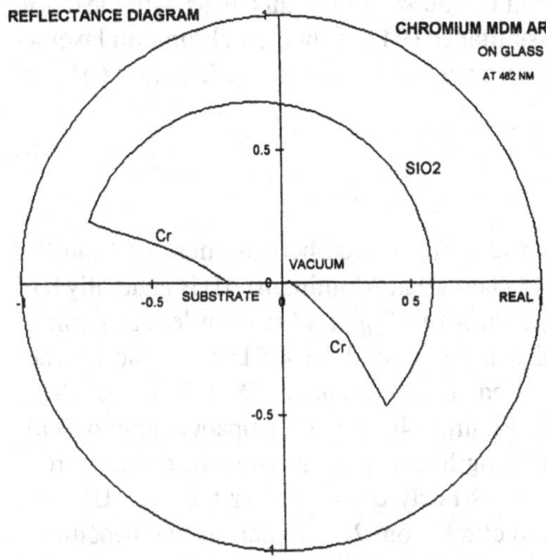

Fig. 1.40 Reflectance diagram (RD) of a MDM (Cr, SiO2, Cr) AR on glass at 482 nm.

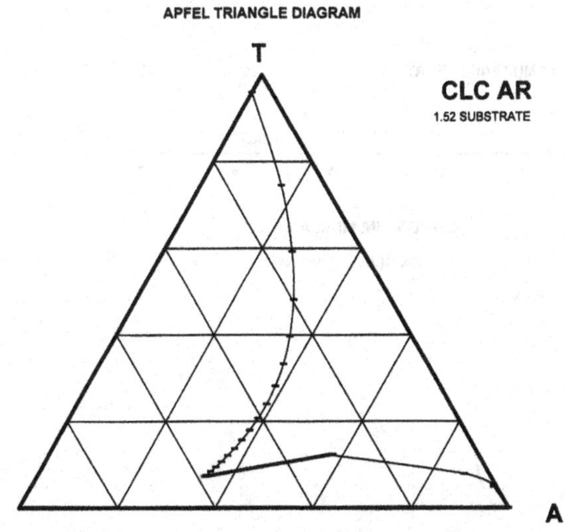

Fig. 1.41 Triangle diagram (TD) of the same coating as shown in Fig. 1.40.

right edge between T and A). This would be an antireflection coating on the chromium layer in this case. The point of intersection of this locus with the 0% R line would show the Potential Transmittance (PT) of the first chromium layer as about 17%. This same coating is illustrated on an Admittance Diagram (AD) in Fig. 1.42.

1.5.1.1 Absorbing Materials

Most of the material properties of n and k used here are derived from the compilations of Palik[31,32]. Figure 1.43 shows that Aluminum (Al) is generally less absorbing than Cr. The Silver (Ag) shown in Fig. 1.44 is even less absorbing. The reflectance of a bulk material can be plotted on a RD from the Fresnel equation. Figure 1.45 shows lines of equal n and k on a RD. On an AD, these lines would form a rectangular grid. Figure 1.46 shows the opaque point or bulk reflectance of Cr as a function of wavelength. It can be seen that the distance from the origin and thereby the reflectance is fairly constant over the near UV and visible wavelengths. It then starts to climb from 2 to 3 microns and becomes a relatively high reflector at longer wavelengths. Cr has a generally neutral gray color with a little less reflectance in the blue. Ni has a similar locus but has a bit more reflection toward the blue. This is probably why Nichrome is favored for neutral density filters, the proper mixture of Ni and Cr should best compensate each other to give nearly constant R and T over the visible spectrum.

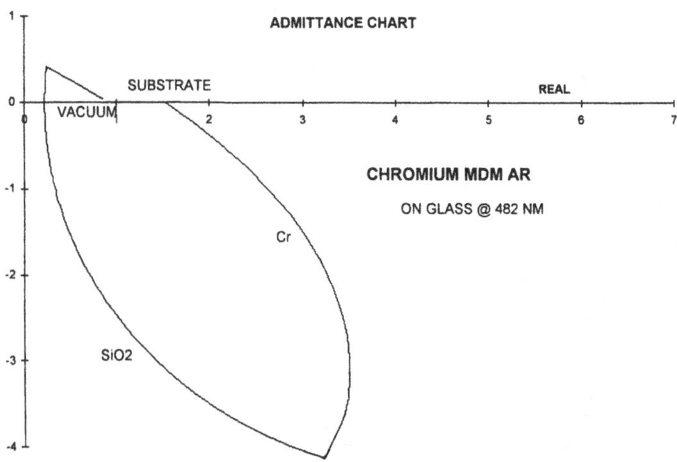

Fig. 1.42 Admittance diagram of the same coating as in Fig. 1.40.

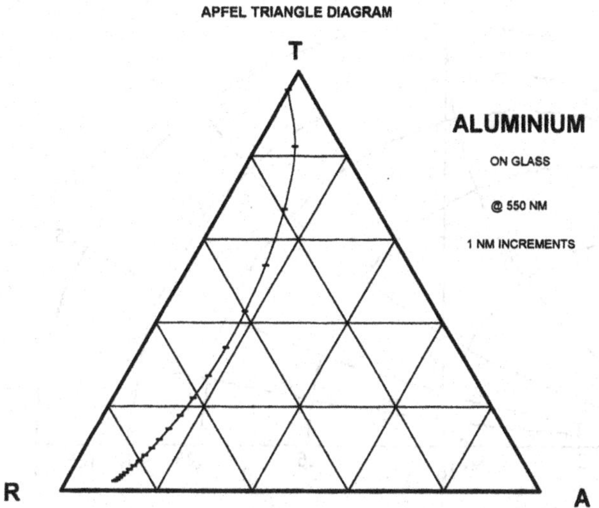

Fig. 1.43 Triangle diagram of a thick layer of aluminum (Al).

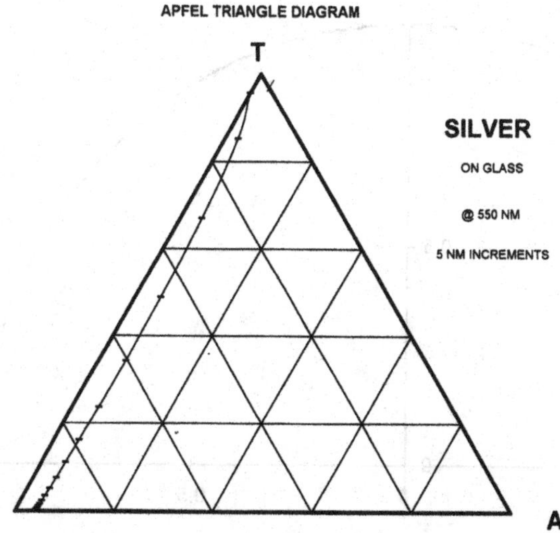

Fig. 1.44 Triangle diagram of s thick layer of silver (Ag).

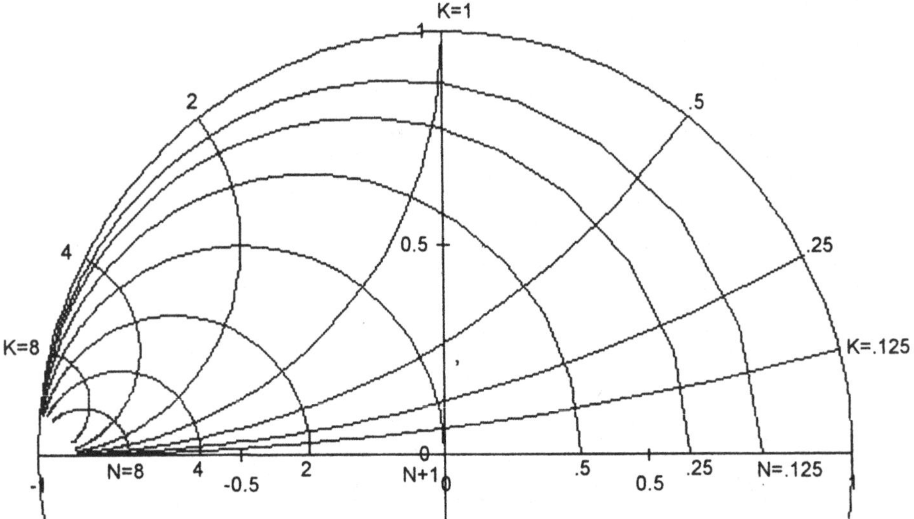

Fig. 1.45 Constant n and k lines on a reflectance diagram.

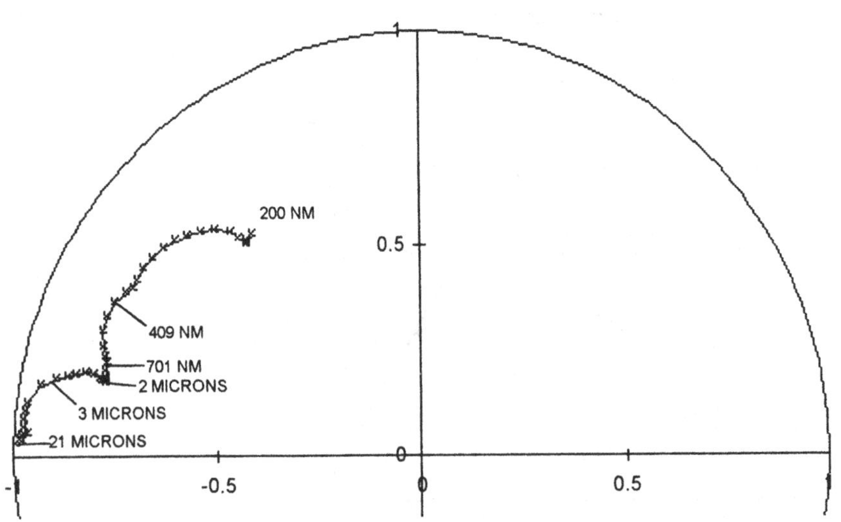

Fig. 1.46 Reflectance of the chromium (Cr) opaque point versus wavelength.

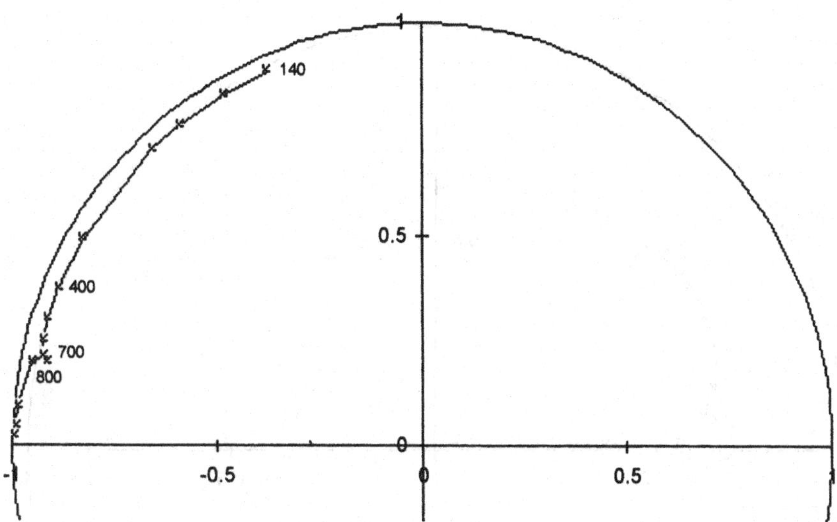

Fig. 1.47 Reflectance of the aluminum (Al) opaque point versus wavelength.

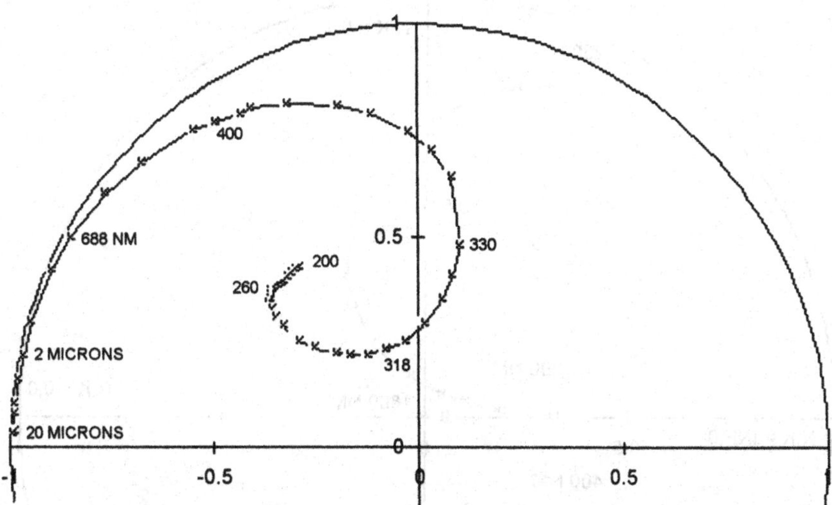

Fig. 1.48 Reflectance of the silver (Ag) opaque point versus wavelength.

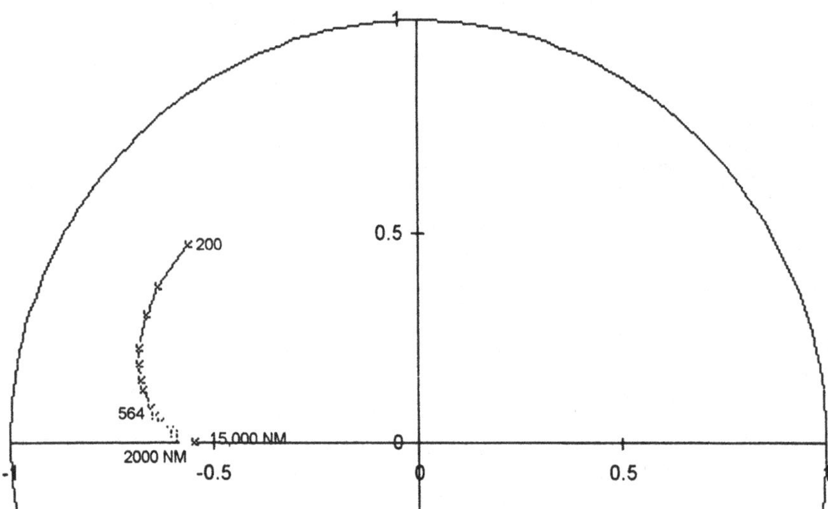

Fig. 1.49 Reflectance of the silicon (Si) opaque point versus wavelength.

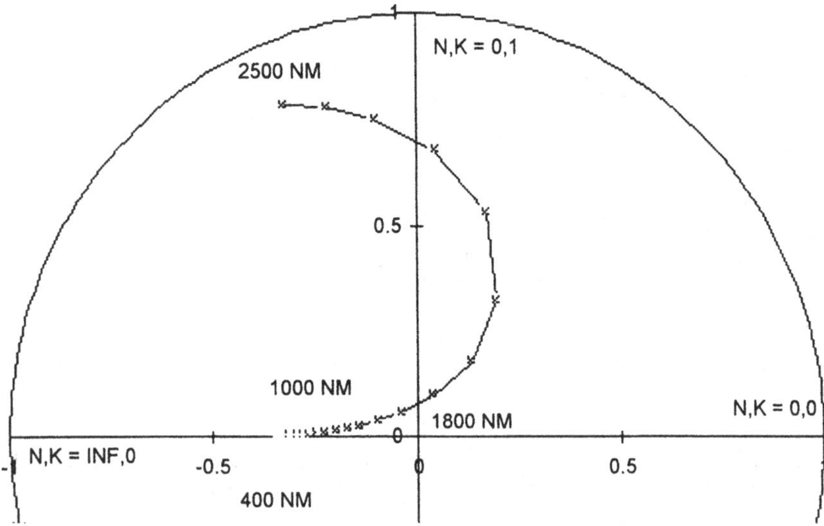

Fig. 1.50 Reflectance of the indium oxide (InO_X) opaque point versus wavelength.

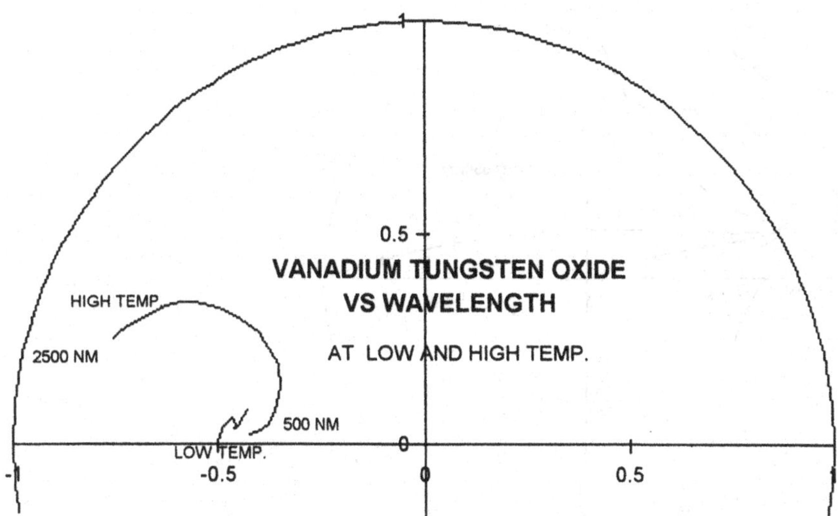

Fig. 1.51 Reflectance of the vanadium tungsten oxide opaque point versus wavelength.

Figure 1.47 shows that Al has a fairly flat spectral reflectance except around the well known dip near 800mm. Figure 1.48 shows that Ag has a minimum reflectance at around 318 nm and then it increases steadily with wavelength. The semiconductor materials of silicon (Si) and indium oxide (InO_x) are shown in Figs. 1.49 and 1.50. The Si has high absorptance in the UV decreasing to none at about 1 micron. Conversely, InO_x has none in the visible and starts increasing from 1 micron to longer wavelengths where it becomes a high reflector.

Figure 1.51 shows the interesting thermochromic properties of tungsten doped vanadium oxide ($V_{1-x}W_xO_2$) as reported by Tazawa, et al.[33] At temperatures below the critical value of 47°C, the material has low absorptance and looks mostly like a high index dielectric. Above this temperature, it behaves more like a metal with increasing reflectance with wavelength. This type of behavior could be useful to make a passive window which would transmit sunlight and infrared warmth when the window was cool and block the heat when the window was hot.

1.5.1.2 Application Examples of Designing With Metals

When viewed on a RD, the locus of the reflection versus film thickness changes predictably from the starting point reflectance and phase. Figure 1.52, 1.53, and 1.54 show such loci for Cr, Al, and Ag. The fact that the metal loci

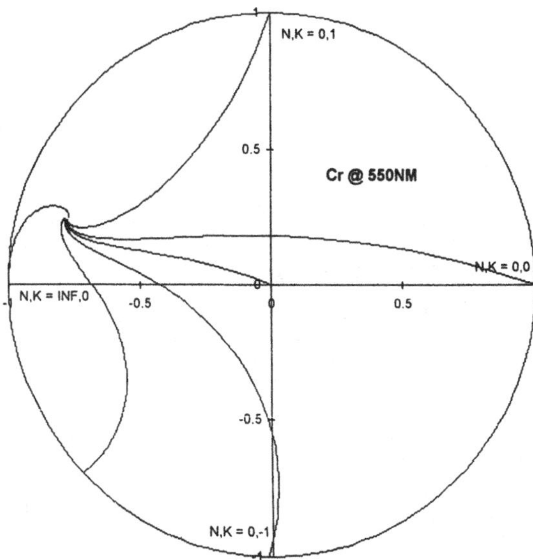

Fig. 1.52 RD showing loci of chromium (Cr) from various starting points.

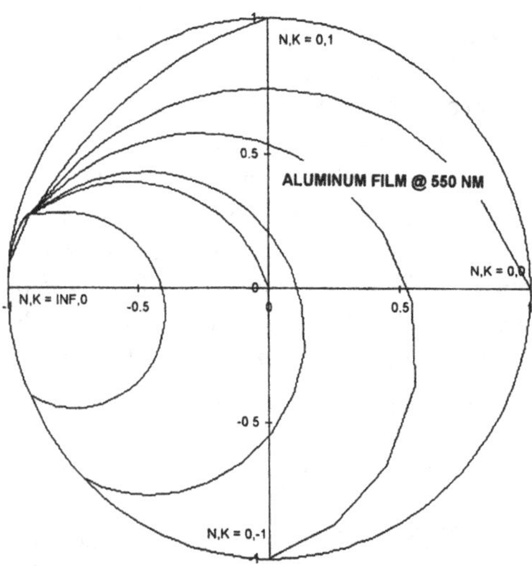

Fig. 1.53 RD showing loci of aluminum (Al) from various starting points.

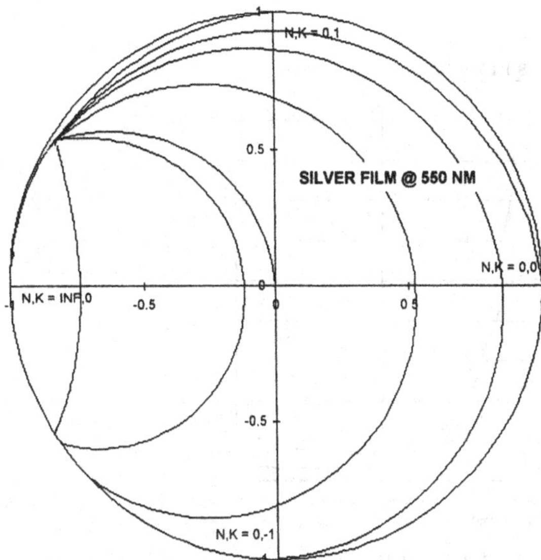

Fig. 1.54 RD showing loci of silver (Ag) from various starting points.

can in some cases move almost in the opposite direction from a dielectric locus may prove useful in future designs where some A can be tolerated. These type of charts can be used in conjunction with those like Figs. 1.14 and 1.15 to design films that combine metals and dielectric materials.

Induced Transmission Filter Example

An example of an induced transmission filter is taken from Fig. 6 of the work by Dobrowolski, et al.[29] Figure 1.55 shows the transmittance of a similar design with and without suppression of the unwanted reflections shown in Fig. 1.56. The higher reflection curve in Fig. 1.56 is the unwanted reflection and the lower curve is the result of the reflection suppression. Figure 1.57 shows how this would look on a TD; the unsuppressed design at 570 nm in the passband ends near the top of the TD with about 90% T and 10% R. The suppression brings the locus to about 50% T and 2% R.

The TD viewpoint as seen in Fig. 1.57 is not very revealing with respect to what is happening to the R and T when the dielectric layers are deposited, just as the RD does not show what is happening to the A and T when there is A present. To overcome this limitations, we have modified the TD to make a Prism Diagram

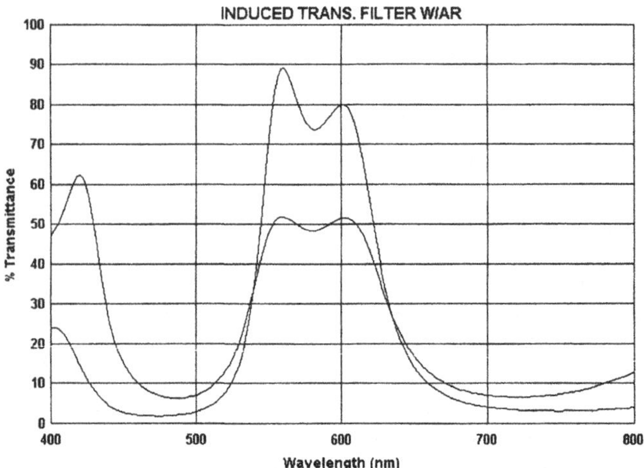

Fig. 1.55 Transmittance spectrum of Dobrowolski, et al.[29] induced transmission design with and without suppression AR at 570 nm.

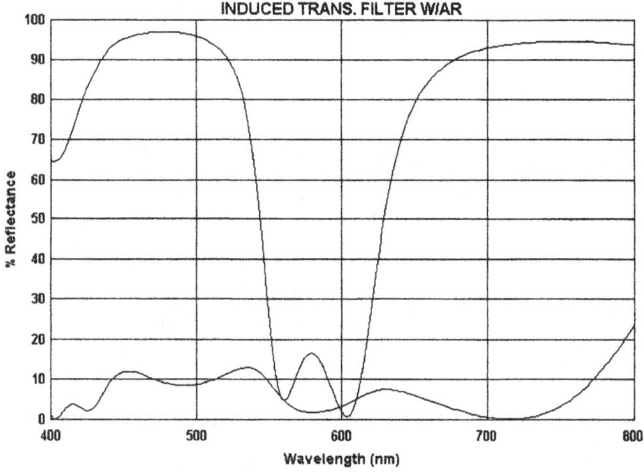

Fig. 1.56 Reflectance spectrum of Fig. 1.55 design with and without AR.

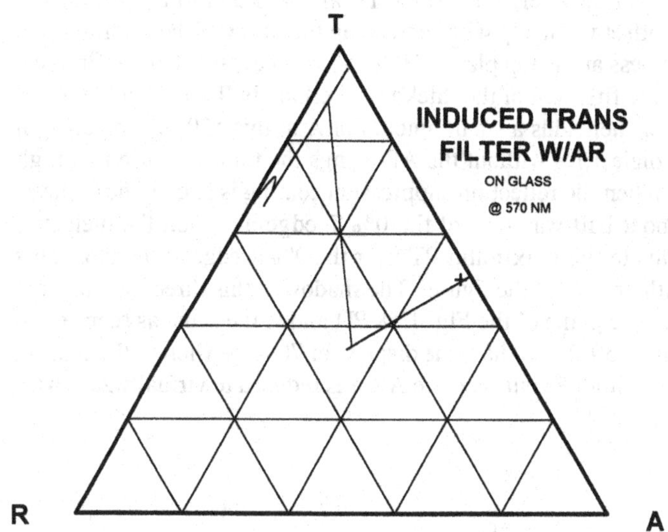

Fig. 1.57 Triangle diagram of Fig. 1.55 design with AR.

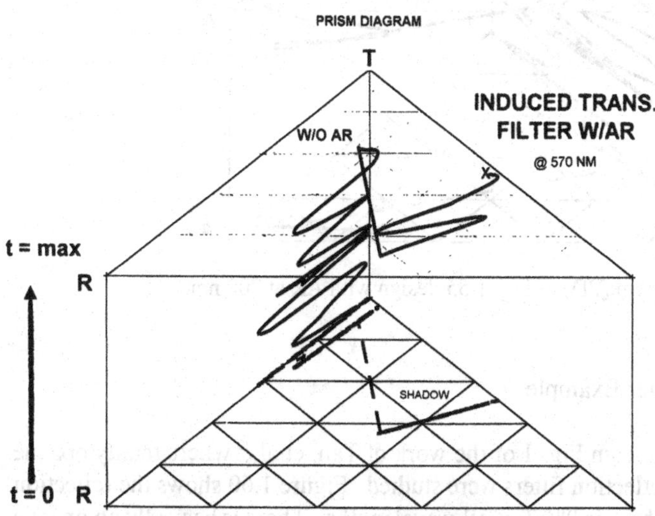

Fig. 1.58 Prism diagram (PD) of Fig. 1.55 design with AR at 570 nm.

(PD) as shown in Fig. 1.58 wherein the same TD as Fig. 1.57 forms the base, but the values are also offset vertically with increasing thickness of the coating up to the maximum thickness at the top plane. Here, we can now see that the first five dielectric layers in the first half of the thickness oscillate in R and T in the 0% A plane. The Ag layer then adds a small amount of A at this 570 nm wavelength and the rest of the dielectrics without the AR brings the transmittance to a high %T and low %A. When the reflection suppression coating is added, the Cr layer moves the locus about half way toward the 0% T edge and then the dielectric layers antireflect this to the maximum PT near the 0% R edge at the top. This point is marked with an "X" in the figure. The shadow of this three-dimensional path is seen on the base plane of the Fig. 1.58 PD and is the same as seen in the Fig. 1.57 TD. Figure 1.59 shows the same display for 700 nm wherein the coating without AR has a very high R, but with the AR it is reduced to virtually zero with a small residual %T.

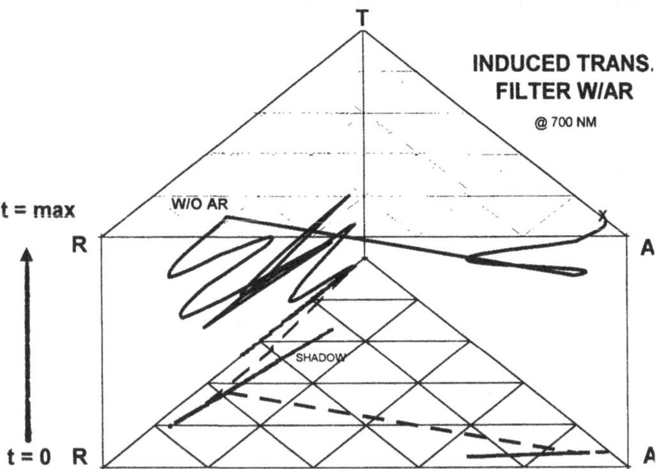

Fig. 1.59 Prism diagram (PD) of Fig. 1.55 design with AR at 700 nm.

NBP Reflection Filter Example

Another example is from Fig. 1 of the work of Tan, et al.[30] where totally opaque narrow bandpass reflection filters were studied. Figure 1.60 shows the reflection of the filter where there is 0% T at all wavelengths. There is basically an opaque Ag layer, dielectric stack, and a final thin layer of Cr. The principal used here is

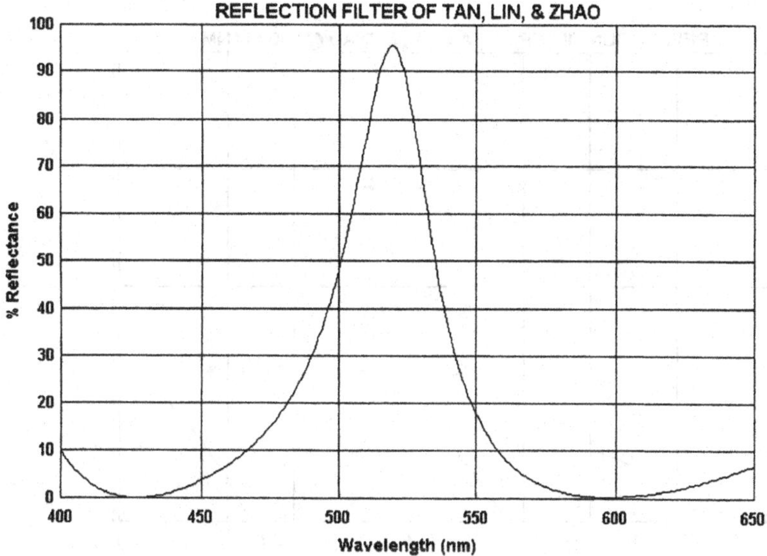

Fig. 1.60 Reflectance spectrum of Tan et al.[30] filter.

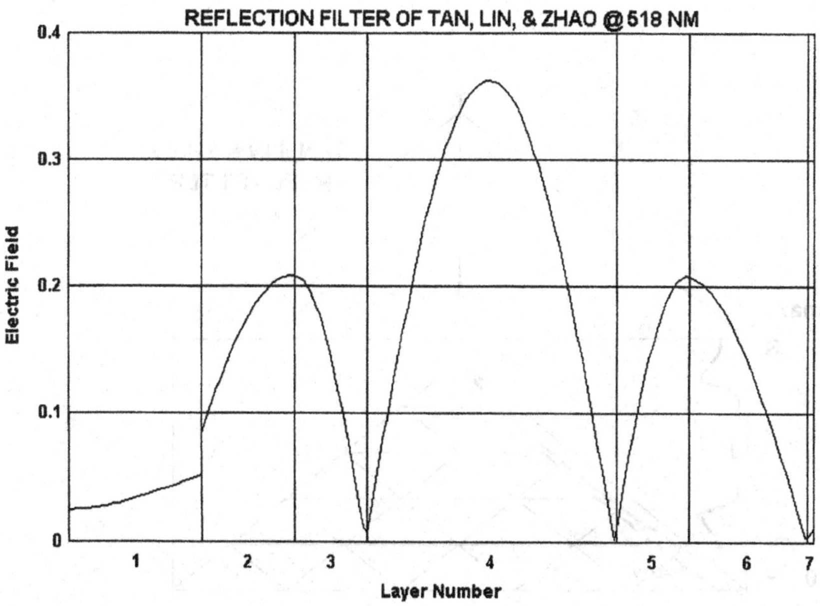

Fig. 1.61 Electric field in coating of Fig. 1.60 at 518 nm.

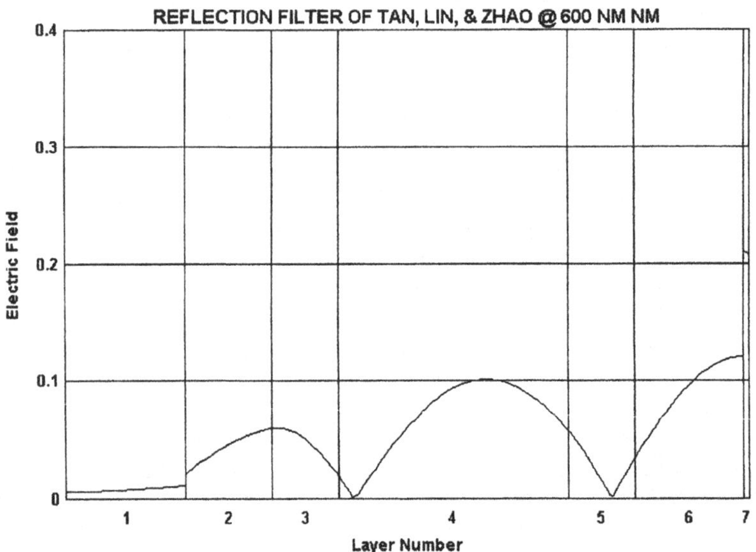

Fig. 1.62 Electric field in coating of Fig. 1.60 at 600 nm.

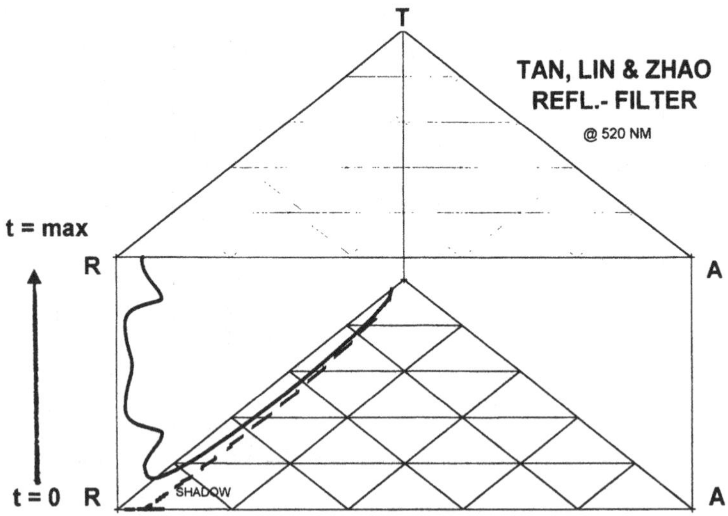

Fig. 1.63 Prism diagram for coating of Fig. 1.60 at 520 nm.

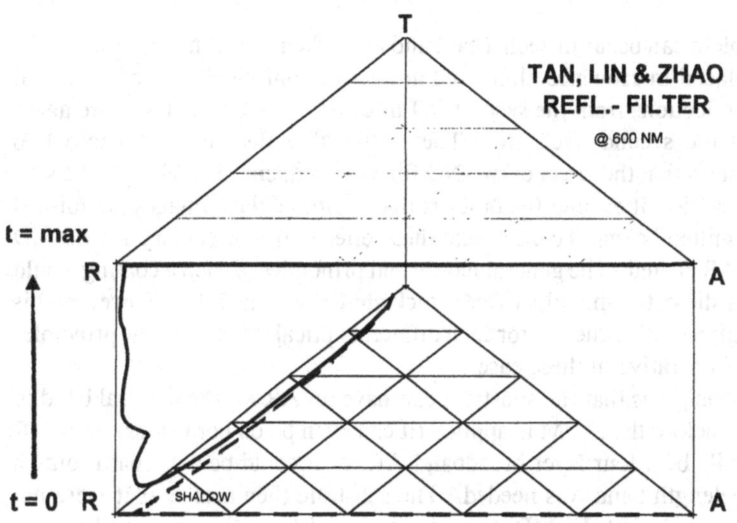

Fig. 1.64 Prism diagram for coating of Fig. 1.60 at 600 nm.

that the stack creates a spacing such that the electric field is near zero at the thin Cr layer for the passband wavelength as seen in Fig. 1.61. Under these conditions, the Cr layer absorbs very little. As the wavelength moves either way from this, the electric field increases rapidly and causes great A in the Cr layer as shown in Fig. 1.62 and thereby low reflection. Figures 1.63 and 1.64 show the PD for this filter in the passband at 520 nm and in the blockband at 600 nm. It can be seen in Fig. 1.60 for 520 nm that the coating ends at very high %R and low %A where there is no %T. In Fig. 1.64 for 600 nm, we see that A goes to 100% and thereby T and R are zero.

AR Coated Variable Neutral Density Filter Example

Neutral density (ND) filters can be made of a variety of metals deposited to a thickness that produces the desired optical density. Aluminum will transmit more in the blue than in the red. Chromium is more neutral than aluminum, but nickel is more neutral than chromium. An alloy of nickel and chromium, "nichrome", is the most neutral of these in the visible spectrum.

Variable ND filters have been produced for decades with either a linear variation along a straight line or a circular filter which varies with rotational position. If two otherwise identical linear or circular filters are positioned with their gradients of density in opposing directions, a nominally uniform density over an area can be achieved which can be varied by moving the filters in opposing directions.

A problem can occur in such a case and also when two or more uniform ND filters are placed in sequence along and at near normal incidence to an optical path. The reflections from the second ND filter return to the first and are again reflected to the second, etc., etc. These "ghost" reflections can cause two problems: one is that the effect of two ND filters in sequence is no longer the sum of their optical densities, and the other is that multiple ghost images are formed in the transmitted beam. To eliminate these effects, it is necessary for the ND filters to be AR coated. The general nature and principles of such a coating would be the metal-dielectric-metal (MDM) stack shown in Fig. 1.40. However, this must be of different thicknesses for each different optical density. A few principles can be found operative in these cases.

One principle is that the substrate can have an AR for the spectral band of interest on it before the MDM is applied (it can even be on both sides). The AR might typically be a four layer broadband AR, or it could be a V-coat if only a narrow wavelength band was needed. The substrate then acts as if it were not there. This means that the MDM can be designed to be the same on both the substrate and air side of the main metal film or as though it were a self supporting and free standing film. Note that the result would then be MDMDM on top of the AR. The reflection amplitude from the top M layer essentially needs to be equal to the reflectance from the M layer below the D spacer which reaches the top and out of phase with it at the center of the AR band. The two reflections will then cancel each other and create the AR effect. The D spacer then can be the same thickness for all densities in the variable ND filter because it is only required to maintain the phase between the reflections from the top and next lower M layers. This is the second principle of interest. The D spacer could be either high or low index, but low index layers seem to give better design results. We believe this to be related to the fact that the low index locus on a circle diagram is more nearly concentric with the origin than a high index locus. From a circle diagram viewpoint, the first metal layer would move out from the origin instead of the substrate index point as in Fig. 1.40. The D layer would travel in approximately a semicircle (QWOT) toward the origin as in Fig. 1.65. The second M layer would pass near the origin from lower right to upper left and terminate at a maximum reflectance amplitude. The second D layer would travel on another larger semicircle. The third M layer would bring the reflectance amplitude essentially to the origin again (the place of no reflection). The higher density filters would have longer metal lines that the lower density filters because this is what causes the absorption for the optical density.

Figure 1.66 shows the reflectance (upper curve) and the transmittance (lower curve) for an 2.0 OD filter (10% T) as shown in the circle diagram of Fig. 1.65.

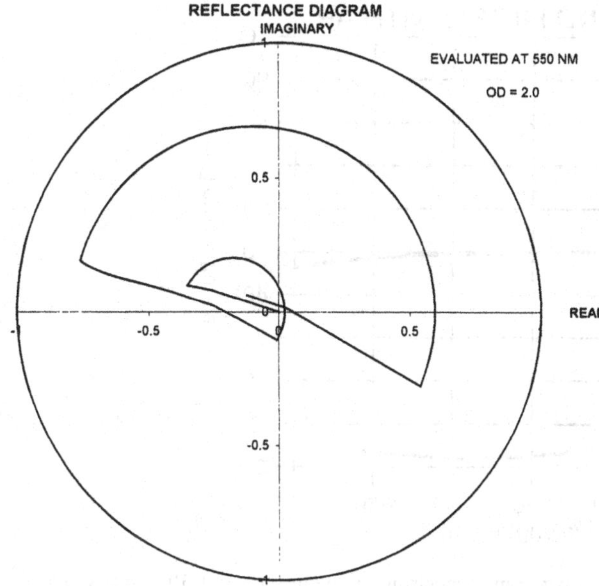

Fig. 1.65 Reflectance diagram of an antireflected neutral density filter at 2.0 OD.

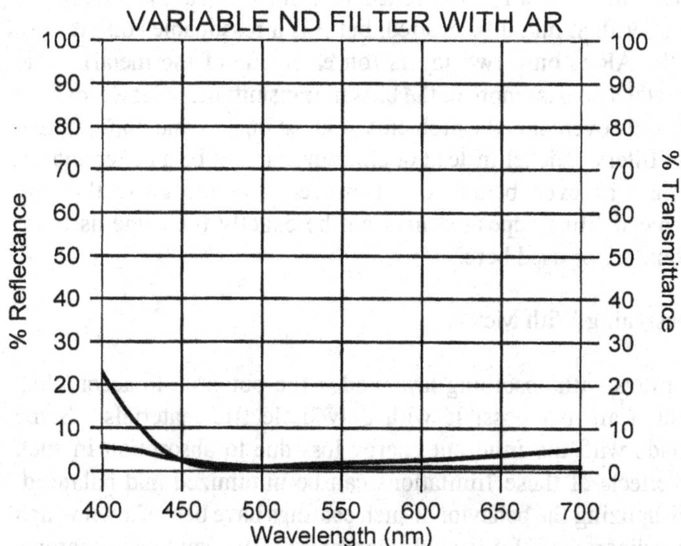

Fig. 1.66 Reflectance and transmittance spectrum of an antireflected filter at 2.0 OD.

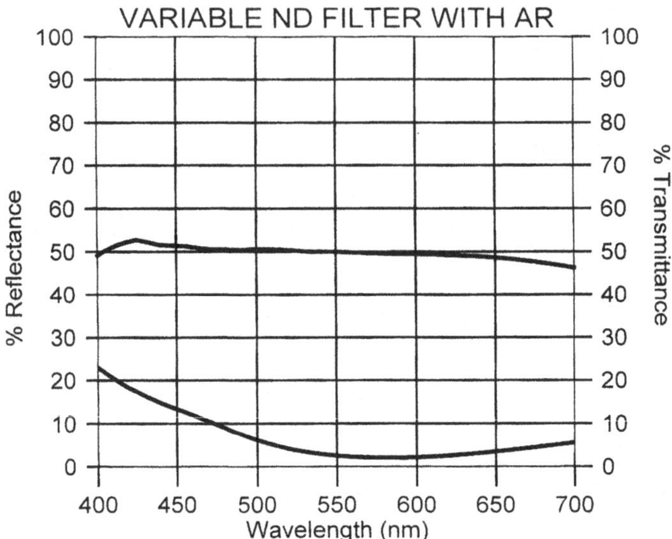

Fig. 1.67 Reflectance and transmittance spectrum of an antireflected filter at 0.3 OD.

Figure 1.67 shows the reflectance (upper curve) and the transmittance (lower curve) for an 0.3 OD filter (50% T). The reflectance of the AR at the blue end of the spectrum is higher than might be desired, but it is a reasonable compromise considering that the AR is only two layers (on each side of the metal). The properties of the metal also pass more of the blue in transmittance. However, this approach will generally overcome the problems of ghost images and non-additive optical densities of filters. Nickel instead of chromium might be a better choice, and nichrome might be even better yet. However, also recognize that the properties of the metal films deposited may not be exactly the same as those reported in the literature or used here.

Conclusions of Designing With Metals

The design of coatings with absorbing layers adds the potential to accomplish characteristics which are not possible with only dielectric materials. Some compromise is made with the inherent energy loss due to absorption in such coatings, but the effects of these limitations can be minimized and balanced. Several ways of visualizing the behavior of such coatings have been shown which include: reflectance diagrams (RD), triangle diagrams (TD), admittance diagrams (AD), and a new prism diagram (PD). These can aid in the understanding of certain types of coatings. We have added, in the Appendix, a selection of TDs and RDs for various materials of possible interest and usefulness that have not yet been shown here.

1.5.1.3 Additional Graphics for Visualization

It is also possible to make reflectance diagrams and prism diagrams, using a spreadsheet program, which are stereoscopic and show what is happening in three dimensions (3D). In some cases, this has been helpful to the author in gaining further visualization and insight concerning the behavior of some coating designs. Figure 1.68 shows a stereoscopic reflectance diagram (SRD) of a high-low stack which starts on glass and spirals clockwise toward the observer with increasing thickness.

Figure 1.69 shows a stereoscopic prism diagram (SPD) of the growth of an opaque film of a metal like Ni starting on a glass substrate.

For many decades, photogrammetric journals have published stereoscopic photo pairs which are printed with a separation less than the typical interpupilary distance of the average observer. Their subscribers have become comfortable with an ability to view the left image with the left eye and the right image with the right eye.

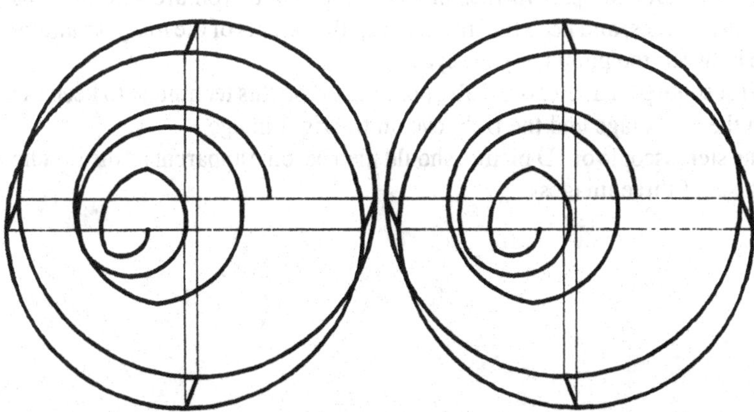

Fig. 1.68 Stereoscopic reflectance diagram (SRD) of a high-low stack which starts on glass and spirals clockwise toward the observer with increasing thickness.

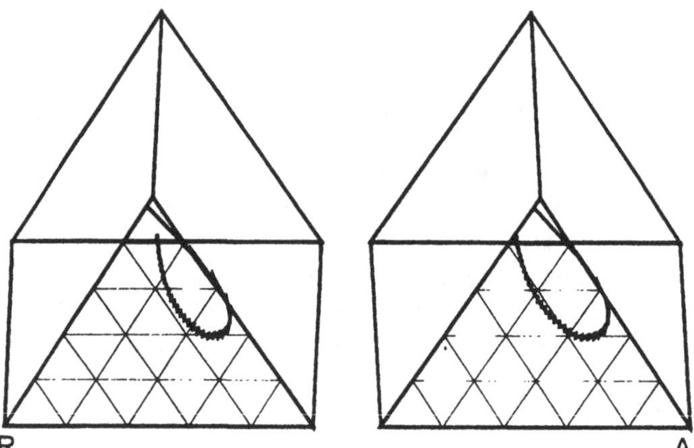

Fig. 1.69 Stereoscopic prism diagram (SPD) of the growth of an opaque film of a metal like Ni starting on a glass substrate.

If you have not done this before, try the following:

1) Position the stereo pair as close as possible in front of your eyes.

2) Slowly move the pair further in front of you until you are able to focus sharply on the images, and keep the line between the centers of the images parallel to the line between the pupils of your eyes.

3) With some practice, it is usually possible to use this technique to keep the left eye on the left image and the right eye on the right image.

4) The stereoscopic or 3D picture should then become apparent as the middle image in a set of three images!

1.6. APPROXIMATIONS OF INDICES AND DESIGNS

We will now touch on the concepts of Herpin equivalent index and Epstein periods. Herpin[11] showed that any film combination is equivalent at one wavelength and angle of incidence to a two film combination. Epstein[12] carried this further to show that a symmetrical combination of layers was equivalent to a single film at one wavelength and angle. We will illustrate those graphically and expand on the concept of approximations of the ideal when the ideal is not directly achievable. We described[13] the possibilities of equivalent index approximations which we will review here for the convenience of the reader.

We discussed above the case of approximating the desired solution shown in Fig. 1.11 and 1.36 by the solution in Fig. 1.35 with only two available real materials. Rabinovitch et al.[14] and others have pointed out that these approximations are only fully equivalent at one wavelength and one angle of incidence. There are various ways to approximate the layer with a combination of two indices that are higher and lower than the index to be simulated. Figure 1.70 shows the principle on the first layer of a three-layer design in graphic form. When a QWOT of medium index 1.65 is deposited on a substrate of index 1.52, the admittance curve follows a semicircular path M from point A to Z in the figure. Any set of layers which bring the admittance from point A to point Z will have the same performance as M at this one wavelength and angle of incidence, but not necessarily at others. The Epstein period is defined as having a central layer of high or low index and two equal and symmetric outer layers of the opposite index.

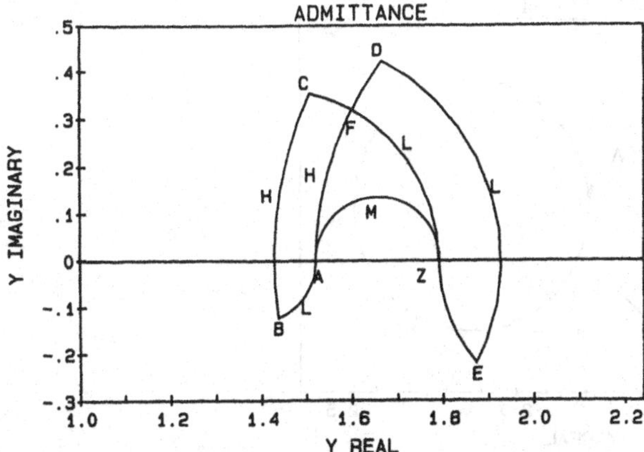

Fig. 1.70 Admittance of a QWOT of index 1.65 (curve M from A to Z), Epstein LHL period from ABCZ, Epstein HLH from ADEZ, and Rock 2-layer Herpin HL from AFZ.

For any given available high and low index materials, there are two solutions to the Epstein period for a given case. These are shown in Fig. 1.19 by the path ABCZ for the LHL period and by ADEZ for the HLH period. We have already mentioned the two layer solution of Rock[5] which is represented by the path AFZ and seen in Fig. 1.35. Herrmann[15] discussed a non-symmetric three-layer solution which was an improvement on the Epstein periods over a broader wavelength region. We showed[13] that the closer the admittance path of the simulation approaches the path of the system to be approximated, the more nearly the same will be the performance of the simulation to the design approximated at all wavelengths and angles, as one would justifiably expect. This is a key point to keep in mind for later sections of this chapter.

By dividing the simulated layer into an arbitrary number of thin layers, the resulting admittance locus can be made to follow the admittance curve to any arbitrary degree of closeness. Figure 1.71 illustrates a 20 layer approximation of the medium 1.65 index first layer. Figures 1.72 and 1.73 show the differences in performance between the approximated design and the various approximations over a broad wavelength range at an angle of incidence of 45 degrees. The simulations were designed to match the 1.65 index at 524nm and normal incidence. It can be observed that the above key point is well demonstrated. Figures 1.74 and 1.75 illustrate how the admittance diagrams change from Fig. 1.70 for the s and p polarizations for each of the approximations at 45 degrees angle of incidence. The Herpin and Epstein concepts are valuable tools as long as their limitations are kept in mind. The concept of the optimum degree of approximation is what we want to emphasize here, however.

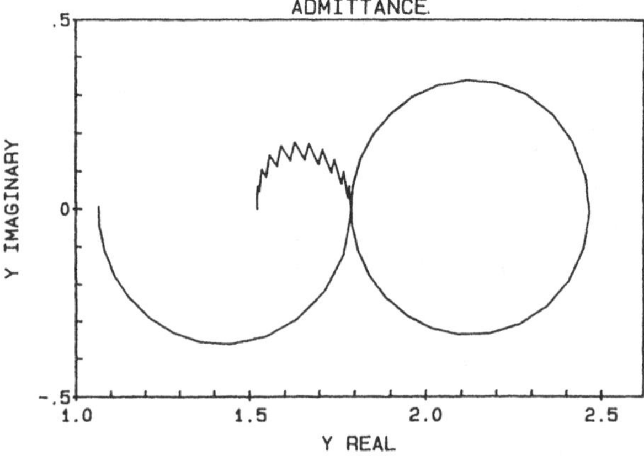

Fig. 1.71 Admittance diagram at 524 nm and normal incidence of the three-layer AR with the 1.65 index layer replaced by and optimized 20-layer Herpin equivalent.

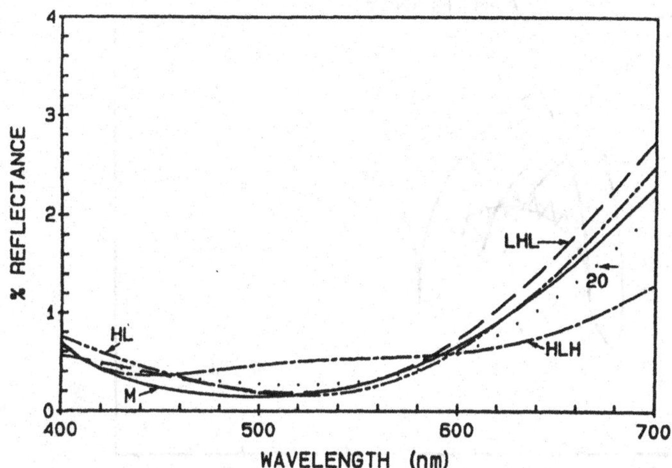

Fig. 1.72 Reflectance at 45° in *p*-polarization of the three-layer AR and the various approximations of Fig. 1.70.

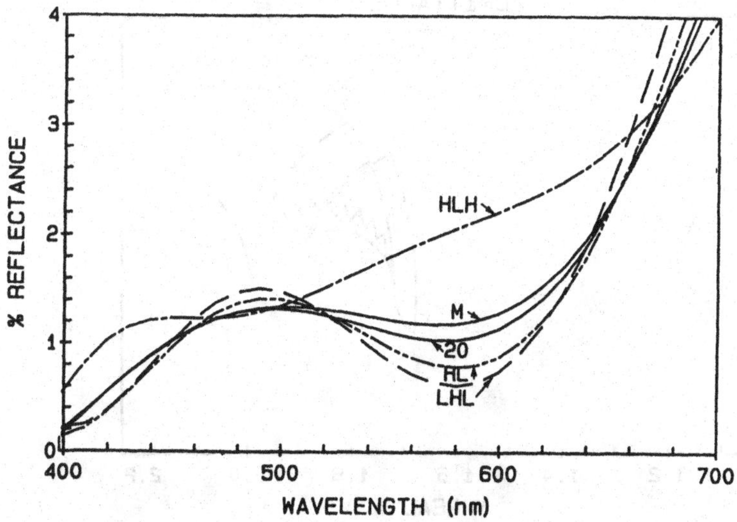

Fig. 1.73 Reflectance at 45° in *s*-polarization of the three-layer AR and the various approximations of Fig. 1.70.

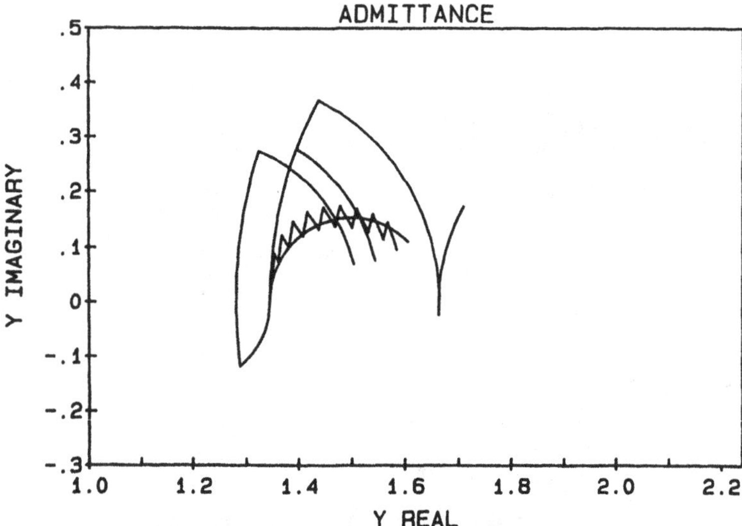

Fig. 1.74 Admittance diagram at 600 nm and 45° incidence in *s*-polarization first layer of the three-layer AR and its approximations in Fig. 1.70.

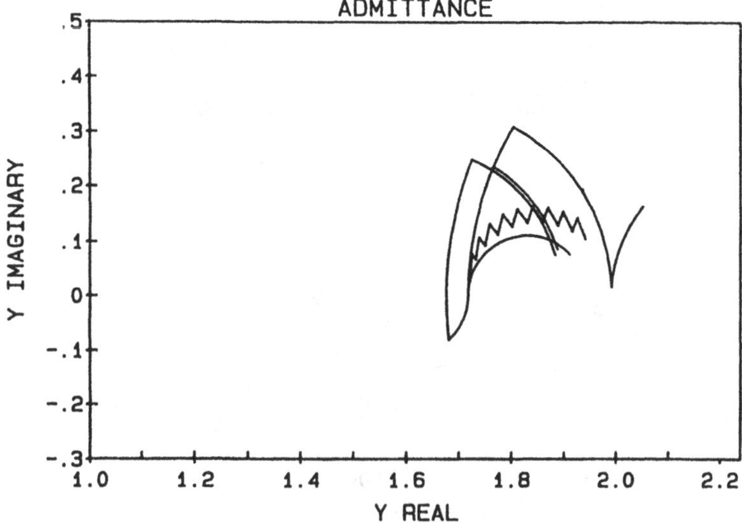

Fig. 1.75 Admittance diagram at 600 nm and 45° incidence in *p*-polarization first layer of the three-layer AR and its approximations in Fig. 1.70.

1.7. INHOMOGENEOUS INDEX FUNCTIONS

The question of what are the underlying principles and "ideal" design for the ultimate broad band AR coating has intrigued us for almost a decade. The question of how to design it, and how to fabricate it has followed us through this period until recent years when the probable answers seem to have finally come together.

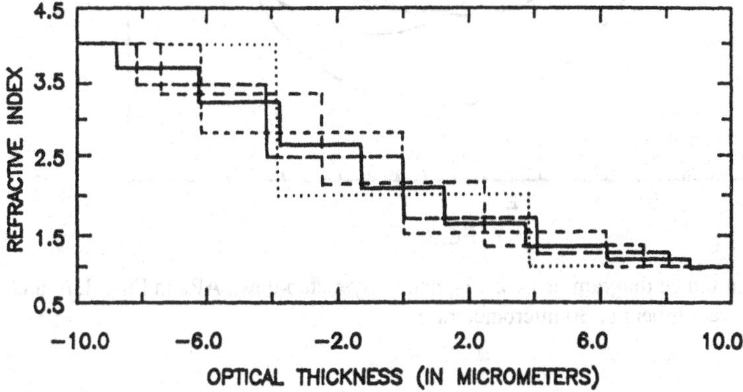

Fig. 1.76 Index profile of step-down coatings of 1, 2, 3, 4, and 7 layers from index 4.0 to 1.0, all giving the same long wavelength limit to the AR band.

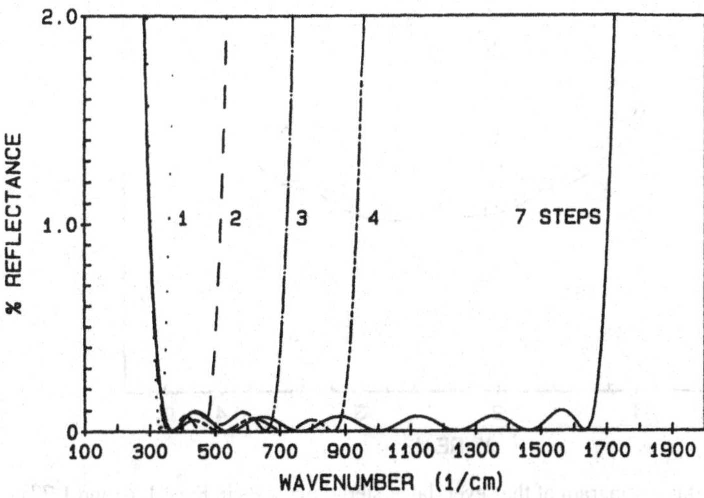

Fig. 1.77 Spectral reflectance of the step-down ARs in Fig. 1.76.

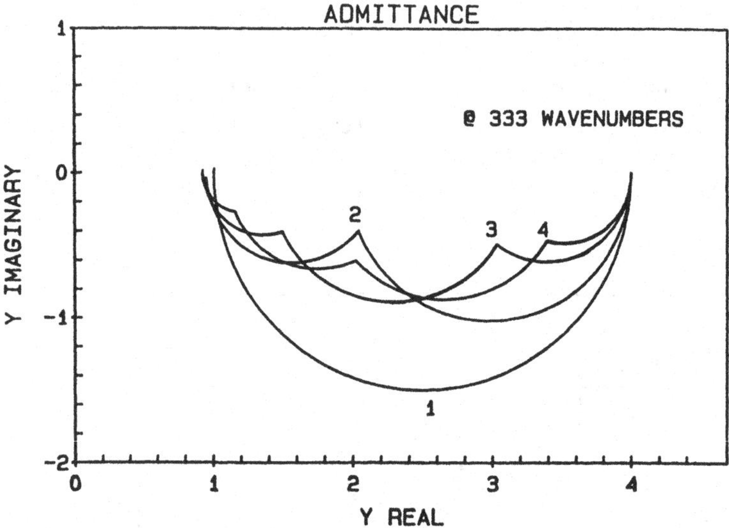

Fig. 1.78 Admittance diagram of 1-, 2-, 3-, and 4-layer step-down ARs in Figs. 1.76 and 1.77 at 333 wavenumbers or 30 micrometers.

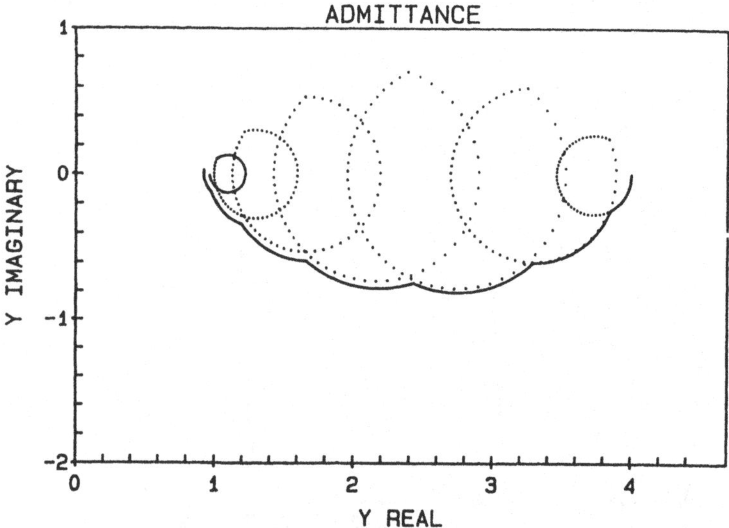

Fig. 1.79 Admittance diagram of the seven-layer step-down ARs in Figs. 1.76 and 1.77 at 333 and 1667 wavenumbers or 6 micrometers.

Jacobsson et al.[16] and Dobrowolski et al.[17] described the merits of step-down layers where the space between the substrate and the ambient medium is divided into approximately equal steps of decreasing index of refraction from the substrate to the medium. The author investigated this in terms of admittance diagrams and reported some of the results in 1989[18]. Figure 1.76 illustrates such a step-down layer from germanium to air with 1, 2, 3, 4, and 7 steps. Note that the overall thicknesses have been adjusted to give the same long wavelength or low frequency limit at 300 1/cm (for 1% reflectance). We will discuss this thickness difference later. Note also that the origin of zero thickness is near the center of Fig. 1.76 and later plots. This causes the profiles in Fig. 1.76 to be more clearly seen as approximations of the ideal shown later in Fig. 1.85. The choice of this origin is also a characteristic of the Fourier transform technique in Sect. 1.7.2. Figure 1.77 shows the performance of each profile in Fig. 1.76. We can observe that the AR band gets wider with an increasing number of steps as the short wave limit (high frequency) moves to the right while the long wave limit is fixed. Figure 1.78 shows the admittance diagrams resulting for the 1-, 2-, 3-, and 4-layer versions at the long wavelength end of the AR band. Figure 1.79 shows the admittance diagram of the seven layer at the longest and shortest wavelengths in the AR band. This figure was something of a Rosetta Stone to deciphering what might be the basics of the AR coating. At the long wave limit of the AR band, the admittance of the "ideal" AR appears to be of an approximately catenary form which seems to osculate the admittance locus of the short wave or high frequency end of the AR band. The admittance locus of the shortwave end has a coiled spring appearance. As the number of layers or steps is increased, the loci become more smooth and the AR bandwidth becomes progressively wider. Figure 1.80 shows this for 24 layers at the long wave end of the band. Figure 1.81 shows the admittance at the high frequency end of the band, and it seems to fit "inside" the admittance of the low frequency end (Fig. 1.80). We shall return to this concept later.

A coincident study of the paper by DeBell[19] and a plot of the index versus thickness of some of his solutions led to Figs. 1.82, 1.83, and 1.84. DeBell looked at the possibilities of inserting additional half waves of alternating high and low index layers between the quarter wave start and end layers of the classical QHQ in order to improve the bandwidth and minimize the reflectance of broadband ARs. He optimized the indices of each layer while holding the thicknesses constant. The index profiles of some of his solutions are shown as broken lines in Figs. 1.82, 1.83, and 1.84. Looking at the index profiles from DeBell's work with homogeneous layers we noticed that these might be taken as approximations of smooth periodic variations of index with thickness, as shown in solid lines. Southwell[20] stated, "Any arbitrary generalized gradient-index interference coating (including homogeneous and inhomogeneous layers) possesses a digital configuration (sequence of thin high- or low-index layers), which is spectrally

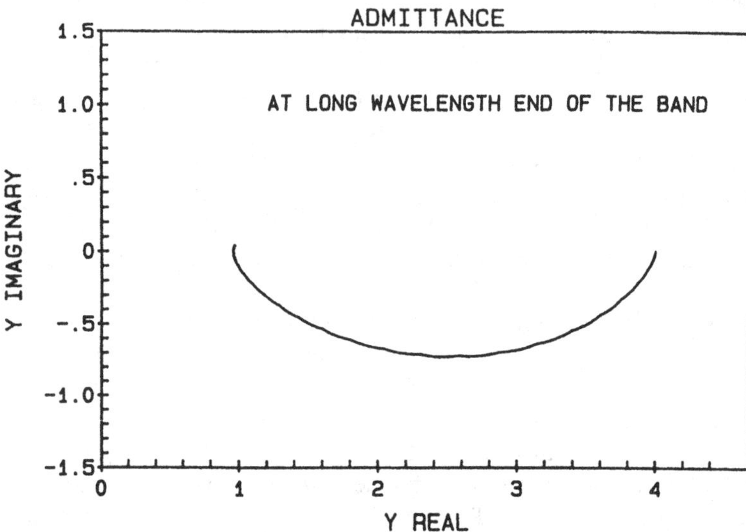

Fig. 1.80 Admittance diagram of 24-layer step-down, near "ideal" AR at the longest wavelength of the AR band.

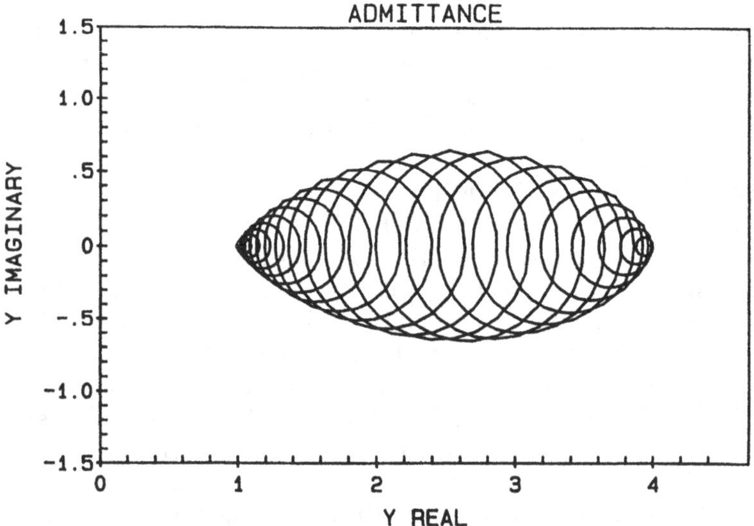

Fig. 1.81 Admittance diagram of 24-layer step-down, near "ideal" AR at the shortest wavelength of the AR band.

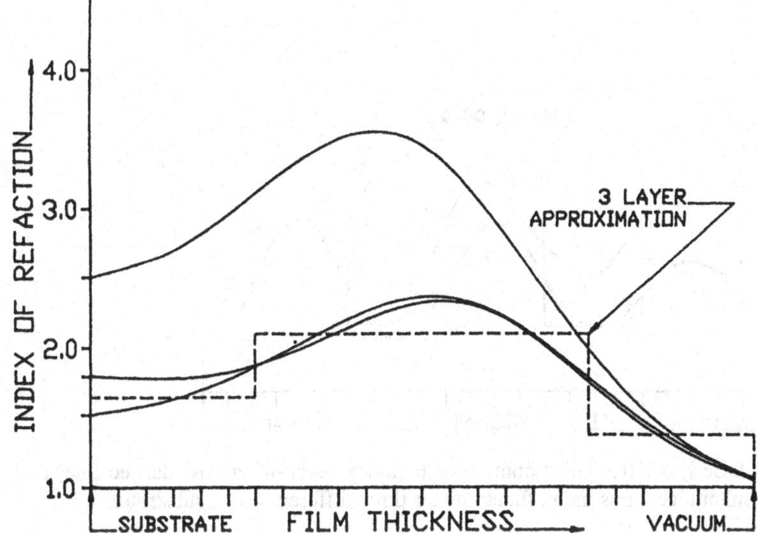

Fig. 1.82 DeBell's QHQ homogeneous index three-layer AR (dashed line) and one-cycle inhomogeneous index functions for three different index substrates.

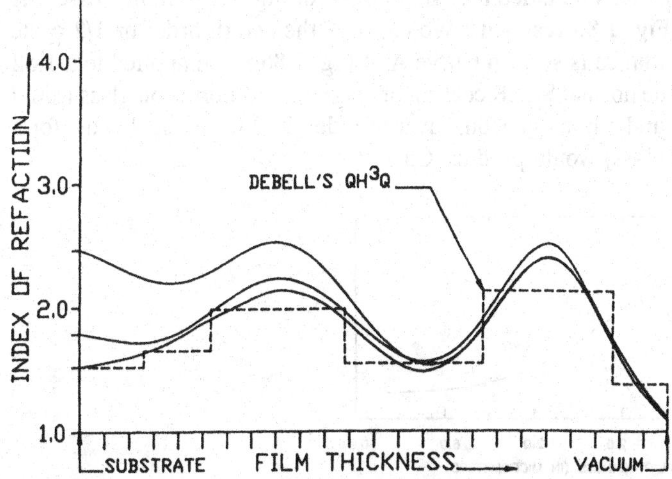

Fig. 1.83 DeBell's QHHHQ homogeneous index five-layer AR (dashed line) and two-cycle inhomogeneous index functions for three different index substrates.

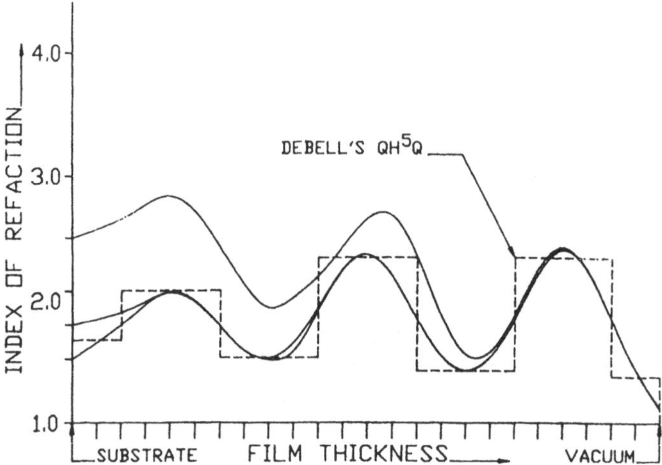

Fig. 1.84 DeBell's QHHHHHQ homogeneous index seven-layer AR (dashed line) and three-cycle inhomogeneous index functions for three different index substrates.

equivalent at all wavelengths. Such digital configurations are found directly from arbitrary index profiles by using a prescribed two-layer high-low equivalent to a thin layer of arbitrary index." Using this basic principle with some modification, we divided the overall thickness of our quasi-periodic functions into 24 layers of equal thickness and optimized the index of each layer with respect to the broadest AR band practical. This included the "step-down" of Fig. 1.76. In that case, the smooth curve in Fig. 1.85 resulted. We call this the zeroth order or 1/2 cycle function. Its admittance is seen in Curve A of Fig. 1.86 when applied to a 1.52 index substrate. The normal SLAR coating of magnesium fluoride on glass would be as in Curve C, and a homogeneous layer of index 1.233 (the ideal value for a SLAR coating on glass) would produce Curve B.

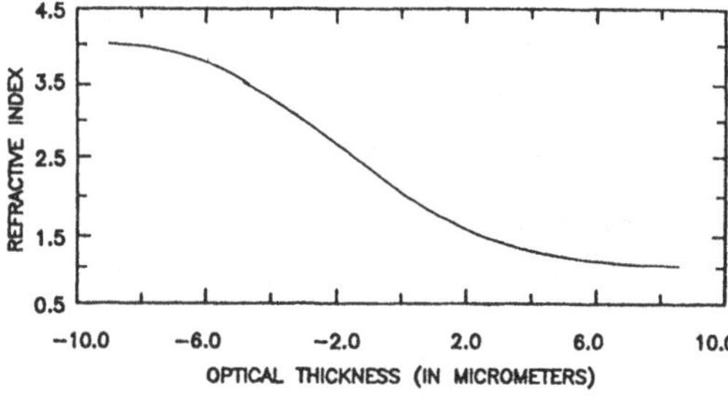

Fig. 1.85 Half cycle inhomogeneous index function from Fig. 1.80 and also generated by the Fourier method.

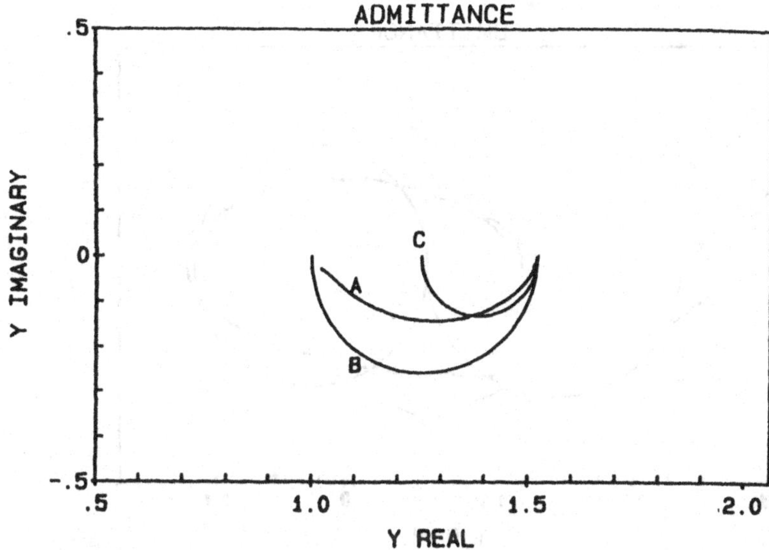

Fig. 1.86 Admittance: curve A is for the "ideal" inhomogeneous function on 1.52 index substrate, curve B is a homogeneous SLAR of index 1.233, and curve C is a homogeneous SLAR of index 1.38.

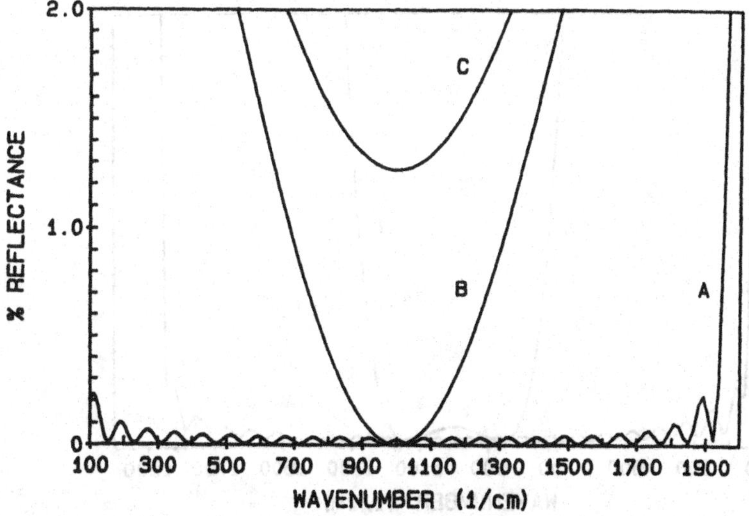

Fig. 1.87 Spectral reflectance of the three coatings in Fig. 1.86.

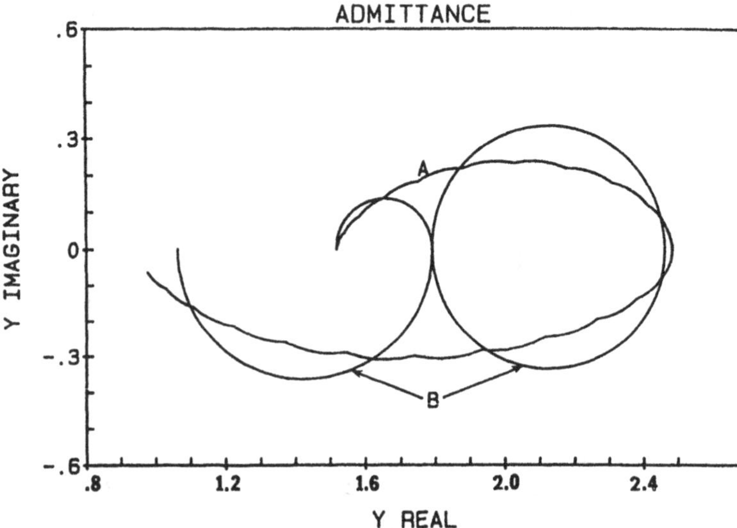

Fig. 1.88 Admittance diagram of the classic three-layer homogeneous layer AR (curve B) compared to the "ideal" one-cycle inhomogeneous function of Fig. 1.82.

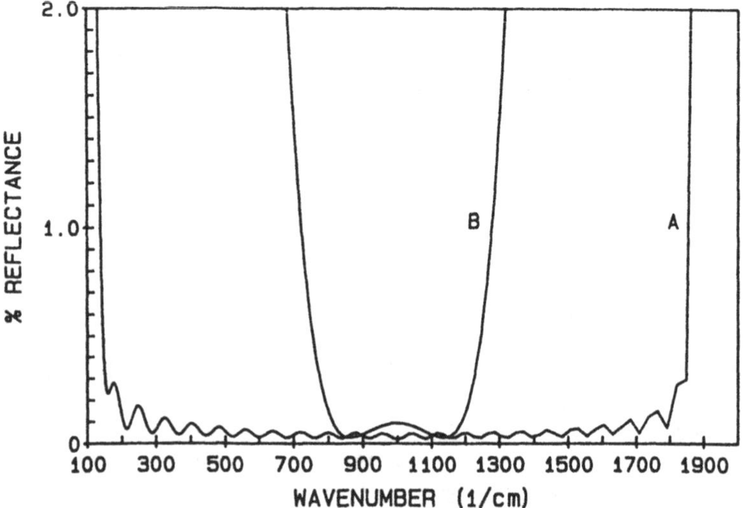

Fig. 1.89 Spectral reflectance of the two coatings in Fig. 1.88.

Fig. 1.87 shows the spectral performance calculated for these single layer approximations of the "ideal" inhomogeneous 1/2 cycle index profile function. The ripples in Curve A and the rise in reflection at the short wave or high frequency end are due to the limitations of 24 layers. The rise occurs when the individual layers approach 1/2 wave optical thickness, and thus become ineffective absentee layers. If we used twice as many layers in the same overall stack thickness, the bandwidth would be twice as wide. This is because the frequency (wavenumber) at which each layer becomes a half wave would be twice as high because the layers would be half as thick. The low frequency (long wavelength) end of the band would remain the same because the overall thickness would be unchanged. Note by comparison of Figs. 1.86 and 1.87 that the closer Curves B and C approximate the admittance of the "ideal" Curve A, the better the result match the "ideal". This was a major point made in Sect. 1.6.

We saw earlier that changes of the angle of incidence from normal will cause the spectral properties to generally shift toward shorter wavelengths or higher frequencies. If a broadband AR coating is needed which will function at increasing angles of incidence, the AR band at normal incidence needs to be "flat" to longer wavelengths (or lower frequencies). This will then allow the coating to be tilted to some greater angle before the long wavelength end of the band encroaches on the AR band of interest. The basic principle for the design of a broadband AR coating to handle large angles is, therefore, to approach a design like that of Fig. 1.85 and curve A of Figs. 1.86 and 1.87. Here the long wavelength end of the AR band must be sufficiently longer than the band of interest at angle.

The performance of any of these "ideal" ARs is ultimately limited by the lowest real index of refraction available as we will discuss in more detail in Sects. 1.7.1, 2.2.3, and 3.2.4.1. The current limits on the indices of real homogeneous materials is approximately the 1.38 of MgF_2. However, a new possibility has appeared, at least as a laboratory curiosity. The photolithography techniques have etched and produced "moth eye" surfaces which have a microscopic texture like "pyramids" which taper from contiguous solid bases to "sharp" points. This then approximates the change in effective index from the substrate to air/vacuum as seen in Fig. 1.85. The index of 1.0 is approximated to the degree that the pyramids come to sharp points. It will also be noted that the pyramids cannot be flat sided, but must curve more like a bullet shape in order to obtain the correct density as a function of thickness from the substrate.

At the time of this writing, we had attempted to contract to have an actual 21 to 67° beamsplitter "coated" in this way for a real application (which cannot currently be done with ordinary coating technology). However, it appears that the technology is not economically viable at this time, and the communications and semiconductor industries offer more commercial interest for related technologies.

Back to the earlier discussions, the techniques used for Figs. 1.80 and 1.81 were applied to the QHQ solution by DeBell for different substrate indices generated the one-cycle function curves in Fig. 1.82. The dotted line is the homogeneous index profile for the classical QHQ broad band AR on a 1.52 substrates previously seen in Figs. 1.11 and 1.36. Figure 1.88 (Curve A) shows the admittance loci (24 layer approximation) for the inhomogeneous profile on a 1.52 index substrate of Fig. 1.82 and the homogeneous (Curve B) solution. Figure 1.89 illustrates the resulting spectral performance of the two. Note in the case of Figs. 1.88 and 1.89 that the three homogeneous layers are an approximation of the "ideal" one cycle function.

Carrying these concepts further to multiple periods, after DeBell's QHHHQ and QHHHHHQ as seen in Figs. 1.83 and 1.84, led us to the reduction of these to similar approximations with thin homogeneous layers[21-23] according to the technique of Southwell[20].

The term "rugate" is defined in *The American College Dictionary* as: "*adj.* wrinkled; rugose." Rugose is defined as: "*adj.* having wrinkles; wrinkled, ridged." The term has come into use in the optical coating community in the last two decades. In this case, it generally is referring to inhomogeneous film structures of which Figs. 1.82 to 1.85 are related examples of "Rugate ARs." The more common references to rugate filters, however, are to structures where there are many cycles of index variation with thickness that have been designed to reflect selected bands much like minus filters. The FBG is also an example of a rugate filter. We will touch on this more in subsequent chapters.

1.7.1. Low Index Limitations

There is a serious limitation placed by indices of real and available materials on what can be achieved by an AR coating over a broad band. This author[22], Aguilera et al.[24], and others have shown that for a given bandwidth, the lowest achievable average reflectance of an AR coating over the low reflectance band is limited by the lowest index available. As we can see in Fig. 1.76, the broader step-down solutions require very low (and unavailable) indices. If 1.35 or 1.45 or such is the lowest index available, the discontinuity from the ideal index profile causes a ripple perturbation in the resulting AR band. There is also another factor in the case of approximating an "ideal" design with a limited number of layers. If the band is made wider when optimizing a given homogeneous layer design, the residual reflectance in the AR band goes higher as seen in Fig. 1.90. Adding more cycles to the ideal inhomogeneous function does not significantly improve the achievable bandwidth. This observation raises the question, "What advantage would a multiple period function have?" It was shown that the higher order "ideal" inhomogeneous index functions tend to spiral into the desired admittance

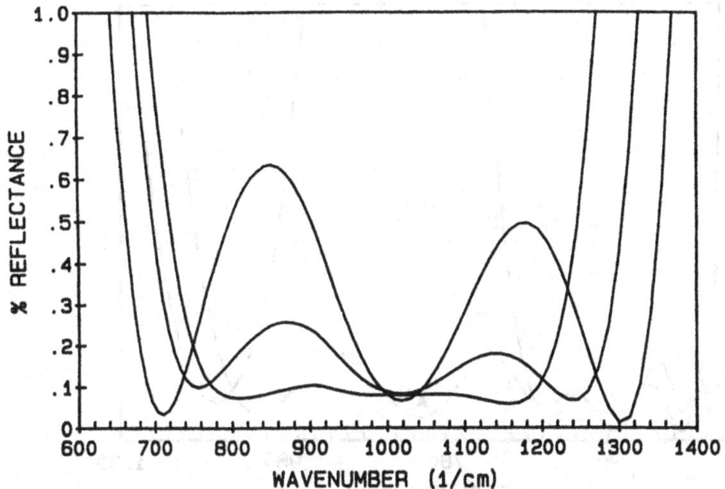

Fig. 1.90 Change in the minimum average reflectance of an AR as the bandwidth of a given homogenous layer design is changed and the design is reoptimized.

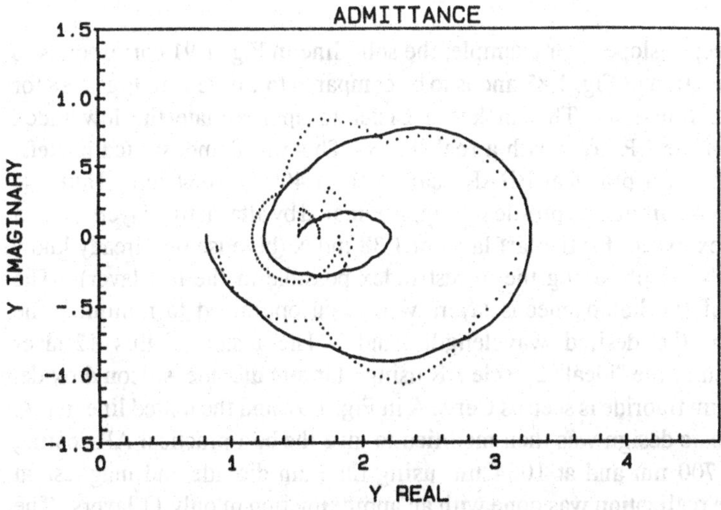

Fig. 1.91 Admittance of the two-cycle inhomogeneous index function (solid curve) and a 12-layer approximation using 2.35, 1.46, and 1.38 index homogeneous layers (dotted line).

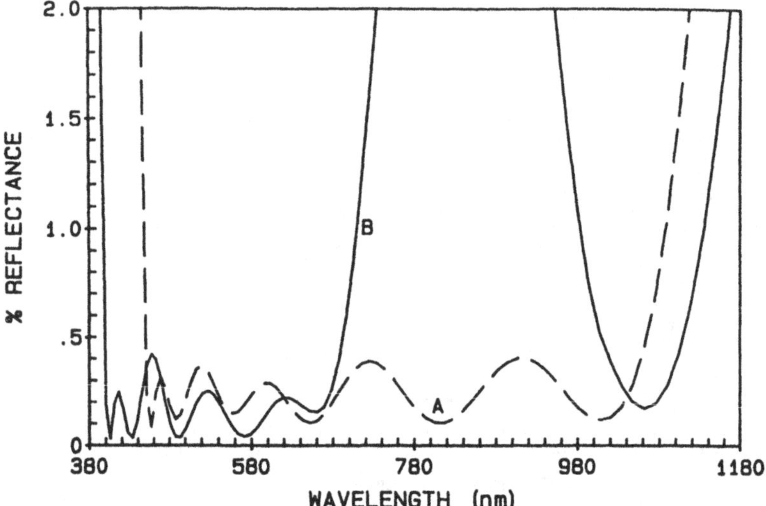

Fig. 1.92 Spectral reflectance of the 12-layer design in Fig. 1.91 (dashed curve) and an 11-layer design optimized for 400 to 700 and 1064 nm.

of 1.0 at a steeper slope. For example, the solid line in Fig. 1.91 corresponds to the 2 cycle function of Fig. 1.83 and is to be compared to Curve A in Fig. 1.88 for the one cycle function. This makes it easier to approximate the low index termination of the AR layer with a real index. This was demonstrated in Refs. 1.22 and 1.23 on a practical BBAR coating from 400 to 1064 nm. Here the inhomogeneous admittance profile was approximated by alternating layers of 2.35 and 1.46 index except for the last layer of 1.38 index (because we already know of the desirability of having the lowest index possible in the last layer). The thicknesses of the homogeneous layers were then optimized to minimize the reflectance in the desired wavelength band. The result of this 12-layer approximation of the "ideal" 2-cycle AR using titanium dioxide, silicon dioxide, and magnesium fluoride is seen as Curve A in Fig. 1.92 and the dotted line in Fig. 1.91. This basic design was then modified to give the best practical AR coating from 400 to 700 nm and at 1064 nm using titanium dioxide and magnesium fluoride. The realization was done with an approximation of only 11 layers. The design result is shown as Curve B in Fig. 1.92. This will be discussed in more detail later in Sec. 7.7.3.

1.7.2. A Fourier Approach

Observing the periodic nature of DeBell's multiple half-wave solutions, as seen in Figs. 1.82 to 1.84, and the subsequent discovery of a family of inhomogeneous index functions for AR coatings were the trigger for the following idea. The Fourier synthesis techniques described by Dobrowolski et al.[25] might lead to these types of periodic inhomogeneous solutions and lend further insight into these matters. The Fourier technique transforms the desired reflectance/transmittance function of frequency (wavenumber) to an inhomogeneous (continuous) index of refraction profile as a function of thickness which will generate a similar R/T function. Although this currently has some limitations, it has the advantage of being a direct synthesis technique requiring no prior knowledge of what the index profile might look like. A collaborative work[26] added to the understanding of the possibilities and limitations, as follows.

Figure 1.85 also represents the "ideal" AR coating which resulted from our studies. Such a smooth function generates an AR range for all frequencies higher than the low frequency (long wavelength) limit. The long wavelength limit is directly proportional to the overall optical thickness of the transition profile, and it is twice that length. That is to say that the profiles of Figs. 1.76 and 1.85 with an optical thickness of 17.5 micrometers provide an AR band starting at about 35 micrometers or 286 1/cm. Note two things about Fig. 1.76 with respect to Fig. 1.85. First, all of the homogeneous layers approximate the smooth inhomogeneous function as closely as possible. Second, the effective width of all the profiles is the same, even though the single layer approximation is only 7.6 micrometers in optical thickness. In combination with a portion of the substrate and the medium, it is the best single layer approximation of this "ideal" inhomogeneous function for the 35 micrometer AR band.

The Fourier point of view explains the high frequency limits seen in Fig. 1.77 as follows. The steps in the homogeneous layer approximations are high frequency perturbations in the smooth low frequency "ideal" profile. These "ripples" in the profile cause a reflection peak at a corresponding high frequency point in the spectral reflectance curve, and thus we get the high frequency limit of the AR band. Note that, to avoid confusion in Fig.1.77, the higher order AR bands which occur at 3x, 5x, etc. of the center band frequency are omitted.

Verly[26] tried to achieve a long wavelength limit for an AR coating with less than the minimum thickness indicated above. Figure 1.93 shows such a solution compared to a three layer step-down. Inference yields the "ideal" profile in this case to be in excess of 8 micrometers, whereas the "too-thin" profile is only about 6.5 micrometers thick. The figure indicates that the "thin layer" solution has approximated the longer smooth profile by adding the spikes on the ends of the layer. These make the average effect over a few micrometers on each end similar

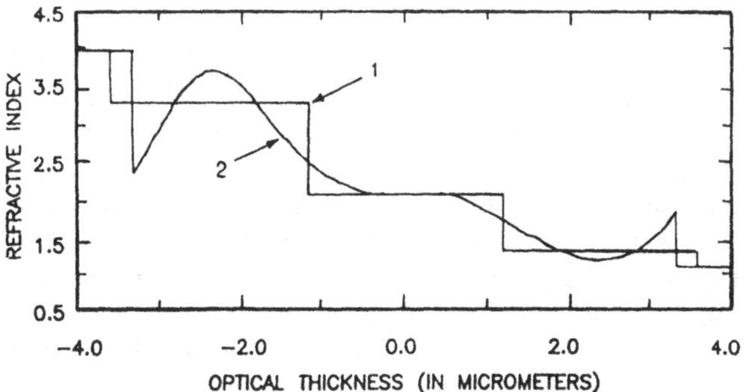

Fig. 1.93 Index profile of three homogeneous layer step-down (curve 1) and a profile of a "too thin" inhomogeneous layer by the Fourier method (curve 2).

to the "ideal". Figure 1.94 and 1.95 show the admittance at the long and short wavelength ends of the AR band for the three layer and the Fourier "too-thin" solutions. The striking similarity of these figures is consistent with all of the foregoing discussion.

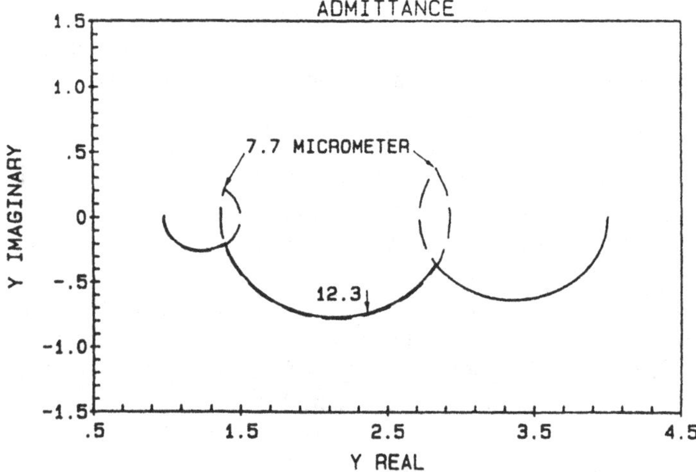

Fig. 1.94 Admittance diagram of the homogeneous layer design of curve 1 in Fig. 1.93 at the longest and shortest wavelengths in the AR band.

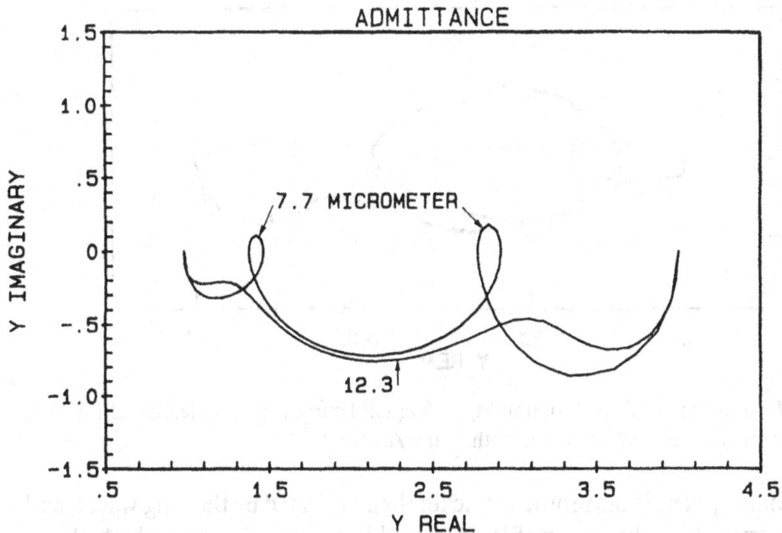

Fig. 1.95 Admittance diagram of the inhomogeneous index profile of curve 2 in Fig. 1.93 at the longest and shortest wavelengths in the AR band.

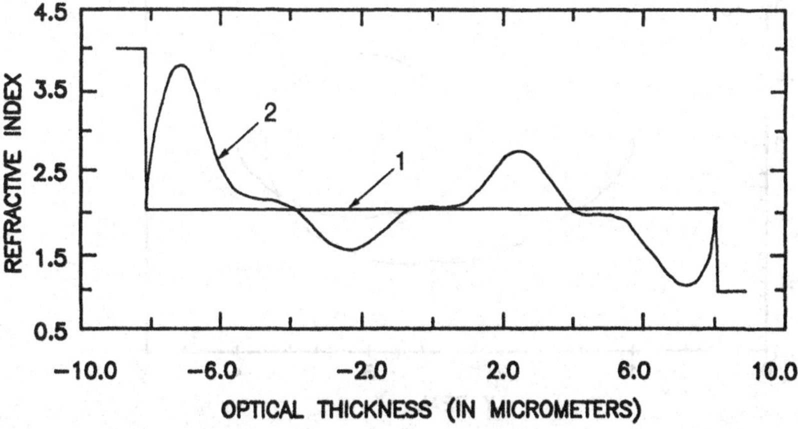

Fig. 1.96 Index profile of a "too-thick" inhomogeneous AR design by the Fourier method (curve 2) and its starting design (curve 1).

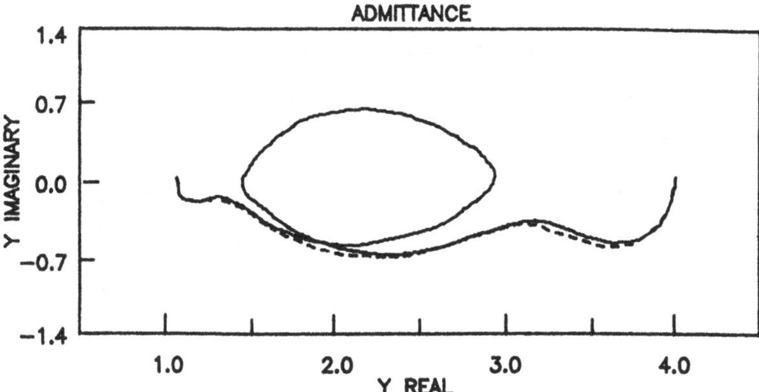

Fig. 1.97 Admittance of the design in Fig. 1.96 (solid curve) and the design of Fig. 1.95 (dashed curve) at the longest wavelength in the AR band.

What happens if the profile is thicker than indicated by the long wavelength limit? Figure 1.96 shows a profile generated by the Fourier method which was intentionally started from a base profile which was approximately twice as thick as it needed to be from the long wavelength limit as discussed above.

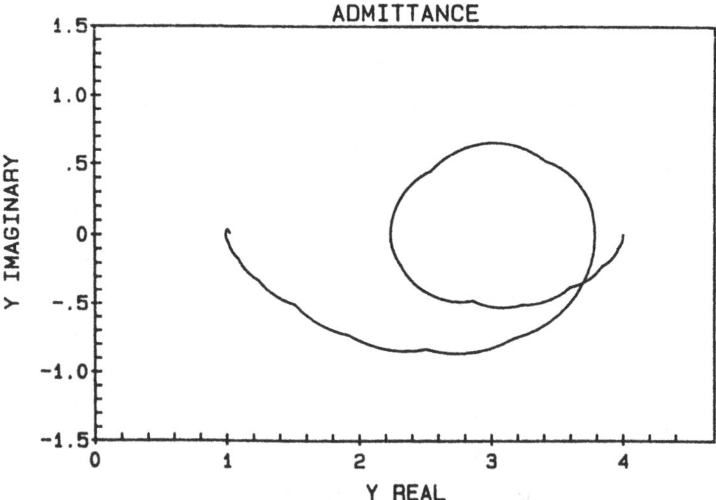

Fig. 1.98 Admittance of a "too-thick" design by the method of varying the index of 24 layers of equal optical thickness.

Figure 1.97 shows the resulting admittance at the long wavelength end of the AR band and in dashed line the same curve from Fig. 1.95 (the too-thin profile). The difference between the two admittances is a loop which has "grown" on the too-thick profile's admittance locus. We can also see in Fig. 1.96 that the portion of the profile between −3 and +3 micrometers might be removed to give a result much like the profile in Fig. 1.93. Figure 1.98 shows a case generated by the author's version of the Southwell method which demonstrates the same effect when compared to Fig. 1.80.

We may draw three conclusions from these observations. The first is that excess thickness is "used up" by the design in the form of admittance loops as seen in Figs. 1.97 and 1.98. The second is that the excess is of no apparent value except to help in overcoming the low index limitations mentioned previously. The third is that the excess loop could be anywhere on the admittance loci and may be more than one loop. This accounts for the findings of Verly[26] that there are an unlimited number of solutions to an AR band requirement if there is excess thickness. These conclusions are not rigorously shown here, but are rather the result of empirical examination of many other related cases. These observations aid significantly in the understanding of the basic phenomena of AR coating designs.

We have expended some effort to date in search of the ideal profile of the index of refraction as a function of thickness. We have referred above to the "ideal" in quotes, because we do not yet know if an ideal profile exists or what it is exactly. However, we do think that we have the "ideal" design narrowed down closer than any current ability to actually fabricate such a coating. The following few specific cases approximate the ideal function to a sufficient degree for all present practical purposes: Verly (private communication) has preferred an Exponential Sine Function of the form, where x is optical thickness,

$$N(x) = A\, e^{-C \sin(x - x_0)} \tag{1.7}$$

The author has started many designs with a Gaussian profile of the form

$$N(x) = A\, e^{x^2} - B \tag{1.8}$$

with the offset B to bring it to the index of the medium in a finite thickness. The

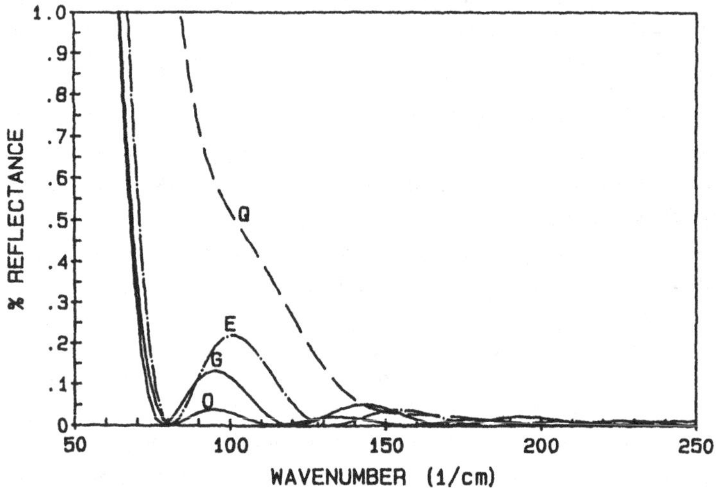

Fig. 1.99 Reflectance at the long wavelength end of the band for the "ideal" index profiles: Q is the Quintic function, E is the Exponential sine function, G is the truncated Gaussian function, and O is the optimized Gaussian function.

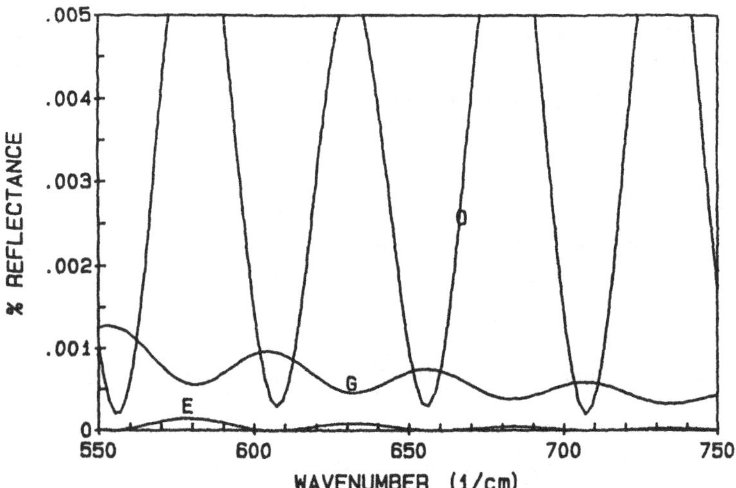

Fig. 1.100 Reflectance as in Fig. 1.99 but toward the center of the AR band and on a highly expanded scale. The reflectance of the Quintic is so low that it is lost on the baseline.

profile given in Eqn. 1.8 also served as a starting point for its optimization over a given bandwidth. We will refer to the result as the Optimized Gaussian.

Southwell[27] reported work with another interesting refractive index profile following the Quintic Function where n_s is the substrate index and n_i is the index of the final (or incident) medium

$$N(x) = n_i + (n_s - n_i)(10x^3 - 15x^4 + 6x^5) \tag{1.9}$$

All of these functions look just like Fig. 1.85 to within a pen width or so, and are hardly worth plotting on that scale to show the differences. We compare the reflectance of each of these four "ideal" layers at the same profile thickness in Figs. 1.99 and 1.100. At the long wavelength end of the band, the Optimized Gaussian has the lowest reflectance because the optimization had been weighted to make it so. The Gaussian and Exponential Sine have intermediate values, and the Quintic rises more rapidly. However, in the middle of the band, the reverse is the case. The Quintic is so low it is hidden in the baseline of Fig. 1.100, whereas the Optimized Gaussian exceeds .005%. Not that .005% is necessarily bad, but this is just to illustrate the differences of the principles employed. What we can observe here is that the usual tradeoffs seem to exist between band width and band depth. It is like a tube of toothpaste, if you squeeze it in one place, it will bulge in another. It may be that the area under the total reflectance band is constant or approximately so. The pursuit of the ideal profile is really somewhat academic and esoteric at this point, but the author believes that (if it exists) it will be totally symmetric when seen in admittance as in Figs. 1.80 and 1.81. The pursuit thus far has been well worth the effort for the author in terms of gaining a better understanding of the underlying principles of AR coatings.

1.8. OPTIMIZATION

We have mentioned the use of optimization many times above without addressing any of its details. One can generally think of an "optimizer" as a black box which takes for its input a starting design, a set of parameters which may be varied, and a set of performance goals, weightings, and constraints. The desired output of this black box is a design which is the best fit to the weighted performance goals that can be achieved with the variables and within the constraints given. The optimizers usually deal with a merit function that is actually a *demerit* function which is reduced to a minimum by the process. The "merit" function might be computed as the sum of the squares of the difference between the targets and the actual spectrum. The program finds the effect of each variable on the merit function and predicts where the best design would be in variable space. This process iterates for a given number of times or until the goals are met. Such "black boxes" are provided as software packages for personal computers by organizations such as lead by Goldstein[34], Noe[35], and Macleod[36]. The starting design, such as might have been obtained through the methods discussed above, would be represented by the index of refraction and the thickness of a given number of layers plus the substrate and media indices. We will discuss some of the other factors and consequences below.

1.8.1. Performance Goals and Weightings

If the goal were to design a four-layer AR for the visible spectrum, one must specify the performance goals as 0.0% reflectance at 50 nm intervals from 400 to 700 nm. This should lead to a good solution such as shown in Fig. 1.8 curve B. In the software which we use[34] (and most likely others), the weighting on each goal or target value is determined by the reciprocal of the tolerance on that target. For example, the above targets of 0.0% for wavelengths of 500, 550, and 600 nm might have tolerances of 0.5% while 450 and 650 nm have 1.0% and 400 and 700 nm have 2.0%. This would give an approximation of weighting for the visual or photopic response of the eye putting the most "pressure" on wavelengths centered near 555 nm. The same design might also result from leaving out the 400 and 700 nm targets entirely. In former times of slower and more expensive computers, keeping the number of targets smaller helped speed up and reduce the costs of the optimization process. These considerations are no longer as much of a concern.

The situation changes when we look at the case of the design in Fig. 2.1 labeled "B = 2.5". This design was optimized from 420 to 1100 nm. If we only used seven targets as mentioned above, we would encounter a problem. The "optimizer" would tend to move the reflection toward 0.0% as much as it could at the targets (only), but the spaces in between the targets would be free to move to much higher values of reflectance. It is advisable to have at least twice as many targets across the band of interest as there are cycles in the ripple pattern, and four times is our usual choice. In this case, that would be about 32 targets. The ripple patterns are more evenly spaced on a frequency or wavenumber (cm^{-1}) scale, therefore it is better to have the targets equally spaced in frequency also.

The art of designing, after the indices and number of layers have been determined (which could be aided by the concepts discussed in Chapters 1, 2, and 3), involves the choices of the magnitude and position of the targets and their weightings in order to achieve the desired results. The author has sometimes found that the original choices need to be modified to apply more or less pressure on some target points to achieve the best balance.

1.8.2. Constraints

There are sometimes occasions where it is desirable to constrain certain aspects of a design such as minimum and maximum layer thicknesses. Some software packages[34] allow great flexibility in the definition of constraints. This is done by bringing design variable and results into a spreadsheet where they can be manipulated and fed back to the optimization part of the program.

We recently used this approach[46] in an investigation of which classes (if any) of AR designs might be improved by design techniques which considered the effects of probable production errors. We further reported on the influence of different error distribution assumptions such as: random errors uniformly distributed within a tolerance range, worst case error distributions, and various sensitivities to errors which might realistically represent those in actual practice. We were not able to discover any class of coatings which gains a significant benefit from the optimization with respect to the expected random errors of production. It was not surprising to find that normal designs result in each layer thickness lying at a minimum of the (de)merit function wherein a movement from the nominal design thickness in any direction will increase the (de)merit. Since this is also the point of the zero first derivative of the (de)merit with respect to thickness, the sensitivity to small errors is at a minimum.

We will discuss the use of this spreadsheet and constrained optimization feature in more detail in Chapter 7 in connection with achieving matching layers for a QWOT stack edge filter which also give specific optical monitoring characteristics.

1.8.3. Global versus Local Minima

We have seen in Fig. 1.7 and 1.8 that there are two solutions to the typical V-coating problem. The optimizer will usually converge on the solution closest to the starting design, or the "local minimum." The other solution in this case might actually be better than the first, depending on the targets and weightings, but the typical optimizer might not find it. The best of all possible solutions is the "global" minimum. One can find discussions of global optimizers in the literature, but they of course must systematically examine the whole universe of parameter space in enough detail to not miss the global minimum.

The analogy would be that a search for the lowest point starting at Salt Lake City would find it to be Salt Lake. A search starting at Barstow, California should find Death Valley, which is lower. However, if one started at Jerusalem, they might find the true global minimum at the Dead Sea.

Therefore, it is well to be alert to the possibility of local minima when using automatic optimizers. This is where some knowledge and understanding can be helpful in evaluating the results of an optimization. It is like the fact that we can know that the Dead Sea is the lowest point on the globe from the collective experience of others, even if we have not been there yet ourselves.

1.8.4. Some Optimizing Concepts

There are a variety of concepts for optimizers which have been used for decades and new ones are being developed all of the time. We will comment briefly on some of the most commonly used that we are aware of, and we will not discuss the "Simplex" or Parabolic Approximation[37] and other methods which have appeared over the years in some applications such as lens design.

Optimization is like letting a ball slide into a bathtub and roll until it rests on/in the drain hole. Its progress will not generally be in a straight line to the drain.

1.8.4.1. Damped Least Squares Optimization

There are many descriptions in the literature such as by Meiron[38,39] which describe the concept of least squares and damped least squares (DLS) optimization. The derivative of the merit function with respect to each variable is calculated. These are used to compute the solution including each variable which would minimize the merit function. The solution is not likely to be the actual minimum because the merit function is not apt to be linear in all of the variables. However, this should move the design in the direction of the minimum (local). There is a risk, particularly as the actual minimum is approached, that the predicted change will overshoot the actual minimum. Therefore, damping is added so that the step taken is shortened to avoid overshooting, if that has been detected. Conversely, the step may be lengthened if the movement continues in the same direction without passing over a minimum. This all done automatically in the software, of course.

1.8.4.2. Flip-Flop Optimization

Southwell[20] introduced a simple but effective optimization scheme which he called "flip-flop" optimization. It stems from the concepts which we discussed earlier concerning index of refraction approximations and also inhomogeneous index functions. As will become more clear in Chapter 2, there is generally an advantage to using the highest and lowest indices practical in a design. Since we have seen that we can approximate any index in between the highest and lowest, there is usually little justification for using more than two indices and the attendant complications.

Southwell's proposed synthesis (design) algorithm is as follows:

(1) Select a total physical thickness for the coating. Divide this thickness into thin layers of equal thickness.
(2) Assign some initial index, either high or low, to each layer. Usually the convergence of iterative solutions depends on starting values, so this step may be important. Here are four suggestions:

(a) Start with all high-index layers.
(b) Start with all low-index layers.
(c) Start with alternating high- and low-index layers.
(d) Start from some known approximate solution.

The first three require no knowledge of thin-film theory, and the fourth attempts to utilize such experience.
(3) Evaluate a merit function based on the desired spectral response. One example is the least-squares sum: square the difference between calculated reflectivity and the desired reflectivity at various wavelengths across the band of interest and add them together. Use the characteristic matrix theory to evaluate the calculated response.
(4) Change the state of each layer (from low to high index or from high to low) one at a time and reevaluate the merit function. If the performance is better in the flipped state, retain the change; otherwise restore it.
(5) If, after testing all the layers (a single pass), the merit function has improved, go to (4) for another pass; otherwise end.

He shows several examples of how well this works.

1.8.4.3. Needle Optimization

We were first introduced by Tikhonravov[40] to "needle optimization" in 1989. DeBell[41] described the method as: "The key idea of this method is to introduce a needle like variation in the refractive index somewhere in the coating design in such a way that it will optimally decrease the value of the merit function. Once this variation, really it is just a very thin (needle like) layer, has been inserted into the existing design the resulting assembly of layers is refined to further decrease the value of the merit function by adjusting the thickness of all the layers. This process may be successively iterated until the introduction of a thin layer of material no longer effects a decrease in the merit function."

Our understanding is that the influence of a very thin layer on the merit function is evaluated through the whole thickness of the design ("scanned" through it). The point of greatest beneficial effect is then chosen to insert the thin layer and reoptimize. From what we will show in Chapter 2, there is likely to be an optimum number of layers for a given overall thickness of coating. This is particularly true for AR coatings. More layers would add more reflection to be dealt with and eliminated; less would reduce design potency. We have seen published cases of needle designs that could be slightly improved by removing one layer and reoptimizing.

There have been more details published about the application of the needle method by Baumeister[42], Sullivan and Dobrowolski[43], Tikhonravov et al.[44], and Furman and Tikhonravov.[45]

1.9. SUMMARY

We have reviewed the principles and shown applications of various useful graphical tools (or methods) for optical coating design: Reflectance Diagram, Admittance Diagram, Triangle Diagram, Prism Diagram, and stereoscopic versions of these. We have shown graphically how unavailable indices can be approximated by two available indices of higher and lower values than the one to be approximated. The basis of ideal antireflection coating design has been shown empirically. The practical approximation of these inhomogeneous index profiles has been demonstrated. Much of the discussions have centered on AR coatings, but most other types have been seen in the perspective of the same graphics and underlying principles. We will show in subsequent chapters that *reflection control* is the basis of essentially all dielectric optical coatings and that transmittance, optical density, etc., are byproducts of reflection (and absorption). We believe the best insight is gained by the study of reflectance. We will also show further that AR coatings, high reflectors, and edge filters are all in the same family of designs. The author has found the graphical tools described in this chapter to be very useful as an aid to understanding and insight, and hopes that they will prove helpful to the reader also.

1.10. REFERENCES

1. J. H. Apfel: "Graphics in optical coating design," *Appl. Opt.* **11**, 1303-1312 (1972).
2. H. A. Macleod: *Thin Film Optical Filters,* 2nd Ed. (MacMillan, New York, 1986) pp. 62-66.
3. A. Thelen: "Equivalent Layers in Multilayer Filters," *JOSA* **56**, 1533-1538 (1966).
4. A. Thelen: *Design of Optical Interference Coatings,* (McGraw-Hill, New York, 1988).
5. F. Rock: "Antireflection Coating and Assembly Having Synthesized Layer of Index of Refraction," U. S. Patent 3,432,225(1969).
6. H. A. Macleod: *Effective Optical Coating Design* (Course Notes, page 31, Thin Film Center, 2745 E. Via Rotonda, Tucson, AZ 85716, 1990).
7. J. H. Apfel: "Optical coating design with reduced electric field intensity," *Appl. Opt.* **16**, 1880-1885 (1977).

8. D. J. Smith: "Comparison of graphic techniques toaid optical thin film interference design," (A), *JOSA A* **4**, 122 (1987).
9. J. H. Apfel: "Triangular coordinate graphical presentation of the optical performance of a semi-transparent metal film," *Appl. Opt.* **29**, 4272-4275 (1990).
10. P. H. Berning and A. F. Turner: "Induced Transmission in Absorbing Films Applied to Band Pass Filter Design," *JOSA* **47**, 230-239 (1957).
11. M. A. Herpin: "Calcul du pouvoir réflecteur d'un systèm estratifié quelconque," *Comptes Rendus Acad. des Sci.*, **225**, 182-183 (1947).
12. L. I. Epstein: "The Design of Optical Filters," *JOSA* **42**, 806-810 (1952).
13. R. R. Willey: "Graphic description of equivalent index approximations and limitations," *Appl. Opt.* **28**, 4432-4435 (1989).
14. K. Rabinovitch and D. Ziv: "Herpin Equivalent Layer at Non-Normal Incidence," *Appl. Opt.* **24**, 312-313 (1985). ·
15. R. Herrman: "Quarterwave Layers: Simulation by Three Thin Layers of Two Materials," *Appl. Opt.* **24**, 1183-1188 (1985).
16. R. Jacobsson and J. O. Martinsson: "Evaporated Inhomogeneous Thin Films," *Appl. Opt.* **5**, 29-34 (1966).
17. J. A. Dobrowolski and F. Ho: "High performance step-down AR coatings for high refractive-index IR materials," *Appl. Opt.* **21**, 288-292 (1982).
18. R. R. Willey: "Antireflection Coatings for Germanium Without Zinc", *Proc. SPIE* **1050**, 205-211 (1989).
19. G. W. DeBell: "Antireflection Coatings Utilizing Multiple Half Waves", *Proc. SPIE* **401**, 127-137 (1983).
20. W. H. Southwell: "Coating Design Using Very Thin High- and Low-Index Layers," *Appl. Opt.* **24**, 457-460 (1985).
21. R. R. Willey: "Rugate Broadband Antireflection Coating Design", *Proc. SPIE* **1168**, 224-228 (1989).
22. R. R. Willey: "Another Viewpoint on Antireflection Coating Design", *Proc. SPIE.* **1191**, 181-185 (1989).
23. R. Willey: "Realization of a Very Broad Band AR Coating," *Proc. Soc. Vac. Coaters* **33**, 232-236(1990).
24. J. A. Aguilera, J. Aguilera, P. Baumeister, A. Bloom, D.Coursen, J. A. Dobrowolski, F. T. Goldstein, D. E. Gustafson, and R. A. Kemp: "Antireflection coatings for germanium IR optics: a comparison of numerical design methods," *Appl. Opt.* **27**, 2832-2840 (1988).
25. J. A. Dobrowolski and D. Lowe: "Optical thin film synthesis program based on the use of Fourier transforms," *Appl. Opt.* **17**, 3039-3050 (1978).
26. R. R. Willey, P. G. Verly, and J. A. Dobrowolski: "Design of Wideband Antireflection Coating with the Fourier Transform Method", *Proc. SPIE* **1270**, 36-44 (1990).
27. W. H. Southwell: "Gradient-index antireflection coatings," *Opt. Lett.* **8**, 584-586 (1983).
28. R. Kashyap, *Fiber Bragg Gratings*, (Academic Press, San Diego, 1999).
29. J. A. Dobrowolski, Li Li and R. A. Kemp, "Metal/dielectric transmission interference filters with low reflectance. 1. Design," *Appl. Opt.*, 34 (25) 5673 (1995).

30. M. Tan, Y. Lin and D. Zhao, "Reflection filter with high reflectivity and narrow bandwidth," Appl. Opt., 36 (4) 827 (1997).
31. E. D. Palik, *Handbook of Optical Constants of Solids*, (Academic Press, Orlando, 1985.)
32. E. D. Palik, *Handbook of Optical Constants of Solids II*, (Academic Press, Orlando, 1991.)
33. M. Tazawa, P. Jin, and S. Tanemura, "Optical constants of V1-xWxO2 films," *Appl. Opt.*, **37** (10) 1858 (1998).
34. F. T. Goldstein, *FilmStar*, FTG Software Associates, P.O. Box 579, Princeton, NJ 08542.
35. T. D. Noe, *TFCalc*, Software Spectra, 14025 N.W. Harvest Lane, Portland, OR 97229.
36. A. H. Macleod, *The Essential Macleod*, Thin Film Center, Inc. 2745 E. Via Rotonda, Tucson, AZ 85716.
37. J. Meiron and G. Volinez: "Parabolic Approximation Method for Automatic Lens Design," *J. Opt. Soc. Am.* **50**, 207-211 (1960).
38. J. Meiron: "Automatic Lens Design by the Least Squares Method," *J. Opt. Soc. Am.* **49**, 293-298 (1959).
39. J. Meiron: "Damped Least-Squares Method for Automatic Lens Design," *J. Opt. Soc. Am.* **55**, 1105-1109 (1965).
40. A. V. Tikhonravov and N. V. Grishina: "New Problems in the Synthesis of Thin Films," *Optical Coatings*, eds. Tang Jinfa and Yan Yixum, (International Academic Publishers, Beijing, 1989).
41. G.W. DeBell: "Optical Thin Film Technology in Russia," *Proc. Soc. Vac. Coaters* **37**, 3-9 (1994).
42. P. Baumeister: "Starting designs for the computer optimization of optical coatings," *Appl. Opt.* **34**, 4835-4843 (1995).
43. B. T. Sullivan and J. A. Dobrowolski: "Implementation of a numerical needle method for thin-film design," *Appl. Opt.* **35**, 5484-5492 (1996).
44. A. V. Tikhonravov, M. K. Trubetskov, and G. W. DeBell: "Application of the needle optimization technique to the design of optical coatings," *Appl. Opt.* **35**, 5493-5508 (1996).
45. Sh. A. Furman and A. V. Tikhonravov: Optics of Multilayer Systems (Éditions Frontières, B. P. 33, 91192 Gif-sur-Yvette Cedex, France, 1992), p. 130.
46. R. R. Willey: "Design of Optical Coatings Taking Consideration of Probable Production Errors," *Proc. Soc. Vac. Coaters* **43**, 230-233 (2000).
47. A. Macleod: "Half-wave Holes, Leaks and Other Problems," *Proc. Soc. Vac. Coaters* **39**, 193-198 (1996).
48. D. Cushing: "Thin Film Interference Filter for 45° Angle of Incidence inside a Glass Prism with Extremely Low Polarization Dependence," *Proc. Soc. Vac. Coaters* **43**, 252-257 (2000).
49. P. Baumeister: "Bandpass design - applications to nonnormal incidence," *Appl. Opt.* **31**, 504-512 (1992).
50. D. Cushing: "Band shape improvement techniques," *Proc. SPIE.* **4094**, (2000).

2

Estimating What Can Be Done Before Designing

2.1. INTRODUCTION

Before designing a coating it is helpful to have some idea of whether the goal of the design is achievable. The ability to estimate performance limits can avoid fruitless design efforts and avoid the neglect of potential performance gains or simplifications. In the case of antireflection coatings, we have collected a broad range of empirical data on optimized designs to provide a formula to estimate what can be expected in typical cases. This is the thrust of this chapter and some information is also provided for estimating the number of layers and other properties of dichroic, bandpass, and blocking filters.

2.2. ANTIREFLECTION COATINGS

In Chapter 1, we addressed the problem of understanding and designing very broadband antireflection (VBBAR) coatings. Here we draw upon those results and further investigations to provide a tool for the engineer or designer to estimate the performance which can be expected from a VBBAR design before the design process is started. To this end, we have fit our cumulative results to an equation for the average reflectance in the AR band as a function of the four major variables. These variables are bandwidth (B), index of refraction of the last layer (L), overall optical thickness of the coating (T), and the difference (D) between the highest and lowest indices used (except for the last layer). It was also found that the minimum number of homogeneous layers required can be predicted. We will show how these formulae were developed, what their limitations are, how to apply them, and what factors are of major and minor importance.

The general predictions are also found to be consistent with the results of two recent AR design "competitions" involving many independent investigators. Some insight with respect to the basic underlying principles of AR coatings can also be gleaned from the results and the process by which they were found.

An AR design problem was posed by Thelen and Langfeld[1] for the thin film conference in Berlin in September of 1992. The problem was to design an AR from 400 to 900 nm that had less than 1% reflectance at 0 and 30 degrees angle of incidence for random polarization over the band. The maximum physical thickness allowed for the film stack was 2000 nm and the choice of indices was limited to given values which approximate the real materials of magnesium fluoride, silica, alumina, tantala, and titania. The maximum physical thickness of titania allowed was 150 nm. The "contest" was to achieve the lowest average reflectance over the band, within the 1% constraint, where the square of the reflectance at 30 degrees in s-polarization was weighted half as much as the square of the reflectance at zero degrees. Thelen[1] reported the results of the many designs submitted. The results correlate well with our earlier work and have added additional breadth and insight as discussed below.

2.2.1. Procedure

To gain empirical design data for a general estimating formula, a series of designs were optimized over fixed bandwidths to give the lowest average reflectance in the band while varying each of the major factors. These empirical results were used to determine the reflectance as a function of these variables. Figure 2.1 shows such a series to find the average reflectance (R) as a function of bandwidth (B) where the L, T, and D were held constant. In effect, this is taking the partial derivative of R with respect to B at specific values of L, T, and D. Bandwidth is here defined as the longest wavelength in the band divided by the shortest wavelength. Figure 2.2 shows the change in R with respect to overall optical thickness (T) while B, L, and D are fixed. We have found it most convenient to use T in the units of one wavelength of optical thickness at the wavelength which is the geometric mean of the wavelengths at the ends of the AR band which define B. However, T correlates even better with the number of periods in a plot of reflectance versus thickness at the longest wavelength in the band (as we will discuss later and show in Fig. 2.8). The four factors B, L, T, and D were found to be the major variables affecting the minimum average reflectance which can be achieved. As shown in Chap. 1, it was found that the substrate index has no major effect on the minimum reflectance possible in the ranges examined as long as T is greater than one half (1/2). The most major influence on the minimum average reflectance and the most restricted in practice is the index of the last layer which

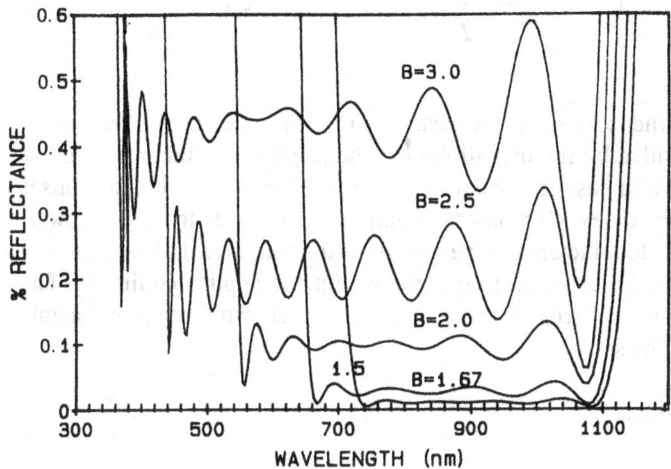

Fig. 2.1 Variation of the percentage of reflectance versus wavelength with bandwidth, while T = 3.0, L = 1.38, and D = 0.89. This illustrates how the minimum average reflectance decreases with bandwidth.

needs to be as low as practical. We have discussed the reason for this extensively in Chap. 1. This is why we mostly use magnesium fluoride as the last layer. It would be desirable from the design point of view to also use the lowest index in the stack in order to have the index difference D as large as possible. However, we have found[2] that our particular process (to date) using magnesium fluoride and titania stacks of more than a few layers has excessive scattering and therefore we use silica instead of magnesium fluoride for the inner layers.

Another observation is that the number of minima in the residual ripple of the AR band is proportional to the bandwidth (B) and the overall optical thickness (T), and it is independent of the indices and number of layers in the design. The number of minima in the band is approximately equal to: $8B/3 + 2T - 4$.

2.2.2. The Formula

The collection of data of R versus B, L, T, and D was empirically fit to functions over the range of the investigation to arrive at Equation 2.1.

$$R_{AVE}\%\,(B,L,T,D) = \frac{4.378}{D}\,\frac{1}{T^{0.31}}\,[e^{(B-1.4)} - 1](L-1)^{3.5} \tag{2.1}$$

The difference in index between the layers of the stack (D) is divided into a constant. One divided by the overall optical thickness (T) is taken to the .31 power. The bandwidth has the constant 1.4 subtracted from it and then it raises e to that power. The index of the last layer minus one is raised to the 3.5 power. The product of these four factors is the estimated minimum average reflectance in percent (%) that can be expected in designs within the applicable limits. The ranges over which these variables have been thus far shown to give reasonable estimates are as follows:

B from 1.4 to 5.0
L from 1.1 to 2.2
T from 1.0 to 9.0
D from 0.4 to 2.8

To set these in perspective, these studies were originally done for the visible and

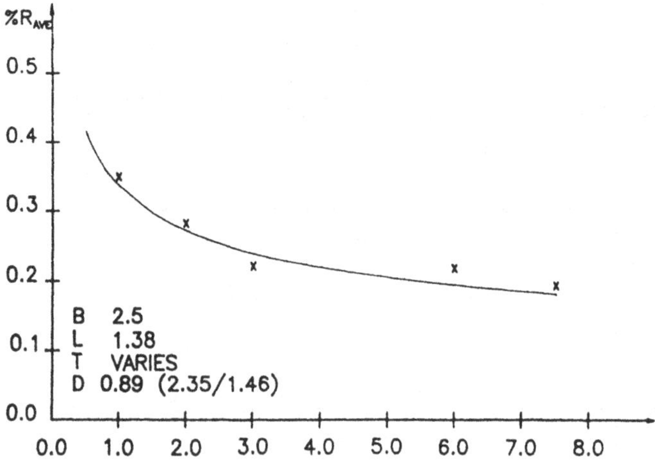

Fig. 2.2 Average reflectance in the band versus optical thickness in cycles (or waves at the geometric mean of the band). The curve is from the formula, while the X's are from empirical data.

near infrared spectral range. B, for example was tested from a 400 to 600 nm (1.5) bandwidth to a 400 to1200 nm (3.0) bandwidth. The lowest real index (L) which we use is about 1.38, but we have studied (in design) the use of imaginary materials down below 1.1 and real materials such as silica up to 1.46. The D-values come from the differences between high index titania at up to 2.58 and down to the low index of magnesium fluoride at 1.38 and combinations of intermediate materials. The overall optical thickness of the stack (T) is given in waves of the geometric mean wavelength of the band which roughly correspond to the cycles shown in Chap. 1. In coatings on high index substrates like germanium where there are several lower and intermediate indices between the substrate and an index of 1.0, we have shown in Chap. 1 that 1/2 cycle or "step-down" coatings seems most advantageous. In the case of visible band materials, there is not a significant choice of lower index materials to make the step-down approach practical. The simplest broadband solutions in the visible are of the classical three layer type which is, in effect, a one-cycle design as we saw in Chap.1. Macleod[3] contributed an interesting design study to the Thelen-Langfeld problem which he evolved toward an optimal one-cycle design before it was even submitted to an optimization process.

More recently we were motivated by the work of Rastello and Premoli[4] to expand the study to cover the infrared band that was the subject of their study and the "competition" reported by Aguilera, et al.[5] This gave further refinement and range to the applicability of the estimation formula. The agreement with the referenced results was satisfactory in the infrared also.

2.2.3. Results

We will now compare the empirical results with the values that would be estimated by Eqn. 2.1. Figure 2.2 shows the reflectance (R) versus optical thickness and Fig. 2.3 shows the variation of R with bandwidth while other factors are held constant. The X's are the empirical results and the curves are the prediction of the formula. The fits show the equation's prediction to be a reasonable approximation of the results obtained by exhaustive design.

Figure 2.4 shows R versus L, the index of the last layer, which points out the advantage of lower indices if they could be found or simulated. Figure 2.5 shows many test cases where the number of layers in the designs were progressively reduced until the results passed through a minimum R. It was counterintuitive to find that the performance improved slightly as the number of layers was reduced, while keeping T constant, down to some minimum. The minimum number of layers is approximately equal to 6T+2 in the visible studies shown in Fig. 2.5 where B=2.5, L=1.38, and D=.89. With the IR examples[7], where B=1.597, L=2.2, and D=2.0, the minimum number is more on the order of 4T+2. Further reduction

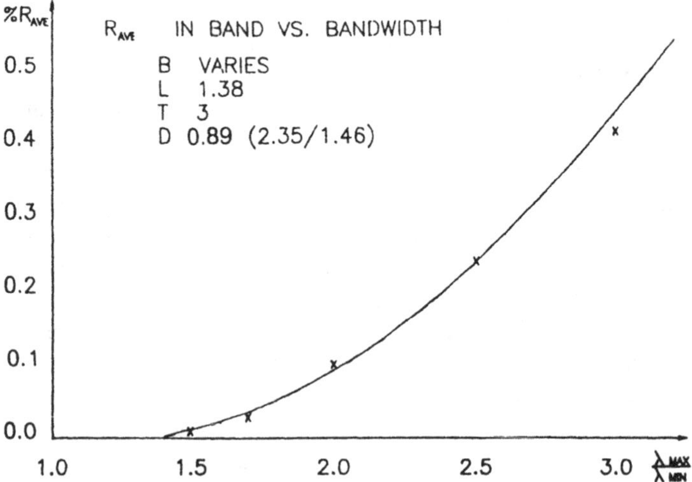

Fig. 2.3 Average reflectance in the band versus bandwidth as seen in Fig. 2.1. The curve is from the formula, while the X's are from empirical data.

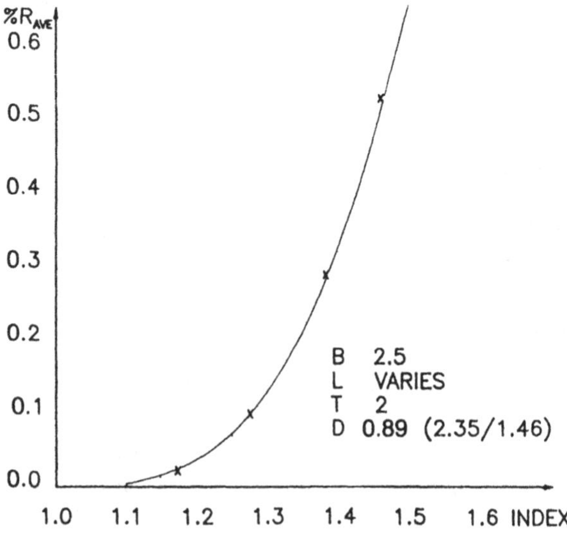

Fig. 2.4 Average reflectance in the band versus index of refraction of the last layer. The curve is from the formula, while the X's are from empirical data.

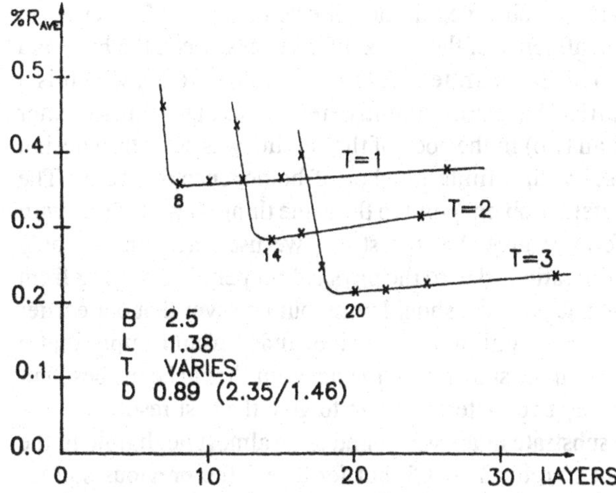

Fig. 2.5 Average reflectance in the band versus the number of layers. This implies that the minimum number of layers for the best results equals 6T + 2. The X's are from empirical data.

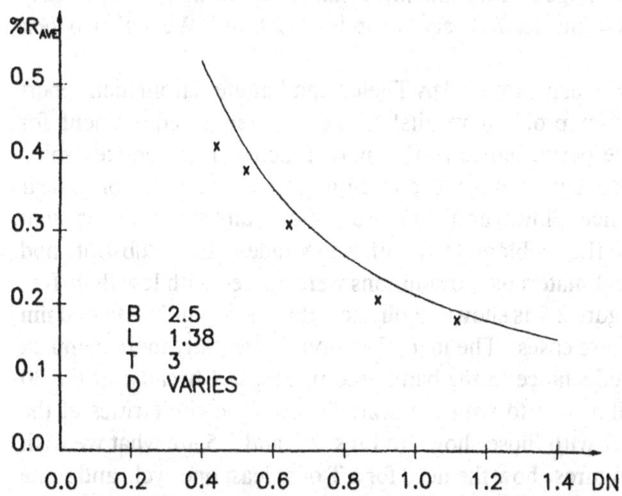

Fig. 2.6 Average reflectance in the band versus the index of refraction difference between the high and low indices of all the layers in the stack except the last. The X's are from empirical data.

in the number of layers below some minimum then causes the achievable results to degrade rapidly which is also observed in the collected results of Thelen[1].

Figure 2.6 shows the influence of the index difference between the high and low indices in the stack. This is separated from the last layer index which is a special case. Another empirical but counterintuitive result was that using a greater variety of indices (more than two) in the body of the coating was actually a design disadvantage in these cases with a finite number of homogeneous layers. The results of the Thelen-Langfeld problem point to the same thing. Two indices with the largest practical difference give the lowest R; we use three indices only because of scattering considerations due to the physical properties resulting from our processes, as mentioned above. We should point out however that our earlier work discussed in Chap. 1 with unlimited choice of index would imply that a smooth variation of index from the substrate to the medium should be the best, but lacking that opportunity, only two materials seem to give the best results.

The influence of the substrate index was found to be almost negligible in the cases studied. This is not true for $T < 0.5$, but for $T > 1$ the previous studies discussed in Chap.1 show that the index profiles first rise to a level higher than that of the substrate before falling toward 1.0. Therefore the starting index has little or no influence on the final results.

As mentioned above, Fig. 2.2 shows the effect of overall optical thickness. We have studied this from other points of view also in Chap. 1, where the effects of extra thickness seemed to have little advantage when any hypothetical indices could be used. The advantage of extra thickness may be in the ability to partially compensate for the lack of the desired very low index last layer. We will also look at this further below.

Figure 2.7 was graciously provided by Thelen and Langfeld from their report on the Berlin AR design problem results[1]. The problem's requirement for including the 30 degree performance in the merit function F introduces some small difference in the results from those of Eqn. 2.1 which is for only near normal angles of incidence. However, the nature of the results are very consistent with Eqn. 2.1. Because the problem dealt with a low index (1.52) substrate and a limited selection of real materials, no solutions were offered with less than $T=1$ of optical thickness. Figure 2.7 is shown in physical thickness, but 350 to 400 nm approximates $T=1$ in these cases. The merit function F is slightly more complex than just the average reflectance in the band used in Eqn. 2.1 because of the 30 degree factor, but similar to it to within a scale factor. The similarities of the results shown in Fig. 2.7 with those shown in Figs. 2.2 and 2.5 are what we wish to point out. All three figures show the need for a T of at least one cycle and some improvement with optical thickness, but less and less effect beyond $T=3$. We have added the number of layers to a few of the key design results on Fig. 2.7. These are consistent with the $6T+2$ estimate for the minimum number of layers.

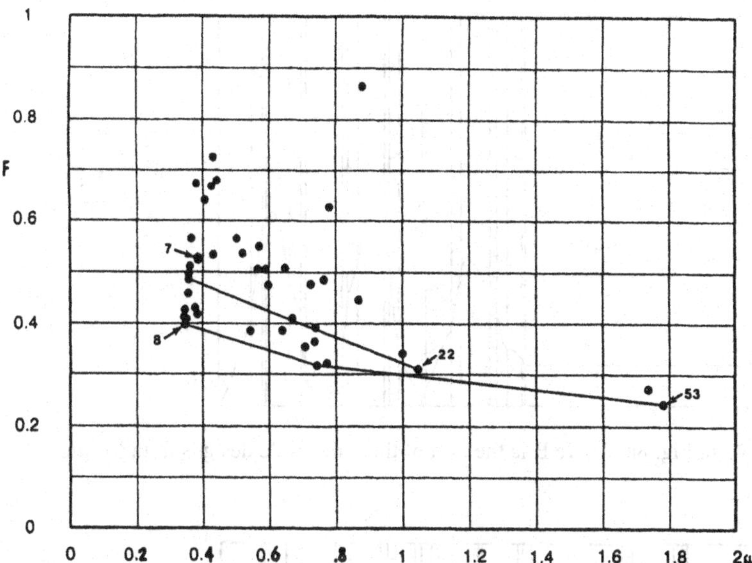

Fig. 2.7 Merit function versus total physical thickness reported by Thelen and Langfeld[1]. The number of layers is shown for selected designs. The upper line connects points of the best designs which used no titania in the design while the lower line connects those that did use titania.

Fig. 2.8a Percent reflectance versus thickness at the longest design wavelength in the band (900 nm) of some of the best designs from the Thelen-Langfeld problem. Curve A is the one-cycle design from the best of the Berlin design problem contributions.

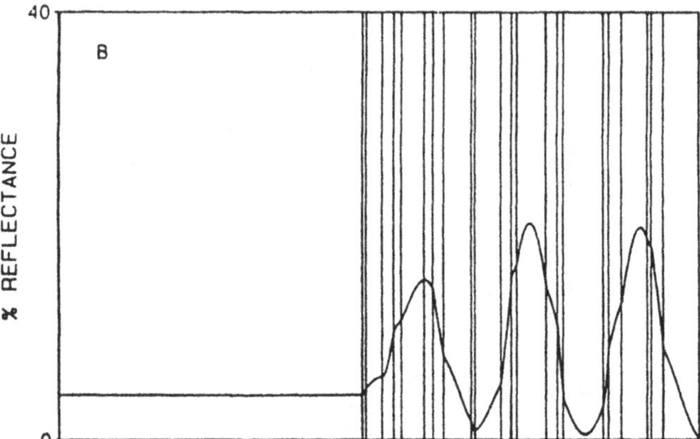

Fig. 2.8b As in Fig. 8a, Curve B is the best of the three-cycle designs from Berlin.

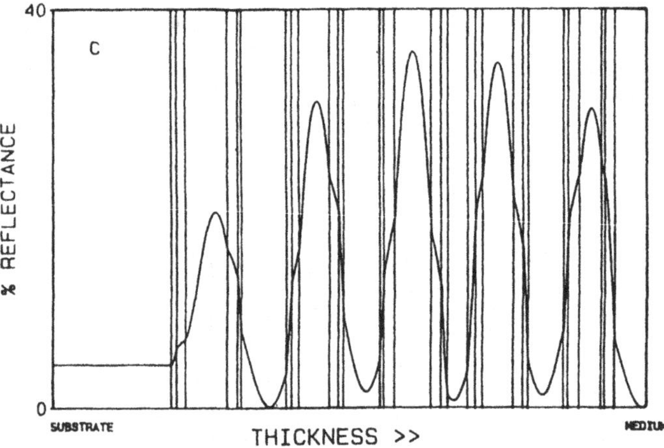

Fig. 2.8c Curve C is an adaptation and refinement of the best of the contributions.

The best result at the thinnest end of the designs had 8 layers and T=1. The best design with a physical thickness of about 1050 nm or a T of about 3 had 22 layers. However, we were able to further optimize that design to be slightly better with 20 layers which is 6T+2. The best design of the contest at about 1780 nm of physical thickness and T=5 had 53 layers. Preliminary work on reducing the number from 53 layers points also to the fact that 32 layers (6T+2) may be enough to get the same results. Note that the best 7-layer design in Fig. 2.7 is not as good as the best

8-layer. This is consistent with Fig. 2.5 also.

The upper line in Fig. 2.7 connects points of designs which used no titania in the design. These cases with the lower resulting D value do not achieve as good a result as those which use near the maximum amount of titania for a higher D.

To further emphasize the characteristics of T, we have plotted in Fig. 2.8 the reflectance versus thickness at the longest design wavelength in the band (900 nm) of some of the best designs from the Thelen-Langfeld problem. These are consistent with the periodic nature pointed out in Chap. 1. The general shape of the periods is similar when plotted as admittance versus thickness and the earlier works show the characteristic curves on the admittance amplitude versus phase plots. All of these appear to point to a natural periodic shape or function which would be the ideal AR design if there were no restrictions on the indices which could be used in an inhomogeneous structure of a given optical thickness.

2.2.4. Summary of Antireflection Coating Estimation

The empirically derived formula allows the estimation of the minimum average reflectance of very broadband AR coatings in the visible and infrared region. The variables with major effects on the results are the lowest available index, bandwidth, index difference from high to low, overall optical thickness, and the number of layers into which the overall thickness is divided. Substrate index is not a significant factor and intermediate indices between the highest and lowest indices available can be a disadvantage. This latter finding was somewhat unexpected.

The general results are somewhat independently confirmed by being consistent with results of the design contributions of many experienced coating designers from around the world to the Berlin AR design problem[1] and that reported by Aguilera, et al.[5] Dobrowolski, et al.[6] have also recently reported further corroboration of this work.

The formula can be used to reasonably predict the best performance that can be expected of a design of homogeneous layers for a given set of materials. The "ideal" number of layers in a design for a given thickness and the number of ripple minima in the band can also be predicted. These estimating formulae are expected to be useful tools for engineers and designers.

2.3. BANDPASS AND BLOCKER COATINGS

We pointed out in Chap. 1 that bandpass, LWP, and SWP filters can be made by properly positioning a QWOT stack or stacks to block or reflect the unwanted wavelengths. Thelen[7] appears to have been the first to discuss "Minus Filters" and he references related work by Young[8] wherein the blocked band is in the middle of two passbands, one on each side. Dobrowolski[9] applied them in detail and Thelen discusses them in his book.[10] It is helpful when working with any of these designs to be able to estimate how many layers will be required to attain the desired reflection/blocking and how wide the blocked band will be. The optical density (OD = log(1/Transmittance)) increases almost linearly with the number of layers in a QWOT stack. The width of the blocking band increases with the ratio of the indices of the high and low index materials in the stack. The relative width of the blocking band is less with higher orders of the reflection band QWOT wavelength. We can use all of these facts to estimate how many pairs of a given material combination will be required to achieve a given result.

2.3.1. Estimating the Width of a Blocking Band

Macleod[11] gives Eqn. 2.2 and its derivation for the estimated half width of the blocking band in the frequency related units of **g** which equals λ_0/λ, where λ_0 is the wavelength of the QWOT stack.

$$\Delta g = \frac{2}{\Pi} \arcsin\left(\frac{n_H - n_L}{n_H + n_L}\right) \tag{2.2}$$

For visible spectrum materials such as TiO_2 and SiO_2, this will give a half width of .138 for $n_H = 2.26$ and $n_L = 1.46$. If such a QWOT stack were centered at 550 nm, the edges of the reflecting band would be at about 474 and 626 nm. In the infrared, where Ge and ThF_4 can be used with $n_H = 4.0$ and $n_L = 1.35$, one might get $\Delta g = .330$. With a stack at 10µm, this would imply band edges at about 6.7 and 13.3 µm. Figure 2.9 shows the relationship of Eqn. 2.2 graphically. When the high and low indices approach the same values, the width of the reflectance band approaches zero. This can be found in fiber Bragg grating reflectors.

As we discussed in Chap. 1, the spectral distributions tend to be symmetrical when plotted versus frequency or g-values. Figure 2.10 is an example with different numbers of layer pairs of index 2.3 and index 1.46 plotted in linear frequency, or *wavenumbers*. The half width predicted by Eqn. 2.2 would be .1434, but the measured width is wider for fewer pairs at an OD of about 0.7. Equation 2.2 predicts a number for the width based on a very high number of pairs. There is an approximation of a pivot point at about the OD = 0.7 or 80% reflectance level. As the number of layer pairs is increased, this point does not change very much, but the slope at the point gets progressively steeper. Figure

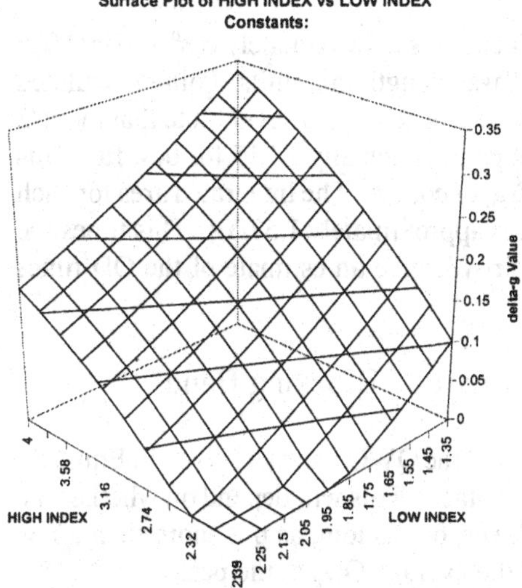

Fig. 2.9 Graphical illustration of the relationship of Eqn. 2.2.

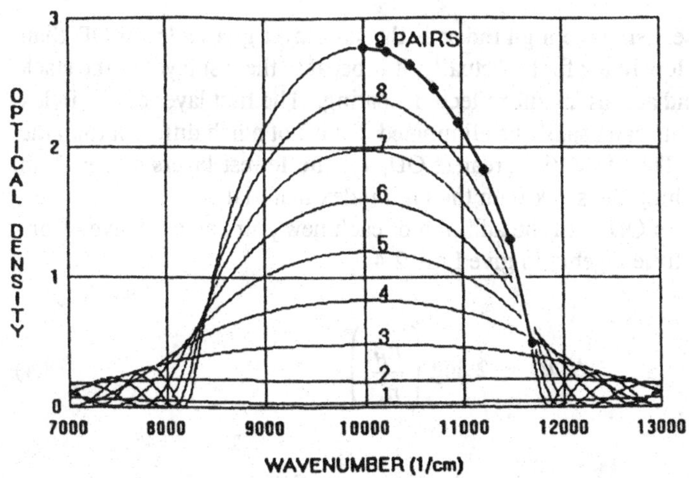

Fig. 2.10 Optical density versus linear frequency (or wavenumbers) with different numbers of layer pairs of 2.3 and 1.46 index of refraction. The values calculated from the approximation $\cos^{0.5}(1.315\delta g/\Delta g)$ are plotted over the right side of the "9 PAIRS" curve.

2.10 also illustrates how linearly the peak OD increases with additional layer pairs after the first few.

Also note that the shape of the OD curve is approximately $\cos^{0.5}(\Pi\delta g/2\Delta g)$ where δg is the distance from the QWOT wavelength in g-units. Points calculated using this formula are plotted to the right in Fig. 2.10. However, note that the $\pi/2$ is not precise and for the example of 9 pairs is actually 1.315 for best fit. This approximation is only valid up to $\delta g = \Delta g$, of course. The integrated area for each pair under this curve from $-\Delta g$ to $+\Delta g$ is approximately $1.57\Delta g$. This times the OD added at the peak by each new pair will give an estimate of the OD times bandwidth contribution of each pair.

2.3.2. Estimating the Optical Density of a Blocking Band

The optical density at the maximum point of the QWOT stack is given in Eqn. 2.3, where p is the number of layer pairs in the stack. This will depend on whether the stack starts with a high or low index layer, but as long as p is more than a few pairs, it gives a good approximation of the average OD_P at the peak.

$$OD_P \approx 2 \log \tfrac{1}{2}\left[\left(\frac{n_H}{n_L}\right)^p + \left(\frac{n_L}{n_H}\right)^p\right] \qquad (2.3)$$

In the typical case, using the high index as the start layer gives a lower OD than starting with the low index first. Actually, it is because the last layer of the stack is of low index and acts as an antireflection coating. The first layer of low index next to the substrate can usually be eliminated if it is not much different than the substrate index. Therefore, the greatest OD_P for the fewest layers comes from starting and finishing the stack with the high index material.

The change in OD with the addition of each new pair can be derived from Eqn. 2.3 using a little algebra to give Eqn. 2.4.

$$\Delta OD \approx 2 \log\left(\frac{n_H}{n_L}\right) \qquad (2.4)$$

This is sufficiently correct as long as there are more than a few layers.

Equation 2.4 combined with the integral over the bandwidth given above allows us to estimate the OD Bandwidth Product (ODBWP) of each additional pair. This is approximated by Eqn. 2.5.

$$ODBWP \approx 1.57\ \Delta g\ \Delta OD \approx 2\ \log\left(\frac{n_H}{n_L}\right)\ \arcsin\left(\frac{n_H - n_L}{n_H + n_L}\right) \tag{2.5}$$

2.3.3. Estimating the Number of Layers and Thickness Needed

As an example of the application of Eqn. 2.5, let us take the case of a requirement for a 99% reflector from 400 to 700 nm using $n_H = 2.25$ and $n_L = 1.45$. How many pairs would we estimate are required for such a design? Such a design would probably consist of gradually increasing or decreasing the layer thicknesses in the stack to give smooth coverage over the reflection band. The real issue is how much reflection needs to be generated by the layer pairs to cover the band. We can work out that this band is a Δg of .273 about a λ_0 of 509 nm, and the OD over the band must be 2.0 (1% transmittance). This gives a total ODBWP of 1.092 required. Using Eqn. 2.5, we find that the ODBWP per layer pair would be about 0.0832. Dividing this into the 1.092 required gives us the estimate that 13.13 pairs would be required or about 26 layers. We also know that each of these layers would average about one QWOT at 509 nm. Dividing the optical thicknesses of .12725 μm by the indices of the high and low layers and multiplying by 13 layers of each index, we get an estimated physical thickness of 1.876 microns. Note that estimation of the physical thickness required is only expected to be applicable where the designs are nearly quarterwave stacks. The estimation of the number of layers to produce a given ODBWP as described above, on the other hand, is not particularly dependent on the thickness of the layers involved.

2.3.4. Estimating More Complex Coatings

If the higher order reflection bands do not come into play, it may be practical to divide the spectral band to be covered into subsections and just consider the sum of the layers needed to meet the ODBWP of each subsection. We have found this to be practical for estimation purposes before starting a design. Angles other than normal incidence will probably add to the number of layers required.

There may be more complex coatings to be estimated where there are multiple bands to be blocked. In such cases, it would be logical to first look to see if higher order reflectances of the QWOT stack can be useful to the requirements.

Macleod[11] also illustrates as in Fig. 2.11 that, for a QWOT each of high and low index materials per pair (equal thicknesses), the reflectance or block band repeats at each odd multiple of a QWOT. The width of each of these block bands is the same in Δg. This then leads to the fact that the 3rd harmonic frequency will have 1/3 the relative width at that *wavelength* as at the fundamental QWOT wavelength. Similarly, the 5th harmonic will be 1/5 as wide. This then allows the creation of narrower bands when needed, but it is at the expense of 3 or 5 times as thick a stack for a given block band wavelength. Using high and low index materials that are closer to each other can accomplish the same thing with thinner stacks if the appropriate materials are usable. The free width between block bands also needs to be considered.

If any of these higher order reflection bands contribute to the reflection needed, they do not add to the layer count because they already exist from some other part of the coating. If, however, a high order reflection band is in a place where transmission is needed, the design would have to be changed to suppress that band. Baumeister[12] showed how different bands can be suppressed by using other than a unity ratio between the thicknesses of the high and low index layers. For example, a 3:1 ratio between the **overall thickness** of the pair to the thinnest layer will add the second and fourth harmonics but suppress the third, sixth, etc., as seen in Fig. 2.12. A 4:1 ratio will add the second but not the fourth, etc., as seen in Fig. 2.13.

Figures 2.14 to 2.16 show the effects of 5:1, 6:1, and 7:1. If we call the ratio A:1, it can be seen that the Ath harmonic of g_0 has a zero value and that the peaks of the harmonics have an envelope which is approximately a sine function of g from 0 to Π. An empirical fit to the data yields Eqn. 2.6.

$$OD_N \approx \left| OD_E\, SIN^{1.2}\left(\frac{\Pi N g}{A}\right)\right|; \qquad N = 1,2,... \qquad (2.6)$$

Here the OD_N is the OD of the peak of the Nth harmonic block band, and OD_E is the OD of the peak achieved by an equal thickness pair stack. This is illustrated in Figs.2.11 through 2.19. It is interesting to note that Eqn. 2.6 gives the correct results even for non-integer values of A such as 4.5, 4.75, and even 1.5 as seen in Figs. 2.17 to 2.19.

These properties can be useful in designing blockers for laser lines with doubled, tripled, quadrupled, etc. harmonics. These series of figures suggest ways to adjust the relative blocking in the harmonics by the choice of the A-value. Other applications using these harmonics may appear in the future in areas such as fiber optics communications filtering.

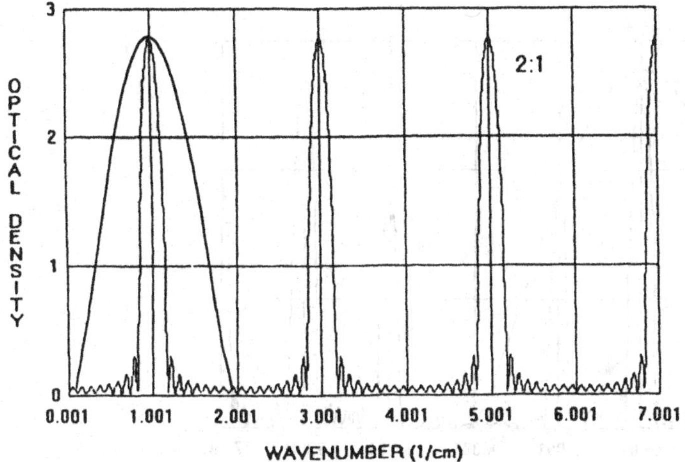

Fig. 2.11 OD versus frequency for a stack of 9 pairs where each of H and L layers is of equal thickness and a QWOT at 1.0 wavenumbers. The reflectance or block band repeats at each odd multiple of a QWOT. The width of each of these bands is the same in Δg. The first cycle of the curve generated by Eqn. 2.6 for this case is superimposed on the plot.

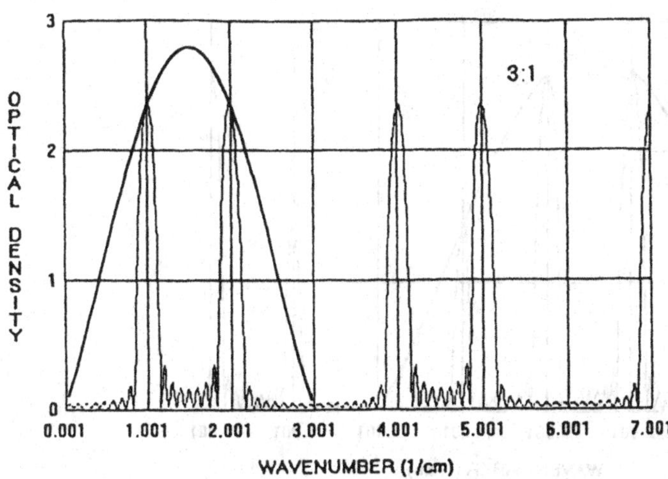

Fig. 2.12 Optical density versus frequency for a stack of 9 pairs that has a 3:1 ratio between the **overall thickness** of the pair to the thinnest layer. This will add the second and fourth harmonics but suppress the third, sixth, etc. The first cycle of the curve generated by Eqn. 2.6 for this case is superimposed on the plot.

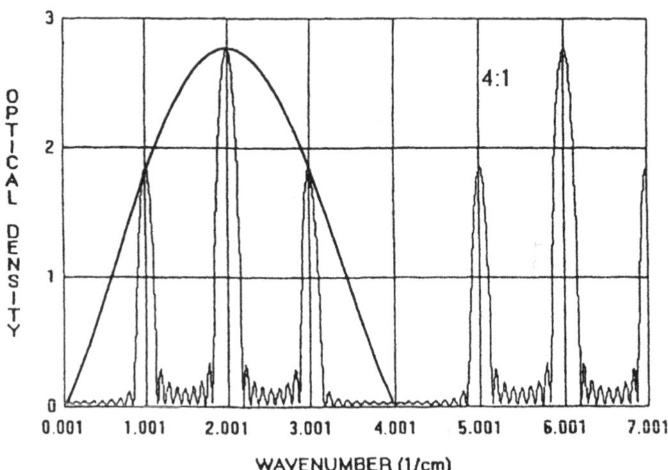

Fig. 2.13 OD versus frequency for a stack of 9 pairs that has a 4:1 ratio as in Figs 2.11 and 2.12. This will add the second harmonic but suppress the fourth, eighth, etc. The first cycle of the curve generated by Eqn. 2.6 for this case is superimposed on the plot.

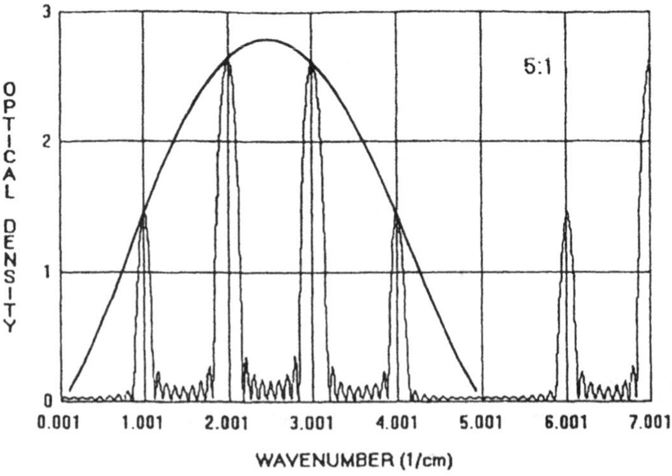

Fig. 2.14 Optical density versus frequency for a stack of 9 pairs that has a 5:1 ratio between the **overall thickness** of the pair to the thinnest layer. This will suppress the fifth harmonic, tenth, etc. The first cycle of the curve generated by Eqn. 2.6 for this case is superimposed on the plot.

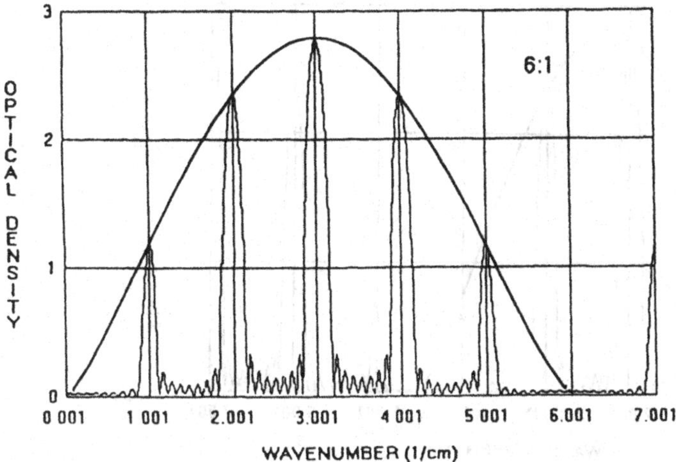

Fig. 2.15 Optical density versus frequency for a stack of 9 pairs that has a 6:1 ratio between the **overall thickness** of the pair to the thinnest layer. This will suppress the sixth harmonic, twelfth, etc. The first cycle of the curve generated by Eqn. 2.6 for this case is superimposed on the plot.

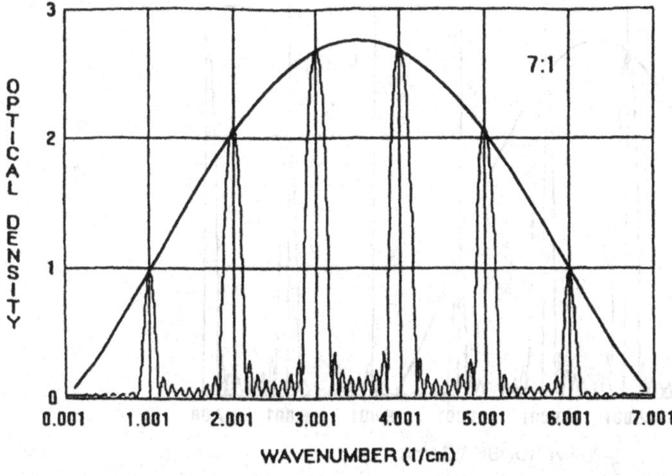

Fig. 2.16 Optical density versus frequency for a stack of 9 pairs that has a 7:1 ratio between the **overall thickness** of the pair to the thinnest layer. This will suppress the seventh harmonic, fourteenth, etc. The first cycle of the curve generated by Eqn. 2.6 for this case is superimposed on the plot.

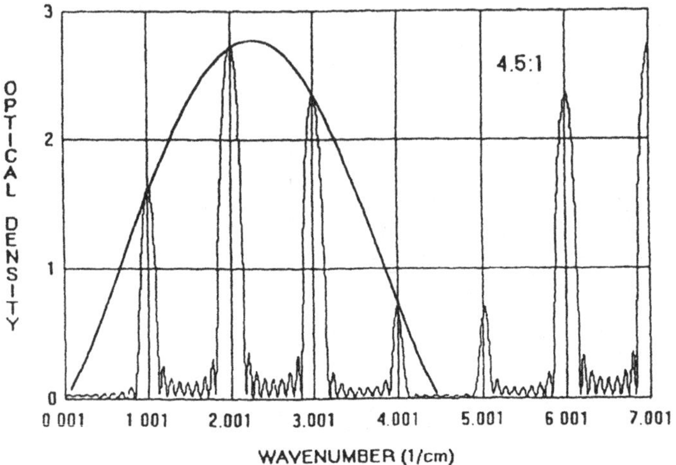

Fig. 2.17 Optical density versus frequency for a stack of 9 pairs that has a 4.5:1 ratio between the **overall thickness** of the pair to the thinnest layer. This illustrates that Eqn. 2.6 gives the correct results even for non-integer values of A such as 4.5. The first cycle of the curve generated by Eqn. 2.6 for this case is superimposed on the plot.

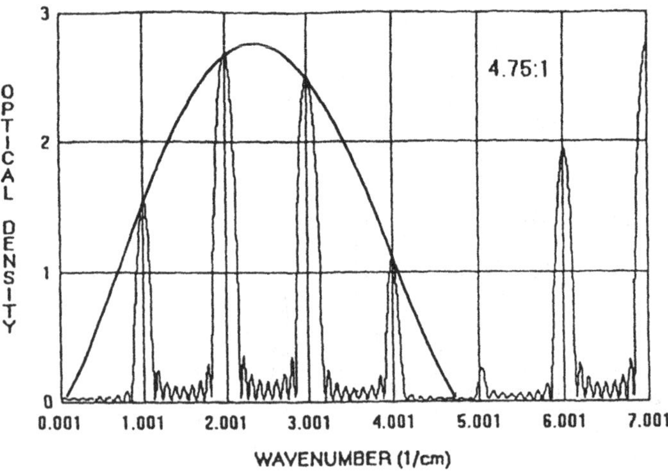

Fig. 2.18 Optical density versus frequency for a stack of 9 pairs that has a 4.75:1 ratio between the **overall thickness** of the pair to the thinnest layer. Note that the minimum of the OD pattern occurs at 4.75 times the fundamental band. The first cycle of the curve generated by Eqn. 2.6 for this case is superimposed on the plot.

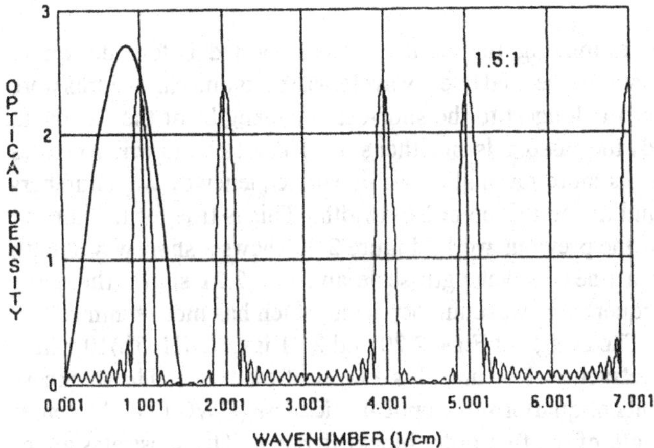

Fig. 2.19 Optical density versus frequency for a stack of 9 pairs that has a 1.5:1 ratio between the **overall thickness** of the pair to the thinnest layer. Note that the minimum of the OD pattern occurs at 1.5 times the fundamental band, and also multiples thereof. The first cycle of the curve generated by Eqn. 2.6 for this case is superimposed on the plot.

The higher harmonics seen in Figs. 2.11 to 2.19 can be attributed to the "square wave" nature of the transitions from high to low index as will be seen in Chap. 3. If the changes in index were smooth sine waves of the same period as the square waves, there would be no higher harmonics, only a single peak as at 1.001 wavenumber in Fig. 2.11. This, then, would be a classic rugate filter used to block only the one line and pass "all" other wavelengths.

2.3.5. Estimating Edge Filter Passband Reflection Losses

We have seen that edge filters which block one spectral region and pass an adjacent region are normally constructed with periodic stacks of high and low index layers of equal quarter wave optical thickness at the center wavelength of the blocked band. A preliminary and a subsequent aperiodic structure of several layers which minimize the reflectance in the passband are usually needed to transition from the effective index of the substrate to the effective index of the periodic structure and from the structure to the final medium. This section addresses the estimation of the possibilities and limitations of these antireflection or matching layers to reduce the reflections before the coating is actually designed. Equations are given for the estimation of the average reflectance in the passband as a function of the number of layers and the width of the passband for both short and long wavelength pass filters and for passbands on both sides of a "minus filter".

In Sec. 2.2 on estimating the results to be expected before designing antireflection (AR) coatings, we used linear wavelength plots and bandwidths were defined as the ratio of the longest to the shortest wavelengths of the AR band. When we went to study the passbands on either side of a block band which creates an edge filter, it appears more rational to use linear frequency or wavenumbers (cm^{-1}) for the plots and the definition of bandwidth. This is true for the current work and possibly for the previous work. Figure 2.20 shows a short wavelength pass (SWP) filter on a linear wavelength scale and Fig. 2.21 shows the same design on a linear frequency or wavenumber scale which has more symmetry in simple design cases. The design of Figs. 2.20 and 2.21 is (.5L 1H .5L)10 where L is a layer index of refraction 1.46 and H is a layer of index 2.2. The thickness of the layers are in units of quarter wave optical thickness (QWOT) at the center frequency or wavelength of the first order blocking band. The substrates are of index 1.52, the medium on the other side of the stack is of index 1.0, and dispersion and absorption are not included in this study. All of the present work considers equal optical thickness of the H and L layers inside the periodic stack.

The design of Figs. 2.20 and 2.21 is a good starting place for a SWP filter. The design of Fig. 2.22, which is (.5H 1L .5H)10, is a good starting place for a long wavelength pass (LWP) filter. We will now define the bandwidth of the AR as the fraction of the frequency range from zero frequency to the center of the blocking band. By examination of Figs. 2.21 and 2.22, we can see that the AR

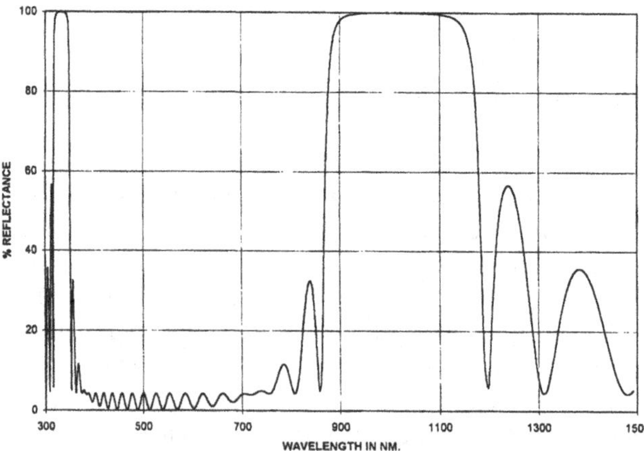

Fig. 2.20. Short wavelength pass (SWP) filter on a linear wave-length scale.

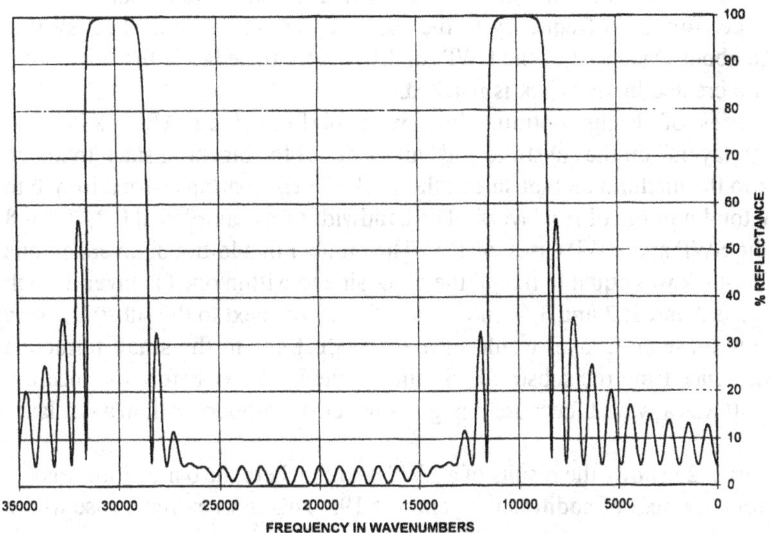

Fig. 2.21. Short wavelength pass (SWP) filter on a linear frequency or wavenumber (cm^{-1}) scale.

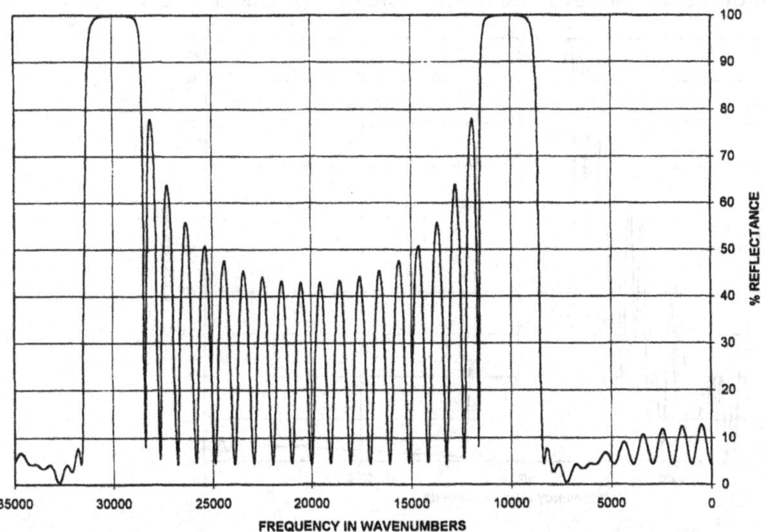

Fig. 2.22. Long wavelength pass (LWP) filter on a linear frequency scale.

bandwidth of a LWP filter in these cases is limited to a maximum of about 0.8 by the distance from zero frequency to the edge of the block band. The SWP is limited to about twice that of the LWP, or 1.6, before the edge of the third order block band created by the stack is reached.

A series of design optimizations were performed on AR coatings or "matching layers" on the substrate and "air" sides of the blocking stack to match the stack to the medium on that side of the stack. These coatings varied from 0 to 19 in the total number of AR layers. The bandwidth was sampled at 0.2, 0.4, 0.8 and, in the SWP and BWP cases, at 1.6. The number of additional layers on one side of the stack was equal to that of the other side to within one (1) layer in each case; such as 2 and 1, 2 and 3, 4 and 3, etc. The layers next to the substrate were always H because an L layer would have little effect due to the small refractive index difference from the substrate. Similarly, the last layer before the medium of index 1.0 was always L because that gives better AR properties than a last layer of H.

Figure 2.23 shows the results of a LWP design where the bandwidth was 0.8 and the total number of additional layers was 19. This is an extreme case where the point of diminishing returns on the number of layers has been approached. It does, however, illustrate a practical limit on what can be done with this type of design at the maximum LWP bandwidth. Figure 2.24 shows a SWP design for the maximum bandwidth using 11 additional layers. When we designed for one half of the maximum SWP bandwidth (0.8), we achieve what is seen in Fig. 2.25. The rest of the samples over the ranges were of this same nature.

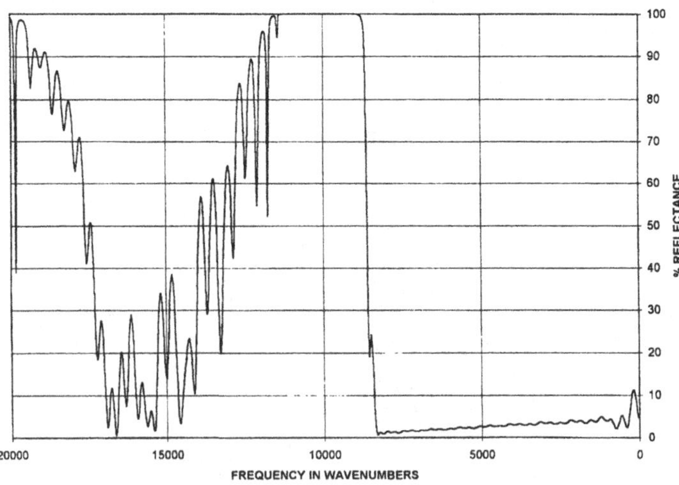

Fig. 2.23. Results of a LWP design where the bandwidth was 0.8 and the total number of additional layers was 19.

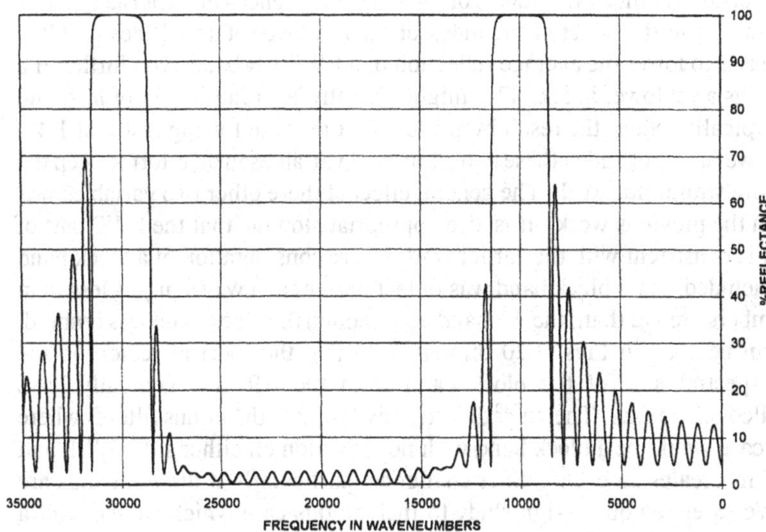

Fig. 2.24. SWP design for the 1.6 maximum bandwidth using 11 additional AR layers.

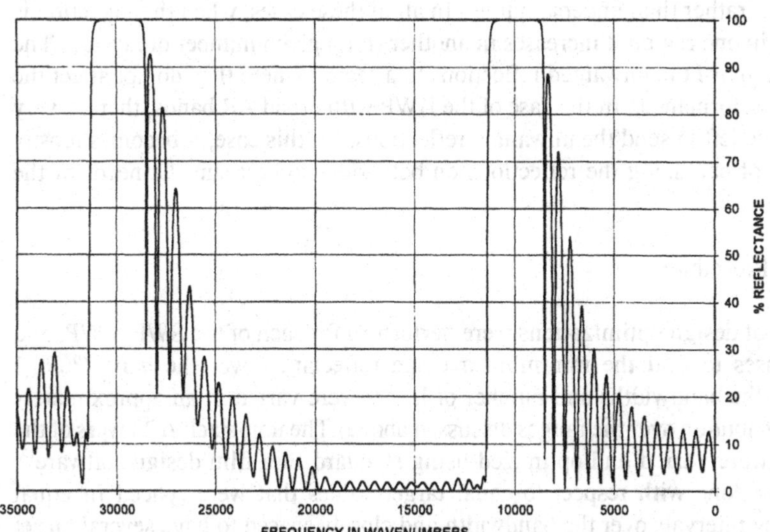

Fig. 2.25. SWP designed for half (0.8) of the maximum band-width using 11 additional AR layers.

Section 2.2 included the effects of the index difference between the L and H materials and also the effect of the index of the last layer of an AR design. It is often possible to lower the average reflection of a LWP AR band even further if a last layer has a yet lower index. This might typically be of index 1.38 or less, and it could typically reduce the result by 1/5 to 1/4 of the result using just L at 1.46. However, we have not added these two additional variables in an effort to keep the results more straightforward. The general effect of these other two variables was covered in the previous work. It is also appropriate to note that the LWP part of this study is consistent with the earlier work where consideration of a block band was not included, but a block band was in fact present at a wavelengths less than (wavenumbers greater than) the AR band even though the block band was ignored.

It can be seen in Figs. 2.20 through 2.25 that the spectral region on the opposite spectral side of the block band from the AR will generally have uncontrolled reflections. Thelen[7,10] did extensive work with "minus filters" where the ideal case would be a block band with no reflection on either side (LWP and SWP). This creates an even greater challenge, particularly if the AR bands are broad. We extended our design study to include this case which we will call a both long and short wavelength pass (BWP) filter. For simplicity in this part of the study, we always set the bandwidth on the SWP side of the BWP filter to two (2) times that of the LWP side. This would not generally have to be the case, but without such a limitation to this study the range of possibilities might lead to confusion rather than understanding. In all of these cases, when the reflection is reduced in one region it increases in another (for a given number of layers). The goal is to move the unwanted reflections to a region where they do not affect the desired performance. In the case of the BWP with broad AR bands, there is very little place left to send the unwanted reflections. In this case, it becomes mostly an issue of balancing the reflections on both sides to best suit the needs of the problem.

2.3.5.1 Procedure

A series of design optimizations were performed for each of the SWP, LWP, and BWP cases to find the minimum average reflectance over the band ($\%R_{ave}$) wherein the bandwidths and number of layers were varied in an approximately even distribution over the ranges discussed above. The non-stack AR layers of the designs were varied and optimized using standard thin film design software[13]. This was done with respect to equal target values that were spaced in equal frequency intervals over the bandwidth and closely spaced to have several target

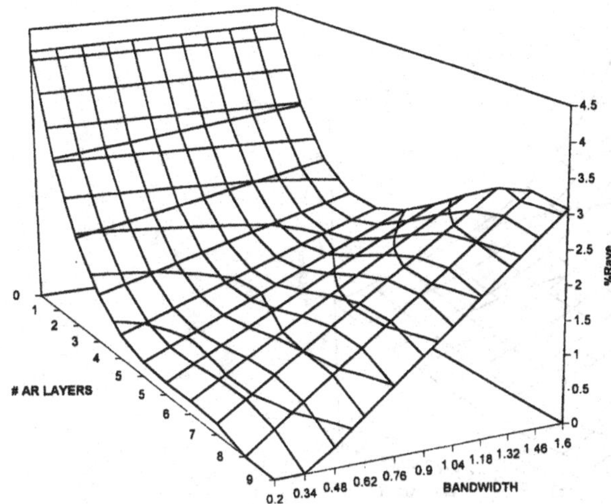

Fig. 2.26. Three-dimensional plot of the resulting minimum predicted $\%R_{ave}$ versus number of AR layers and bandwidth for the SWP filters.

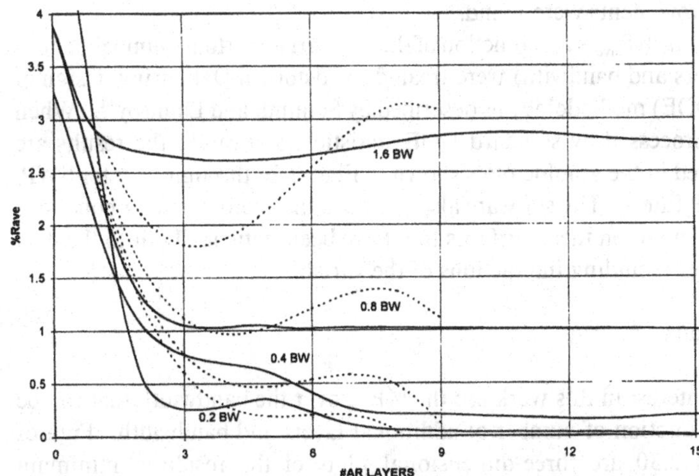

Fig. 2.27. Solid lines are $\%R_{ave}$ for SWP filters from the individual detail designs versus layers and bandwidth. The dotted lines are the values predicted by the equations.

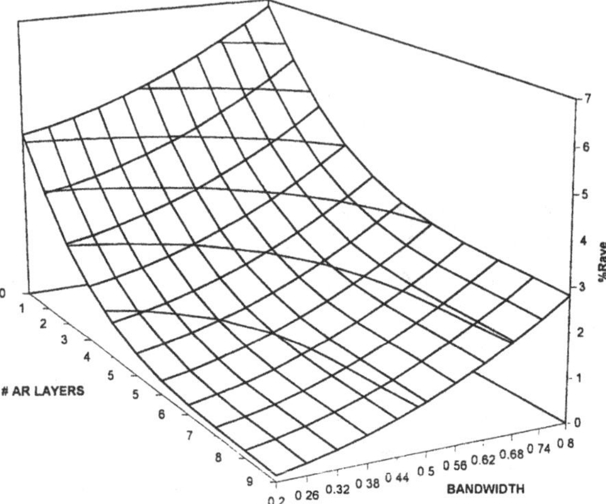

Fig. 2.28. Three-dimensional plot of the resulting minimum predicted $\%R_{ave}$ versus number of AR layers and bandwidth for the LWP filters.

points on each ripple in the spectral result. When an optimum was reached with the NOL Gradient Methods of the software[13], the result was reoptimized with Levenberg-Marquart algorithm to see if it could be further optimized. Usually, only small improvements were found.

The resulting $\%R_{ave}$'s as a function of the two variables (total number of non-stack (AR) layers and bandwith) were treated as Historical Data using design of experiments (DOE) methodology as described by Schmidt and Launsby[14]. When this data was processed by standard DOE statistical software[15], the results are readily illustrated in the graphic plots shown in Figs. 2.26 through 2.30 for SWP, LWP, and BWP filters. The software also provides the coefficients for equations to calculate any point on these surfaces that have been statistically fit to the data to the third order including interactions of the variables.

2.3.5.2 Equations

The results of interest in this work are the $\%R_{ave}$ over the bandwidth that can be achieved as a function of number of additional layers and bandwidth. Figures 2.26, 2.28 and 2.30 are three-dimensional plots of the resulting minimum predicted $\%R_{ave}$ versus number of layers and bandwidth for the SWP, LWP, and BWP designs. This analysis allows us to generate the following equations to estimate the minimum $\%R_{ave}$ to be expected as a function of the number of layers, N, and the bandwidth, B:

$$\%R_{ave}(SWP) = 3.29586 - .49358N + .8634B + .01769N^2 + .03682NB^2 \tag{2.6.1}$$

$$\%R_{ave}(LWP) = 2.1678 - .7247N + 5.0606B + .0441N^2 - .0007N^3 \tag{2.6.2}$$

$$\%R_{ave}(BWP) = 8.71509 - 2.17882N + 15.907B + .16413N^2 - .00409N^3 + .3327NB - 15.38B^2 \tag{2.6.3}$$

These equations are approximations and will generate some small negative values for large N and small B, but these can be taken to mean near zero values for R_{ave}. A spreadsheet can be easily set up to take the input of N and B and display the predicted $\%R_{ave}$.

In Figs. 2.27 and 2.29, we can see the $\%R_{ave}$ from the individual detail designs versus the number of AR layers as solid lines for the different bandwidths.

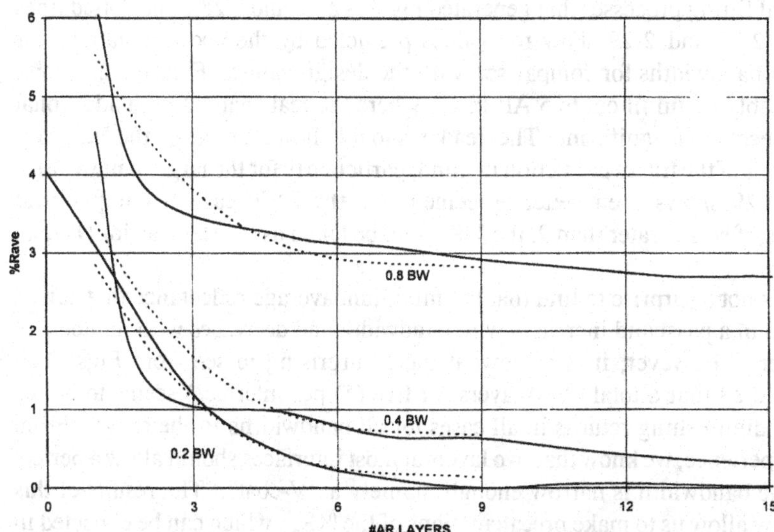

Fig. 2.29. Solid lines are $\%R_{ave}$ for LWP filters from the individual detail designs versus layers and bandwidth. The dotted lines are the values predicted by the equations.

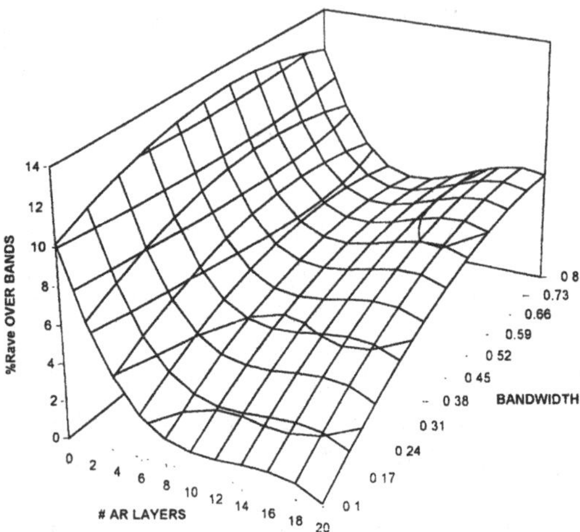

Fig. 2.30. Three-dimensional plot of the resulting minimum predicted $\%R_{ave}$ versus number of AR layers and bandwidth for the BWP filters.

After a total of about 5 AR layers in the SWP case of Fig. 2.27, there is no significant improvement in the $\%R_{ave}$. The LWP case is slightly more gradual, but reaches a point of diminishing returns at total of about 9 AR layers. As a result of this observation, the raw data from 0 to 9 layers are all that were included in the statistical fitting processes that generated Figures 2.26 and 2.28. The dotted lines in Figs. 2.27 and 2.29 show the values predicted by the above equations for different bandwidths for comparison with the design values. Figure 2.27 shows a reasonably useful fit out to 5 AR layers where the real changes with additional layers become insignificant. The reader should, however, keep the accuracy limitations of the filters prediction in mind, particularly for the larger bandwidths. Figure 2.29 shows even better agreement for the LWP cases. For practical purposes, if N is greater than 9, the $\%R_{ave}$ can be taken as the same as for N equal to 9.

It is not a surprise to find that the minimum average reflectance that can be achieved in a passband increases with bandwidth and decreases with numbers of AR layers. However, it is somewhat more surprising to see from Figs. 2.26 through 2.28 that a total of 10 layers (or five (5) per interface) seems to be the point of diminishing returns in all cases for all bandwidths in the range. From other experience, we know that two layers at most interfaces should allow a perfect AR if the bandwidth is narrow enough, namely a "V-coat". The results of this work now allow us to make precalculations of the $\%R_{ave}$ which can be expected in the passband of edge and minus filters when using the common materials like SiO_2 and TiO_2.

2.4. DICHROIC REFLECTION COATINGS

The estimation of general coating spectral shapes such as color correction filters may be reasonably approximated by the above methods in most cases. In the case of dichroic filters for color separation, etc., the above approach should be satisfactory. However, it is also of great interest to know how many layers are needed to achieve a certain edge slope between the pass and block bands. This is often the determining factor rather than OD in such filters.

The steepness of the side of an edge filter is in inverse proportion to the number of layers or pairs. The spectral distance from the high to the low transmittance region is usually the important factor for the designer. This might be from 80% to 20% T (about .1 to .7 OD) or some other choice of limits. If we call the spectral distance *dg* and the peak density at the QWOT wavelength OD_P, the effect of steepness may be approximated by Eqn. 2.7. We know the Δg from Eqn. 2.2 and the OD_P from Eqn. 2.3.

$$dg \approx \frac{1}{2}\left(\frac{\Delta g}{OD_P^{\ 1.74}}\right) \approx \frac{1}{2}\left(\frac{\arcsin\left((n_H - n_L)/(n_H + n_L)\right)}{\pi \log \frac{1}{2}\left((n_H/n_L)^P + (n_L/n_H)^P\right)}\right) \tag{2.7}$$

The 1/2 factor (found empirically) would need to be changed if the specific OD range from high to low transmittance or reflectance were changed from the 80% to 20% used above. Since we know that the peak OD is directly proportional to the number of layers in the stack, the *dg* will be inversely proportional to the number of layers (to some power). This shows how adding layers for steepness has a strong effect at a low total number of layers, but a weak effect if there are already many layers. The wavelength of the edge multiplied by dg gives the approximate change of wavelength from 20 to 80% reflectance. Figure 2.31 illustrates this for 5, 10, 15 and 20 layer pairs of L and H indices of 1.45 and 2.25 on a spectral plot. Figure 2.32 shows actual data measured from Fig. 2.31 plotted for comparison with a curve generated by the formula. The formula estimates a slope which is less steep than the real case for many layers and more steep for less than seven (7) layer pairs. However, it does show clearly the great number of layers needed for a steep edge and the "point of diminishing returns." In the region of 10 to 20 layers, this should be an adequate engineering estimate of performance before designing.

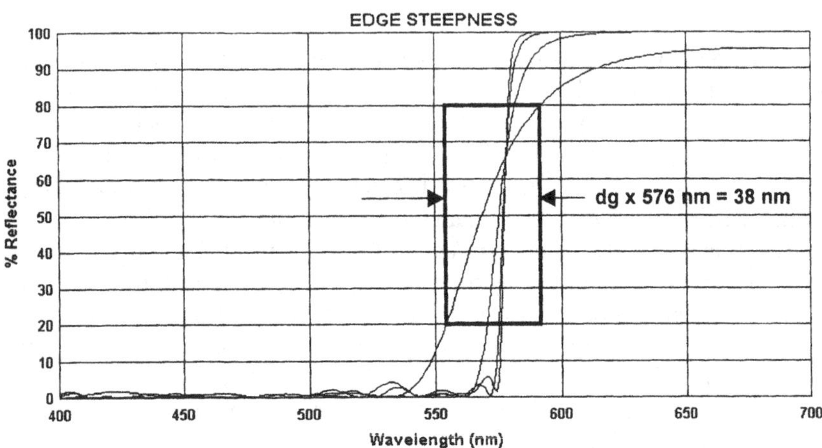

Fig. 2.31 Spectra of filter edge for 5, 10, 15 and 20 layer pairs of indices 1.45/2.25.

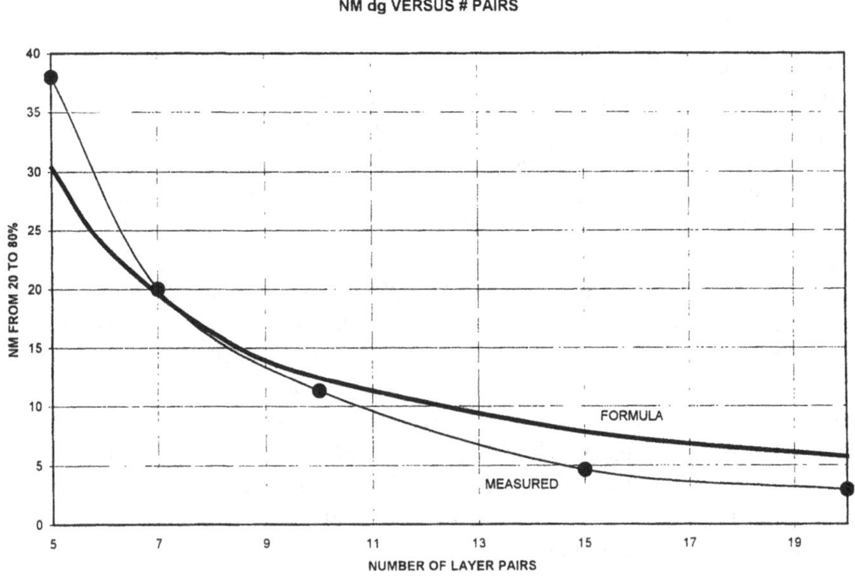

Fig. 2.32 Actual data measured from Fig. 2.31 plotted for comparison with a curve generated by Eqn. 2.7.

2.5. DWDM FILTERS

It is practical to estimate the pass band width and blocking band width of a DWDM filter as a function of refractive indices, spacers, number of layer pairs, and number of cavities before designing the filter. This can be a practical design guide as to which parameters should be used to gain a desired result. We will now describe the derivation and resulting formulas for these estimates. The successful production of a final design then depends on the process stability, monitoring techniques, and related errors which will be discussed in a later section.

Observing that the three-cavity filter seen in Fig. 1.26 is generally well suited to the typical 100 Ghz requirements, we performed a systematic investigation of the likely design space using design of experiments methodology (DOE). Four variables were considered: indices of the high and low index materials (2), the number of layer pairs in each mirror, and the number of half waves in the spacer layers. However, the indices were combined to form what was deemed to be two more meaningful variables: the difference in index ($n_H - n_L$), and the average index ($(n_H+n_L)/2$). The extremes of the sampling ranges in a Box-Wilson or CCD configured DOE were:

#Layer-Pairs	5 to 13
#Spacer-HWs	1 to 5
Index-Difference	.31 to .87
Average-Index	1.615 to 1.895

The most likely cases are well within these ranges. The experimental values of the 25 cases of this DOE are seen in Table 1. Figure 2.33 shows the results of the regression analysis.

Table 1. The experimental values of the 25 cases of this DOE.

Factor Row #	#LAYER PAIRS	SPACER HWs	INDEX DIF.	AVERAGE INDEX	PASS BANDWIDTH	Factor Row #	#LAYER PAIRS	SPACER HWs	INDEX DIF.	AVERAGE INDEX	PASS BANDWIDTH
1	7	2	0.45	1.685	3 82	*13*	11	4	0.45	1 685	0.32
2	7	2	0.45	1.825	5.26	*14*	11	4	0.45	1.825	0.52
3	7	2	0.73	1.685	0.38	*15*	11	4	0.73	1.685	0.0074
4	7	2	0.73	1.825	0.64	*16*	11	4	0 73	1.825	0.016
5	7	4	0.45	1.685	2.76	*17*	9	3	0.59	1.755	0 32
6	7	4	0.45	1.825	3 88	*18*	5	3	0.59	1 755	4.88
7	7	4	0.73	1.685	0.25	*19*	13	3	0.59	1.755	0 022
8	7	4	0.73	1.825	0.42	*20*	9	1	0.59	1 755	0.5
9	11	2	0 45	1.685	0.45	*21*	9	5	0.59	1.755	0 234
10	11	2	0 45	1.825	0 72	*22*	9	3	0.31	1 755	5.24
11	11	2	0.73	1.685	0 0112	*23*	9	3	0.87	1.755	0.016
12	11	2	0.73	1.825	0.025	*24*	9	3	0.59	1.615	0.18
						25	9	3	0.59	1 895	0 52

Multiple Regression Analysis

Y-hat Model

Factor	Name	Coeff	P(2 Tail)	Tol	Active
Const		0.22070	0.2011		
A	#LAYER-PAIRS	-1.04402	0.0000	1	X
B	SPACER-HW'S	-0.15270	0.1066	1	X
C	INDEX-DIF.	-1.10118	0.0000	1	X
D	AVE-INDEX	0.17343	0.0698	1	X
AC		0.75498	0.0000	1	X
AA		0.51692	0.0000	0.920	X
CC		0.56117	0.0000	0.920	X

Rsq	0.9587
Adj Rsq	0.9417
Std Error	0.4390
F	56.4098
Sig F	0.0000

Source	SS	df	MS
Regression	76.1	7	10.9
Error	3.3	17	0.2
Total	79.4	24	

Factor	Name	Low	High	Exper
A	#LAYER-PAIRS	7	11	9
B	SPACER-HW'S	2	4	3
C	INDEX-DIF.	0.45	0.73	0.59
D	AVE-INDEX	1.685	1.825	1.755

Prediction

Y-hat	0.2207
Std Error	0.439025691

99% Prediction Interval

Lower Bound	-1.09637707
Upper Bound	1.537777073

Fig. 2.33 The results of the regression analysis of the data in Table 1.

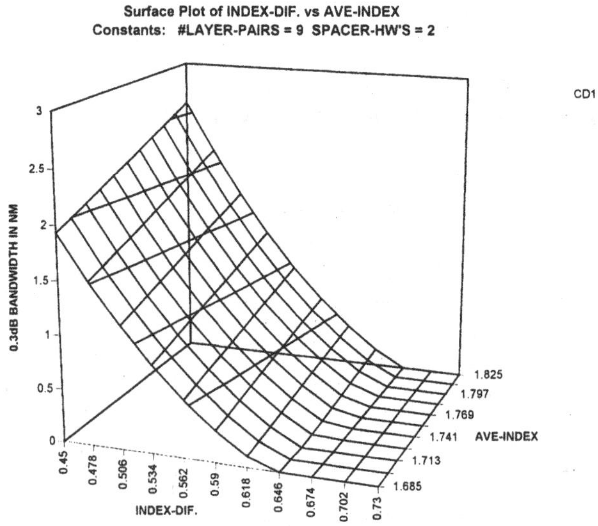

Fig. 2.34 Shows that the 0.3 dB bandwidth (BW) is a strong function of the index difference but the average index has little effect.

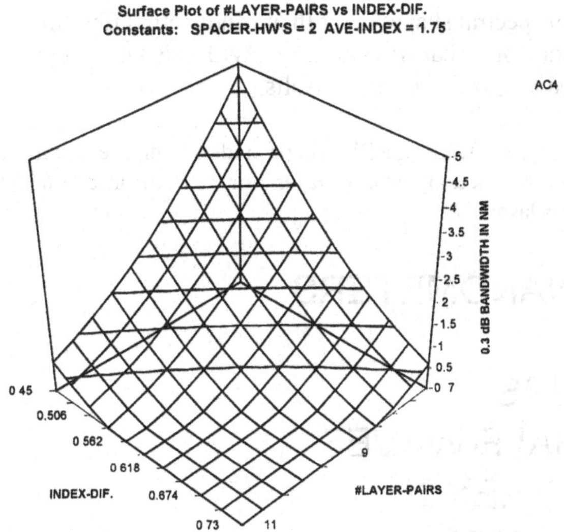

Fig. 2.35 Shows that the BW is a strong function of the difference of index and of the number of layer pairs.

Figure 2.34 shows that the 0.3 dB bandwidth (BW) is a strong function of the index difference from n_H to n_L, but the average index of n_H and n_L has little effect. Figure 2.35 shows that the BW is a strong function of the difference of index and of the number of layer pairs. Figure 2.36 shows that both the average index and the number of half waves in the spacer layers do not have a strong effect on the BW. It is therefore advisable to obtain the gross features of the design (rough BW and blocking) by index difference (which is usually fixed by other considerations) and the number of layer pairs in the mirrors. The fine details of the BW can then be adjusted with the number of half waves in the spacers. The average index is also usually fixed by other considerations. Therefore, the design is first set as close as possible to the required results with the number of layer pairs and then refined by the number of half waves in the spacers.

Table 2, which was derived from the information in Fig. 2.33, gives equations and coefficients from which the BW can be predicted for three-cavity filters by knowing the high and low indices, number of layer pairs per mirror, and the number of half waves in the spacer layers. This is easily entered in a spreadsheet program for routine calculations in the design process. It could also be useful to find the difference between n_H and n_L from an actual design run. Since the index of n_L is likely to be close to 1.45 or 1.46, the index n_H can be determined to the accuracy that n_L is known. This does however require that the results are good (have the right spectral shape) other than bandwidth differences due to the index being different from what was assumed. With this knowledge, the design can then be adjusted to give the desired results.

Table 2 Equations and coefficients from which the BW can be predicted for three cavity filters by knowing the high and low indices, number of layer pairs per mirror, and the number of half waves in the spacer layers.

BANDWIDTH IN NANOMETERS
= Const
+ A * #LAYER-PAIRS
+ B * #SPACER-HALF-WAVES
+ C * INDEX-DIFFERENCE
+ D * AVERAGE-INDEX
+ AC * #LAYER-PAIRS * INDEX-DIFFERENCE
+ AA * #LAYER-PAIRS ^ 2
+ CC * INDEX-DIFFERENCE ^ 2

	0.3dB	20dB
Const	40.42101	99.05472
A	-4.43899	-11.1974
B	-0.1527	-0.39881
C	-65.9173	-150.719
D	2.477614	5.069167
AC	2.696339	6.397634
AA	0.12923	0.339522
CC	28.63107	63.3527

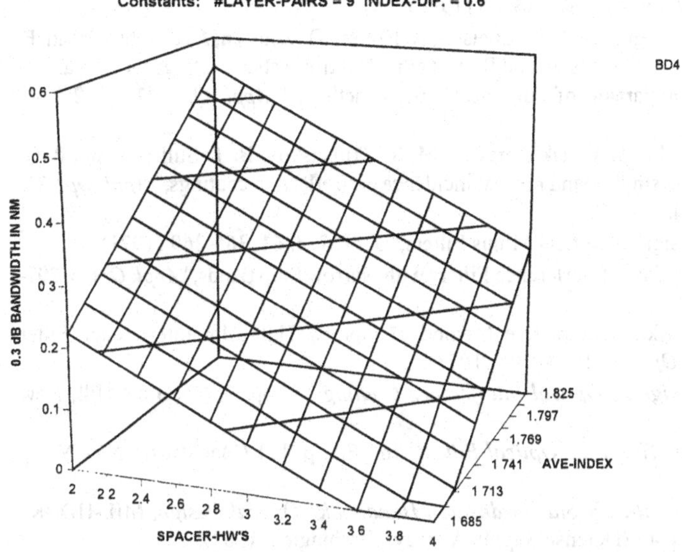

Fig. 2.36 Shows that both the average index and the number of half waves in the spacer layers do not have a strong effect on the BW.

2.6. SUMMARY

It has been shown that there is quite a bit that can be estimated about most coating designs even before the design is started. The estimating process also provides guidance for the design process. It might be said that the estimating process is the first order design as in lens design, and the computer optimization is the rigorous completion of the design process.

2.7. REFERENCES

1. A. Thelen and R. Langfeld: "Coating design problem," *Proc. SPIE* **1782**, 552-601 (1992).
2. R. R. Willey: "Realization of a very broad band AR coating," *Proc. Soc. Vac. Coaters* **33**, 232-236 (1990).
3. A. Macleod: "Design of an antireflection coating for glass over the region 400 nm to 900 nm," *Proc. SPIE* **1782**, 602-611 (1992).

4. M. L. Rastello and A. Premoli: "Continuation method for synthesizing antireflection coatings," *Appl. Opt.* **31**, 6741-6746 (1992).
5. J. A. Aguilera, J. Aguilera, P. Baumeister, A. Bloom, D. Coursen, J. A. Dobrowolski, F. T. Goldstein, D. E. Gustafson, and R. A. Kemp: "Antireflection coatings for germanium IR optics: a comparison of numerical design methods," *Appl. Opt.* **27**, 2832-2840 (1988).
6. J. A. Dobrowolski, A. V. Tikhonravov, M. K. Trubetskov, B. T. Sullivan, and P. G. Verly: "Optimal single-band normal incidence antireflection coatings," *Appl. Opt.* **35**, 644-658 (1996).
7. A. Thelen: "Design of Optical Minus Filters," *J. O. S. A.* **61**, 365-369 (1971).
8. L. Young: "Multilayer Interference Filters with Narrow Stop Bands," *Appl. Opt.* **6**, 297-315 (1967).
9. J. A. Dobrowolski: "Subtractive Method of Optical Thin-Film Interference Filter Design," *Appl. Opt.* **12**, 1885-1893(1973).
10. A. Thelen: *Design of Optical Interference Coatings,* Chap. 7, (McGraw-Hill, New York, 1988).
11. H. A. Macleod: *Thin Film Optical Filters,* 2nd Ed., p. 171 (MacMillan, New York, 1986).
12. P. Baumeister: *Military Standardization Handbook, Optical Design,* MIL-HDBK-141, Chap. 20, p 45 (Defense Supply Agency, Washington, 1962).
13. *FilmStar Design*™, FTG Software Associates, Princeton, NJ (1998).
14. S. R. Schmidt and R. G. Launsby: *Understanding Industrial Designed Experiments*, Sec. 3.8, (Air Academy Press, Colorado Springs, 1994).
15. *DOE KISS*, Ver. 97 for Windows, (Air Academy Associates and Digital Computations, Inc., Colorado Springs, 1997).

3

Fourier Viewpoint of Optical Coatings

3.1. INTRODUCTION

We observed the interesting periodicity which develops in thicker AR coatings as developed in Sec. 1.7.2 in our collaboration with Dobrowolski and Verly[1,2]. We will now explore some of the principles that relate to the Fourier approach which contribute to the understanding of thin film designs in general and the effect of additional thickness on ARs in particular by empirical and comparative means.

We showed in Chap. 1 that the lack of very low index of refraction materials is the major limitation to achieving a very low reflectance coating over a very broad spectral band. It was also shown that additional coating thickness can be employed to make up for this deficiency to a certain extent. Thicknesses which are an order of magnitude thicker than the minimum necessary for a reasonable very broad band AR coating can reduce the reflection to about one half that of the minimum thickness case. This result is empirically predictable to a satisfactory degree as we have seen in Chap.2, but the underlying reasons for this have not been clear. The Fourier viewpoint adds to the understanding of variations from these ideal designs.

3.2. FOURIER CONCEPTS

A history of the development of the application of Fourier transform techniques to thin films and its mathematical foundations might best be found in the paper by

Dobrowolski and Lowe[3]. Here, however, we will attempt to make the concepts clear using graphics and without recourse to mathematics. The state of the art at the present time in coating design is based on **analysis** and adjustment. By this we mean that it is possible to fully analyze what a specific configuration of layers and indices will give in terms of reflectance, transmittance, absorptance, polarization, etc. We can then make small changes in the parameters and see if the results are in the direction of better or worse with respect to what we desire. We can optimize such a design with respect to defined design goals or targets by multiple iterations of these steps. This process is the backbone of thin film design at this time. However, what is really desirable is to be able to **synthesize** a design. By this we mean: to directly put together a design which meets the requirements without iteration/optimization. The Fourier technique put forth by Dobrowolski[3] and his predecessors, Delano[4] and Sossi[5], and his successors, Bovard[6-8], Southwell[9], and Fabricius[19], has the promise of being just such a synthesis process. However, as we shall illustrate, there is still some work to be done before a true synthesis version is achieved. Because of the approximations which must be used at present, a preliminary synthesis by the existing techniques must then be optimized for a final solution. We will later describe these approximations and limitations that will hopefully be overcome in the future. The real potential of the synthesis technique seems to be for cases where intuition and experience do not indicate how best to design for some complex reflectance profile. It would be highly desirable in these cases to have a tool to take the required reflectance profile and directly synthesize a solution.

The published works referenced are heavy in the mathematical description of the technique. For some of us, the mathematics do not paint a lucid picture of the principles. We will put forth here a brief graphical illustration to lend some intuition and understanding about the Fourier techniques. Although mathematics may be the engine which propels the vehicle, we need not be an engine designer or automobile mechanic in order to drive the vehicle to where we want to go. In this section, we will give a "Driver's Ed" version of what makes the "car go."

3.2.1. Background

We reviewed the concepts of the Fresnel reflection amplitude coefficient (r) in Sec. 1.2, Eqn. 1.1. We will later show empirically that this is the "coin of the realm" in which we should deal when working in the Fourier domain. It is not magic, but "it **is** all done with mirrors." Reflectance amplitude is the key. This is the reflection caused by a discontinuity of index of refraction or admittance at an interface between two homogeneous media. In the case of inhomogeneous media with no discontinuities, the concepts still apply, but we must deal with the rate of change of index or admittance. If there is no change, there is no reflection. For the rest of this discussion, we will assume homogeneous media and no absorptance or dispersion, for simplicity's sake.

Figure 3.1a shows the index of refraction versus position in space or optical thickness (t) in the direction of the propagation of the light. Figure 3.1b shows the reflectance amplitude (in math: a Kronecker delta) versus the same position (t) that this discontinuity produces. Figure 3.1c shows the reflectance intensity (R = rr*) versus spectral frequency (f = 1/wavelength) which this produces. Note that it is the same for all frequencies since there is no dispersion. This is similar to a pure sharp electrical pulse, like lightning, that produces "white" noise at all frequencies.

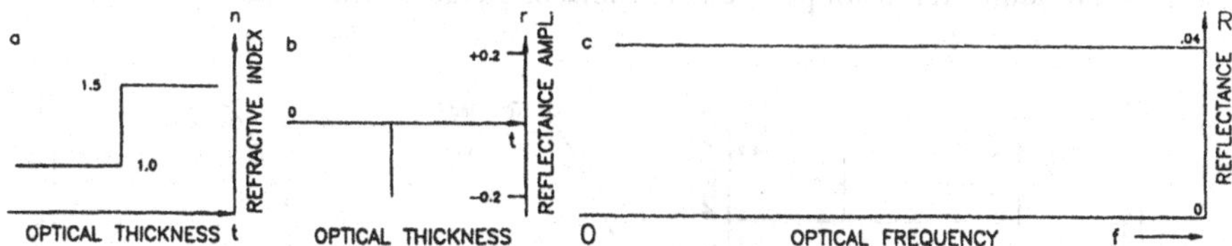

Fig. 3.1. Fourier view of reflection at a discontinuity of refractive index. Part a, the index profile, shows the index of refraction versus optical thickness in the direction of the propagation of the light. Part b shows the reflectance amplitude profile versus the same position that this discontinuity produces. Part c shows the resulting reflectance intensity versus spectral frequency.

If we now take a thin slab of material of index 1.5 surrounded by an index of 1.0, we know that we will get interference between the reflections from the first and second interface surfaces. Figures 3.2a-c show this index profile and r vs. t, plus the resulting first cycle of the R vs. f spectrum. This is the familiar result of a higher index single layer coating where the r at the first and last interfaces are of equal and opposite amplitudes. The r and R go to a maximum at a frequency for which the thickness is one QWOT and back to a minimum when the frequency is twice that (or t = halfwave OT). The pattern repeats at all integer multiples of this latter frequency. Note that the first minimum would occur at half the

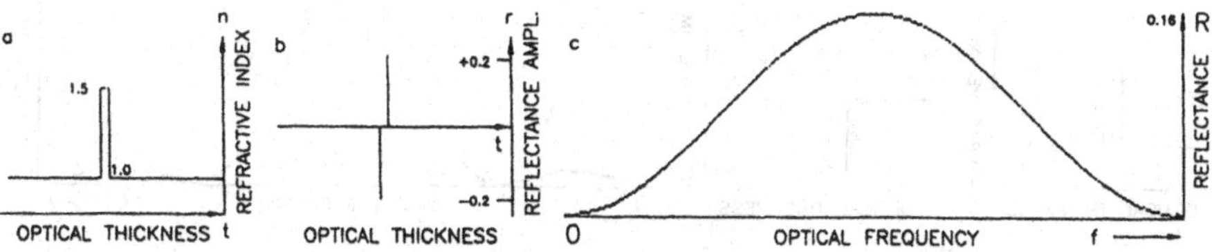

Fig. 3.2. Fourier view of reflection at a thin slab of material of index 1.5 surrounded by an index of 1.0. As in Fig. 3.1, parts a-c show this index profile and r vs. t, plus the resulting first cycle of the R vs. f spectrum.

frequency if we doubled the thickness of the slab as illustrated in Figs. 3.3a–c. Similarly, it would occur at 2f if t were cut in half, etc.

The c subfigures in this series are all the Fourier transform (squared) of the b subfigures. That is to say that the reflectance amplitude versus optical thickness is Fourier transformed to the reflectance amplitude (and squared to give the R we are used to seeing) versus optical frequency. A key point is that the Fourier transform is reversible. If we transform the c-data, it will give us back the b-data. This implies that we should be able to define what we want in a c-curve and transform it to the reflectance vs. thickness which would produce that result. It is not yet quite that easy, but the state of the art is getting close. We will first show a few additional examples to give more of a feel for the results.

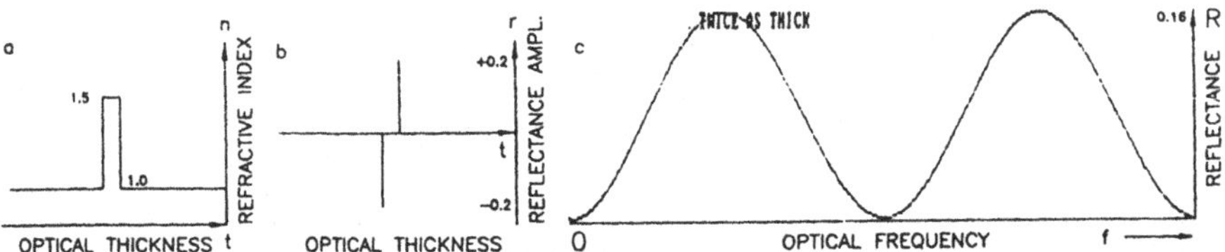

Fig. 3.3. Fourier view of reflection at a slab of material as in Fig. 3.2, but twice as thick. Note that the first minimum occurs at half the frequency.

If we space two slabs apart by an optical thickness equal to their own, we get the results shown in Fig. 3.4. The interactions of the various reflections and phases add to produce the results shown. Note that the scale of R has quadrupled because the reflection amplitudes have added to give 2r which converts to 4R. It may further be noticed that there are three of the lowest frequency components developed by the interferences between each of the three adjacent pairs which have the same frequency spectrum as Fig. 3.2 but differing phase relations to each other. There will also be two interferences between the 1st and 3rd interfaces and

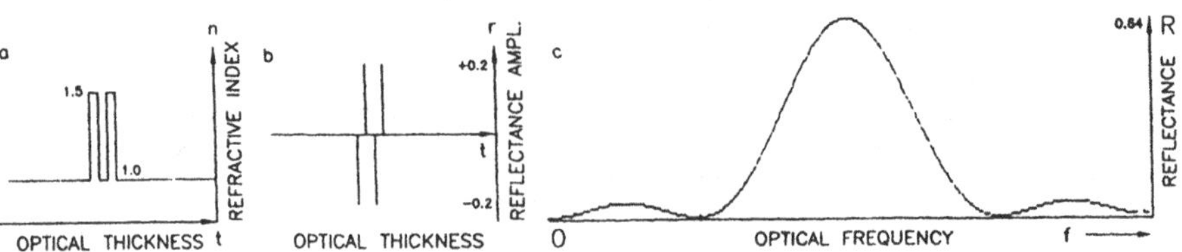

Fig. 3.4. The result of two slabs spaced apart by an optical thickness equal to their own. Note that the scale of R has quadrupled because the reflection amplitudes have added to give 2r which converts to 4R.

the 2nd and 4th which will have minima twice as often as the "fundamental" period. There is one more interference between the 1st and 4th interface which will have a minimum three (3) times as often as the fundamental. Keeping track of all these reflectance amplitudes, frequencies, and phases can quickly lead to confusion as can be seen from this most simple three-layer example. However, we do not have to concern ourselves with this in general, as the Fourier transform algorithms will do all of that for us. The point to remember is that the reflections from each surface interact with the reflections from each other surface through their amplitudes and relative phases.

Figure 3.5 shows the effect of eight (8) equally spaced reflections as would

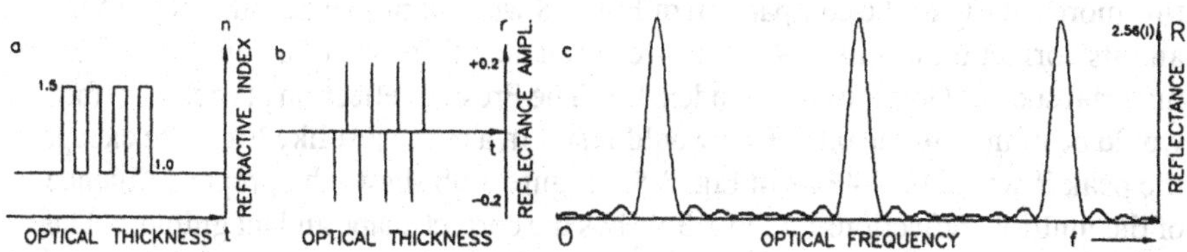

Fig. 3.5. Fourier view of the reflection effect of eight (8) equally spaced reflections as would be found in a 7 QWOT stack.

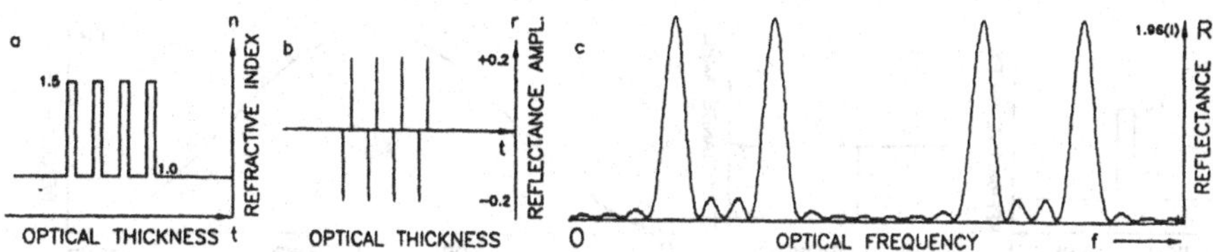

Fig. 3.6. Fourier view of the reflection of a stack where the ratio of the thickness of a pair to the thickness of the first layer is 3:1; note the similarity to Fig. 2.12.

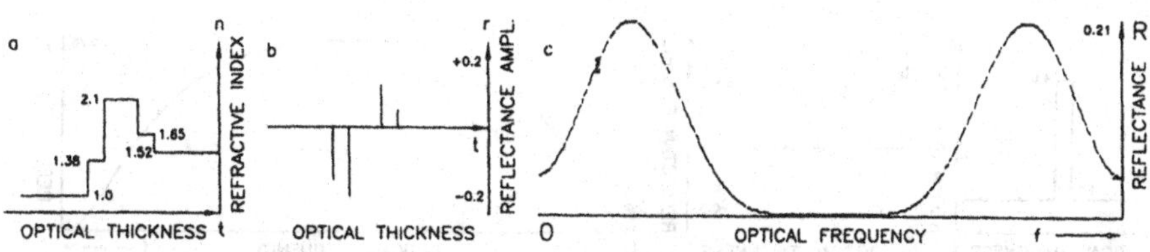

Fig. 3.7. Fourier view of the reflection of the common three-layer broad band AR where the AR band actually lies between two higher reflection peaks.

be found in a 7 QWOT stack. The pattern starts to look consistent with Fig. 2.11 of a QWOT stack in the last chapter. Figure 3.6 shows a stack where the ratio of the thickness of a pair to the thickness of the first layer is 3:1; note the similarity to Fig. 2.12. Figure 3.7 shows how the common three-layer broad band AR would look in this scheme where the AR band actually lies between two higher reflection peaks.

3.2.2. Some Limitations

If we look closely we notice a problem. In Figs. 3.5c and 3.6c, the peaks in R are greater than the physical limit of 1.0. The problem is that the b subfigures shown have not taken the multiple internal reflection (MIR) into account. We can see this more clearly in the comparison of Fig. 3.8 without proper account of the MIR and its correction in Fig. 3.9. Here we simulate a thin slab of a very high index material such as Germanium of index 4.0. The Fresnel reflection at each interface would be of magnitude 0.6. This would result in a curve just like Fig. 3.2c except the peak R would be 1.44 as in Fig. 3.8. Figure 3.9b shows the proper influence of the multiple reflections and Fig. 3.9c has the correct shape and magnitude. The reflection from the first interface is just as before. The first reflection from the second interface encounters the 1st surface on its way back, and only 0.4 of it is

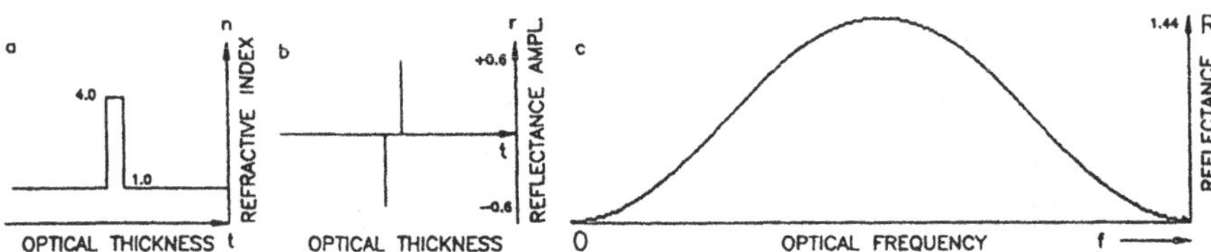

Fig. 3.8. Fourier view of the reflection of a thin slab of a very high index material such as Germanium of index 4.0 where the multiple internal reflections (MIR) have not been taken into account.

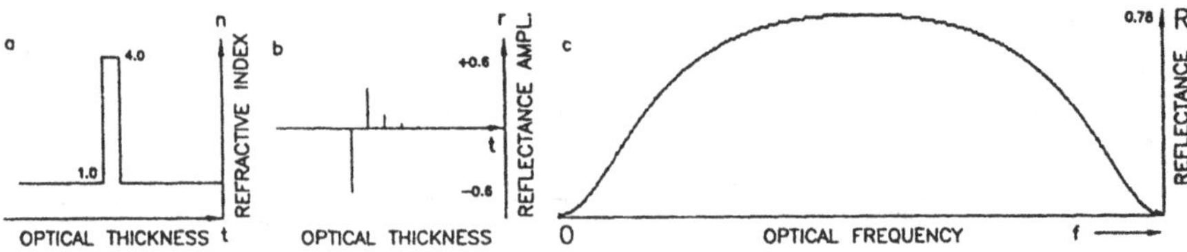

Fig. 3.9. Fourier view of the reflection of the case in Fig. 3.8 where the proper influence of the multiple reflections has been taken into account.

transmitted to interfere with the first surface reflection, a value of $0.4 \times 0.6 = 0.24$. This is to be compared to the 0.6 expected when we ignored multiple reflections. This is not the end of it, however. The flux continues to reflect back and forth inside the slab letting 0.4 of its remaining strength transmit through whichever interface it is reflecting from at that instant. This results in a rapidly decaying series of reflectance pulses coming back in the original reflection direction. This is similar to the laser "ring-down" technique for measuring very high reflectance coatings. When the interface reflectances are low, this problem is not so obvious, but for high reflectances the problem of neglecting multiple reflections becomes severe. Even in the cases of Figs. 3.5 and 3.6 the problem has become apparent.

Most of the material published to date has been restricted to small reflections in nonabsorbing and nondispersive media. It is apparent from the above examples why the small reflections limitation is imposed. The technology to date does not incorporate the effects of multiple reflections; other than that, it works very nicely. Bovard[8] addressed the problems of the distortions which occur. As we have seen above, the shape of the results are distorted as well as the magnitudes. This is illustrated in Fig. 3.10 where the true result A is compared with the distorted result B for a 7 layer QWOT stack. Bovard and Verly et al.[10] have worked with various forms of the "Q-function" in an attempt to correct the distortion. The Q function is the spectral description of the amplitude of Q versus frequency which when transformed will yield the proper index of refraction versus thickness function to produce the spectrum. The Q functions tried have been various combinations of R and T (transmittance intensity). As a result of our work, we concluded[11] that the proper function is the second Q function used by Verly et al.[10] as given in Eqn. 3.1.

$$Q(f) = [1 - T(f)]^{1/2} = [R(f)]^{1/2} = r(f) \tag{3.1}$$

This then is simply Q(f) = r(f)! However, this does require that we properly account for the multiple reflections. We have not yet been able to deduce how this can be properly done, nor has anyone else to our knowledge. It is hoped that someone with the proper mathematical background and insight will be able to work this out soon, or else prove (if so, regrettably) that it cannot be solved.

We believe that this resolution as to the cause of the distortion problem was not apparent until our work.[11] Our approach was empirical and has not been proven with mathematical rigor, but the evidence is convincing. The two-interface case of Germanium was reasonably easy to handle. Each of the multiple reflections with its magnitude and phase was identified manually and calculated to the point where the contributions of any higher order became insignificant. The summations were made which resulted in Fig. 3.9b, and the transform of that was taken and squared to give Fig. 3.9c. The results agree with the rigorous calculations given by the accepted matrix algorithms of analysis. When more than two interfaces are considered, this approach quickly becomes unmanageable. However, we were able to carry the three interface problem to a high enough order to convince ourselves that the answer was universal. The three-interface problem is shown in Fig. 3.11 where no account has been taken of the MIR. Figure 3.12 shows the comparison of our work, curve B, to 15th order with the rigorous matrix solution A. The results show that still higher orders are needed to reach the point of insignificance.

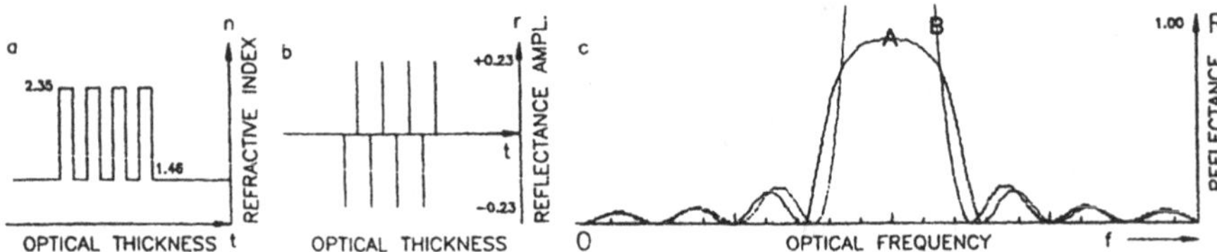

Fig. 3.10. Shape distortion in curve B, as in Fig. 3.5 for a 7 layer QWOT stack, without proper account of multiple reflections as compared to the correct case in curve A.

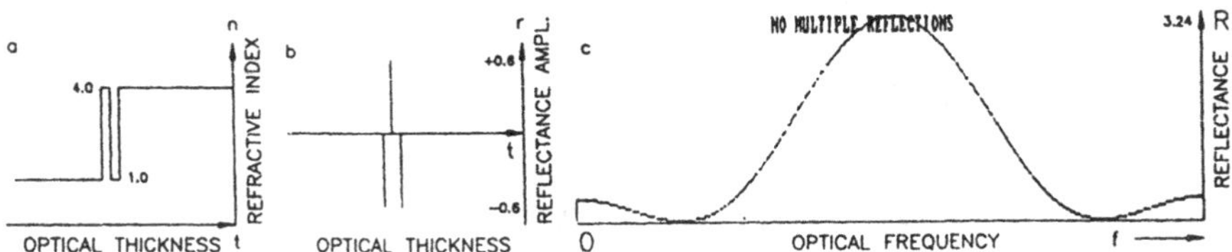

Fig. 3.11. The three-interface problem where no account has been taken of the MIR.

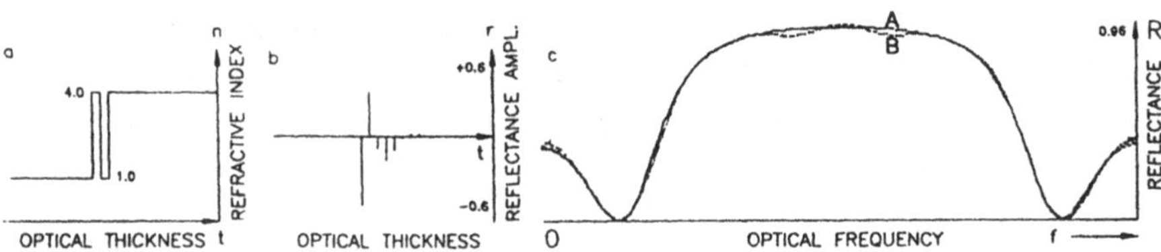

Fig. 3.12. Fourier view of the reflectance of the case in Fig. 3.11 where MIR is included to 15th order in curve B as compared to the rigorous matrix solution show in curve A.

3.2.3. A Method to Determine the Multiple Reflections

We will reproduce here the scheme which we used to develop the values for each reflection. We do this in the hope that some more astute mathematician can find a ready way to properly incorporate these recursively related multiple reflections into the form which can be Fourier transformed forward and back. The early work of Pegis,[12] Hodgkinson and Stuart,[13] and Kaiser and Kaiser[14,15] may prove helpful in such a pursuit. Liddell[18] has a review of Fourier transform synthesis methods and some potentially useful views on the recursive problem.

We worked only with QWOT layers to keep the phase relations as simple as practical, but the method could be extended for general optical thicknesses. Figure 3.13 shows a portion of a reflectance/transmittance tree that can be used to find the

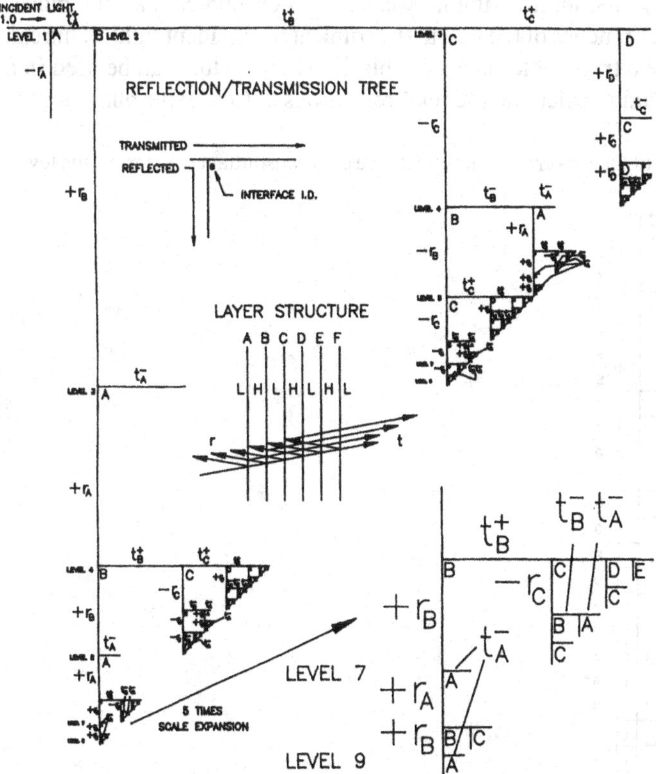

Fig. 3.13. Illustration of a portion of a reflectance-transmittance tree that can be used to find the r and t factors that make up each multiple reflection. Fine detail is expanded to show the principles and fractal form, but otherwise is not intended to be legible.

r and t factors that make up each multiple reflection. The chart represents rays that are transmitted through an interface as horizontal lines and those that are reflected as vertical lines. The nomenclature is taken from Macleod,[16] where the r factors are reflectance amplitude and the t factors are transmittance amplitude. The superscript plus signs on the t factors indicate a ray continuing through the stack in its original direction, the minus signs are for transmittance through an interface while the ray is traveling in reverse. The minus signs on the r factors indicate the phase change at a low- to high-index interface. This binary tree allows one to account for all of the reflections and transmissions until they emerge from the film stack as a final reflection (t_A^-) or transmission (t_B^+ in the case of only two interfaces, A and B). That is, each t_A^- is the terminus of a ray path, which results in a multiple reflection to be added (with phase) to the other reflections from the film. The factors defining the amplitude of such a reflection are found by starting at the t_A^- and multiplying it by each r or t encountered as one moves back down (up) the branches of the tree to the trunk at the incident light. This can be used to generate a truth table such as Table 1. These factors can be used in a spread sheet program to calculate the total reflections at successive phase levels.

Table 1. Truth table of the power of each reflectance and transmittance factor versus level.

	r_A	r_B	r_C	r_D	t_A^+	t_A^-	t_B^+	t_B^-	t_C^+	t_C^-	LEVEL
1	−1										1
2		1			1	1					3
3	1	2			1	1					5
4	2	3			1	1					7
5	3	4			1	1					9
6			−1		1	1	1	1			5
7	1	1	−1		1	1	1	1			7
8		−1	2		1	1	1	1			7
9	1		2		1	1	2	2			9
10	2	2	−1		1	1	1	1			9
11	1	−2	2		1	1	1	1			9
12	2	2	−1		1	1	1	1			9
13		2	−3		1	1	1	1			9
14	1	−2	2		1	1	1	1			9
15	2	2	−1		1	1	1	1			9
16	1	1	−1		1	1	1	1			7
17				1	1	1	1	1	1	1	7
18	1	1		1	1	1	1	1	1	1	9
19		−1	−1	1	1	1	1	1	1	1	9
20			1	2	1	1	1	1	1	1	9
21		−1	−1	1	1	1	1	1	1	1	9
22	1	1		1	1	1	1	1	1	1	9

These then can be used as the *r* input data for the Fourier transform, which will yield the correct reflectance spectrum. Figure 3.12 is such a result to the 15th order, but it shows residual ripples because even higher orders are needed to give a correct result in this case of such high reflectance at each interface.

The potential of the Fourier technique seems to be great enough to justify further investigations to overcome these limitations.

3.2.3.1. Using Fourier Techniques for DWDM Filters and Rugate Filters

When applying Fourier techniques to DWDM filters, the limitations of not including the effect of multiple reflections overwhelm any usefulness that we have been able to find thus far. The general patterns are correct, but the widths and amplitudes are not correct, as we found above. The simple Fourier transform approach might be more useful for FBG work because the reflectance amplitude of each cycle is so low that multiple reflections should not be a major factor. However, it is not yet clear to us as to what insight might be gained for FBGs.

The whole field of rugate filters similarly relates to Fourier techniques. Rugates might, in some respects, be considered intermediate between ordinary homogeneous designs and FBG designs. Both rugates and FBGs are more nearly sinusoidal and lack sharp changes in index with thickness. One of the major concepts is to avoid sharp discontinuities in the index versus thickness profile in order to suppress higher harmonic reflection bands. Therefore, the understanding of rugate filters is aided considerably by an intuitive feel for Fourier transforms, but the fine details may be obscured by the incomplete mathematics available at this time. Greenham et al.[20] show examples of concept and fabrication of rugate filters.

3.2.4. Overcoming Low Index Limitations with Thickness

We have shown that the ideal AR coating, when any and all indices are available, would be a smooth inhomogeneous "step-down" in index from the substrate to the medium. The form of the index profile is approximately a Gaussian decay from the substrate to the medium. When the medium is a vacuum or air and the lowest available index is represented by a real material such as MgF_2 at index 1.38 rather than values very close to that of the medium, the discontinuity from the smooth step-down profile causes a reflection residual that cannot be overcome by adjustments in the rest of the smooth profile. Additional thickness and the appropriate index profile can be used to reduce this residual reflection, but not eliminate it entirely. We will discuss some observations and findings on these effects from the Fourier point of view.

3.2.4.1. Low Index Limitations

If the AR requirement were for only a narrow band such as the classical "V" coat, the lowest index available would be of no consequence. It is, in fact, possible to design a coating to antireflect a single wavelength with layer materials whose indices are both higher than that of the substrate. However, as the desired bandwidth increases and the reflectance requirement decreases, the effects of the lowest index material becomes much more important. As we showed in Chap. 2, the minimum average reflectance intensity (R) in a moderate bandwidth goes approximately as $(L-1)^{3.5}$, where L is the lowest index available (and used as the last layer). It can be seen that for an uncoated substrate over an infinite bandwidth the R would be just the Fresnel reflectance or $((1-n)/(1+n))^2$ where n is the substrate index. As n approaches 1.0, R approaches $((n-1)/2)^2$. (Note how this is consistent with the $r = dn/2n_{ave}$ which can be used for inhomogeneous functions and was the basis of Bovard's[7] integrations and those of his predecessors.)

The ideal AR coating for the broadest band (a smooth "step-down") is shown in the lower curve of Fig. 3.14 for such a fictitious coating on a germanium substrate. If the lowest index available is greater than 1.0, then there will be a step discontinuity between it and the medium which causes a Fresnel reflectance. The upper curves in Fig. 3.14 show these effects of 1.282 and 1.400 for minimum indices. Figure 3.15 shows these same results over the AR band. The right end of Fig. 3.15 would be flat lines to infinity but for an artifice of the calculation technique. The right end is due to the fact that we had to calculate the results using a finite number of homogeneous layers (30-40) to simulate a smoothly varying homogeneous coating transition. The right end is the frequency where

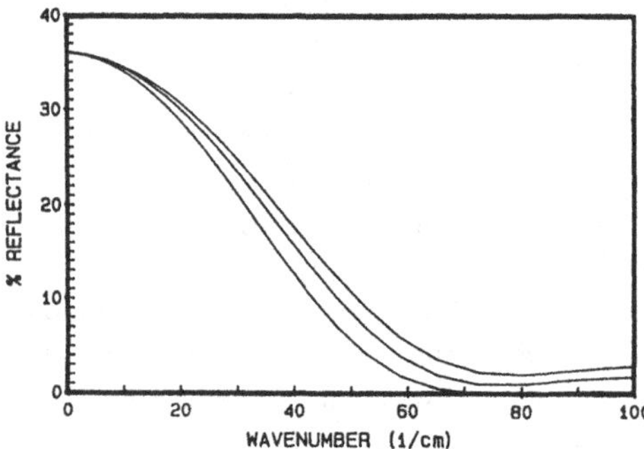

Fig. 3.14. A fictitious ideal AR coating on a germanium substrate for the broadest band is shown in the lower curve where the lowest index available is 1.0. The upper curves show the effects of 1.282 and 1.400 as the minimum available indices.

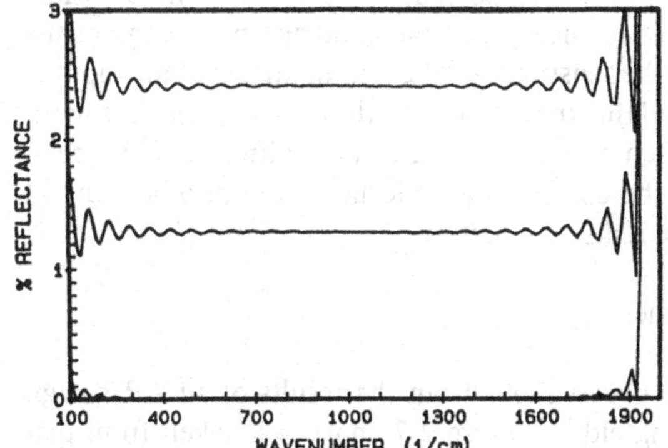

Fig. 3.15. The same results as Fig. 3.14 over the AR band. The right end would be flat lines to infinity but for an artifice of the calculation technique (See the text).

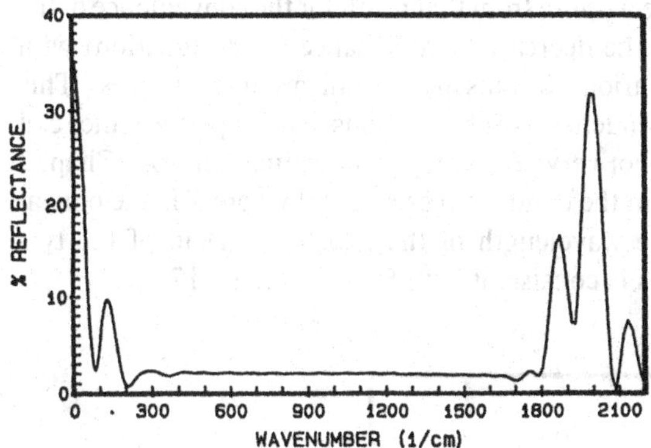

Fig. 3.16. The result of an optimization attempt made to improve on the index profile of a smooth step-down function which was truncated at the index value of 1.38. This is essentially the same in the AR band as the truncated unoptimized form.

those finite layers become one halfwave of optical thickness. The minimum reflectance values in the band are essentially the Fresnel reflectance levels for the minimum index values. They are actually somewhat less (≈80%) which we conjecture to be caused by the fact that the simulated band is not infinite and therefore allows some reduction in the limiting reflectance.

An optimization attempt was made to improve on the index profile of a smooth step-down function which was truncated at the index value of 1.38. Figure 3.16 shows the result which is essentially the same in the AR band as the truncated unoptimized form. We conclude that the minimum index value determines the lower limit on reflectance in a broad band and no change of the inhomogeneous profile between the substrate and the minimum index can improve the result (within the same overall thickness film). As the bandwidth is narrowed, however, it is possible to approach zero reflectance as in the limit of a "V" coat. As more thickness is added to the coating, the reflectance can also be reduced somewhat, which is what we shall discuss next.

3.2.4.2. Benefit of Extra Thickness

We discussed in Chap. 2 some observations from the results of an AR design contest run by Thelen and Langfeld.[17] Figure 2.7 above was taken from that report. These results are somewhat disappointing in that they confirmed that an order of magnitude increase in thickness is needed to cut the reflectance in half. The same phenomenon of reflectance versus thickness can be observed in a plot arrived at from different approaches. Verly, et al.[2] refined and extended the work of reference 1 and we reproduce a figure from that result for the convenience of the reader as Fig. 3.17. It shows the decrease in reflectance (merit function) with increasing film thickness for various designs and optimization techniques. This confirms from yet a third independent set of observations that AR performance can be improved by thickness, but not very efficiently. Our estimation from Chap. 2 is that the average reflectance in the band reduces as $T^{-0.31}$ where T is the optical thickness in full waves at the wavelength of the geometric mean of the two extremes of the AR band. This is consistent with Figs. 2.7 and 3.17.

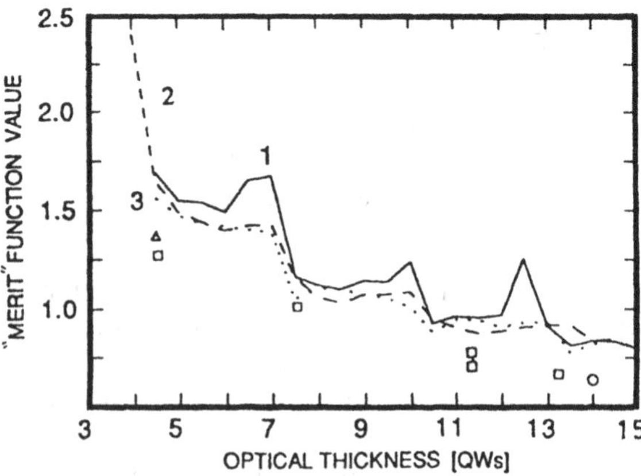

Fig. 3.17. A figure reproduced from Verly, et al.[2] shows the decrease in reflectance (merit function) with increasing film thickness for various designs and optimization techniques.

It also becomes apparent that there is a minimum thickness needed of approximately one wave of optical thickness at the above mentioned geometric mean wavelength for best results. This is consistent with the classical QHQ three-layer AR coating which is one wave of optical thickness at 530 nm for a band from 400 to 700 nm.

It is clear that there is benefit to be gained by additional thickness beyond the minimum, but why this is true has not been well understood. We will next discuss some concepts which should aid in understanding the underlying principles.

3.2.4.3. Observations of Extra Thickness Characteristics

Figure 3.18 uses the same design as the lower curve in Fig. 3.14, but it is scaled and plotted to match the general band for the Berlin design problem of 400 to 900 nm (11111 to 25000 cm^{-1}). This represents the ideal step-down index profile of index from the substrate at 1.52 to the medium at 1.00. If any and all indices were available between these extremes, the coating could be "perfect" from the low frequency limit at 11111 cm^{-1} to infinite frequency.

It is interesting to note that the low frequency end of the spectral frequency plot looks similar to the shape of the index versus optical thickness profile. Figure 3.18 represents the best that could be done if any indices could be used; the rest of the figures will show what can be achieved when one is limited to real materials.

Figures 3.19 through 3.22 are adapted from real designs resulting from the Berlin problem submissions. Figure 3.19 is the best of the minimum thickness designs and is the 8-layer point in Fig. 2.7. Figure 3.20 is the best of the designs

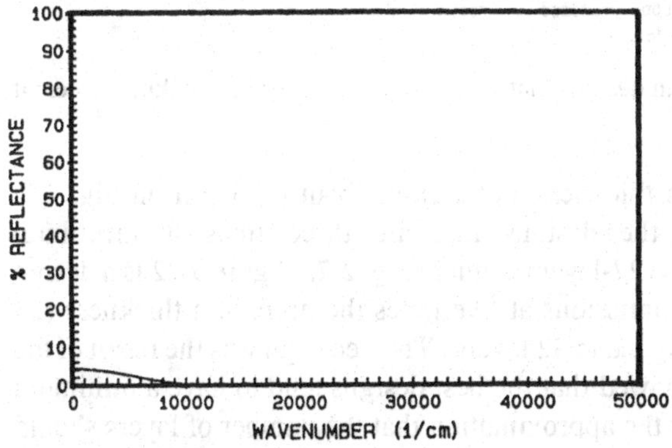

Fig. 3.18. The same design as the lower curve in Fig. 3.14, but scaled to match the general band for the Berlin design problem of 400 to 900 nm (11111 to 25000 cm^{-1}). This represents the ideal step-down index profile of index from the substrate at 1.52 to the medium at 1.00 if any indices could be used.

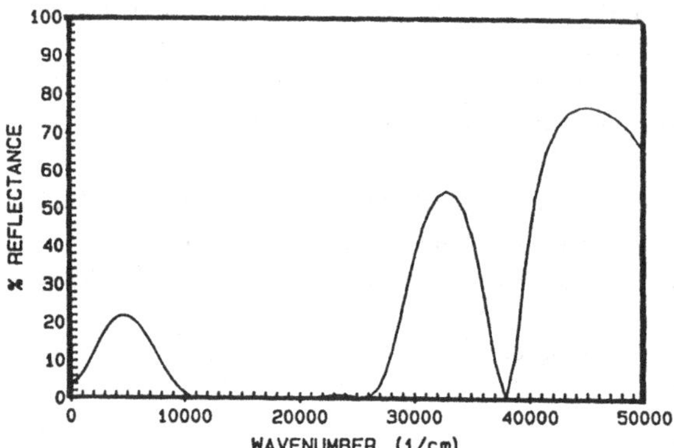

Fig. 3.19. The best of the minimum thickness designs from the Berlin problem, the 8-layer point in Fig. 2.7.

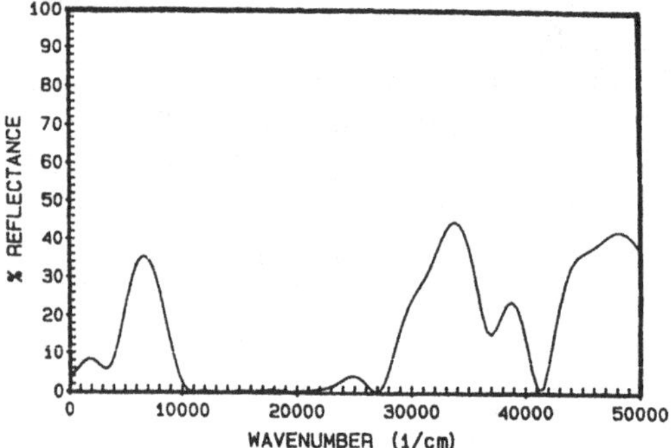

Fig. 3.20. The best of the Berlin designs that were twice the minimum thickness, seen at about 0.73 μm on Fig. 2.7.

that are twice the minimum thickness and seen at about 0.73 μm on Fig. 2.7. Figure 3.21 is the best of the designs that are three times the minimum thickness and identified as the 22-layer design in Fig. 2.7. Figure 3.22 is a design similar to the best of the submissions at five times the minimum thickness (53 layers), but it has been redesigned to 32 layers. This redesign was the result of the Chap. 2 studies where we showed that the best designs tend to have a minimum number of layers dictated by the approximation that the number of layers should be 6T+2 (in the visible) where T is the minimum thickness. Note that the 8-layer design of Fig. 3.19 fits this and the 22-layer design of Fig. 3.21 is close to the 20 predicted by the formula.

Notice that the number of maxima and minima (periods) to the left of the AR band in each of Figs. 3.19 to 3.22 is equal to the number of times that the coating

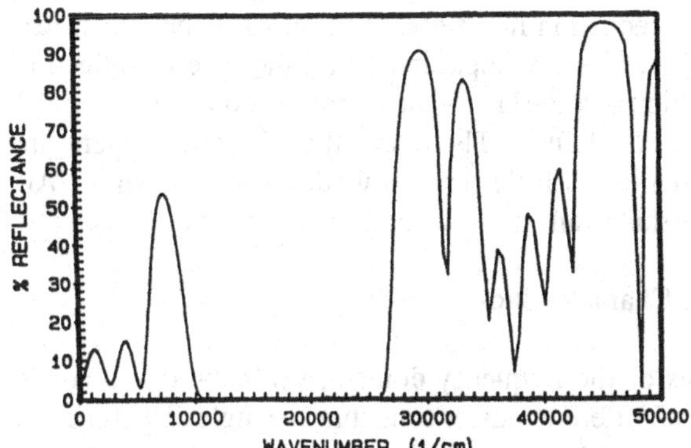

Fig. 3.21. The best of the Berlin designs that were three times the minimum thickness, the 22-layer design in Fig. 2.7.

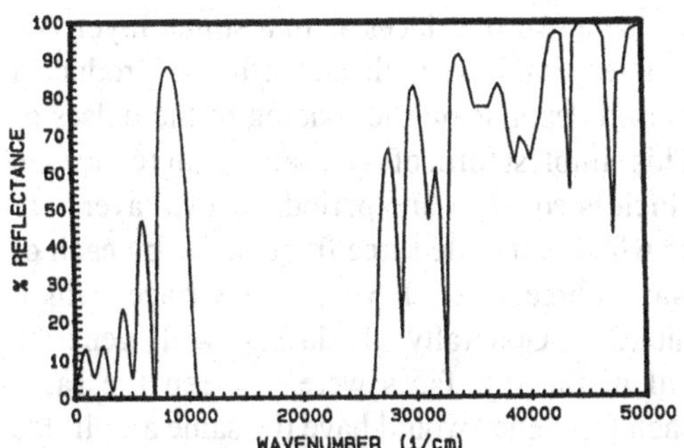

Fig. 3.22. A design similar to the best of the Berlin submissions at five times the minimum thickness (53 layers), but redesigned to 32 layers.

is thicker than the minimum (T). From our previous reports[1,2], it can also be seen that the profiles index versus thickness have the same number of extrema. In this area to the left (low frequency/long wavelength) side of the AR band are the "universal" shapes from which we hope to gain some understanding. These shapes will be generally the same for any coating designs in the same classification. The right sides of the figures, on the other hand, will generally be unique to the specific details of an individual design.

In comparing Figs. 3.19 to 3.22, it can be seen that the reflectance peak to the left of the AR band increases with increasing optical thickness of the design. In effect, the AR is a short wave pass filter (SWP) where heavy emphasis is placed on the antireflection properties for the passband. The ripples to the left of the peak are also just like those on the non-antireflected side of any QWOT-stack blocker used for SWP.

It would appear that the reduction in reflection in an AR band, when real homogeneous materials are used, is accompanied by an increase in reflection outside of the band. The addition of more layers and therefore more interfaces will add reflectance over the infinite band. Therefore, the additional layers are increasing the reflectance in areas of little concern while reducing it in the AR band. We will next consider this result from the Fourier viewpoint.

3.2.4.4. Fourier View of the Characteristics

If we think in Fourier terms or the frequency domain, we know that a single interface between two media of different indices constitutes a single delta function in the reflectance amplitude **r** versus optical thickness. Here $\mathbf{r} = (n_1 - n_2)/(n_1 + n_2) = \Delta n/2n_{ave}$ or the Fresnel amplitude reflectance of the interface. This spike is like a lightning bolt in time that generates some of all time frequencies. In the reflectance versus thickness case, the spike generates a reflectance which is equal over all spectral frequencies. If we have two interfaces as in a single layer AR coating, two spikes are generated. These interfere with each other to produce a single sinusoidal function whose period depends on the spacing of the pulses or the optical thickness of the layer. This simplest form of AR coating can reduce the reflectance over a narrow band which is equal to one period. If two layers are involved, there are three interfaces which generate three frequencies by each of three interfaces taken two at a time. Three layers have four interfaces which generate a total of six frequencies. Generally, L layers will generate $(L+1)!/2(L-1)!$ frequencies. If all of the interfaces were between the same materials such as TiO_2 and SiO_2, each frequency would have the same amplitude and differ only in phase. For example, if all the layers were of equal thickness as in a QWOT stack, all the frequencies would be in phase and add for maximum reflectance at the QWOT wavelength. The highest frequency will be generated by the two interfaces which are most widely separated in optical thickness. This points to the fact that additional thickness in and AR coating provides higher frequencies which can be used to reduce residual reflectance in the AR band. The optimization process basically is adjusting the phases of the various frequency components to achieve the desired reflectance results as closely as possible within the constraints.

In Chap. 2, we found that the number of ripples in an AR band were approximately $8B/3 + 2T - 4$. Here B is the ratio of the longest wavelength to the shortest wavelength and T is the optical thickness as defined above. This observation shows that the ripples are a linear function of bandwidth and thickness. If we increase the thickness while keeping the B constant, we will increase the number of ripples (higher frequencies). This is consistent with concepts put forth above.

The whole process can be related to the Fourier synthesis of a square waveform by adding together appropriate amplitudes and phases of the sinusoidal harmonics of the fundamental frequency of the square wave. As more harmonics are added, the ripples reduce in amplitude and increase in frequency until the result is not distinguishable from a true square wave. This seems to be the same case for a broad band AR coating design. There is also the possibility of applying this same concept to the design of broadband high reflectors.

3.2.4.5. Conclusions with Respect to More Thickness

We have shown empirically that the lowest available index limits what can be achieved in a broadband AR coating design because the residual discontinuity between it and the medium produces a Fresnel reflectance similar to that of an uncoated interface of the same material. The benefits of additional thickness in the coating beyond the minimum for a good AR have been shown from the author's work in Chap. 2, Fourier design,[2] and a collection of many designers' work as collected by Thelen and Langfeld[17] We have discussed some viewpoints that should be of help in understanding the phenomena by which additional thickness improves a broadband AR coating and why it quickly reaches a point of diminishing returns.

3.3. DESIGNING A VERY BROAD BAND AR COATING

Given all the foregoing discussions of principles, estimates, viewpoints, etc., how do we go about actually designing a very broad band AR coating? Much of this discussion is also applicable to many other types of coatings. It is assumed that we know the index of the substrate and the medium. The AR is then to be a good impedance match (matching layers) between the substrate and the medium over the required spectral band. The next choice to make is the coating materials which are a function of the spectral band and the coating facilities, processes, and capabilities to be used. We have shown above that the lowest index possible is advantageous for the last layer, and also that the greatest spread between high and low index is beneficial. Once we have chosen the materials and know the desired AR bandwidth, the only choice is the overall coating thickness. If we have a required R_{AVE}, we can look at Eqn. 2.1 to see if a thickness of one T-unit will meet the requirements. If it will, we have only to design the filter. If it will not, there is more to do. The first question might be whether three T-units would be enough to make it meet the requirements. We don't consider much more than 3T to be practical. If that doesn't meet the requirements, we need to see if a lower index and/or a larger difference between high and low indices is practical. If not, the requirements cannot be expected to be met by any design.

If it is found that the requirements can be expected to be met by a given choice of materials and thickness, Fig. 2.5 offers a guide to the number of layers that are expected in a final design. However, we have not often found it easy to go directly to an optimum by starting with that number of layers. Our preferred approach is to start with many more layers and narrow the design down to the minimum number. We have seen from Figs. 3.19 to 3.22 that the AR band is generally flanked on each side by two high-reflecting bands which are usually ignored. We use this fact to start with a QWOT stack just beyond the short wavelength or high frequency side of the AR band. The number of lair pairs is chosen so that the overall thickness of the stack is about the number of T-units chosen above. We assume that this starts with many more than the optimum number of layers.

The optimization then is done by varying all of the layers. Some layers will become thicker and some thinner. A few layers will eventually become very thin. As this occurs, we remove such a layer and reoptimize with the smaller number of layers. Additional layers will become very thin and be removed until the optimum (fewest) number of layers is reached for a given set of conditions. This "weeding out" of layers occurs from the basic principle in AR coatings that more layers cause more reflections that need to be dealt with or removed. The optimization process naturally minimizes the thickness of these "unwanted" layers.

We check after some optimization has been done to see if the T-value is converging on the planned value. We do this as in Fig. 2.8a, b, and c by evaluating the reflectance versus thickness at the longest wavelength in the AR band. If a 3T design was desired and we find we have a 4T, we will remove the layers associated with one of the 4 "humps" and then reoptimize the design. If we wanted 3T and found only 2T, we would copy the layer pattern of one of the humps and add it to the design before reoptimizing.

Elements of this procedure may be useful for some beamsplitter designs also.

3.4. CONCLUSIONS

We have attempted to aid the understanding of the Fourier viewpoint of thin film design and also the nature of AR and other designs from that viewpoint. The Fourier tool holds the promise of becoming a true **synthesis** technique in the future, but still needs additional work and insight to reach that goal. In the meantime, Fourier techniques have already added to our understanding and can be an aid to the solution of complex spectral profile problems.

3.5. REFERENCES

1. R. R. Willey, P. G. Verly, and J. A. Dobrowolski: "Design of Wideband Antireflection Coating with the Fourier Transform Method," *Proc. SPIE* **1270**, 36-44 (1990).
2. P. G. Verly, J. A. Dobrowolski, and R. R. Willey: "Fourier-transform method for the design of wideband antireflection coatings," *Appl. Opt.* **31**, 3836-3846 (1992).
3. J. A. Dobrowolski and D. Lowe: "Optical thin film synthesis program based on the use of Fourier transforms," *Appl. Opt.* **17**, 3039-3050 (1978).
4. E. Delano: "Fourier Synthesis of Multilayer Filters," *JOSA*. **57**. 1529-1533 (1967).
5. L. Sossi: "A method for the synthesis of multilayer interference coatings," *Eesti NSV Tead. Akad. Toim. Fuus. Mat.*, **23**, 229-237 (1974).
6. B. G. Bovard: "Derivation of a matrix describing a rugate dielectric film," *Appl. Opt.*, **27**, 1998-2005 (1988).
7. B. G. Bovard: "Fourier transform technique applied to quarterwave optical coatings," *Appl. Opt.*, **27**, 3062-3063 (1988).
8. B. G. Bovard: "Rugate filter design: the modified Fourier transform technique," *Appl. Opt.*, **29**, 24-30 (1990).
9. W. H. Southwell: "Spectral response calculations of rugate filters using coupled-wave theory," *JOSA,A*, **5**, 1558 (1988).
10. P. G. Verly, J. A. Dobrowolski, W. J. Wild, and R. L. Burton: "Synthesis of high rejection filters with the Fourier transform method," *Appl. Opt.* **28**, 2864-2875 (1989).
11. R. R. Willey: "Resolution of two problems in the Fourier analysis of thin films," *Appl. Opt.* **32**, 2963-2968 (1993).
12. R. J. Pegis: "An exact design method for multilayer dielectric films," *JOSA*, **51**, 1255-1264 (1961).
13. I. J. Hodgkinson and R. G. Stuart: "Thin-film-synthesis algorithm for realization of refractive indices and Airy summation," *JOSA*, **72**, 396-398 (1982).
14. H. Kaiser and H.-C. Kaiser: "Mathematical methods in the synthesis and identification of thin film systems," *Appl. Opt.*, **20**, 1043-1049 (1981).
15. H. Kaiser and H.-C. Kaiser: "Identification of stratified media based on the Bremmer series representation of the reflection coefficient," *Appl. Opt.*, **22**, 1337-1345 (1983).
16. H. A. Macleod: *Thin Film Optical Filters* (Macmillan, New York, 1986), pp. 27-31.
17. A. Thelen and R. Langfeld, "Coating design problem," in *Thin Films for Optical Systems*, K. H. Guenther, ed., *Proc. SPIE*, **1782** (1992).
18. H. M. Liddell: *Computer-aided Techniques for the Design of Multilayer Filters* (Adam Hilger Ltd, Bristol, 1981) pp. 111-112.
19. H. Fabricius: "Gradient-index filters: designing filters with steep skirts, high reflection, and quintic matching layers," *Appl. Opt.*, **31**, 5191-5196 (1992).
20. A. C. Greeham, B. A. Nicols, R. M. Wood, N. Nourshargh, and K. L. Lewis: "Optical interference filters with continuous refractive index modulations by microwave plasma-assisted chemical vapor deposition," *Opt. Eng.* **32**, 1018-1024 (1993).

4

Typical Equipment for Optical Coating Production

4.1. INTRODUCTION

In the previous chapters, we have looked at how to design coatings to give the desired reflection properties for a given application. This would, however, be nothing but an interesting academic exercise unless we can actually produce what was designed. In most of our experience, the design is the easy part. It is all just numbers in a computer or on paper until you fabricate it.

There are a great variety of ways to deposit optical coatings. We will describe in some detail what we believe to be the most typical in today's optical coating industry. We will then touch on some of the old and newer alternatives by comparison. The most typical equipment in today's optical coating industry is what is affectionately called the "box coater." This has overtaken the "bell jar coater" in the last decade or two, but the requirements and functions are mostly the same except for the usually greater size of the box coater. Tanks and drum coaters generally fall into the same category of *batch coaters* as contrasted to "continuous" processes such as web coaters and in-line coaters which we will also touch on later.

Currently, the most common process is Physical Vapor Deposition (PVD) as contrasted with Chemical Vapor Deposition (CVD), Plasma Enhanced CVD (PECVD), Dip Coating, and others. In PVD, the materials to be deposited are typically heated in a vacuum until they create a sufficiently high vapor pressure to transfer the material from the source to where it condenses on the cooler substrate to be coated. Let us also place sputtering in the class of PVD, but here the material is "vaporized" by ion bombardment rather than heat (per se). We will discuss some of the other processes subsequently.

It is hoped that this approach will be helpful to the greatest number of readers, and we will refer the reader to other sources for more details on some of the more diverse approaches and applications.

4.2. GENERAL REQUIREMENTS

There are many businesses throughout the world which produce optical coatings in batches of lenses, mirrors, and filters. They may have from one to dozens of bell-jars and box coaters in which they deposit these coatings. This equipment ranges from relatively simple to systems that are as automated and sophisticated as that of any industry. We will describe the requirements and equipment from the most basic to start with and then evolve to the more complex.

Figure 4.1 illustrates the simplest case of PVD. There is a chamber from which "all" the gas has been removed to create a vacuum. The substrates to be coated are somehow held so that the surfaces to be coated face the material source. The material is heated to a temperature at which there is sufficient vapor pressure of the material to cause atoms of the material to radiate from the source and condense on the substrate. This would be like steam from a tea kettle condensing to a film of water on a colder plate held in the steam. The atoms in a vacuum travel in straight lines because there are few gas molecules in their path with which to collide.

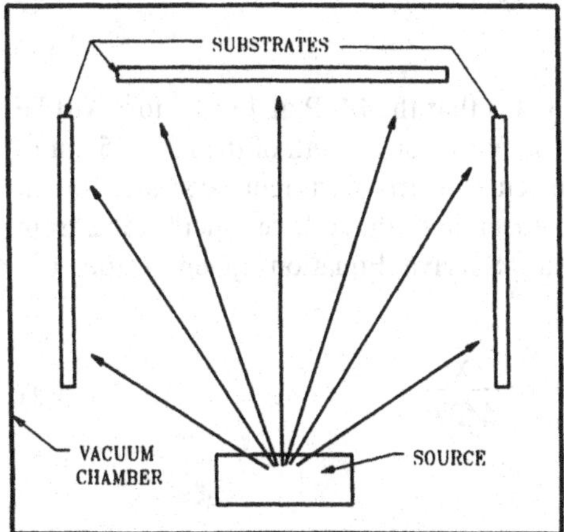

Fig. 4.1. The simplest case of PVD. The material is heated to a temperature at which there is sufficient vapor pressure of the material to cause atoms of the material to radiate from the source and condense on the substrate.

4.2.1. The Vacuum

If there was a small amount of residual gas in the chamber to collide with, the evaporated molecules which collided with the gas would go in any and all directions and might also lose some of their energy (heat). If there were large amounts of gas, the molecules might actually lose so much energy by multiple collisions as to condense to solid particles before reaching any surface. Such high pressure depositions can result in powdery films or even dust which falls to the bottom of the chamber. Therefore, all of our discussions assume that the vacuum in the chamber is such that the collisions are few and do not significantly affect the result as compared to a "perfect" vacuum.

4.2.1.1. How Good a Vacuum Is Needed?

The average path at a given pressure and temperature that a molecule can be expected to travel before colliding with another is the "mean free path" (MFP). Figure 4.2 shows the relationship of the MFP to pressure at 25°C. It also shows the molecular rate of incidence and the time to form a monolayer on the surface. This helps us to get some idea of how good a vacuum is needed. Equation 4.1 can be used to calculate the MFP at room temperature.

$$MFP = \frac{5.0x10^{-3}}{P\ Torr}\ cm \qquad (4.1)$$

We can see from Fig. 4.2 or Eqn. 4.1 that the MFP at 1×10^{-5} torr will be $5.0\times10^{+2}$ cm or 5 meters. Pulker[1] points out that at 1/100th of the MFP (5 cm in this case) we can expect 99% of the molecules to arrive without a collision, but at the MFP only 37% would have had no encounter. Pulker's comment stems from what is known in thermodynamics as the "Survival Equation" given in Eqn. 4.2.

$$\frac{N}{N_0} = \exp\left(-\frac{x}{MFP}\right) \qquad (4.2)$$

N is the number of molecules from a starting number N_0 that have not collided after a distance x. About 90% have not collided by 0.1MFP, 50% by 0.69MFP,

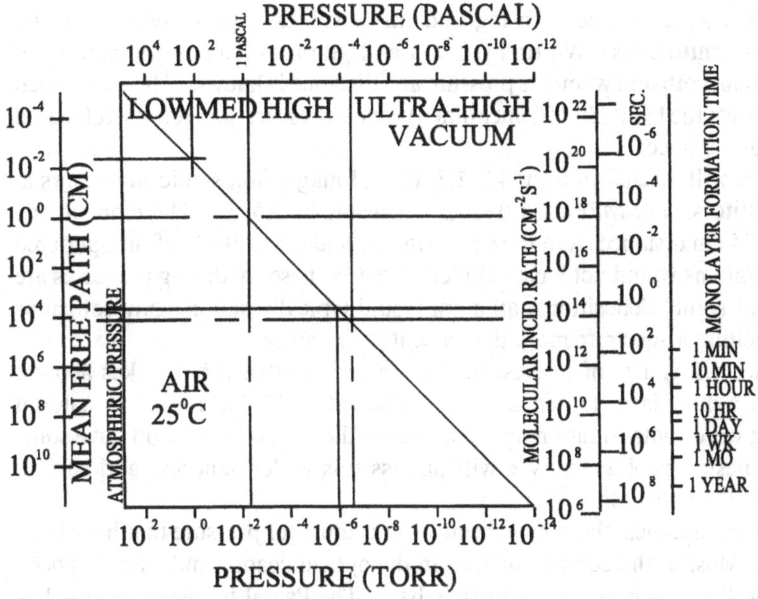

Fig. 4.2. The relationship of the Mean Free Path (MFP) to pressure at 25°C. It also shows the molecular rate of incidence and the time to form a monolayer on the surface.

and only 13.5% by 2MFP. This implies that collisions in a one meter size box coater are not significant at 1×10^{-6} torr but would be a factor to consider at 1×10^{-5} torr. At 1×10^{-5} torr, from the forgoing, we would expect 50% of the molecules to have had a collision by 3.45 meters. However, the 1.5×10^{-4} torr which might be used in reactive oxygen depositions of SiO_2 and TiO_2, is another story. Equation 4.1 gives a MFP = 33.3 cm at this pressure. Therefore, at 100 cm or 3MFP (about the size of a 1 meter box coater), only 5% of the molecules reach the other side of the chamber without collisions. As we will discuss later under uniformity masking, small changes in pressure at these levels can have major effects on material distribution.

$$\% \ UNCOLLIDED \ ATOMS = 100 \exp\left(- \frac{X(cm)\ P(Torr)}{5x10^{-3}} \right) \tag{4.2}$$

We can combine the information of Eqns. 4.1 and 4.2 to find what percent of the atoms would travel a given distance at a given pressure without having a collision with another atom. Equation 4.2.1 is illustrated in Fig. 4.3 where the opposing sides are labeled with log and linear values of pressure in Torr and distance in centimeters. We find this a useful graph to look up the percent of atoms without collision when the pressure and distance is known. The bold dotted line represents the MFP (36.8% uncollided) versus pressure, and it is useful to find the MFP at a glance.

As we will discuss in Sect. 4.2.2.3, typical magnetron sputtering occurs at tens of millitors. The MFP for 1.0×10^{-2} torr would be 0.5 cm. Therefore, at one inch or 2.54 cm distance from the sputtering cathode, only 0.6% of the sputtered molecules/atoms would not have collided. Clearly these sputtering processes are not "line of sight" depositions; an atom would typically suffer many collisions before reaching a substrate more than a centimeter away.

Figure 4.2 is also of interest in that we can see that it only takes about a second for a monolayer to deposit on a surface at 1×10^{-6} torr! If the material depositing were contamination and it stayed on the surface, we might have some serious adhesion problems. We will discuss this under adhesion and surface cleaning in later chapters.

Over the decades, there have been various units for pressure that have been preferred. Most of the equipment seen in the optical coating industry has been gauged for Torr, millitorr, or millibar (mbar). The Pascal has been specified as

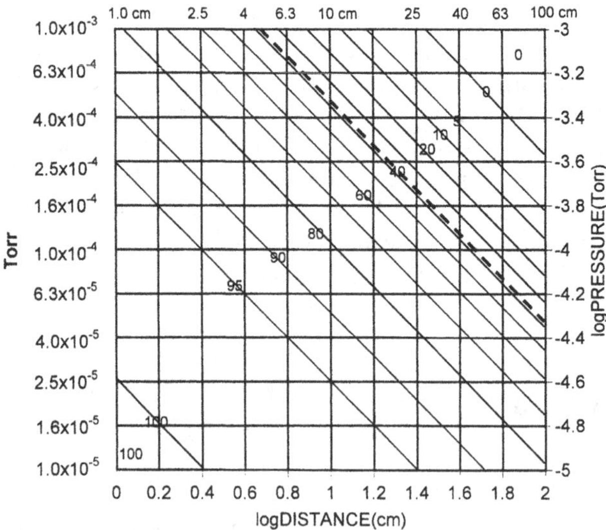

Fig. 4.3. Percent of uncollided atoms versus pressure and distance traveled. Dotted line is the mean free path (MFP) versus pressure.

the preferred unit by some journals. Table 1 provides an easy look-up table for conversion from Torr to mbar or Pascal and what the mean free path would be at that pressure. The table is structured in two parts for the convenience of the user, one which starts on the left with Torr, and the other which starts with Pascal.

Table 1. Look up table for conversion from Torr to mbar or Pascal and what the mean free path would be at that pressure.

Torr	mbar	Pascal	MFP (cm)
Atmos.760	1013	101300	6.6E-06
1	1.33	133.3	0.005
0.10	0.13	13.30	0.05
0.01	1.3E-02	1.33	0.50
1.0E-03	1.3E-03	0.13	5.00
1.0E-04	1.3E-04	1.3E-02	50.0
1.0E-05	1.3E-05	1.3E-03	500.0
1.0E-06	1.3E-06	1.3E-04	5.0E+03
1.0E-07	1.3E-07	1.3E-05	5.0E+04
1.0E-08	1.3E-08	1.3E-06	5.0E+05
1.0E-09	1.3E-09	1.3E-07	5.0E+06

Pascal	mbar	Torr	MFP (cm)
101300	1013	760	6.6E-06
100	1	0.75	6.7E-03
10	0.10	7.5E-02	0.07
1	0.01	7.5E-03	0.67
1.0E-01	1.0E-03	7.5E-04	6.66
1.0E-02	1.0E-04	7.5E-05	66.6
1.0E-03	1.0E-05	7.5E-06	666.4
1.0E-04	1.0E-06	7.5E-07	6.7E+03
1.0E-05	1.0E-07	7.5E-08	6.7E+04
1.0E-06	1.0E-08	7.5E-09	6.7E+05
1.0E-07	1.0E-09	7.5E-10	6.7E+06

4.2.1.2. How Can We Create the Vacuum Needed?

Books are written[2] on the subject of how to generate vacuums, but we will here summarize some of the technology that is most common in the optical industry at this time. Figure 4.4 shows a typical box coater vacuum pumping system. The chamber has a door through which parts and fixturing can be inserted and removed. In the case of bell jars, the jar is raised to provide such access. The door or the jar has a vacuum tight seal when closed. The box or the base plate of the bell jar has two pumping port valves, the high-vacuum valve and the roughing valve. When the chamber has been loaded and the door closed, it is at atmospheric pressure. The types of pumps which can achieve the low pressures (high vacuum or HiVac) required for PVD cannot operate at atmospheric pressure. It is therefore necessary to lower the chamber pressure sufficiently before the HiVac pump is brought into action. A mechanical rotary piston positive displacement pump is commonly used to transfer most of the gas from the chamber out to the atmosphere. With the chamber isolated by the roughing and HiVac valves closed, the mechanical pump is turned on and the foreline valve is opened to the HiVac pump. The HiVac pump is typically a fractionating diffusion pump which we will describe in more detail later. It needs to be warmed up to operating temperature while being mechanically pumped. When it is warm

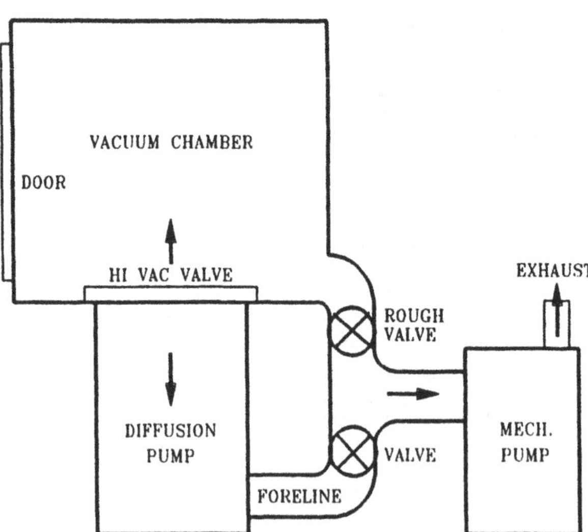

Fig. 4.4. A typical box coater vacuum pumping system. The chamber has a door through which parts and fixturing can be inserted and removed.

enough and pumped down to about 10^{-2} torr, the foreline valve can be closed (for a short while) and the roughing valve opened to the chamber. The mechanical pump then reduces the chamber pressure to about 10^{-1} torr where the diffusion pump can be used. First the rough valve must be closed and then the foreline valve opened to clear the diffusion pump of any gas which has built up while the chamber was being rough pumped. When the foreline pressure is back to 10^{-2} torr or less, the HiVac valve is *slowly* opened to the chamber. If the valve is opened quickly, the gas load on the diffusion pump would be too great and undesirable effects such as oil backstreaming would happen (discussed below). The diffusion pump is then evacuating the chamber to the 10^{-6} torr region in preparation for coating. The mechanical pump is evacuating the foreline of the diffusion pump to something on the order of 10^{-2} torr unless the process uses induced gas or the substrates or chamber walls and fixtures are outgassing significantly.

A "holding pump" is sometimes added to a system like that above. Its function is to keep the diffusion pump foreline pressure low even if the DP is isolated such as when the HiVac valve is closed and the mechanical pump which is roughing the chamber is valved off from the DP. This can be a very small pump by comparison to the mechanical pump. It can be particularly useful if the roughing cycle times are so long that the pressure in the isolated DP rises too high.

We have seen cases where an automated system without a holding pump was roughing a chamber, but the sequence had to be interrupted by the control system because the foreline pressure rose too high. This caused the roughing valve to close, the foreline valve to open, and the foreline to be pumped until some low pressure was reached. The valves were then automatically reset and the roughing continued. Such behavior is usually indicative of a system with problems. The problems might include: leaking chamber, leaking HiVac valve, very dirty chamber, very dirty DP oil, etc.

Note that it undesirable to rough the typical chamber much below 10^{-1} torr (100 microns). This is because at lower pressures the gases flowing down the roughing line to the pump cease to be in viscous flow where it drives the pump vapors back into the pump. At the lower pressures, the molecules of mechanical pump oil may be in molecular flow and some can travel up the roughing line into the chamber. Viscous flow would be like a river that flowed so rapidly that no fish were able to swim upstream against it. It will be seen below that the cross-over pressure to change from roughing to HiVac is a compromise between upper and lower pressure limits. Hablanian and Caldwell[3] have reviewed in detail how the correct cross-over pressure can be determined. The dashed lines in Fig. 4.5 show that the pressure at which the transition from viscous to molecular flow regime occurs depends on the mean dimension of the openings involved. The 10^{-1} torr (100 microns) value mentioned above would be an appropriate lower limit for a 2-inch diameter roughing line, whereas a 4-inch line might go to 0.5×10^{-1} torr

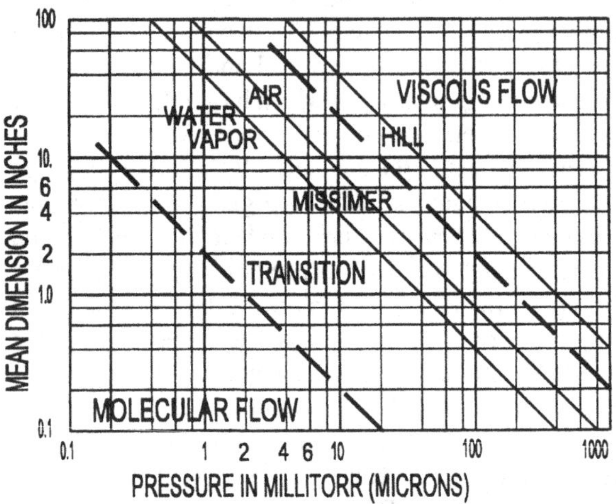

Fig. 4.5. A chart of the pressures and mean gas conduit dimensions where flow changes from viscous to molecular. Dashed lines are typical, see text for other lines.

(or 50 microns), but a 1 inch line goes into transition at 200 microns. The criteria for the beginning of molecular flow is stated on page 64 of Roth[2] to be when the MFP is the same as the diameter of the pipe or vessel and 110 times as great for viscous flow. The region between is referred to as Transition Flow. We can find from Eqn. 4.1 that the diameter of the pipe in centimeters would be $D < 5\times10^{-3} / P$ torr for molecular flow and two orders of magnitude greater for viscous flow.

The dashed lines of Fig. 4.5 represents Roth's[2] criteria. Hill[4] on page 5 has somewhat different numbers for air at room temperature that is "not in viscous flow." Hill's line is the unbroken line above the upper dashed line of Roth. A third view is supplied by Missimer[64] for air and water vapor (which is a lower line than air) which are below Roth's upper line. It can be seen that there is some room for interpretation as to where viscous flow goes into transition flow. This is also a function of the gas species and the temperature. The safer choices for crossover decisions would be the upper Roth line or even Hill's line.

The times for the pumping cycles and the pressures vary considerably with the size and nature of the chamber and pumps. A typical system might rough down to the 10^{-1} torr cross-over pressure in 3 to 10 minutes depending on how clean the chamber is. The HiVac part of the cycle is even more variable depending on the temperature and pressure required, and the cleanliness of the system and conditions of the pumps and chamber seals. This might range from 5 minutes to over an hour.

Types of Pumps

There are a great variety of vacuum pumps as can be seen by Fig. 4.6. We have heavy-boxed the types which we will discuss. The two main branches of the Vacuum Pump Tree in Fig. 4.6 are the Gas Transfer and Entrapment vacuum pumps. The gas transfer pumps transfer the gas from the inside of the vacuum system and exhaust it to the outside. The entrapment pumps trap the gas within the vacuum system by freezing the gas on a cold surface, "physiochemically" trapping it in a porous solid such as graphite, or chemically reacting it with a solid (gettering) to reduce the vapor pressure to some insignificant level. The entrapped gas will eventually saturate or fill the capacity of its trap and the system will have to be emptied or regenerated. A cryopump is regenerated by allowing it to warm up to room temperature while being rough pumped. This evaporates and drives off the entrapped gases. There is some danger, if a saturated cryopump warms up with all of its ports closed, that it could build a large internal pressure and explode. Pressure relief valves are built into the pumps to prevent this from happening.

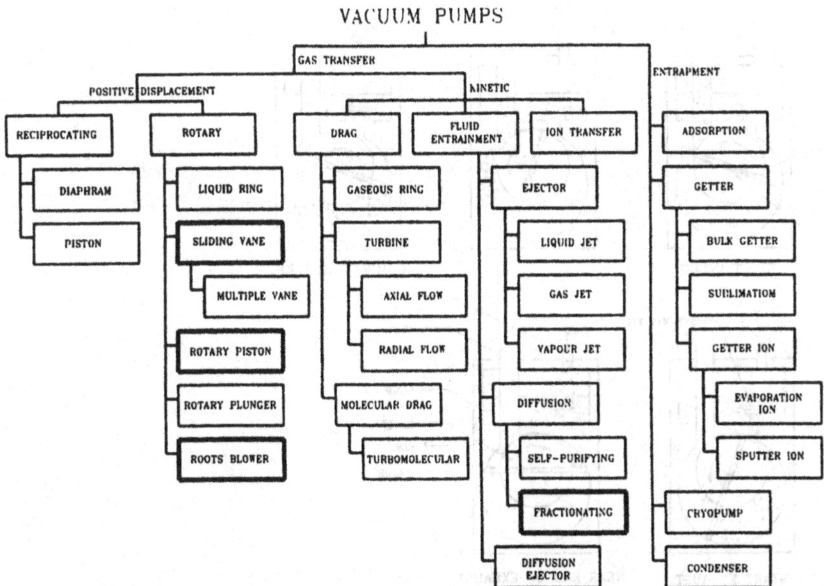

Fig. 4.6. A family tree of the great variety of vacuum pump types available. The types discussed in the text are heavy-boxed.

The gas transfer pump branch of Fig. 4.6 divides further into Positive Displacement and Kinetic vacuum pumps. The positive displacement pumps encapsulate a volume of gas (as in a valved piston and cylinder of an automobile engine) and transfer it from one part of the system to another. Kinetic pumps essentially sweep or knock the originally random motioned gas molecules along in a specific direction with "brooms" or other forcefully directed solids, liquids or gases.

Mechanical Pumps

The most typical mechanical pumps in the optical industry are Rotary pumps of the Positive Displacement type, namely the Sliding Vane and Piston varieties. The piston type can be related to the usual automobile engine but different in detail. The sliding vane type is illustrated in Fig. 4.7. The rotor is closely fitted with two or more vanes that form a seal with the stator cavity. A reservoir of mechanical pump oil supplies the rotor and vanes for lubrication and sealing. The vanes slide in the rotor to maintain a seal and sweep gas behind the vane from the inlet aperture (when exposed) into the stator volume. After about 180° of rotation in Fig. 4.7c, the pump has isolated a volume of gas at the inlet pressure. As the rotor

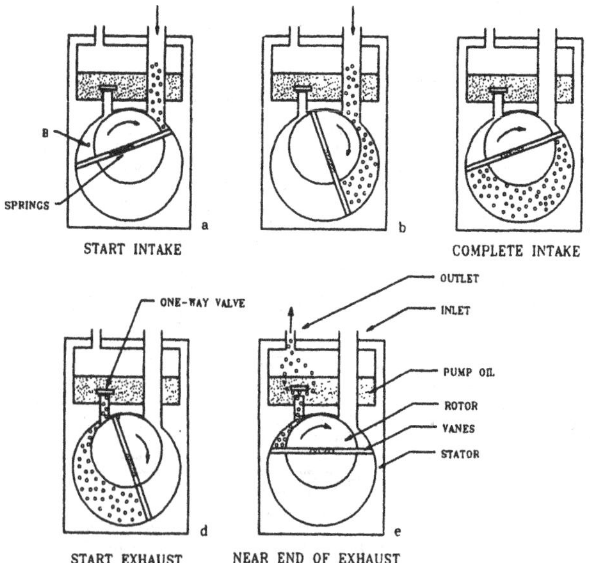

Fig. 4.7. The sequential action of the sliding vane type pump. The rotor is closely fitted with two or more vanes that form a seal with the stator cavity.

progresses further, this volume is compressed until the vane reaches the exhaust port where the gas is expelled. In the case illustrated, the gas passes through a one-way valve and the oil reservoir to the outlet. The cycle then repeats as the second rotor vane advances. Single stages of such a pump might achieve pressures of 10^{-2} torr because of the high compression ratios possible. Two such stages in series or "Duo" pumps may reach nearly 10^{-3} torr, but are limited ultimately by the vapor pressure of the pump oils. In the early stages of roughing a chamber when the gas load is great, the exhaust from the pump contains a significant amount of oil mist. The exhaust should pass through a filter (like an automobile carburetor air cleaner). The exhaust should be vented outside, not into a coating area where oil vapors would have very negative effects on optical film quality and durability.

One problem with such pumps is that water vapor can condense before the gas is expelled to the outlet. The condensate tends to stay in the pump oil and build up. Two practices are used to overcome this problem. The pump oil is allowed to run hot at about 50 to 70°C to help drive off contaminating liquids and gases. The other is called "gas ballasting." The B opening in Fig. 4.7a can be opened to the atmosphere for ballasting. It is positioned such that the mixture reaches the ejection pressure before the vapor condensation takes place. When gas is ballasted, the ultimate pressure capability of the pump is decreased because some of the compression ratio is lost. The rigorous procedure would be to rough the chamber with the ballast open and then close it when the gas load is low and the ultimate pressure is desired. In many cases, however, the ultimate mechanical pump pressure is not needed and/or the process gas load is such that ballasting is desired all of the time. Pump oil reservoirs are typically designed to be large enough to handle some contamination before it becomes so concentrated that the oil needs to be changed. When necessary, the oil is drained and replaced.

Diffusion Pumps

The fractionating diffusion pump (DP) is illustrated in Fig. 4.8. The specialized diffusion pump oil is heated to evaporation in the boiler at the bottom. The vapors pressurize and rise in concentric cylinders that feed downward spraying jet nozzles. Three stages of jets are illustrated. The vapors from the upper jets sweep in a downward direction any molecules that wander into their path from above. This increases the pressure below the stream as compared to that above. The second stage below the top jets does the same thing, and similarly the bottom stage. The pressure differential between the top and the bottom of the three-stage pump is then approximately the cube of the differential at one stage. This allows us to go from a chamber pressure of less than 10^{-6} torr to a foreline pressure of about 10^{-2} torr.

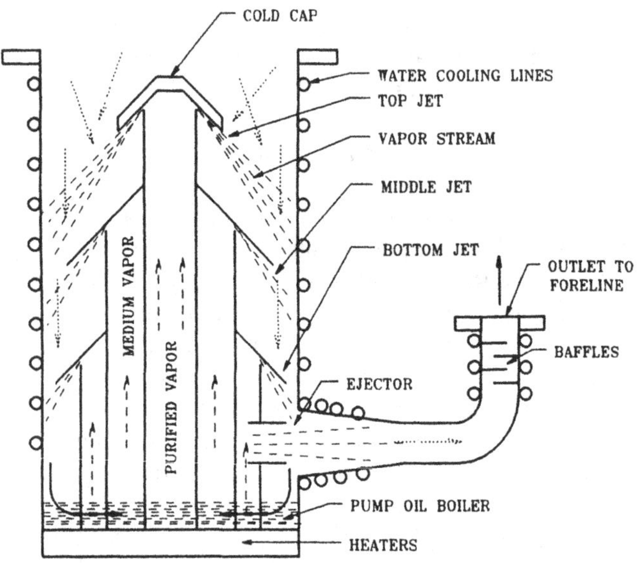

Fig. 4.8. A typical configuration of a fractionating diffusion pump (DP).

Figure 4.9 shows the action in the diffusion region in more detail. The massive and highly directional vapor jet molecules collide with the randomly moving gas molecules and transfer momentum to them in the general direction of the vapor molecules. The vapor molecules are generally much more massive than those of the gases being pumped. The vapor molecules then hit the cooled outer pump wall where they condense and run down by gravity back into the boiler. The pressure in the area below the jet is therefore higher than the pressure above the jet because of the kinetic momentum transfer action of the diffusion pump.

Referring again to Fig. 4.8, the gases compressed down below the bottom jets are ejected out of the exit elbow from the pump by a horizontal vapor stream. The exit section is cooled and has baffles so that the pump oil vapors are condensed and return to the boiler and do not leave the pump with the "compressed" gases going into the foreline.

The fractionating DP is designed to deliver the purest high boiling point vapors to the top jets where the lowest pressure (highest vacuum) is desired. The pump oils tend to absorb gases and also to break down to less massive and lower boiling point components. The pumps are designed so that the returning oils which have run down the outer wall of the DP flow toward the center of the boiler through passages between the outer and inner stack cylinders. The outer space is cooler than the center, but the lower boiling point fractions evaporate in this space and supply the lowest jets where the pressure is highest and the least harm is done.

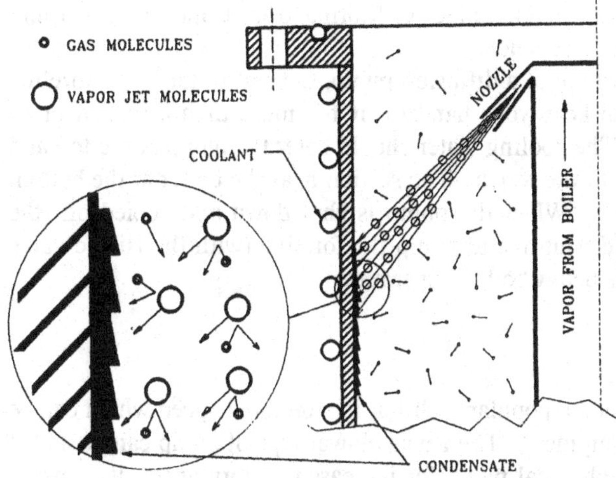

Fig. 4.9. The action of the molecules in the diffusion region of a diffusion pump in more detail. The massive and highly directional vapor jet molecules collide with the randomly moving gas molecules and transfer momentum to them in the general direction of the vapor molecules, thus sending them downward.

The oil which reaches the middle space is hotter and boils off the intermediate vapors which feed the middle jets. The oil which reaches the center is the purest and supplies the top jets.

It is important to not admit too great a gas load to the pump[3] or the top jets will be overwhelmed and "break down." When this happens, there tends to be increased backstreaming of pump oil up into the chamber. The rule of thumb is to keep the foreline pressure down below 3×10^{-1} torr as the HiVac valve is opened. This calls for either a slowly opening valve or a throttled bypass path during the transition from roughing to high vacuum operation.

Backstreaming under normal circumstances is still a major concern. Pulker[1] has a discussion (pp.150 to 152) of the need for careful design of cold caps on the tops of pumps and chevron baffles to minimize backstreaming to tolerable levels.

Chevron baffles between the DP and the chamber are at least water-cooled. More commonly they are cooled with a refrigeration system or liquid nitrogen (LN2). Since most of the pumping task is to get rid of the water vapor in the chamber, LN2 traps or the equivalent refrigeration system add greatly to the pumping speed of the system. We have found the refrigeration systems such as those supplied by Polycold™ of San Rafael, California to be the most desirable and economic way to operate in production.

In the normal operation of a diffusion pump, the exit water at the foreline should be just too hot to keep your hand on it for more than a second or so (approximately 50°C). The cooling water should enter the pump at the top and that area should be cool to the touch. The section near the boiler at the bottom will be too hot to touch. When the pump is shut down and cooled off, the mechanical pump should continue to pump the foreline (with the HiVac valve closed) until the elbow is below body temperature.

Roots Blowers

We have not yet mentioned a popular addition for roughing speed which can be made to the pumping equipment. The Roots blower type of pump can be added just upstream of the mechanical pump for increased performance. It is also a positive displacement type pump and is illustrated in Fig. 4.10. The technology

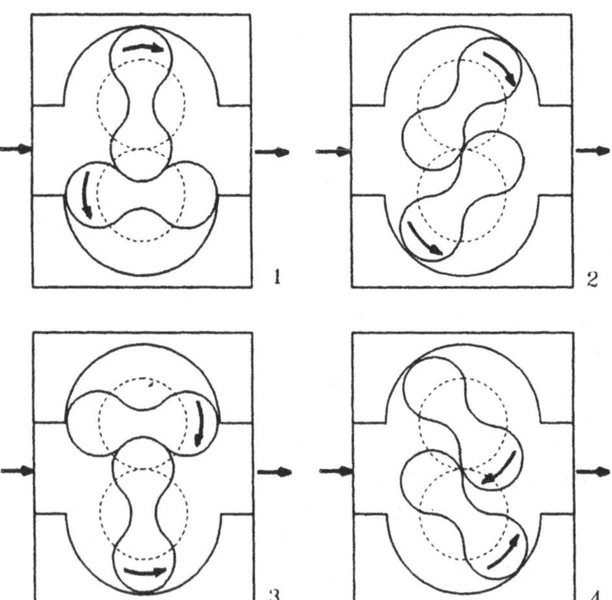

Fig. 4.10. The sequential action (frames 1-4) of the Roots blower type of pump, also a positive displacement type.

has been used for over a century. The two lobes of the pump rotate in opposite directions such that their surfaces are in almost rolling contact, but they do not actually make contact. The outer orbits of the lobes sweep volumes of the captured gas through the stator part of the pump. The geometry of the lobes captures gas from the inlet and releases it to the outlet. Actually the clearances between the lobes themselves and between the lobes and the casing are about 1/4 mm. One rotor is driven directly by a motor shaft and the second is slaved to it by two gears. Speeds are typically from 1000 to 3000 rpm which accounts for the high pumping speed capability.

When backed by the typical mechanical pump, a Roots pump can attain 5×10^{-4} torr. They can be used at the start of a roughing cycle or are sometimes turned on at 15 torr; the best pumping speed is found between 10^{-1} and 10^{-2} torr. According to Hill,[4] it is common practice to use a smaller vane or piston mechanical pump with about 1/10 the pumping speed of the blower. The Roots blower and mechanical pump combination offers increased pumping speed at a somewhat increased cost. They tend to also help reduce effects of mechanical pump oil molecules getting back into the chamber during the last stages of roughing when the transition from laminar to molecular flow occurs.

Cryopumps

As alluded to above, a surface in a vacuum chamber which is cold enough to freeze and trap water vapor can be an effective pump. The most common form is the LN2 trap above the DP. It is also not uncommon now to find LN2 coils run inside the chamber itself in what is called a Meissner trap. This captures the water vapor and freezes it on the coils and any attached plates. If for example the Meissner trap was a half meter diameter plate on an inside surface of the chamber, it would act like a "perfect" half meter diameter pump for water vapor. That is to say that any water molecule that reached the plate is trapped the same as if it had gone through a half meter diameter aperture into a perfect vacuum. The conductance or effectiveness of the Meissner is 100% because it is unobstructed in the chamber as opposed to the typical LN2 chevron baffle which is usually around a corner and down in the throat of the diffusion pump stack below the HiVac valve. The disadvantage of the Meissner is that it must be warmed and defrosted before the chamber is opened, which takes a few minutes and some complexity. The LN2 baffle below the HiVac valve does not have this problem since it is not exposed to atmosphere when the chamber is open, but it too must be defrosted from time to time.

Our practice with respect to the cold trap above the DP is to defrost as often as each night if the water vapor load has been high. The HiVac valve is closed, the ballast to the mechanical pump opened, and the Polycold is turned off or LN2 cleared from the trap. We have found that it takes about one hour for all of the ice to melt and the residue to be pumped away. If there is a lot of condensed water in the trap, the foreline pressure may rise above 3×10^{-1} torr. This will shut off the DP power on an automated system, or it should be manually shut off if this happens. The recommended alternative is to shut off the DP power to start with. When the foreline pressure is back down to 10^{-2} torr, the system can be restarted. If the DP has cooled off, it may take a half hour or more to warm up to operating temperature again. It is sometimes practical to use a shorter defrost cycle between coating runs by turning off the refrigerator but not the DP while the chamber is cooling, venting, unloading, and reloading. This may avoid the DP shut down due to overpressure in the foreline (actually in the top jets of the DP) and keep the ice buildup under control.

LN2 at 77°Kelvin is not cold enough to effectively pump most of the gases other than water vapor which need to be pumped such as nitrogen, oxygen, etc. However, liquid helium at 4°K can be effective. What is normally referred to as a Cryopump is a refrigeration unit at near liquid helium (LHe) temperatures which freezes out most gases and acts as the HiVac pump for the system. LHe, of course, cannot freeze out helium; but there is a solution for this. Cryopumps typically have a matrix of cooled activated charcoal which adsorbs helium, hydrogen, and neon which are not trapped on the Cryopump cold surfaces.

Cryopumps as an alternative to DPs have the advantage that they use no oils which might backstream or otherwise contaminate a chamber. They still must work with some sort of roughing pump to reduce the chamber and the pump pressure itself to where the Cryopump can operate. If a conventional mechanical pump is used for this, great care in the system design and operation are needed, or it can cause chamber contamination during roughing and otherwise.

Since a Cryopump works on the basis of temperature, it is important to avoid heat loads on the pump. If a process in the chamber uses high temperatures such as 300°C, special care must be taken in the system design to keep the radiant heat from the pump. The additional heat baffling may seriously reduce the conductance and therefore the effective pumping speed.

Cryopumps are entrapment pumps and will eventually become saturated. When this happens, they must be warmed up and the trapped gases exhausted. Modern systems tend to automate this regeneration process to do it unattended when the system is not in use. Vacuum system designers must carefully consider the gas loads of the processes to be used when deciding if the time between regenerations is practical.

There has been an ongoing debate since the advent of Cryopumps between DPs and Cryopumps. Both are still viable contenders with merits and limitations. It mostly depends on the details of the applications.

Other Pumps

As can be seen from Fig. 4.6, there are many types of vacuum pumps. There has been increased interest in recent years in Turbomolecular pumps. These are like jet turbines and act like a DP in that the molecules are swept by solid vanes rather than the vapor molecules of the DP. The advantage of the Turbomolecular pump is that it can be more oil-free.

Many years ago, oil-free pumping could be achieved by a combination of an adsorption pump and a getter ion pump. The chamber was roughed by opening it to an adsorption volume filled with something like activated charcoal cooled to LN2 temperatures. Most of the gases were adsorbed by this as a rough pumping action. The chamber was then opened to the HiVac getter ion pump. A gettering material had been freshly evaporated or sputtered on a surface which would then capture atoms/molecules by gettering.

The semiconductor industry often drives the technologies which become useful to the optical coating industry because of the usually more stringent purity requirements and often more extensive equipment requirements.

4.2.2. Evaporation Sources

In this section we will deal with the ways that material is normally evaporated for PVD. The most common methods for bringing materials to evaporation temperatures have been resistance sources. Here an element made of typically tungsten, tantalum, or molybdenum holds the material to be evaporated and is heated by passing a large electrical current through it. Next most common in the last several decades has been the electron beam source. The "E-gun" bombards a material with high energy electrons until it heats to the point of vaporization. Sputtering appeared over a century ago as a technique for making mirrors, but in recent decades has been more highly developed for semiconductor applications and also has some optical applications. In this case, the atoms of material are knocked out of the source (like billiard balls in a "break") by energetic ions of usually inert gas such as argon. The argon is ionized and highly accelerated (like a cue ball) by a few kilovolt electrical potential toward the material to be sputtered. Some processes are now being used which introduce the material to the chamber as a gas (sometimes initially as a liquid) which then reacts and/or condenses to produce a deposit. These processes tend to fall in the category of CVD and PECVD, which have existed for over a century but recently have become more common in the semiconductor industry than the optical industry.

4.2.2.1. Resistance Sources

Figure 4.11 shows a few examples of resistance sources. The simplest might be the coil of tungsten wire like a filament in a light bulb as shown in Fig. 4.11a. This is a typical source for evaporating aluminum. One or more rods, wires, or J-shaped "canes" of aluminum are laid in the center of the coil. An alternative is to hang small *U* or *V* shaped pieces of aluminum wire on the bottom of each turn of the filament. If the filament is heated with the right temperature versus time profile, the aluminum melts and wets the filament before it evaporates significantly. The power is usually then applied quickly to evaporate the aluminum rapidly so that the depositing film does not have much time to react with the residual gas in the chamber. This high rate has been found to be important in giving the highest reflectance and purity of the aluminum. Some other metals may be evaporated this way. The coils may also be made of multistranded wire.

We have seen cases where coiled filaments have been used successfully to evaporate SiO and MgF_2, but care must be taken in such processes to avoid having the material fall out of the filament before it evaporates.

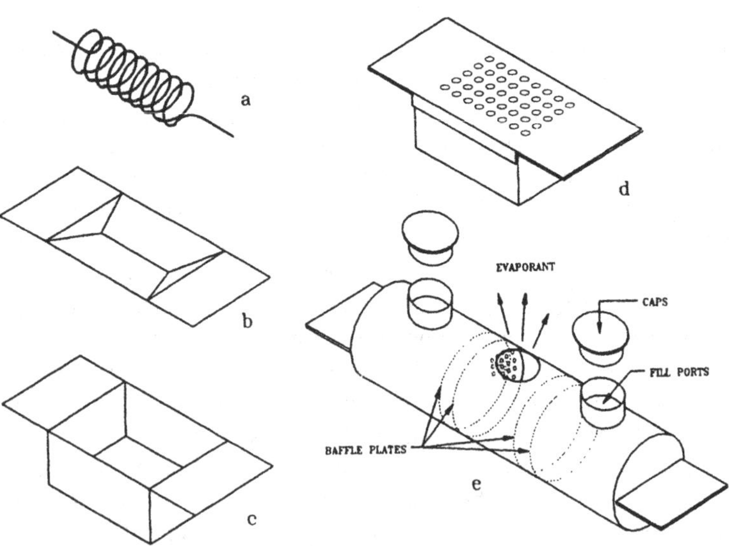

Fig. 4.11. A few examples of resistance sources: (a) filament of tungsten wire, (b) dimpled "boat," (c) larger volume folded or welded boat, (d) boat with baffles to prevent direct paths for "sparks," (e) the large capacity R. D. Mathis Co.[5] "beer can" boat.

Materials such as MgF_2 are often evaporated in a dimpled "boat" or something like Fig. 4.11b. A strip of flat refractory metal is formed to have a depression or dimple which will hold a material when it melts to be a liquid and is raised to evaporation temperature. Such a configuration is simple and inexpensive if a small charge of material is all that must be evaporated.

When larger amounts need to be evaporated, a larger volume boat can be fabricated as in Fig. 4.11c. These configurations can sometimes be folded sheet stock of the refractory metals so that they are seamless or else the joints are welded. The volumes that can be handled this way are mostly limited by the available drive power and the ability of the boat to handle the power without excessive hot spots or weak points.

Some materials such as SiO and ZnS are best evaporated from covered boats with a labyrinth path that prohibits solid particles from being blown toward the substrates. SiO from an open boat tends to send off sparks like a fire up a chimney. The boat might be as in Fig. 4.11d where there is typically another baffle, under the top shown with the holes in it, which prevents a direct path from the material to the outside of the boat. SiO actually sublimes. If there are fine particles, they may be carried by the rush of vapors.

We have had extensive experience with the R. D. Mathis Co.[5] "beer can" boat as shown in Fig. 4.11e. This is a large capacity source for materials such as SiO. The two ends have compartments which are loaded with SiO and then capped. There are baffle plates or bulkheads with perforations which are misaligned such that there is no direct path out for vapors or particles. Such a boat with a capacity for over 100 grams of SiO material might operate at 2.0 volts and 600 to 800 amps for a 1 to 2 nm/second deposition rate in a 1 meter box coater.

In more recent times, we have found the type of source shown in Fig. 4.12 to be recommended by Mathis and more economical than the "beer can" presumably because it is easier to construct. We have obtained equally satisfactory results and lifetime with it.

The larger boats such as seen in Fig. 4.11d and e are heated to between orange and yellow hot when depositing SiO. The radiation losses are a major factor in power consumption and may be a source of excess chamber heat. These larger boats most often have one or two outer layers of radiation shields to reflect as much heat back into the source as practical. The shields have as little conductive contact as possible with the heated boat.

We found it was advantageous to have flexibility in one end of the power electrodes and clamps to the boats. The thermal expansion of the boat can be significant in a large boat. This can induce large stresses and major damage to the boats if no room for flexure is provided. Tungsten, tantalum, and probably the

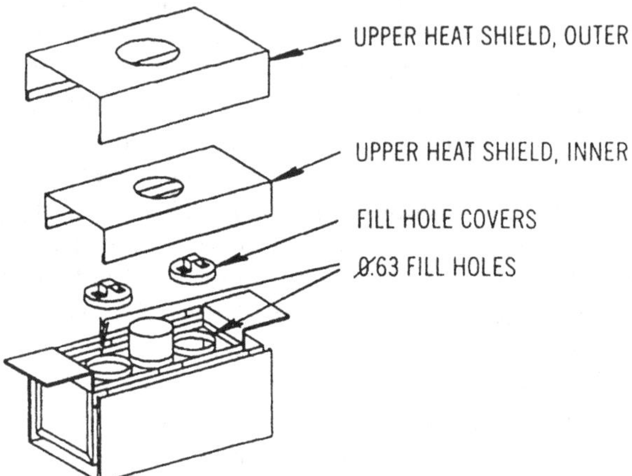

Fig. 4.12. Source recommended by Mathis Co.[5] and more economical than the "beer can" presumably because it is easier to construct.

other refractory metals become much more brittle after being heated and cannot tolerate much flexure in themselves.

There are a myriad of other resistance source designs for special situations and materials. Alumina coated boats, for example, are available for materials which tend to react with the boat metal. For a more extensive study of what is available, see the Balzers[6], Mathis[5], or similar evaporation source catalogs.

4.2.2.2. Electron Beam Sources

A significant attribute of an electron beam source (E-gun) is its ability to concentrate a large amount of power in a small surface independent of the type of material being heated. Another major factor is that the material being evaporated may provide its own barrier to chemical reactions/interactions with its holder. It is also possible to reach higher temperatures than the melting and evaporation points of refractory oxides and even the refractory metals normally used for resistance boats.

The schematic illustration of a typical E-gun is found in Fig. 4.13. The material to be evaporated is held in a cavity called a pocket or crucible which may or may not have a liner. A beam of electrons impinges on the material at several kilovolts and at some fraction of an amp. A 15 kilowatt (KW) power supply for the gun with a range of 4 to 10 kilovolts (KV) and up to 1500 milliamperes (mA)

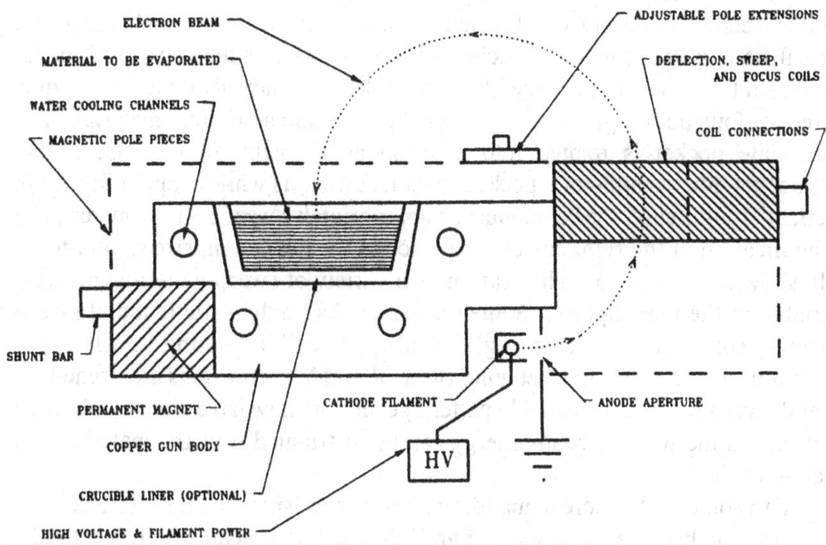

Fig. 4.13. Schematic of a typical electron beam source (E-gun).

of current capability is not uncommon in a modern box coater. Water cooling is supplied to the crucible to remove the heat which escapes the source material by conduction through the material, liner, and E-gun body. When no crucible liner is used, the source material is in direct contact with the copper crucible. With most materials, the cooling keeps the edges of the source material at a low enough temperature so that it does not react or interact with the copper, only the more central part of the source material is at evaporation temperature. This is what is meant by the material acting as its own barrier.

In a case such as the evaporation of TiO_2 and related oxides, we prefer to use a molybdenum liner as illustrated in Fig. 4.13. The liner reduces the conductivity so that the material retains more of the heat and remains at a more uniform temperature at or near a molten state throughout the material. Sometimes we have even seen quartz granules put under the liner for more insulation. We also choose to evaporate hafnium metal from an insulating vitreous carbon liner. With proper oxygen IAD, this is our preferred process to get HfO_2 (which we will discuss further in Chap. 5). Materials like SiO_2 have a much lower thermal conductivity and also sublime instead of melting so the heat conducted to the copper is much less. As a result, a liner may only be an advantage for mechanical handling.

A great variety of E-gun configurations are available. A rotating hearth crucible version is one of our favored versions because uniform heating and thermal homogenization can be achieved with a simple single sweep of the E-beam from the center to the edge of the crucible. Another advantageous form is a three- or four-pocket gun where each pocket can hold a different material. The appropriate pocket is rotated into position as needed. In the case of the multipocket gun, rotation of a pocket about its own axis while evaporating is not practical. In this case, the beam must be appropriately swept over the material by the application of the right current sequence to the deflection, sweep, and focus coils shown in Fig. 4.13. There are now a variety of sweep pattern generators available in the marketplace. Some older guns had only a single radial sweep deflection coil available. This might be adequate with a rotating hearth, but is very limiting with a multipocket gun. Both "X and Y" deflections are needed for general sweep patterns. Available pattern generators now have almost unlimited flexibility in the way the beam energy can be distributed over the material as a function of time.

With some of the more demanding performance specifications for coatings, uniformity can become a problem. If an E-gun pocket is deep with respect to its diameter (as is typical of a multipocket gun), the change of height of the material as it is depleted from the pocket by evaporation can cause changes in the deposition distribution and therefore the coating uniformity. For this reason, we prefer a large diameter, totally molten pocket of material whose material height does not change as much from full to "empty".

The basic functioning of the E-gun is described briefly below, see Hill[4] for an even more detailed description and history. The electrons are generated by thermionic emission from a tungsten filament. The filament is heated by an AC or DC current. The filament is also at an electrical potential of several KV below ground potential. As shown in Fig. 4.13, the emitted electrons are repulsed by the high negative potential and are attracted and accelerated to the ground potential of the anode aperture. The electrons then pass through the opening in the anode with several kilo-electron-volts (KeV) of energy (a high velocity and kinetic energy). As much as 1000 ma of E-beam current may be provided in the typical E-gun. There would be 10 KW of drive power applied at 10 KV and 1000 ma.

In the absence of obstructions and any magnetic fields, the electrons travel in straight lines. However, in a magnetic field, the electron experiences a force which is perpendicular to its direction of motion and the magnetic field. In a uniform magnetic field, the electron will travel in a circular path whose radius depends on the strength of the magnetic field and the velocity of the electron. An ionized atom would also travel in a circle in a magnetic field, but the masses of atoms are so many orders of magnitude greater than electrons, that the curves of the trajectories of atoms are not noticeable by comparison. A Pierce-type E-gun does not use a magnetic field and therefore the E-beam travels in a straight line from the cathode filament to the source material. This can cause some problems with evaporated source material coming back toward the E-gun, but such guns are still used when very high power guns are needed.[4] The E-gun illustrated in Fig. 4.13 uses a strong magnetic field perpendicular to the plane of the paper to bend the electron beam through 270° from the cathode to the crucible. This configuration protects the cathode assembly and high voltage leads from deposition of evaporant material and also from any debris which might fall on the E-gun from above. Earlier versions with 90° and 180° deviations have become much less common because of the advantages of the 270° system.

As illustrated in Fig. 4.13, the basic magnetic field is supplied by a large permanent magnet and magnetic pole pieces on both sides of the crucible which conduct the field lines so as to produce a nearly uniform field in the region where the electrons travel. Since the radius of the electron path is a function of the velocity and therefore the electron volts of the electron, the magnetic field must be changed if the KV setting is changed on the power supply. The field strength can be changed by changing the amount of field that is shunted from the permanent magnet by various shut bars as seen in Fig. 4.13. Adjustable pole extensions can be used to shape the field somewhat. The fine position of the beam on the material in the crucible can then be controlled by the variable magnetic field components supplied by the currents through the coils of the deflection, sweep, and focus assemblies illustrated.

We once had a very frustrating experience where the beam in an E-gun would function and sweep well at the start of a production run, but would eventually move toward the inside edge of the crucible and have no excursion left in the sweep pattern. We finally found that the deflection coils were heating up and loosing their effect. The heat was coming from the crucible through the copper and possibly also from the filament block. When additional cooling was supplied to the deflection coils, the problem disappeared.

The alignment of the cathode filament and the anode can be sensitive because the trajectory of the electrons and thereby the tangent of the circle of their orbit is changed by this. Even though the 270° configuration helps keep the filament and high voltage area clean, debris falling into that area from a very flakey process can cause high voltage shorts which shut down the process. We have heard rumors of people who have devised a means to provide a discharge pulse of high voltage to blast away shorting flakes without having to break the vacuum. This would need to be done in a way that would not be damaging if the short were something other than a small flake of material.

One of the characteristics of the E-gun is that the beam can potentially be focused to a small spot on the material. As a result, the watts per square centimeter can be made very high which should allow almost anything to be vaporized. The beam can also be defocused to spread the energy. Materials like TiO_2 will tend to spit molten droplets like a volcano if the intense beam moves quickly from a hot area to a cool part of the material. Our practice has been to heat TiO_2 in a rotating molybdenum lined crucible with a radial sweep from edge to center. A large source was typically 75 mm in diameter by 20mm deep. When starting a new charge, the bottom of the liner was covered with material pellets, the chamber pumped below 1×10^{-5} torr, crucible rotation set at about 1 rpm, and the E-gun power brought up slowly until the pellets started to melt. If TiO_2 is the starting material, oxygen will be released as the material melts and stabilizes at some value of TiO_x, perhaps Ti_2O_3. We melt the charge for several minutes with the power at a level that drives the gas off at up to 5×10^{-4} torr. After a few rotations of the crucible, the melt is molten liquid under the E-beam and yellow hot but not molten over the rest of the crucible. Note that this is too bright to be viewed with the naked eye, thus we view it with crossed polarizing filters or welder's glass. When the melt is uniform and outgassed until the pressure drops below 1×10^{-4} torr, the process is stopped, cooled, and vented. The crucible is loaded with another layer and the melting-in repeated until the charge is to the top of the liner. This melting-in process can be tedious, but the consequences of entrapped gas pockets in the melt can cause the loss of a coating run due to explosive spatter and spitting of liquid material on the substrates. In particular cases, it may be possible in a production run to melt-in a new charge of material under the evaporation shutter with production parts in the chamber. This does have some risks, however, of leaked and multiple bounce material getting on the substrates, effects of the gases driven off, and inadequate melt-in of the materials.

Silicon dioxide (SiO_2), or fused silica, is a useful and commonly used low index material. Although this is sometimes referred to as "quartz", natural quartz "is a disaster" according to Russell Hill in a private communication. He says it is believed that natural quartz has an included gaseous component (perhaps hydrogen) which expands when heated and causes eruptions. Fused silica has some favorable attributes such as low index and sealing capability, but it also has some problems when it comes to getting a satisfactory evaporation. Its evaporation from an E-gun is quite different from that of TiO_2. It does not melt to a liquid, but it sublimes. It does not have very high thermal conductivity so that hot spots tend to develop. A liner is of little value except to hold a charge if granular material is used. If solid discs or blocks are used, they can usually rest on the bottom of the crucible unless they just need to be raised to a higher level in the crucible. Since the material will not melt or heat very uniformly, there is no melt-in phase. At best, one might run a new charge under the shutter for a short time to "burn off" any dust or small particles of SiO_2 to reach a more steady state condition.

Our usual practice with solid discs of SiO_2 has been to rotate the crucible at about 1 rpm with a center-to-edge radial beam sweep. The operator has to be very attentive to keep the power level adjusted for uniform erosion of the evaporation material. Holes and groves tend to be unstable in that they evaporate more quickly than the surrounding material and become progressively deeper and more severe. This "tunnelling" effect causes the angular distribution of the evaporated material to change from what it had been from a flat surface. This in turn changes the distribution of material deposited on the substrates which can result in the unsatisfactory coating of some parts or even a whole load of substrates.

Granular SiO_2 material is sometimes used. Here the facets of the granules sublime under heating and some granules fuse together with others forming a crust somewhat like a snowflake on top of the unfused "sand". Some coaters use a "soft" beam of electrons at lower KV and defocus the beam to a broad spot. A beam with too much energy concentration can cause the smaller granules to "blow all over the chamber". The more difficult materials like SiO_2 require a lot of experimentation and experience and attention to get reasonable results. Such processes do not yet lend themselves to unattended automation in any but the simplest of cases.

Electron beam guns have added significantly to the capabilities of coaters to evaporate almost any inorganic material. Each material, however, may have its own peculiarities which need to be worked out as to the evaporation conditions. This constitutes an area of art and skill in many of today's coating processes.

Ion Plating

It is difficult to separate equipment (the subject of this chapter) from process (the subject of Chaps. 5 and 6) in some of these discussions. We will discuss several equipment configurations here which are driven by the processes with which they are used. This is the usual case of "form follows function."

Mattox[36] gives the description and history of the ion plating technology for which he has been well known since 1963. The essence of it is the bombardment of the surface before and during deposition by "atomic-sized energetic reactive or inert species." The key elements are ionizing the gases **and** vaporized depositing materials and accelerating these ions to the substrate(s) by a negative bias applied to the substrate with respect to the source of the ionized material. The source of material vapor has typically been an E-gun, but can also be thermal, sputtering, arc, or even chemical vapor. Ion plating technology precedes by many years the extensive application of IAD, which achieves similar results in ionization and acceleration by a different set of details in the equipment. Ion plating and the related "Ivadizer"process had been seen mostly in the metal processing industry and had not found much application in optical coatings directly until more recent times.

Reactive Ion Plating

The commercial application of ion plating to optical coatings has most notably been seen in the introduction of the Balzers BAP 800 system described by Pulker[40] and later by Guenther[45,60] and also Edlinger and Pulker[46]. It is illustrated in Fig. 4.14 and variously referred to as Reactive Ion Plating (RIP), Low Voltage RIP (LVRIP), and Reactive LVIP (RLVIP). A normal box coater is modified to have a hot cathode plasma source which operates with argon gas. The E-gun is electrically isolated from ground and biased as the anode with respect to the plasma such that the plasma is attracted to the molten/evaporating material in the E-gun. This ionizes the evaporant material. The substrate/holder is electrically isolated and naturally acquires a negative bias with respect to the E-gun from the plasma so that the ions are accelerated to the substrate. The energy of the depositing ions/atoms plus that of the plasma activated gas are key to the properties of the deposited films. Source drive voltages are typically lower and currents higher than the ion sources in common use at this time. This generally produces very dense films with compressive stress as described by Guenther[43,45,47,49] and Pulker[40,41,44,46]. The principal difference between this and IAD is likely to be the ionization and acceleration of the depositing material and not just the ion beam gas.

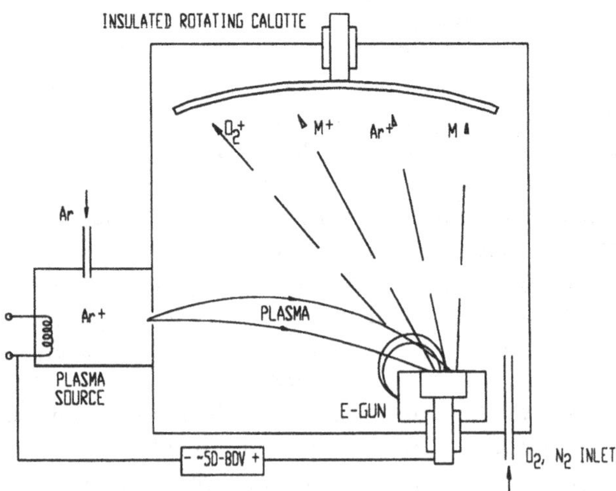

Fig. 4.14. System variously referred to as Reactive Ion Plating (RIP), Low Voltage RIP (LVRIP), and Reactive LVIP (RLVIP).

Because the chamber pressures are higher in RIP than IAD, the uniformity characteristics approach those of sputtering, but the source to substrate distances are those of the box coater. Guenther[45] points out that the low voltage aspect is important to reduce film resputtering and roughening effects caused by the impingement of higher voltage ions on the growing film. It has been well demonstrated to provide densification to eliminate humidity shifts and further to result in a vitreous-amorphous[45] state beyond that of other coating processes.

Many papers have been published on the application[41-51] of LVRIP to optical coatings by authors such as: Balasubramanian[49], Buehler[44], Conrath[42], Dobrowolski[50], Dubs[48,51], Edlinger[41,44,46], Emiliani[44], Fellows[43], Guenther[43,45,47,49,60,61,62], Gürtler[42], Kimble[62], Han[49], Himel[62], Hora[48,51], Jeschkowski[42], Jorgensen[49], Lee[49,61], Plante[50], Piegari[44], Pulker[40,41,44,46], Ramm[41,44], Sullivan[50], Tafelmaier[51], Waldorf[50], Willey[43], and Zarrabian[61].

Advanced Plasma Source

The advanced plasma source (APS) is probably best described in the paper by Zöller, et al.[29,52] and is illustrated in Fig. 4.15. The plasma source is in the center of a box coater. It is a large area LaB_6 cathode, an anode tube, and a solenoid magnet.

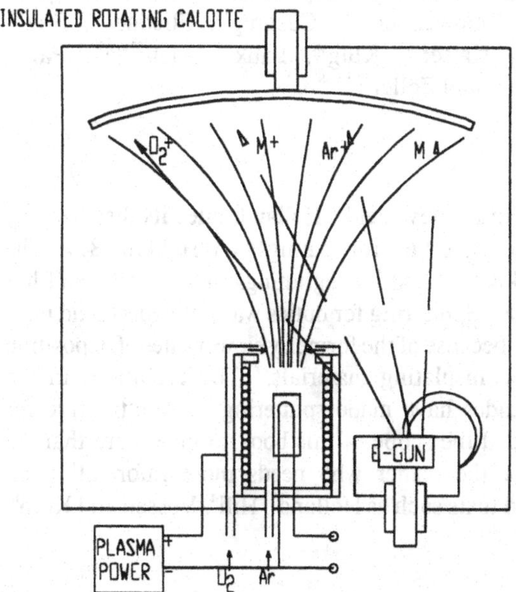

Fig. 4.15. Advanced plasma source (APS).

The cathode is heated by a graphite filament. The cathode generates copious electrons and the applied potential between the cathode and anode gives rise to a plasma in the Ar gas supplied to that space. The magnetic field lines of the solenoid are axial to the assembly and limit the electron motions in the radial direction much like the fields of a magnetron sputter source. This enhances the ionization of the Ar. The electric and magnetic fields drive the ions upward into the chamber. This can be seen to be very similar to other IAD/plasma sources in function but different in its efficiency of ion production. The source is designed to operate at powers up to 8kW. Reactive gasses such as oxygen are introduced by a "shower ring" above the plasma source. It is stated that the plasma activates and partially ionizes the reactive gas. It is further stated that the evaporating species are also partially ionized by the widespread plasma in the chamber. It would appear, however, that this effect would be weaker than that of the LVRIP equipment/process.

The plasma source is electrically isolated from the chamber and a self bias develops between the source and substrates as in the LVRIP system. This generates the acceleration of the ions to the substrate for IAD and ion plating. It is stated that the bias voltage depends on: discharge voltage, magnetic field strength, and chamber pressure. The anode tube is water cooled to minimize the thermal load on the substrates.

Various papers have been published on the application[41-51] of APS to optical coatings by authors such as: Beißwenger[53,54], Cushing[52], Fliedner[54], Friz[57], Götzelmann[52-55,58], Hagedorn[58], Kaiser[56], Klug[58], Laux[57], Matl[52-55,58], Pan[59], Schallenberg[56,57], Uhlig[56], Zhou[59], and Zöller[52-55,58].

4.2.2.3. Sputtering Sources

Sputtering is both an old field and a new field. Holland[7] cites its discovery by Grove in 1852 and application to deposit mirror coatings by Wright in 1877. The optical coating industry as we know it had its beginnings in the 1930s and has expanded rapidly since the 1940s. Sputtering for optics was not widespread until recent decades. This is probably because of the formerly slower rates of deposition and difficulties with depositing insulating materials. The evolutions in the technology over the recent decades have made sputtering more attractive for various applications. It is beyond the scope of this book to give more than an overview of sputtering, but for the reader who needs more information on sputtering there are several good texts such as Holland[7], Hill[4], Vossen and Kern[8], and Bunshah[9].

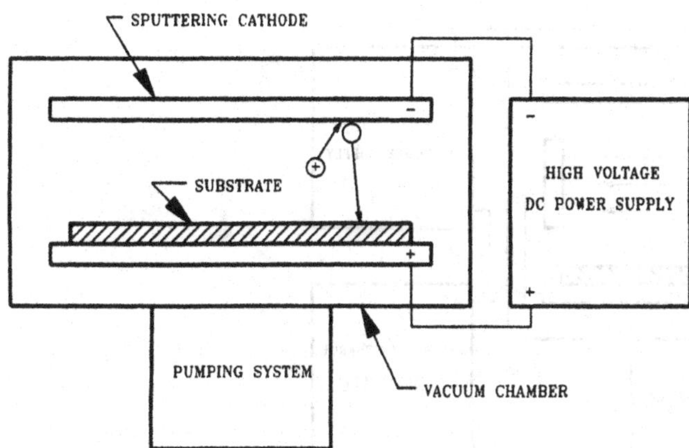

Fig. 4.16. Schematic of the simplest (diode) sputtering chamber.

The most basic configuration for sputtering is the glow discharge tube which is like a "neon" light. Figure 4.16 shows a schematic of the simplest sputtering chamber. A potential is applied to the two electrodes (anode and cathode) in a low pressure gas. At the appropriate pressure and potential, a stable plasma discharge is formed. The positively charged ions formed are accelerated to the negative cathode. The ions collide with the cathode and transfer momentum to atoms near the surface of the cathode. These atoms are knocked out of the surface like billiard balls by the cue-ball in a "break". The liberated atoms then travel until they collide with gas atoms or a solid surface such as a substrate, fixture, or container wall where they condense. This is usually referred to as Direct Current (DC) Diode Cathodic Sputtering.

One advantage of sputtering is that the sputtered atoms have an energy of several eV to hundreds of eV's, whereas thermally evaporated materials usually have an energy of about 0.1 eV. This gives the sputtered atoms more mobility to move and consolidate as they condense on a surface. The film growth comparison might be like the difference between growing frost and freezing rain in terms of a robust, fully densified coating. Another advantage seems to be that the process is a glow discharge that bombards and cleans the substrate surface and promotes better adhesion. These advantages have contributed to the resurgence of interest in sputtering in the past decade or two. However, the evolution of Ion Assisted Deposition (IAD) in the last decade offers some competition to sputtering with the same sort of benefits. Still another advantage is that multi-component films can

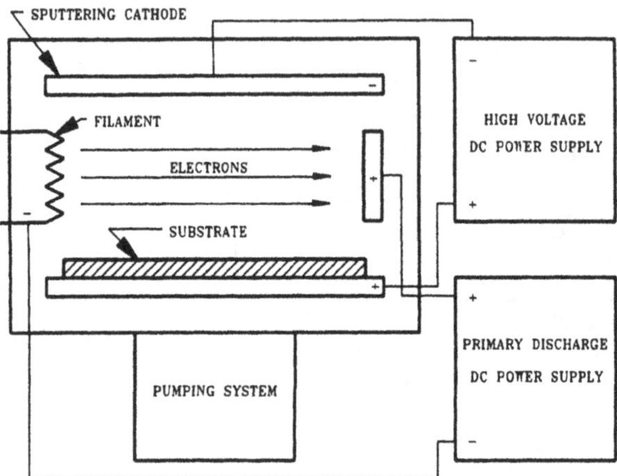

Fig. 4.17. A Triode Sputtering System with a filament source of electrons to generate more ions and thereby greater deposition rates.

be deposited from a cathode of mixed materials such as an alloy. Some special attention to the composition of the target is needed particularly if the materials of the mixture have different sputtering rates. The whole issue of sputtering rates of various materials is covered in detail in the references.

Hill[4] illustrates many of the variations in sputtering equipment as do Vossen and Kern[8], and we will briefly discuss a few of them. The simple diode sputtering system is limited in deposition rate by the number of ions that can be generated by the natural processes. By adding a source of electrons from a heated filament in a Triode Sputtering System as in Fig. 4.17, more ions can be generated by the extra electron excitation and thereby greater deposition rates can be achieved.

The problem with sputtering insulating materials is that a charge quickly builds up on the nonconducting material which is the cathode to be sputtered and the cathode then repels the sputtering ions. If radio frequency (RF) rather than DC power is supplied to the electrodes, this problem can be overcome and insulating materials can be sputtered very well. The RF sputtering system might look schematically similar to the triode system in Fig. 4.17 except that the high voltage power supply would be an RF power supply and matching network. Matching networks are needed to tune to the chamber so that most of the power goes into the plasma and is not reflected back to the power supply. The frequency of operation needs to be above 10 MHz and has typically been at the FCC allocated frequency of 13.56 MHZ. There are many details of RF sputtering that must be attended to and are found in the references[4,6-9].

DC sputtering typically operates at pressures of tens of millitorr whereas RF sputtering might operate as low as 0.2 millitorr. The higher pressures tend to have more problems with gaseous impurities included in the films and less reproducibility of film properties. As we can see from Eqn. 4.1 and Fig. 4.2, the mean free path at the best RF pressure is about 4 cm wherein 63% of the atoms have had collisions before they reach the substrates. At the DC pressures, the mean free paths are in millimeters. The depositions are done with spacings between the cathode and the substrates that are of this order, and therefore the sputtering equipment has been quite different from a box coater. It is clear that there is a lot of gas scattering and the sputtered material is strongly influenced by collisions. The material distribution patterns from deposition are more nearly akin to electroplating than vacuum deposition. Charge distributions and fluid flow around the parts to be coated or plated come into play.

Since sputtering is due primarily to momentum transfer, the heavier atomic weight gases are more effective. One usually uses an inert gas to avoid chemical reactions from the sputtering. The choices of inert gases are He (atomic weight 4), Ne(20), Ar(40), Kr(84), and Xe(131). Although Ne is just a little heavier than nitrogen(14) and oxygen(16), Ne and He are too light for practical sputtering. Argon is the more common gas used because it is economical. Krypton and xenon could be more physically effective, but the economics of the situation usually would not justify their use.

Many metal oxides need some extra oxygen added to the process to get a stoichiometrically satisfactory deposited film. A percentage of oxygen may be added to the sputtering gas for this purpose. If a metal is sputtered and oxidized to deposit a metal oxide, more oxygen is required. This sort of sputtering and equivalent versions producing nitrides are referred to as reactive sputtering. The hard gold-colored coatings now in common use for machine tools and decoration are a titanium nitride (or related materials) produced in this way.

The efficiency of converting the sputtering gas to the ions is low for DC Diode Sputtering, better for Triode Sputtering, and much better for RF Sputtering. This is why the gas pressures can be lower and deposition rates higher as we move up the scale of complexity. Magnetron Sputtering improves the situation even more. Figure 4.18 illustrates the principle. A properly shaped and intense magnetic field near the sputtering cathode creates a region where electrons in the plasma are trapped or at least forced to spend a longer time. The $E \times B$ forces cause the electrons to move in spirals instead of straight lines. Under the right field conditions, they can even be trapped by being reflected by the converging magnetic field lines close to the magnetic poles. As the electron path spirals more and more tightly toward the south pole as shown in Fig. 4.18, its direction can be reversed or reflected back toward the north pole where it is reflected again. The principle is the same as the origin of the Van Allen radiation belt in the magnetic field of the Earth. This increases the frequency of collisions of the electrons with

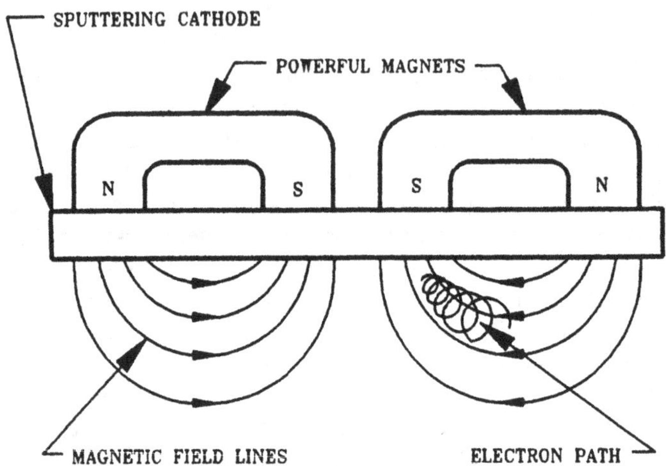

Fig. 4.18. Magnetron Sputtering principle. Electrons are constrained to spend more time near the cathode and thereby cause more ionization.

the gas and thereby the ionization efficiency. As a result, the gas pressures can be lower and/or the sputtering rates much higher than simple DC Diode, Triode, or RF Sputtering. Hill[4] shows many configurations and the theory of the electron confinement and sputtering details. The magnetron addition to sputtering with its increase in rate, decrease in pressure and thereby increase in mean free path has moved sputtering to the point of more interest in the optical coating industry.

Arcs can be a source of problems in sputtering systems, particularly in reactive sputtering where the metal target may build up an oxide insulating layer beyond the edges of the most actively sputtered area. An arc is effectively a small lightning bolt to the charged area of the target or other parts of the system. These can cause surface damage to the target and/or substrates and also create particulate matter which degrades or destroys the usefulness of the substrates. Scholl[10,11] describes progress in sputtering power supply designs which suppress the arcs before they can cause damage. Such power supplies can be very valuable in stabilizing a previously unstable process.

Another area of concern when depositing insulating layers is the "disappearing anode." The surfaces which act as the anode for the sputtering cathode can become coated with dielectric just like the substrate. As this happens, the effectiveness of the sputtering decreases due to this "disappearing anode."

Sellers[67] discusses this problem and its solution in some detail. He describes things such as high aspect cavities and reentrant surfaces that do not get coated with the dielectric. Sieck[68] discusses the effects of anode location on deposition profiles and uniformity. Scholl, et al.[84] gives another solution to the anode problem using an AC power supply and a novel anode and circuit configuration.

Pulsed and Dual Magnetron Sputtering

There have been extensive developments in the past decade in the use of AC Magnetron Sputtering reported by Glocker[15] wherein about 10KHz frequencies are used. This is sometimes also found under the heading of Pulsed Magnetron Sputtering (PMS). Figure 4.19 illustrates the principle. By alternating the voltage in see-saw fashion between two sputter sources, conductors that quickly form thin insulating layers when reactively sputtered can be continuously sputtered at rates equivalent to DC sputtering. This also eliminates the disappearing anode problem because the alternation between the two sputtering sources causes the one which is not sputtering at any instant to act as an anode. This bipolar or Dual Magnetron Sputtering (DMS), which is also called "Twin-Magnetron" sputtering by Leybold, is described by Schiller, et al.[69,70]. These have been applied to large sheet and roll coaters by Schiller, et al.[72-74], Hoetzch, et al.[75], Rettich, et al.[77], Heister, et al.[78,83], and Schilling, et al.[79], Tachibana, et al.[81], but also to some optical products such as AR coatings and multilayers by Kirchoff, et al.[71], Beisenherz, et al.[80], by Strumpfel, et al.[76], Milde, et al.[81], Lishan, et al.[85], and Fahland, et al.[86]

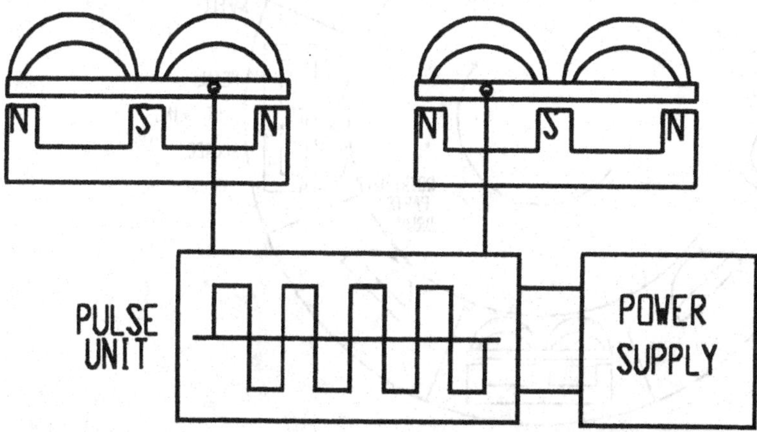

Fig. 4.19. Pulsed Magnetron Sputtering (PMS) or "Twin-Mag" Sputtering.

MetaMode™ and MicroDyne™ Sputtering

The MetaMode™ process/equipment was early described by Seeser, et al.[87] and recently reviewed by Mattox[88]. Figure 4.20 shows the general concept. A metal target is sputtered in relative isolation from the gas with which the metal film will be reacted to form oxide or other compound. The parts move into a reactive area past a gas conductance isolation baffle. The reactive plasma converts the metal to the compound. The parts continue to move into a new sputtering regime where a similar cycle can occur. The figure shows a four cycle per revolution version of such a system.

There is some limitation on the thickness of the film that can be deposited in any one cycle so that the plasma can react with the full depth of the newly deposited film. The scheme, in general, avoids the cathode poisoning by reaction with the oxygen or other active species.

The MicroDyne™ process and equipment was described by Boling, et al.[89] Its general layout is shown in Fig. 4.21. Targets sputter metal which are reacted with the active gas with the help of a microwave plasma. The active gas is controlled optically to avoid target poisoning.

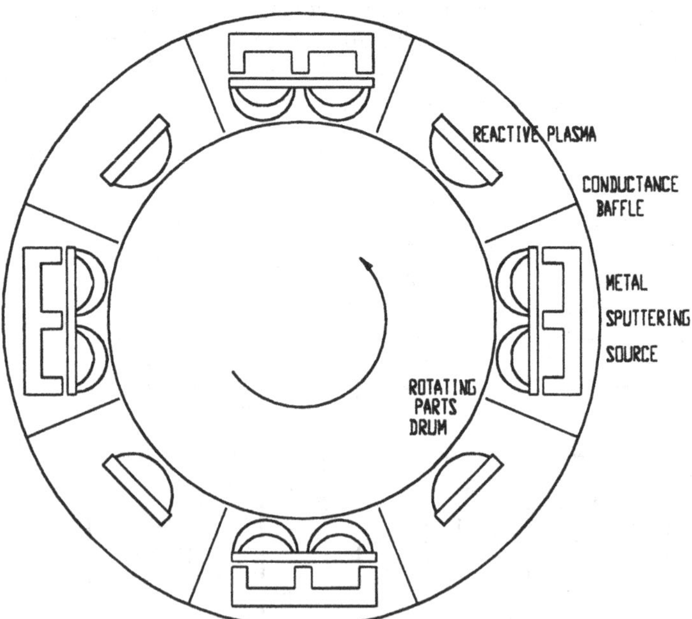

Fig. 4.20. MetaMode™ process/equipment general concept.

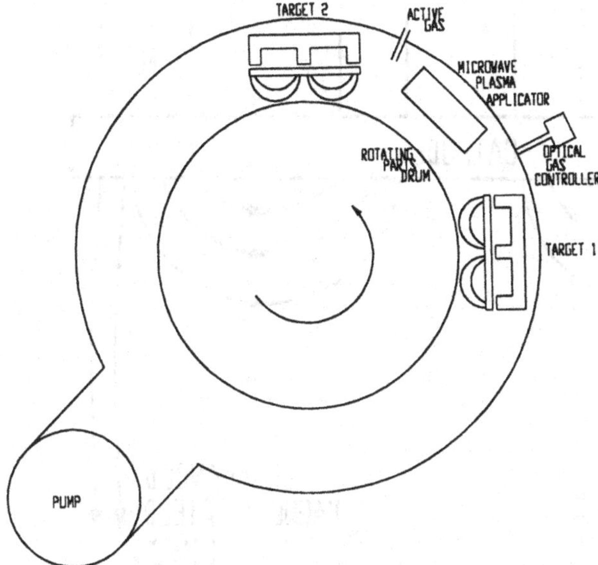

Fig. 4.21. MicroDyne™ process and equipment.

Unbalanced Magnetron Sputtering

Unbalanced magnetron sputtering (UBMS) was introduced in 1986 by Window and Savvides[27,28]. Figure 4.22 shows the difference between UBMS and the balanced (BMS) case illustrated in Fig. 4.18. The essential difference is that the center magnets are not as strong as those at the edges thus leaving magnetic field lines that radiate away from the magnets and the cathode. Sproul[30,31] has worked in this area extensively and describes the functioning of UBMS in detail. Basically some energetic electrons that would have been trapped near the cathode in the BMS case are now able to move away from the cathode toward the anode. However, they cannot travel directly because of the magnetic fields, and they therefore have more ionizing collisions with the gas atoms in the path. This causes a secondary plasma which acts like ion-assisted deposition for the film growing on the substrate. Krug, et al.[32] described some early applications of the concept.

More than one UBMS cathode can be used in multiples of two wherein the magnets with alternate cathodes are of opposite polarization to those of the adjacent cathode. This is so that the field lines connect from cathode to cathode and trap the electrons for the desired action. Figure 4.23 shows a configuration similar to that reported by Münz[33,35] for the commercial application to hard coatings for machine tool parts, etc.

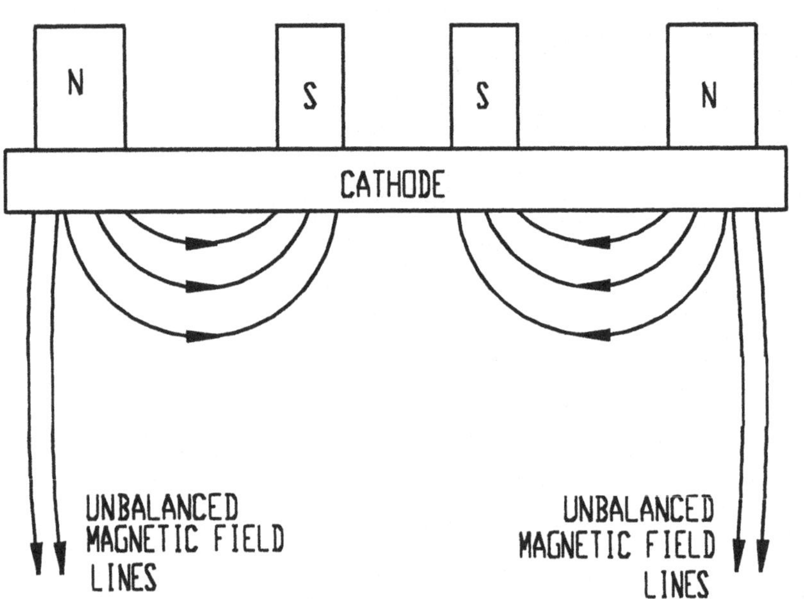

Fig. 4.22. Difference between UBMS and the balanced (BMS) case illustrated in Fig. 4.18.

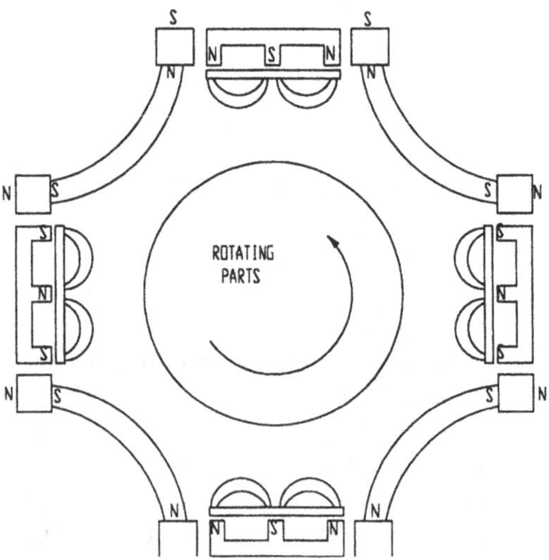

Fig. 4.23. UBMS Configuration similar to that reported by Münz[33,35] for the commercial application to hard coatings for machine tool parts, etc.

Ion Beam Sputtering

Another type of sputtering which has recently come into play (2000) on more than a laboratory scale is Ion Beam Sputtering (IBS). We cover ion beams in more detail in the section on that subject, but we will mention its application to sputtering here. An ion source or gun can provide highly accelerated ions in a focused, diverging, or collimated beam. These techniques can operate at lower pressures more like evaporation, and therefore the chambers might look more like a box coater than a sputtering chamber. Typically a near-collimated beam is aimed at a sputtering target at about a 45° angle of incidence as shown in Fig. 4.24, and the sputtered atoms are condensed on a substrate placed more normal to the target on the "reflected" side. A Kaufman or other type of gridded ion gun can provide a source of high eV ions with a highly controlled and narrow range of energies. This arrangement avoids much of the substrate heating and substrate gas interaction characteristics of the other sputtering processes mentioned. Several different target materials can be mounted so that they can be positioned alternately in the sputtering ion beam as shown in Fig. 4.24 for alternating material depositions.

IBS has been a good investigative tool with very controlled conditions, and some very specialized films can be done this way on a small scale. IBS is not ordinarily well suited for large area coatings, but the DWDM applications fit well because of the small area of the parts to be produced. There have recently been a large number of systems delivered to coat DWDM filters with IBS. Harper's chapter in the book by Vossen and Kern[8] goes into some detail concerning IBS.

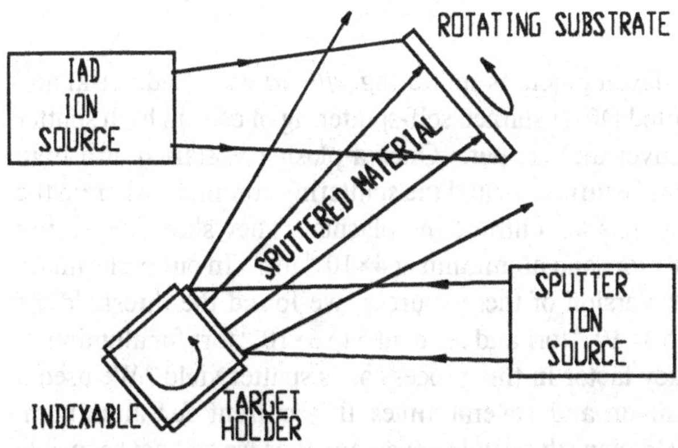

Fig. 4.24. Typical dual ion beam sputtering (DIBS) configuration.

Until recent times, the deposition rates for IBS have been at least an order of magnitude lower than that of evaporative processes. We understand that the DWDM oriented machines are now about 1/3 as fast as conventional evaporation. This is probably determined primarily by the ion power available for sputtering.

The DWDM machines also have another ion beam aimed at the substrates which is used for ion assisted deposition (IAD) with all of its inherent advantages as described in later chapters. Such systems are referred to as dual ion beam sputtering systems (DIBS) and are illustrated in Fig. 4.24.

In-Line Sputtering Systems

Simpler coatings on sheet glass such as architectural and automotive glass are done on a massive scale in large continuous in-line sputtering coaters. Such systems are built with a goal of achieving a low cost per unit area of coating. It is not unusual to think in terms of *acres per year* of coated substrate. Another advantage is that sputtering can occur in any orientation with respect to gravity which allows the substrates to travel face-up on a conveyor and be sputter coated from on top. The parts fixturing is minimal in this case. Advances by the development and use of C-MAG™ sputtering by Airco Coating Technology[12-14,68] have made the reactive sputtering of dielectrics practical on a large scale. The semiconductor and compact disc industry use sputtering extensively and often the substrates and cathodes are held with their surfaces in a vertical position. This has the advantage that the effects of falling debris are less on both the substrates and the cathodes. Hill[4] reports that aluminum has been deposited for semiconductors at 20 nm/sec or more over 150 mm semiconductor wafers such that a 1 μm film is put down in less than one minute.

Sputtering Without Gas

Another interesting development is sputtering *without gas*. Radzimski and Posadowski[16,17] have reported DC-sustained self-sputtering of certain high sputter yield materials such as silver and copper. Once a plasma is started at a high enough current, the gas can be turned off and the sputtering continues wherein the sputtered metal acts as the gas to continue the plasma. They show sputtering copper and silver at 2×10^{-5} torr and aluminum at 4×10^{-4} torr. In our preliminary tests of a 10 cm diameter version of these sources, we found the threshold for starting silver at about 2 to 3×10^{-4} torr and at about 5 to 6×10^{-4} torr for aluminum. High power density is a key factor in this process as is sputter yield. We used 5 amps at 800 volts for silver and several times that current for aluminum. Aluminum has a lower yield than silver and copper, and thus we had not been able to operate with no gas, but with a relatively small amount compared to previous techniques. This has an appeal for mirror coating where large throw distances are required without major gas scattering effects.

Sputtering is in some respects more complex and in others less complex than evaporation in a high vacuum. Although sputtering is not yet a common way to do general and versatile optical coating, it is gaining consideration for large scale, dedicated production processes, and should be watched for future developments.

4.2.2.4. Filtered Arc Sources

Arc vaporization and deposition have been around for some time and its history is reviewed by Mattox[36]. The cathodic arc source is the most widely used. An arc of high current and low voltage vaporized the cathode surface creating vapor, plasma, ions, and ***globules*** of the cathode material. The energetic vapor, plasma, and ions are all favorable attributes for robust coatings. Martin, et al.[37] state that the "species emitted from the cathode spot are both highly energetic (~40 to 60 eV per charge state) and highly ionized (~80% in the case of Ti)." The globules, on the other hand, are highly detrimental to most results. This would be mostly unacceptable for optical coatings and are similar to "spatter" which will be discussed in the chapter on materials. Martin[37,63] describes the magnetic plasma duct filtering system which they used to bend the metal ion species via a magnetic field to separate them from the globules which do not deflect. The deposition of

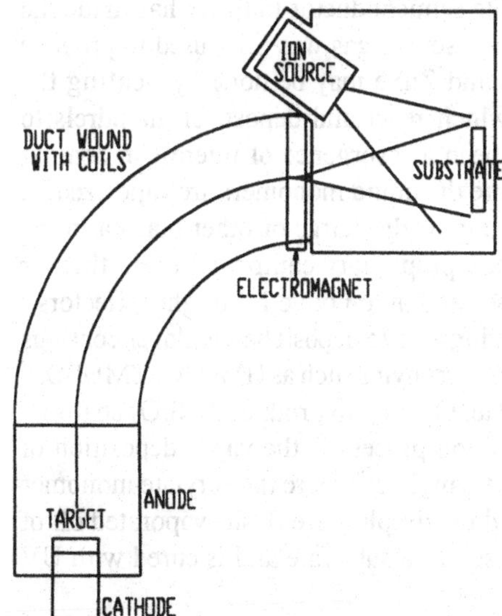

Fig. 4.25. Filtered Arc system with an added IAD source.

the filtered metal ions occurs at the exit of the duct. The substrate can be biased to accelerate the ions and provide a similar effect to that of IAD. Baldwin and Falabelle[38] reported on a commercial adaptation of this technology. Fulton[39] described a system with an added IAD source as illustrated in Fig. 4.25 to modify the depositing films. He reports on Amorphous Diamond-Like-Carbon (A-DLC) and Al_2O_3 films with impressive hardness and tribological properties.

The area covered by such systems is limited as compared to that of a box coater, so that the applications other than research and development at the moment seem somewhat limited. All the above reports mention scanning of the output beam by magnetic deflection (somewhat the converse of an E-beam gun sweep system) in order to obtain wider substrate coverage and uniformity.

4.2.2.5. Chemical Vapor Sources

It is difficult to separate equipment from process in many cases. We will touch more extensively on the myriad of CVD, PECVD, and other processes in Chap. 5. We will confine ourselves for the moment to a few comments about the sources for introducing chemicals for CVD into a coating chamber. There is also a somewhat nebulous line between CVD and reactive evaporation and sputtering where the gaseous reactant (such as oxygen, nitrogen, etc.) is participating in a chemical reaction with the PVD.

CVD by its name is a vapor process and the sources are usually introduced into the chamber from bottled gases. The semiconductor industry has made the most extensive use of CVD and a common source gas was SiH_4 used to produce SiO_2. However, the production of ZnS and ZnSe may be done by heating the elemental solids to create the vapors which react and deposit on mandrels to develop substrate material. There has been a resurgence of interest in coating polymers onto substrates by PECVD where the liquid monomers are vaporized on entry to a chamber and polymerized by a glow discharge or other plasma in the chamber. Balzers[6] now apparently offers proprietary equipment to do this for protective coatings on ophthalmic lenses and automotive headlight reflectors. Much work has been done in the web coating area to deposit barrier layer coatings by related techniques. The monomers with acronyms such as HMDSO, TMDSO,[18] and TEOS[19] are fed into a plasma with He, O_2, etc., to produce an SiO_2 coating.

Another emerging chemical deposition process is the vapor deposition of acrylate thin films described by Shaw and Langlois.[20] Here the acrylate monomer is fed through an ultrasonic atomizer and the droplets are flash evaporated off of a heated surface. The monomer condenses on a substrate and is cured with UV light or an electron beam.

We will discuss some of these processes in more detail in Chap. 5 on processes.

4.2.3. Fixturing and Uniformity

Holding the parts to be coated or fixturing can be simple or complex depending on the needs. In the era of bell jars and no substrate movement, the coating of lenses and prisms has been done with very simple adjustable rails between which the parts lay and were held in place by gravity. Coating odd-shaped parts and surfaces may require very complex fixturing. Getting the coating where it is needed and keeping it from where it is not allowed are often difficult issues. A substrate which is a hemispheric dome can be difficult to coat uniformly on the inside and outside. Keeping the coating from the faces of prisms which are not to be coated can be a problem. We will focus our attention on the box coater in this discussion. Uniformity of the coatings on the parts is often a major consideration, and the number of parts which can be coated in one batch is usually of importance. The fixturing approach may quite different between large volume repetitive coating processes and coatings and products that change from batch to batch. The degree of flexibility required and the overhead of the tool changing time influence the choice of tooling or fixturing.

4.2.3.1. Fixturing Configurations

Figure 4.26 shows two types of single rotation part holders or *calottes*. The calotte normally will be as large a diameter as the box coater can hold and rotates near the top of the chamber about a vertical axis. The parts are attached to the calotte with the side to be coated facing downward toward the sources. The uniformity of the coating depends on the angular distribution of the material from the source (or its "cloud" shape), the position of the source with respect to the calotte, and the radius of curvature of the calotte. In some cases, the uniformity of the coating from a properly curved calotte with well positioned sources is acceptable. If not, masks can be added to improve uniformity as discussed in Sec. 4.2.3.4. When two or more materials are being deposited, the sources may not generate the same angular distribution of material which further complicates the uniformity problem. One solution to this which we have often used is to mount the sources on opposite sides of the chamber so that there is less effect of the mask for one material on the deposition of the other material. Another solution that we have seen is to use separate masks for each material which are positioned in the path to the substrate only when the particular material with which they are associated is being deposited. A flat calotte does not generally provide good uniformity without

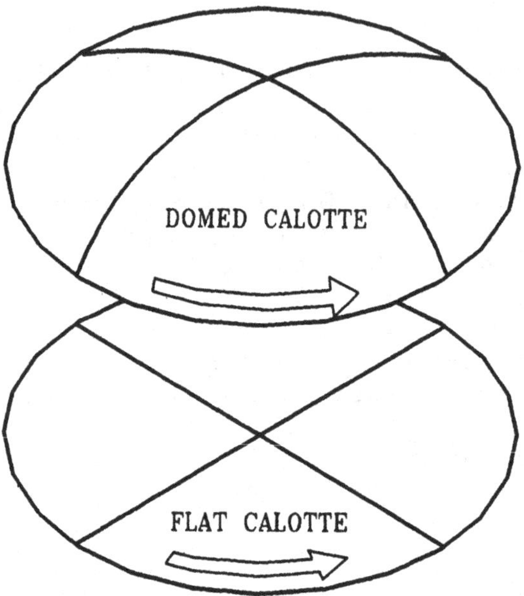

Fig. 4.26. Two types of single rotation part holders or calottes.

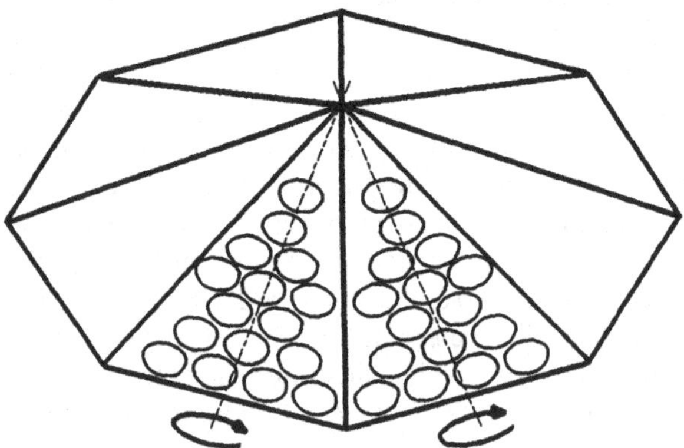

Fig. 4.27. A calotte with rotatable or "flip" segments so that the second side of the lenses can be coated in a single evacuation cycle, typical of ophthalmic coaters.

masking, but the most flat parts can be fixtured and coated on a flat calotte. Large flat parts coated on a domed calotte cannot take much advantage of its uniformity benefits because the parts cannot conform to the domed surfaces. Some coaters for eyeglass lenses use a variation between the dome and flat calotte in the use of a number of flat triangular segments as illustrated in Fig. 4.27. These usually can be "flipped" over by some clever mechanisms without breaking the vacuum so that the second side of the lenses can be coated in a single evacuation cycle.

Planets or double rotation parts carriers as shown in Fig. 4.28 are a major help to uniformity, but they reduce the total area of parts which can be coated in a given size chamber. Figure 4.28a shows three planets rotating about a common center (isometric view). Musset, who had studied uniformity extensively,[21,22] found that the number of flat planets in a system for the most efficient use of the available space should be about 4.5 planets. This implies that 4 or 5 planet configurations are near optimum. The choice will then depend on the expected parts sizes and mix. In some planetary drive designs, the planets can be tilted in (or out) as seen in Fig. 4.28b. This can improve the uniformity in some cases. The extreme of tilted planets is the Knudsen configuration shown in Fig. 4.28c. Here the planets are domed and rotate such that their inner surfaces behave as though they were confined to the surface of a hemisphere. This could provide uniform coating on parts attached to the planets from a source at the center of the hemisphere which propagates material equally in all directions. It also would provide uniformity from a Lambertian or cosine source distribution placed at the bottom of the extended sphere from the hemisphere. The Knudsen configuration

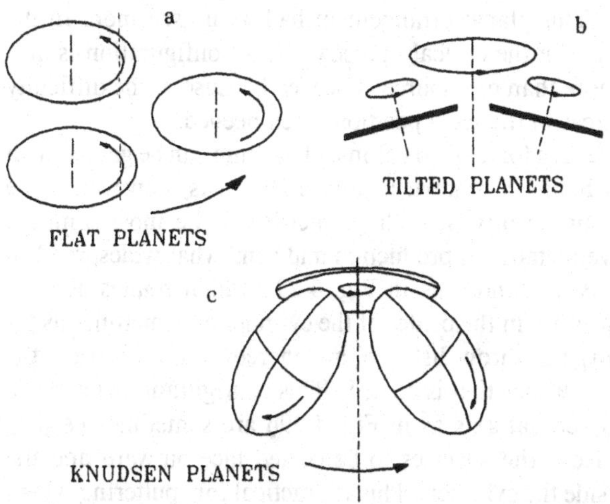

Fig. 4.28. Planets or double rotation parts carriers can help uniformity, but reduce the total area of parts which can be coated in a given size chamber.

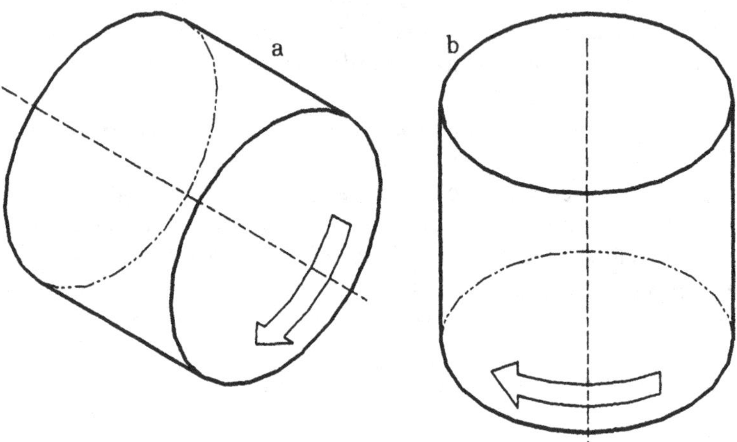

Fig. 4.29. Large coated surface area for a given volume can be gained with cylindrical parts holders.

has a fairly high packing density of the planets, but gravity cannot be depended upon to hold the parts in the planets. Some other means to retain the substrates in the planet is required. This planet arrangement had been used more in the semiconductor industry than in the optical industry. This configuration is also more problematic when more than one source is needed because of the difficulty of having each of two sources in the ideal position when needed.

A large coated surface area for a given volume of vacuum can be gained with a rotating cylinder parts holder as shown in Fig. 4.29. It is, however, more difficult to obtain precise uniformity in such geometries. The most common configuration in decorative metalizing production and somewhat widespread in simpler optical coatings is a cylinder as in Fig. 4.29a which rotates about a horizontal axis. The sources are in the center of the cylinder or sometimes as far below center as possible (to gain "throw distance" and thereby uniformity) and the inside of the cylinder is the surface that is coated. This configuration and also a cylinder rotating about a vertical axis as in Fig. 4.29b are sometimes used in sputtering applications. Here the surfaces to be coated face outward and the sputtering sources are outside the cylinder. This is practical for sputtering where the source to substrate distance is small and long uniform sources can be placed vertically. It clearly would not work for molten evaporating materials.

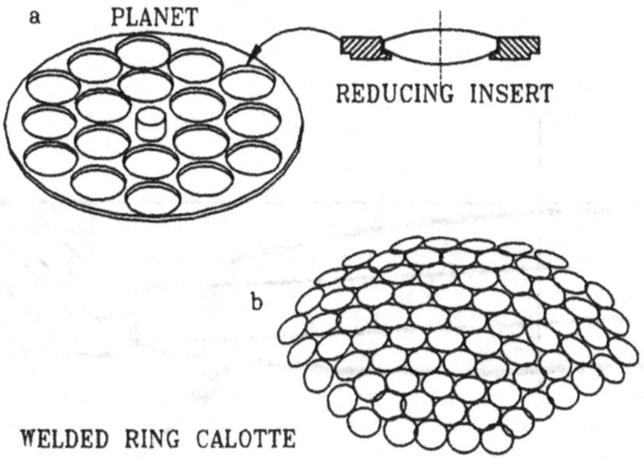

Fig. 4.30. More detail of a typical planet (a), and (b) an adaptation of the holder ring where rings are laid over a mandrel and welded or braised where they touch.

The most common configuration for optical coating of the past decade or two has been the planetary system in a box coater. Figure 4.30a shows more detail of a typical planet. They usually can be quickly mounted and dismounted from the chamber for the loading and unloading of parts. Various planets are made with appropriate sized holes with narrow ledges where the lenses are held by gravity. Smaller diameter lenses can be adapted to larger holes in the planet by reducing inserts as illustrated. The inserts do not allow as dense a packing of parts as when a specific planet is made for that diameter part. However, the economics of the trade-off in flexibility and tooling cost need to be examined for specific cases.

Figure 4.30b shows an interesting adaptation of the holder ring. When large quantities of repetitive parts of the same diameter are to be coated, rings like the insert are made with just enough extra diameter for the strength needed. The rings are then laid over a mandrel like the domed calotte and welded or braised where they touch each other. This can produce a domed calotte with maximum packing density, and it is easier to produce than by trying to machine into a dome. It is not inconceivable to use modern CNC programming and machining to do the same thing, however.

4.2.3.2. Protection from Back Coating

Some evaporated materials may deposit a thin unwanted "overspray" layer on the back of the parts being coated. SiO_2 and ZnS are examples. The ZnS at elevated temperatures has a low "sticking" coefficient. At above 150°C, ZnS acts as though

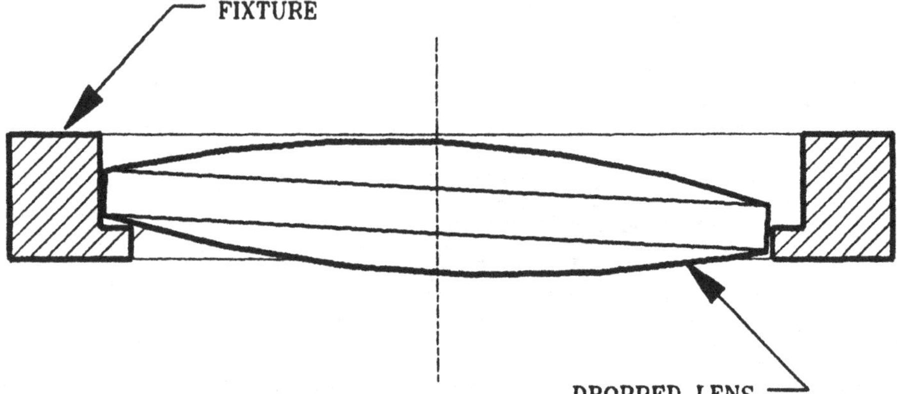

Fig. 4.31. A more common failure where the fixturing heats more rapidly than the substrate and allows the part to drop partially into the opening and then crack upon cooling.

it bounces off of the walls and may be found many places beyond the line of sight paths from the source. This contributes to the reputation of ZnS as a "dirty" material to work with. SiO_2 exhibits a similar behavior but to a much lesser degree. Temperatures and mean free path will have a major influence on these various processes. In many cases, it is desirable if not necessary to block the stray material from depositing on the backs of the parts. One approach is to cover the back side of a planet filled with parts with aluminum foil or a polymer sheet such as Kapton which will not outgas or deteriorate at the process temperatures. Kapton tape has been used to hold covers in place when needed. Another approach is to have a more permanent sheet metal cover which attaches on the back of a planet or calotte to block the unwanted coating. Fortunately, many materials and conditions do not require this "covering-your-backside", but one should be alert to the possible problem.

4.2.3.3. Fixture Temperature Considerations

We have seen many difficulties with damaged or destroyed parts due to differential expansion between the parts and the fixturing. The difference in the coefficient of thermal expansion of glass or other substrates and steel or aluminum fixtures can be large. The rate of heating and cooling of different materials under the influence of different processes and heating can be quite different also. If a process ran at 300°C for example, an aluminum (or steel) fixture might cool much more rapidly than the glass substrate. The lens could be "squeezed to death" even if there were enough clearance at high and low temperatures when both part and fixture were at the same temperature. A more common failure is illustrated in Fig. 4.31. Here the fixturing heats more rapidly than the substrate and allows the part to drop partially into the opening to be coated. When the combination cools back to room temperature, the part is broken by being trapped at the edges in a space where it would not ordinarily fit.

4.2.3.4. Uniformity

The choices in fixturing as described in the previous section are partially dictated by the uniformity requirements of the products. A calotte or drum (cylinder) can hold more parts than a planetary fixturing system, but will not provide as uniform a coating. Tilted planets can be adjusted for more uniformity than untilted planets. We should interject here that the adjustments of planet tilt and height can be critical. We have found that a few millimeters of mismatch in the height of planets or their tilt can make a surprising difference in uniformity from planet to planet. The planet scheme in the Balzers BAK760's that we have used is quite adjustable, but must be done very carefully. If the adjustable fixturing is removed from the chamber for cleaning, a readjustment is usually required for best uniformity. The non-tiltable, non-adjustable planet systems that we have worked with do not have this characteristic, but they also do not have the versatility. This probably falls in the category that: "if it is variable, it will vary" of "if it is adjustable, it will need to be adjusted."

Source Positioning

The positioning of the sources and the angular distribution of the evaporating materials has a major influence on uniformity. Musset[21,22], Stevenson and Sadkhin[91], and others have reported on techniques to measure the distributions from sources and decide where best to position the sources. Musset[21] mentions the lack of reproducibility of the distribution of SiO_2 from an E-gun as we discussed in Sec. 4.2.2.3, and we will show how to solve in the process-related chapter below. The angular distributions will vary from material to material and with how the materials are evaporated. For example, the distribution of TiO_2 from an E-gun will vary with the sweep pattern, rotation speed, background pressure, etc.

On his pages 70 to 73, Hill[4] points out that the physical reality of E-gun source distribution can be considerably different from the theoretical ideal. One cause is the high pressure just over the crucible when the evaporation rate is high. There is not molecular flow in this region. The real system behaves more like a virtual source at some distance above the crucible, and it has a broader material distribution more like a ball than the cosine (Lambertian) distribution that might be predicted from the diameter of a flat molten pool of material.

Note that the chamber pressure becomes very important to uniformity in reactive processes where the O_2 partial pressure may be at 0.5 to 2 x 10^{-4} torr. The mean free path can be less than the distance from the source to the substrates at these pressures. Therefore, a significant number of depositing atoms have undergone some gas scattering before reaching the substrates. This changes the uniformity as a function of pressure. In work where very close uniformity is required, there can even be concern as to the reproducibility of the pressure gauge and control readings.

We believe that the uniformity calculations are approximate at best, even if they are based on measured distributions from sources. The source positioning should be done on the basis of available information at the time of chamber layout and the uniformity then corrected by masks as necessary. If the source positions can easily be movable, source position can be adjusted for best uniformity before resorting to masks. This is may be more easily done with resistance sources than with E-guns that have rotational feed-throughs in the chamber's baseplate. Sometimes the rotary drive for the E-gun is connected by a chain drive or could have adjustable multiple idler gears so that the position can be adjusted.

Masking

Masks can be used to reduce the amount of material which arrives at selected regions of the substrates. Figure 4.32 illustrates a plan view of a chamber with a calotte (flat or domed) with leaf-shaped masks. The sources are near the baseplate

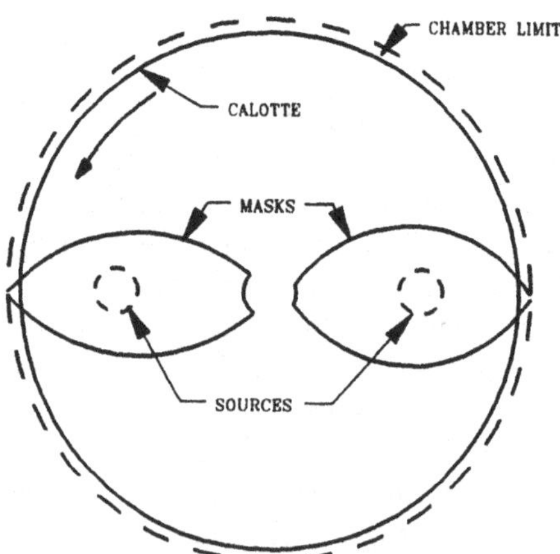

Fig. 4.32. Plan view of a chamber with a calotte (flat or domed) and leaf-shaped uniformity masks.

of the chamber, while the calotte is near the top of the chamber. The masks are usually best as close to the substrates as is safe enough to avoid collisions of the substrates with the masks. The variations with chamber pressure will increase with increasing distance from mask to substrate. Because the sources do not often have identical material distribution patterns, the masks over each source are not identical. There is some effect of both masks on the uniformity from one source, but the mask farthest from the source usually has a much smaller effect. Placing the mask in the position directly over its source has the most effect for the smallest mask because of its proximity to the source. We have heard of the use of a single mask midway around the circle between the sources. This could be trimmed by adjusting the edge nearest the source of concern.

There are books, papers, and software which describe how to calculate the shape that a mask "should have." However, our experience is that such calculations can be a waste of time. The reason for this is that the effects discussed above and of material "bounce" from the chamber walls and fixtures cannot be well accounted for and will change with the state of those surfaces. We have found major changes in the required mask shape from a freshly foiled wall, to a foil after some layers of coating, and to walls with no foil. The effects may also be from radiant heat transfer differences and not as much a case of whether the atoms "bounce" from the walls.

Our approach to developing a mask shape is empirical. First, witness substrates are located so as to represent all of the area to be made uniform. It is usually sufficient to cover a radius from the center to the edge of a planet or calotte. However, we have seen significant differences along a diameter of a flat calotte of 1 meter diameter which had a warp of 1 cm from one end of a diameter to the other. This much difference from the mask to the substrates is the major factor, but the difference in distance to the source may be significant also.

Second, a measurable thickness of material is deposited with no masks in place. When testing a high index material like TiO_2 on a glass substrate, we usually use about 4 QWOTs in the visible. This gives a spectral curve from 400-800 nm which has several peaks or crossings of a given % transmittance which can be measured and compared amongst the witnesses for uniformity. When testing a low index like SiO_2 on glass, the change in reflectance or transmittance with wavelength is small. Our practice has been to apply a QWOT of high index like TiO_2 first and then 4 QWOTs of the low index material. This generates a high contrast spectrum where the first QWOT has little influence on the results even if it is not perfectly uniform.

Third, insert a first approximation mask and repeat the test. The approximation might even be as crude as a rectangle which is expected to be large enough to provide the material for the final mask. If the same thickness of material is deposited as without a mask, one can then make some calculations as

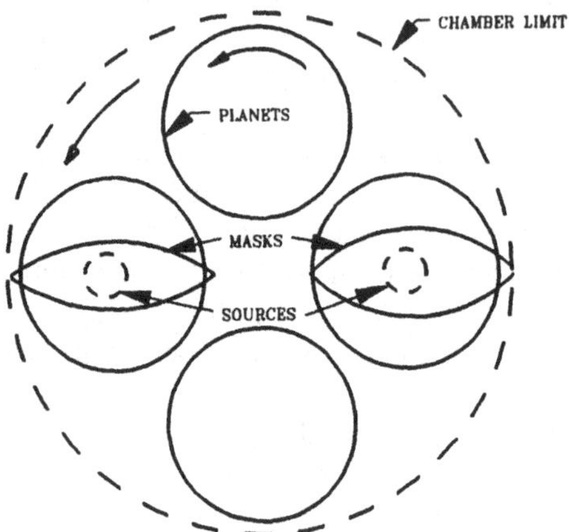

Fig. 4.33. Masks for planets which are fixed in space while the planets move over them.

to how much of the trial mask would need to be reduced at each distance from the calotte center. The favored source position without any mask (if practical) is such that the center of the calotte and the edge receive about the same thickness of material without a mask. Wherever the unmasked material is the thinnest is where the final mask width will approach zero. Everywhere else the mask will be wider to reduce the thickness of the deposit to match the thinnest radial zone on the calotte. If the material were deposited from the source uniformly over the whole calotte (which it is not), the change between the width of the trial mask at a given radius from the center of the calotte and the final mask could be worked out proportionally from the trial mask and no-mask tests. However, much more material is deposited in the vicinity of the mask than elsewhere. As a first approximation, we assume the change is linear from no-mask to the trail mask and adjust the next trial mask widths by zonal radius accordingly. This can require several iterations if the goal is better than 1% uniformity over the whole calotte. If masking for two or more materials is needed, it is well to adjust the masks alternately or in sequence so that errors due to mask interactions are minimized.

What has been said will also generally apply to masking for planets. This is illustrated in Fig. 4.33. In both cases, we have found that the masks need to be mounted in a way that the position is repeatable to within better than about 2 mm. These numbers are predicated on controlling the edge of a 500 nm filter to better than 5 nm or 1%. More stringent or more relaxed requirements will change the difficulty and tolerances proportionately.

The application of source position and masking to a domed calotte and drums or cylinders operates on the same principles and differs only in detail.

Ramsay, et al.[65] and Netterfield[66] have described work using rotation masks between the source and substrates somewhat like cooling fans. These systems have some flexibility by the adjustments of position provided but are also more complex than fixed masks. Netterfield and Ramsay[90] also report using such rotating masks to deposit films that corrected the flatness and optical path of Fabry-Perot spacers to better than $\lambda/400$.

Ion Sources and Uniformity

The reader should be cautioned that the application of ion-assisted deposition to a process can change the uniformity of optical and physical thickness. The ion beams have a distribution much the same as a material source. Fortunately, our experience and that reported by Gibson et al.[92] is that there seems to be a saturation effect in IAD, and excess ion density does not have much effect. The most weakly bombarded areas need enough IAD to get the desired results, and the most heavily bombarded areas will not be damaged. However, it does change the resulting thickness of the deposit and therefore must be taken into account. We tune the uniformity masks with all of the process as close to final values as practical including IAD.

We should also caution the reader that different substrate thicknesses and/or material types can change the effective uniformity. This is most likely due to differing temperatures having an effect on the net deposition rate and film density. Therefore, final mask adjustment should be based on tests with parts that are as nearly "real" as practical.

4.2.4. Temperature Control

It was discovered over the years, as we mentioned above, that many optical coating materials were more dense and durable if deposited at elevated temperatures. MgF_2, for example, has much better durability and much less shift with change in humidity (high packing density of the deposit) if deposited at 300°C rather than room temperature. Similar things can be said for many materials including TiO_2 and SiO_2. However, as we mentioned earlier, ZnS loses its "sticking coefficient" as the temperature gets up to 150°C. Germanium also has some adverse effects if deposited above 200°C as we shall discuss in the process chapter. In this section, we will discuss how heat is usually applied to or removed from a process and how it is sensed and controlled.

4.2.4.1. Heating

The expanded use of IAD may eventually eliminate the use of added heat in most optical coating processes, but we will discuss the heating methods that are still common. It is unusual for an optical coating chamber to be used at much over 300°C. Special considerations of insulators such as Teflon and bearing behavior are needed for temperatures above this range. Conduction and convection are not effective in a vacuum; therefore radiant heating is the normal way to heat substrates.

Probably the most common heat sources now are "quartz" heat lamps with fused silica cover classes to keep the coatings from the lamps. The filaments may operate at over 2000°K. The envelopes transmit well from the visible to about 5 μm, beyond that the heat is radiated from the hot envelope. With infrared transmitting substrates such as Ge, ZnSe, ZnS, etc., most of the heating probably comes from the visible spectrum absorption. Glass substrates absorb from about 2.5 μm out, so that most of the heating may be due to the 2.5 to 5 μm region which would be near the peak of 700 to 2000°K blackbody sources. The protective cover windows become coated and need to be cleaned and sometimes replaced due to breakage. It is sometimes practical to clean these flat window surfaces by scraping the coating off with a razor blade. Bead blasting has also been used successfully. There may be some change in the angular distribution of the heat due to the diffuse scattering of a bead blasted surface, but we have not noticed any obvious effects.

A one meter box coater might need about 10KW of heater power to reach 300°C in a reasonable time. The condition of the chamber walls will have a major effect on the time for heating and the ultimate temperature. An unfoiled stainless steel chamber wall shield that has been bead blasted for cleaning will absorb about 50% of the energy falling on it. If the shields are water-cooled, this heat will be removed. A chamber wall or shield which has been freshly covered with aluminum foil will reflect about 90% of the heat and only absorb 10%. The foil and/or shield may not make good thermal contact with the wall, and therefore not even all of the 10% is lost heat. As the foil becomes coated, its average reflectance will probably drop and the heat balance will change. This may be the source of some process changes noticed from a "dirty" chamber to one which has just been cleaned.

Nichrome wire heaters have been used in various forms. We have worked with several Leybold Heraeus A1000 chambers with heaters of nichrome wire tightly coiled on a ceramic core. They operate at a much lower temperature than a quartz lamp, a less than orange glow. These have been surprisingly durable. They can be bead blasted for cleaning and rewound with fresh nicrome wire when necessary. Such heaters are generally positioned like quartz lamps near the base plate of a box coater illuminating the substrates from below.

Another approach which we have observed but not used from Balzers is an array of rod and coil heaters of the general type used in domestic ovens, electric stove-top "burners," and electric hot water heaters. The array is placed above the calotte in a box coater and radiates downward on the back of the substrates in the calotte. The calotte would need to be open in the back to expose the substrates to the radiation from the heaters. It would seem that the back side heating is desirable in avoiding coating on the heaters, but also obviates its use with a planetary system.

4.2.4.2. Cooling

The only common source of cooling for box coaters is water cooling, other than cryopumps and Meissner traps. The chamber walls are usually water cooled for hot processes such as 300°C so that the work area is bearable. Crystal monitor heads must be kept cool, but more importantly at a constant temperature. Some coating facilities use a separate small water circuit at 100°F to feed the crystals water at a constant temperature. It is easy to thermostat the small reservoir at above room temperature and add a little heat to the water as needed to maintain the feed temperature. The E-guns and high current feeds for the resistance sources need water cooling. Some ion guns for IAD need water cooling.

The source of cooling water can be critical. If crystal cooling lines which typically have a small bore become clogged, the crystal readings become poor to useless and a batch of parts can be lost. We will touch further on this in the section on utilities.

One practice with the chamber cooling water is to turn it off at the end of a process when the chamber has cooled to some temperature appropriate for venting but not as low as ambient. When the vented chamber is opened it is still warm enough to minimize the condensation of water vapor inside the open chamber. Some facilities will provide a separate hot water circuit and valving to heat the chamber at venting. We have not found this hot water necessary even in the high humidity of Florida as long as the above procedure was used. In cases where the cooling water is well below room temperature, leaving the cooling water on with the chamber open can cause visible water condensation inside the chamber.

4.2.4.3. Temperature Sensors and Control

The almost universal temperature sensor used in a box coater is a thermocouple. The electronics used for converting its voltage to temperature and controlling the power to the chamber heaters is generally available as "off-the-shelf" equipment from suppliers such as Omega Engineering of Stamford, Connecticut. We will not belabor the details of the sensor and control, but touch on their application.

The sensor is usually placed close to the substrates since it is the substrate temperature that is most important to control. The sensor leads might come through a central hole on the axis of a calotte or planetary drive and have some height adjustability. They might come up from the base plate or through a side port of a box coater to the vicinity of the substrates. The reading at the junction point of a thermocouple is only an approximation of the substrate temperature at best since it is not imbedded in the surface of the substrate. It is a relative reading, but some effort is needed to have the readings as reproducible as practical because some processes are quite sensitive to temperature.

We have had a situation with an infrared coating using many germanium and ThF_4 layers. The process was running well until the chamber was cleaned. After cleaning the chamber, the next run of parts exhibited absorption. A week's effort of tests and checks was wasted before we discovered that the thermocouple had been mispositioned by about 10 cm from its normal distance to the center of the chamber. We knew that the Ge would show absorption losses if deposited at above about "190°C" as indicated on the thermocouple (in its normal position). This mispositioning caused more heat to be applied than normal to get the reading up to its usual setting, and the product therefore showed absorption. This experience convinced us of the need to locate the sensor reproducibly.

Some facilities will place a block of glass (about 5 to 10 cc) over the thermocouple with a hole drilled into its center for the junction. This adds a thermal mass to the temperature measurement somewhat simulating a lens in the fixturing. We have also had success without such a mass. It appears that the sensor comes to equilibrium with the radiant heat of the surrounding chamber and the residual gas molecules that impinge on it. Figure 4.2 implies that more than 10^{14} molecules are incident on a square centimeter in one second at any pressure above 10^{-6} torr.

We have seen pyrometers used to sense the temperature of the substrates through infrared windows in a chamber, but even this is not an absolute measure of the temperature. Temperature sensitive paints and other materials can be used to calibrate the substrate-to-thermocouple tooling factor. Small high limit recording mechanical dial thermometers are sometimes placed in the fixturing for the parts to do this same type of calibration. Within a single process chamber, we believe that the critical element is reproducibility of temperature, it does not matter what the absolute value is. The process temperature would have been adjusted for the results needed from that chamber. However, when a process is to be transferred to another chamber, the absolute values would be more helpful. Because obtaining the absolute values is not often practical, it is necessary to test for optimum temperature and upper and lower limits of the process in the new chamber.

Temperature controllers are now typically of the "PID" type. Here the power applied to the heaters can be proportional to the error between the desired set-point and the actual temperature. The gain and damping or phase lag of the control can be set so that power to the heaters does not oscillate on the one hand or act too sluggishly on the other. Usually the desired settings give a slight overshoot of the temperature which settles to the set value without oscillation. This is "critically damped" and reaches the set-point value to within some tolerance in the minimum time.

When E-guns, resistance sources, and some ion sources are operated at high power levels, most of the power appears in the chamber as heat and less power is needed from the heaters to maintain a given temperature. We have had processes at over 200°C that became overheated by the depositions using 6 to10 KW of E-gun power, and the process had to wait for the chamber to cool down to proceed at the set-point temperature. Controllers may allow an upper and lower limit to be set to the permissible temperature band and inhibit the process until those limits are satisfied. Excess heating due to sources is a common problem when plastics are to be coated. Heat removal may be desirable. In web or roll coaters, there is often a water-cooled drum behind the surface being coated to keep the plastic below a critical temperature. A Meissner trap should also help keep a chamber cooler, even though its principal function is additional water vapor pumping speed.

4.2.5. Process Control

As of 1995, the technology existed to control most of the processes used in the optical coating industry automatically. Some processes are restricted if the requirements are stringent. For example, the uniformity problem we mentioned with SiO_2 from an E-gun might defeat some performance requirements without manual attention. Almost any new coating chamber will have at least an automatic pumpdown sequence where the door is closed, a button is pushed and the pumping sequence proceeds until the chamber is at the specified pressure. Pumps are turned on and valves opened and closed as needed. This capability has been available for decades and avoids operator errors which could contaminate a chamber with pump oils.

It is not a big step to add a control to turn on the heat for a given set-point temperature as soon as the pressure is appropriate. Ion sources which require a set gas flow in SCCM can be automatically turned on when the pressure and temperature set-points are satisfied, and the voltages and currents applied as soon as the gas flow is established. Once the ion source is turned on and given a specified time to stabilize, the gas backfill can be turned on to bring the process pressure to an automatically controlled level, typically on the order of 1×10^{-4} torr of oxygen. The E-gun and resistance sources can then be automatically energized to a preparatory soak power level to warm up the materials before evaporation. After an appropriate soak time, the power to a source can automatically be ramped up to deposition levels and the shutter opened to deposit the first layer. The deposition can be continued until it is automatically terminated by closing the shutter when the programmed physical or optical thickness has been reached. Chapter 7 will deal in detail with thickness monitoring and layer termination. The next source is then ramped up automatically as the previous source is ramped down to the idle/hold power level. The second shutter is opened and closed automatically to deposit the specified thickness of the second material. The power ramping and shutter cycles are repeated according to the controller programming until the layer sequence is complete. The controller can shut down the heat and wait for a temperature set-point to be reached before automatically venting the chamber. The high vacuum valve is automatically closed at the appropriate point and the cryobaffle can be defrosted while the chamber is vented as we mentioned in Sec. 4.2.1.2. When the chamber door opens automatically after venting, the parts can be unloaded and the chamber reloaded for the process to begin again. This degree of automation exists now and has become economically viable and reasonably reliable. Establishing robust processes to take advantage of the automation is not a trivial task, but it has been done.

There are other controls which can be worked into the automation of process control. One example would be a Residual Gas Analyzer (RGA). An RGA can measure the partial pressure of specific gases in the chamber. It could be integrated into the automation to keep the partial pressure of oxygen at a fixed level in the chamber during a layer deposition. The example above controlled the O_2 make-up gas to keep the *total* pressure constant. If there were significant other gases present in the chamber which varied in a less predictable or controllable way, the RGA might be an improvement. Usually, the partial pressure of other gases can be kept low enough that the total pressure control is adequate.

There are a variety of system architectures for box coater automation in the field today. One approach has been for a single computer to poll all of the sensors for status, decide what each control should do next, and issue commands to all of the controls. Another approach is to have several subordinate controllers or "PLCs" to which the central computer delegates more involved responsibilities. The single computer is like a one man company, it is limited in the capacity of work which it can handle and the frequency with which it can attend to each required function. The delegated, multiple PLC system can expand to handle larger tasks like a company of many employees who receive delegated tasks from their coordinator. The current availability of less expensive computing power and PLCs has made the delegated approach most practical now. A given PLC will have full responsibility for a task such as optical monitoring for thickness control.

It will accept a thickness command, monitor the thickness, and return to interrupt the higher level controller when it is time to close the shutter and terminate the deposition of the layer. The PLC would also report any unusual occurrences or error as exceptions as soon as they happen.

Some modules of even an unautomated box coater system may have control functions like a PLC. A gas pressure controller as used for the oxygen background will admit O_2 to the chamber until it has reached the set-point pressure. It will control the valve with a PID loop to maintain that pressure even if other factors disturb the pressure. A gas flow controller will maintain a flow rate and a temperature controller maintains a set-point in the same way. Crystal monitor systems have built-in computers and offer extensive control capabilities. They typically will have outputs to open and close deposition shutters. Therefore, the single computer systems are actually a limited form of the delegating PLC type system because the various modules take some delegation of responsibility. The main issue is whether the single computer can give the attention as frequently as it is needed to all of the remaining details without help from one or more other computers (PLCs). It may be that the newer high-speed computers will change the balance again toward a "single" computer.

The major electronic challenge in automated control systems and even some of the modules is to make the systems immune to electrical noise spikes. E-guns sometimes "arc" which creates strong electromagnetic pulses. This is essentially lightning-in-a-box! Some ion sources can have similar arcing, but worse than that is the fact that ions in the chamber can promote E-gun arcing. The high voltage leads to the E-gun must be well shielded to ground or they will attract positive ions to their very high negative potential (6 to 10 KV). The success of an automated system is highly dependent on how insensitive it can be made to these noise sources. The PLC manufacturers have had to deal with this problem. Smaller noise effects may come from motors being turned on and off.

Another set of problems to deal with are how to save a process when something fails in a system such as a crystal, high vacuum gauge, or the *electrical power* to the facility. Crystal controllers can have two or more crystals on one monitor and some logic to determine when a crystal has failed or is about to fail. We have not always been satisfied with this capability however, as we will discuss further in the process chapter. The typical pressure indicator for high vacuum is the ionization gauge (IG). These have a filament which will eventually burn out. They burn out more frequently when the process includes oxygen partial pressure. The IG tubes usually have a spare filament (until one burns out). An automated chamber control system should be able to sense the IG failure and halt the process such that the spare filament can be connected and the process continued. If the spare filament has also failed, the chamber would have to be vented to replace the IG unless it can be valved off and replaced or there is a whole spare IG also on the chamber.

In the Florida environment, there are frequent electrical storms in some seasons where the power may go off for an instant or sometimes for minutes. Any reasonable system will have been configured so that the valving for the pumps and chamber go to a fail-safe condition when power is lost. It is more challenging, however, to design the process control system to avoid losing a coating run when this happens, and to make the restart procedure such that it "doesn't take a rocket scientist" to do it. It is important to have at least the control computer powered by an uninterruptable power supply (UPS) so that no data or status is lost at a power failure. We made the demonstration of this restart capability an acceptance requirement for the last two systems that we purchased.

Process control and automated coating systems have matured significantly in the past decade. The technical advances and cost reductions in computer and PLC hardware and in software have made a great improvement in what can be done for a given investment. We believe the critical task at this time is to develop to best system architecture and choose the right hardware and software for the control system. The application specific software should be much easier to develop and therefore less expensive than in the past. It is probably an unusual new system that would not have long-term economic benefits by having fairly automated process control.

4.3. TYPICAL EQUIPMENT

We will define typical equipment in this case as what we would purchase from an equipment manufacturer to use in the production of visible and infrared optical coatings for commercial and military optical instrument applications. Other requirements would be the flexibility to quickly change tooling and to coat a wide mix of parts and sizes up to about 450 mm diameter in planets. For the production of a more limited range of specialized parts, some items might be eliminated or others might have to be added. Figure 4.34 illustrates a typical system.

The chamber would be large enough to accommodate a 1200 mm (48") diameter calotte and have a four-planet drive interchangeable with it. Smaller chambers down to 760 mm might be considered, but the coating capacity increases with size more rapidly than the equipment or operation costs. Unless there is never expected to be a need for the larger system or other considerations such as space constraints prevail, a 1000 to 1200 mm system seems to be the preferred size.

The chamber would have one door in the front which actually breaks at 1/4 to 1/3 of the distance from the front to the rear of the chamber as shown in Fig. 4.35. This is to allow easier access to the back of the chamber without having to

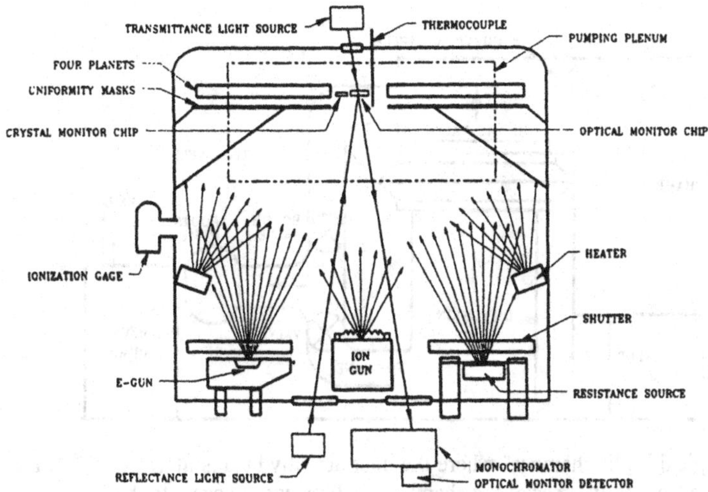

Fig. 4.34. A typical optical coating system with the components which are commonly used.

lean into it as far. The door section would usually have no hardware mounted in it, but it would be water-cooled as would the walls and top of the chamber. This configuration is sometimes referred to as a "split chamber." The door should have one or more viewport windows positioned for user convenience and to allow as much as practical of the chamber and process to be examined while under vacuum. It is particularly important to be able to see all of the sources. The windows should have manually operated shutters which protect the windows from coating buildup and replaceable glass inserts inside the vacuum window to catch the coating when the shutters are open. Chambers are sometimes built with a rear door, and the front door jam is flush with and sealed to a clean-room wall. This allows the chamber to be cleaned from outside of the clean-room to avoid contamination of the room. This configuration is more common in the semiconductor industry, and rarely found in the optical industry.

Before the accoutrements are added to the chamber, it is basically a nearly cubic box with or without rounded edges and corners with a pumping system as illustrated in Fig. 4.35. The plenum for the pumping stack will typically be at the back and as high in the chamber as possible. A chevron baffle prevents coating material from getting to the high vacuum valve. The plenum is in this position because of the height of a 500 mm diameter DP which is appropriate to a 1200 mm box. The pump also has a cryobaffle on top of it and under the high vacuum plate valve. When 100 to 200 mm is left below the DP for servicing, it is difficult to get the level of the plate valve below about 1700 mm. Many chambers of this capacity have been built which are so high as to require a hole in the floor for the

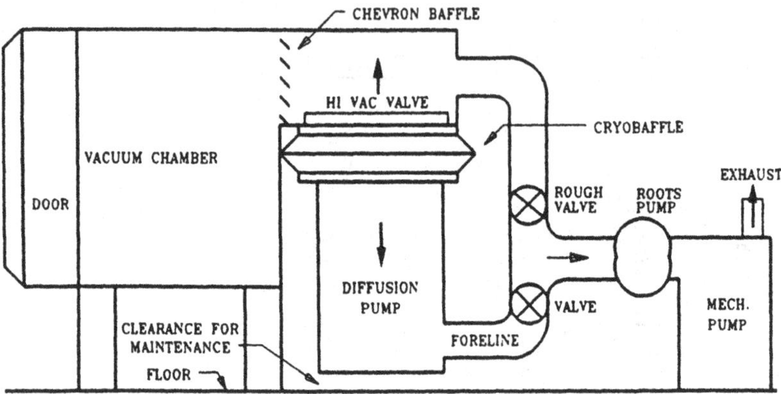

Fig. 4.35. A typical "split chamber" where the door actually breaks at 1/4 to 1/3 of the distance from the front to the rear of the chamber to allow easier access to the back.

DP or a false floor to comfortably reach the working parts of the chamber. Our experience is that there is no need for a chamber of this capacity to be that high. Although some percentage of production chambers may have cryopumps instead of DPs, a careful study of the intended temperatures and materials of production should be made before choosing that alternative.

The pressure measurement equipment will include a high pressure gauge such as a thermocouple in both the chamber and in the foreline. An ionization gauge will be in the chamber. A gas pressure control will allow reactive evaporation with oxygen as discussed in Sec. 4.2.5. The ion source will include a gas flow controller to maintain the required gas flow rate in SCCM through the ion source. A pressure safety interlock is usually included to prevent high voltage from being applied to the E-gun or the ion source until an appropriate vacuum is achieved.

The heat is typically supplied by several banks of quartz lamp heaters to provide about 10 KW when needed. The lamps are controlled by a PID loop system.

In past decades, glow discharge electrodes and power supplies were part of a normal system and used for precleaning. Ion sources do this cleaning task better and have obviated the need for the glow discharge equipment. A broad beam ion source is rapidly becoming common in the optical coating chamber for substrate cleaning and even more for IAD.

A minimum of two resistance sources is recommended. Since one may be soaking one material while evaporating another, independent power and control for the sources are common. Additional water cooled feedthroughs with switching for the power and control allows more sources in the chamber at a small additional cost. One E-gun is sometimes enough, but two is not uncommon. Although a single E-gun can have 3 or 4 pockets for different materials, there are some disadvantages as mentioned in Sec. 4.2.2.3. An "X and Y" sweep pattern generator is highly advisable if the gun pockets are not rotating, but these generators are not yet in common use. Pocket rotation with a radial sweep is typical. Each source needs a shutter. The space available on the baseplate becomes limited if many sources are installed. We have seen a single shutter used to alternately cover two adjacent sources to save space, but that can complicate some operations.

Almost all modern chambers will have at least one crystal control for rate and physical thickness. Dual crystal heads are becoming common and more than one vender offers a "six-shooter" crystal controller and head. It has been determined recently that almost all optical coaters (we know of only one exception) use optical thickness monitors to determine when to cut a layer for a given optical thickness. As we discuss in the monitoring chapter, crystal thicknesses are an approximation to the desired optical thickness, but are difficult to make reliable enough for most optical requirements beyond the simple coatings. Figure 4.34 shows a typical optical monitor scheme where a monitor chip can be viewed in either reflected or transmitted light. The light is then passed through a monochromator and/or a filter to the detector. The detector signal is then processed to maximize the signal to noise ratio and used by the control system to determine the optical thickness. The processing electronics is usually some form of lock-in amplifier to reduce the effective noise bandwidth to the order of 1 Hz. If the optical cuts are performed by an operator, the signal is fed to a chart recorder for his use. Even if the cuts are automatic, we prefer a chart recorder record for performance supervision and troubleshooting. We do not have any hard data, but we believe that much less than half of the *automated* optical coating processes/systems use optical monitoring at this time, but much more than half of optical coatings of more than 4 layers are done with optical monitoring. Eyeglass coatings may be adequate with only careful crystal monitoring.

We discussed control systems in the previous section. As of 1995, a typical "fully automated" box coater of the type described with optical and crystal monitoring as cut system options, two E-guns, an ion gun, and generally as in Fig. 4.34 can be purchased for about $500,000 (US). In past decades, there have been many organizations who tended to purchase "vanilla" box coaters and install many of the accoutrements themselves. This put such groups as much in the equipment-building business as in the coating business. It seems to be more practical in the recent decade to rely on the equipment manufacturer to integrate all of the necessary components and subsystems. This allows the optical coater to concentrate on processes and production. His customers only pay for the results, not his equipment-building expertise. One may argue that certain unique or proprietary processes require the user to also be the builder, and that his equipment expertise is reflected in the results. The development time required to make automated systems for one facility is hard to justify as compared to spreading it over a product line of coaters for the equipment manufacturer. In the 1970s when Dale Morton and his associates at Texas Instruments developed state-of-the-art automated coaters, it was probably justified because the equipment builders were not then very sophisticated in control systems. Today such an approach is much more questionable. However, we have also seen a problem on the other side of the coin. More than one European equipment manufacturer has taken an attitude that what they have is what you should want, and if it does not meet your requirements, you should change your requirements. They do not sell a lot of equipment to knowledgeable people on that basis except for very standard applications.

We had a painful experience in the past wherein we purchased several box coaters, and after the acceptance of the equipment, it took many months to get even simple processes into production in the new systems. The next time a system was purchased, the acceptance test was a four-layer AR process that we used extensively in production. The day after acceptance we were actually turning out production parts! We have seen system purchase specifications with 50 pages of detail on how the system should be built. We have also seen responses to requests for quotation which list hardware details almost down to the pedigree of the nuts and bolts. Both of these are contrary to our philosophy. We have satisfactorily purchased more than one $500,000 system with a two-page specification. The acceptance test was structured so as to demonstrate all of the performance which was required to meet the production goals of the equipment and be quickly adapted to the actual production process. The issue is the same as with management: delegate the task of *what* results need to be accomplished to someone who is competent to determine *how* to meet the goals and then inspect the results. The latter is because people tend to "do what you *inspect*, not necessarily what you *expect.*" Therefore, from our point of view, the performance is what we need. If we specify *how* it is to be achieved and it does not work, that is *our* responsibility, not the builder's. On the other side, we are not particularly concerned with the nuts and bolts of how the builder does it as long as the overall approach is viable and the builder is credible, which is ultimately proved at the acceptance tests.

4.4. ALTERNATIVE APPROACHES

When it comes to large area coating on flat surfaces for automobile windows and architectural applications, large in-line sputtering systems are preferred.[12-14] The coating of polymer sheets for window treatment is done on roll or web coaters. The major application for web coaters is in the food packaging industry where they are used to deposit SiO_2 and Al_2O_3. This makes the plastic film more impervious to oxygen and water vapor which keeps the food fresh longer in the package. However, some large capacity systems are currently depositing 6 or 7 layers for ARs, solar control, and other applications.

The MetaMode™ system described above was in a vertical axis drum configuration as in Fig. 4.29b. There is also increasing application of horizontal drum versions of the box coater as in Fig. 4.29a for some routine applications of AR coatings. However, the box coater is still the "typical equipment" for versatile optical coating at this time.

4.5. UTILITIES

The typical box coater has an array of utilities required for its operation. It requires at least electricity, cooling water, and compressed air. Liquid nitrogen or a PolyCold™ is usually needed. The compressed air is normally used to drive valves and no significant volume is needed. The power, cryocooling, and water, however, are major considerations in terms of both installation and operation.

A typical 500 mm diffusion pump will require about 15 KW of three-phase power. Chamber heaters may draw as much as 15 KW and E-guns another 15 KW. The resistance sources, mechanical pumps, calotte or planet drive, and ion gun may need another 5 KW. A PolyCold™ instead of LN_2 may draw yet another 5 KW. These total about 40 KW if eveything is running and over 20 KW even when the system is on but it is idle. At $ .10 per kilowatt hour, this would cost about $4.00 per hour under full operation and $2.00 per hour when idle.

There is often a debate as to whether a coater should be shut down overnight if there is only a one- or two-shift operation. In the 1990s, the idle cost was about $16.00 per shift to keep it ready to use. It takes about one hour to shut down and one hour to start up a chamber. If a chamber cost $500,000 and was amortized over 5 years, its capital cost would be about $100,000 per year or $50 per hour on a one-shift usage basis. If it were turned on and off every day, 2 hours or $100 worth of use would be lost to save $32 of idle power on a one shift operation or $16 for two-shift usage. For a weekend shutdown of a one-shift operation, the savings would be for 48+16 hours or $128. This two hour loss could be obviated if someone turned the equipment on before the production shift and off afterward without significant cost, because they were doing something else in the facility anyway. Careful planning may be in order for any given production operation. Our choice would be to try and have enough work for such a chamber to run it three shifts and only shut it down on weekends for maintenance if necessary. If one is running a single shift and has no security between times, it might be wiser to shut down overnight.

The biggest problems we have experienced in this area have been major water leaks developing when no one was around. We installed moisture sensors on the floor in the chamber area with an alarm to alert security. It is also important to have a floor drain to avoid flooding effects as much as possible.

Liquid nitrogen usage can be frustratingly high in such a chamber if it is brought to the system in 75 liter dewars. The logistics can become labor-intensive and expensive. If a large tank of LN_2 is plumbed directly to the chamber and filled from a tank truck, the logistics are not a problem. However, the vacuum jacketted pipe between the tank and the chamber costs about $100 per foot to install. Studies and our experience tend to show that a PolyCold™ refrigerator substitue for LN_2 is more satisfactory and less expensive on a life-cycle-cost basis. The LN_2 will allow a system to reach a somewhat lower ultimate pressure, but it is not necessary for most applications.

Cooling water can be a surprisingly large problem. At one time, we knew of two chambers running three shifts using city water for cooling and discharging it down the drain. This is not even legal in some areas now, but the cost turned out to be about $2000 per month! This is about $4 per hour. In Florida, the best source of cooling turned our to be deep wells to water at about 24°C (76°F). Wisnewski[23] described how this cooling system was designed and constructed with a heat exchanger and a closed loop treated water system circulating to four chambers and their supporting PolyCold cryobaffles. This system provided cooling to the chambers at or below 26°C (79°F) year round. Some lore is propagated by some equipment manufacturers that coolant at lower temperatures than this is required. We have operated box coaters with DPs from four different builders at these temperatures without noticable ill effects. Missimer,[24] Gordon,[25] and Zahniser[26] reviewed cooling towers which are the most common approach to obtaining water cooling in most parts of the USA. Any of these systems should bring the operational costs of cooling down to a small fraction of the electrical costs.

The installation of the necessary facilities and utilities is usually a significant effort and investment. The power distribution and wiring, and transformers to get the right match between local voltages and coater requirements are more than the usual production facility's needs. The cooling water system needs special attention in construction and maintenance. Certain biological and chemical masses seem to grow in closed loop cooling systems and clog the cooling lines. The first to get plugged are the smaller lines that feed the crystal monitors. If evaporation source or ion source cooling lines become clogged, damage can occur. The coolant usually flows through a variety of materials including several different metals. It appears that galvanic action can occur and cause some metals to erode and build deposits and scale. We have had an aluminum cooling passage corrode and leak due to such actions. The pH of the coolant appears to be important and certain inhibitors may be appropriate, but unfortunately we do not have any detailed advice on ideal and universal solutions.

We have found that the maintenance of four or five coating systems and support facilities can almost be a full-time job for a skilled maintenance technician. When one looks at all of the opportunities for something not to work correctly, it is surprising that the optical coating industry is as successful as it is.

4.6. REFERENCES

1. H. K. Pulker: *Coatings on Glass*, (Elsevier, New York, 1984) p.144.
2. A. Roth: *Vacuum Technology*, (North Holland, Amsterdam, 1976).
3. M Hablanian and K. Caldwell: "The Overload Conditions in High Vacuum Pumps," *Proc. Soc. Vac. Coaters Techcon* **34**, 253-258 (1991).
4. R. J. Hill: *Physical Vapor Deposition*, (Temescal, Berkeley, 1986) p. 10.
5. R. D. Mathis Co., 2840 Gundry Ave., Long Beach, CA 90806.
6. Balzers Group, 8 Sagamore Park Road, Hudson, NH 03051-4914.
4. L. Holland: *Vacuum Deposition of Thin Films*, (Chapman and Hall Ltd., London, 1966).
8. J. L. Vossen and W. Kern, eds.: *Thin Film Processes*, (Academic Press, Inc., Orlando, 1978).
9. R. F. Bunshah, ed.: *Deposition Technologies for Films and Coatings*, (Noyes Publications, Park Ridge, NJ, 1982).
10. R. A. Scholl: "A New Method of Handling Arcs and Reducing Particulates in DC Plasma Processing," *Proc. Soc. Vac. Coaters Techcon* **36**, 405-408 (1993).
11. R. A. Scholl: "Advances in Arc Handling in Reactive and Other Difficult Processes," *Proc. Soc. Vac. Coaters Techcon* **37**, 312-316 (1994).
12. J. J. Hofmann: "D C Reactive Sputtering Using a Rotating Cylindrical Magnetron," *Proc. Soc. Vac. Coaters Techcon* **32**, 297-300 (1989).
13. B. P. Barney: "3" C-MAG™ Sputter Deposition Source Development," *Proc. Soc. Vac. Coaters Techcon* **33**, 43-48 (1990).
14. M. W. McBride: "New Coaters Employing DC Sputtering of SiO_2 for the Production of Optical Components," *Proc. Soc. Vac. Coaters Techcon* **33**, 250-256 (1990).
15. D. A. Glocker: "AC Magnetron Sputtering," *1995 Technical Digest Series* **17**, 110-112 (Optical Society of America, Washington, 1995).
16. W. M. Posadowski and Z. J. Radzimski: "Sustained self-sputtering using a direct current magnetron source," *J. Vac. Sci. Technol. A* **11**, 2980-2984 (1993).

17. Z. J. Radzimski and W. M. Posadowski: "Self-Sputtering with DC Magnetron Source: Target Material Consideration," *Proc. Soc. Vac. Coaters Techcon* **37**, 389-394 (1994).
18. E. Finson and J. Felts: "Transparent SiO_2 Barrier Coatings: Conversion and Production Status," *Proc. Soc. Vac. Coaters Techcon* **37**, 139-143 (1994).
19. S. Menichella, C. Misiano, E. Simonetti, L. DeCarlo, and M. Carrabino: " PE-CVD Hardening and Matching Coating for Opthalmic Plastic Lenses," *Proc. Soc. Vac. Coaters Techcon* **37**, 37-40 (1994).
20. D. G. Shaw and M. C. Langlois: "A New High Speed Process for Vapor Depositing Acrylate Thin Films: An Update," *Proc. Soc. Vac. Coaters Techcon* **36**, 348-352 (1993).
21. A. Musset and I. C. Stevenson: "Obtaining Uniformly Thick Films in Coating Chambers," *Proc. Soc. Vac. Coaters Techcon* **31**, 203-209 (1988).
22. A. Musset: "Uniformity of Coating Thickness on the Insides of Rotating Cylinders," *Proc. Soc. Vac. Coaters Techcon* **33**, 243-245 (1990).
23. T. E. Wisnewski: "Low Cost Chiller System for Coating Equipment," *Proc. Soc. Vac. Coaters Techcon* **33**, 274-277 (1990).
24. D. J. Missimer: "Removal of Excess Heat in Vacuum Deposition Operations-An Overview," *Proc. Soc. Vac. Coaters Techcon* **33**, 269-273 (1990).
25. S. J. Gordon: "A Closed Loop Cooling System for Small Coating Plants," *Proc. Soc. Vac. Coaters Techcon* **33**, 278-283 (1990).
26. D. J. Zahniser: "Practical Consideration in the Design and Operation of Cooling Towers for Use with Vacuum Process Equipment," *Proc. Soc. Vac. Coaters Techcon* **33**, 284-286 (1990).
27. B. Window and N. Savvides: "Charged particle fluxes from planar magnetron sputtering sources," *J. Vac. Sci. Technol. A*, **4**(3), 196-202(1986).
28. B. Window and N. Savvides: "Unbalanced magnetron ion-assisted deposition and property modification of thin films," *J. Vac. Sci. Technol. A*, **4**(3), 504-508 (1986).
29. A. Zöller, R. Glötzelmann, H. Hagedorn, W. Klug, and K. Matl: "Plasma ion assisted deposition: a powerful technology for the production of optical coatings," *SPIE* **3133**, 196-203 (1997).
30. W. D. Sproul: "Unbalanced Magnetron Sputtering," *Proc. Soc. Vac. Coaters Techcon* **35**, 236-239 (1992).
31. W. D. Sproul: "Advances in Reactive Sputtering," *Proc. Soc. Vac. Coaters Techcon* **39**, 3-6 (1996).
32. T. G. Krug, S. Beisswenger, and R. Kukla: "High Rate Reactive Sputtering with a New Planar Magnetron," *Proc. Soc. Vac. Coaters Techcon* **34**, 183-189 (1991).
33. W. D. Münz: "Unbalanced Magnetrons: Their Impact on Modern PVD Hard Coating Equipment," *Proc. Soc. Vac. Coaters Techcon* **35**, 240-248 (1992).
34. W. D. Münz: "Production of PVD Hard Coating Using Multitarget UBM Coating Machines," *Proc. Soc. Vac. Coaters Techcon* **36**, 411-418 (1993).
35. W. D. Münz and I. J. Smith: "Wear Resistant PVD Coatings for High Temperature (950°) Applications," *Proc. Soc. Vac. Coaters Techcon* **42**, 350-356 (1999).
36. D. M. Mattox: "The Historical Development of Controlled Ion-assisted and Plasma-assisted PVD Processes," *Proc. Soc. Vac. Coaters Techcon* **40**, 109-118 (1997).

37. P. J. Martin, R. P. Netterfield, A. Bendavid, and T. J. Kinder: "Properties of Thin Films Produced by Filtered Arc Deposition," *Proc. Soc. Vac. Coaters Techcon* **36**, 375-378 (1993).
38. D. A. Baldwin and S. Falabella: "Deposition Processes Utilizing a New Filtered Cathodic Arc Source," *Proc. Soc. Vac. Coaters Techcon* **38**, 309-316 (1995).
39. M. L. Fulton: "Ion-Assisted Filtered Cathodic Arc Deposition (IFCAD) System for Volume Production of Thin-Film Coatings," *Proc. Soc. Vac. Coaters Techcon* **42**, 91-95 (1999).
40. H. K. Pulker: "Modern Optical Coating Technologies," *Proc. SPIE* **1019**, 138-147 (1988).
41. J. Edlinger, J. Ramm, H. K. Pulker: "Stability of the Spectral Characteristics of Ion Plated Interference Filters," *Proc. SPIE* **1019**, 179-183 (1988).
42. K. Gürtler, U. Jeschkowski, E. Conrath: "Experiences with the reactive low voltage ion plating in optical thin film production," *Proc. SPIE* **1019**, 184-188 (1988).
43. K. H. Guenther, C. W. Fellows, R. R. Willey: "Reactive Ion Plating– A Novel Deposition Technique for Improved Optical Coatings," *Proc. Soc. Vac. Coaters Techcon* **31**, 185-191 (1988).
44. M. Buehler, J. Edlinger, G. Emiliani, A. Piegari, and J. Ramm, H. K. Pulker: "All-oxide broadband antireflection coatings by reactive ion plating deposition," *Appl. Opt.* **27**, 3359-3361 (1988).
45. K. H. Guenther: "Recent progress in optical coating technology: low voltage ion plating deposition," *Proc. SPIE* **1270**, 211-221 (1990).
46. J. Edlinger and H. K. Pulker: "Ion currents and energies in Reactive Low-Voltage Ion Plating (RLVIP), preliminary results," *Proc. SPIE* **1323**, 19-28 (1990).
47. K. H. Guenther: "Recent advances in low voltage ion plating deposition," *Proc. SPIE* **1323**, 29-38 (1990).
48. M. Dubs and R. Hora: "Production of Stable Interference Filters by Low Voltage Reactive Ion Plating," *Proc. Soc. Vac. Coaters Techcon* **34**, 223-228 (1991).
49. K. H. Guenther, C. Lee, X-F. Han, K. Balasubramanian, and G. J. Jorgensen: "Weather Resistant Aluminum Mirrors with Enhanced UV Reflectance," *Proc. Soc. Vac. Coaters Techcon* **35**, 165-168 (1992).
50. A. J. Waldorf, J. A. Dobrowolski, B. T. Sullivan, and L. M. Plante: "Optical coatings deposited by reactive ion plating," *Appl. Opt.* **32**, 5583-5593 (1993).
51. M. Dubs, R. Hora, and H. Tafelmaier: "Reduced Set-Up Time and Improved Production Yield of Dielectric Filters with Reactive Ion Plating," *Proc. Soc. Vac. Coaters Techcon* **37**, 25-30 (1994).
52. A. Zöller, R. Götzelmann, K. Matl, and D. Cushing: "Temperature-stable bandpass filters deposited with plasma ion-assisted deposition," *Appl. Opt.* **35**, 5609-5612 (1996).
53. S. Beißwenger, R. Götzelmann, K. Matl, and A. Zöller: "Low Temperture Optical Coatings with High Packing Density Produced with Plasma Ion-Assisted Deposition." *Proc. Soc. Vac. Coaters Techcon* **37**, 21-24 (1994).
54. M. Fliedner, S. Beißwenger, R. Götzelmann, K. Matl, and A. Zöller: "Plasma Ion Assisted Coating of Ophthalmic Optics," *Proc. Soc. Vac. Coaters Techcon* **38**, 237-241 (1995).

55. A. Zöller, R. Götzelmann, K. Matl: "Plasma ion assisted deposition: investigation of film stress," *Proc. SPIE* **2776**, 207-211 (1996).
56. H. Uhlig, U. B. Schallenberg, N. Kaiser: "Shift-free narrow band filters for the UV-B region," *Proc. SPIE* **2776**, 342-352 (1996).
57. M. Friz, U. B. Schallenberg, and S. Laux: "Plasma Ion Assisted Deposition of Medium and High Refractive Index Thin Films," *Proc. Soc. Vac. Coaters Techcon* **40**, 280-291 (1997).
58. R. Götzelmann, A. Zöller, W. Klug, K. Matl, H. Hagedorn: "Plasma Ion Assisted Deposition: An Innovative Technology for High Quality Optical Coatings," *Proc. Soc. Vac. Coaters Techcon* **40**, 320-323 (1997).
59. J. J. Pan, F. Q. Zhou, and M. Zhou: "High Performance Filters for Dense Wavelength-Division-Multiplex Fiber Optic Communications," *Proc. Soc. Vac. Coaters Techcon* **41**, 217-219 (1998).
60. K. Balasubramanian, X. F. Han, K. H. Guenther: "Comparative study of titanium dioxide thin films produced by electron-beam evaporation and by reactive low-voltage ion plating," *Appl. Opt.* **32**, 5594-5600 (1993).
61. S. Zarrabian, C. Lee, K. H. Guenther: "Emission spectroscopy of reactive low-voltage ion plating for metal-oxide thin films," *Appl. Opt.* **32**, 5606-5611 (1993).
62. T. C. Kimble, M. D. Himel, K. H. Guenther: "Optical waveguide characterization of dielectric films deposited by reactive low-voltage ion plating," *Appl. Opt.* **32**, 5640-5644 (1993).
63. P. J. Martin, A. Bendavid, and H. Takikawa: "Ionized plasma vapor deposition and filtered arc deposition: processes, properties, and applications," *J. Vac. Sci. Technology A* **17**, 2351-2359 (1999).
64. D. J. Missimer: "Optimizing surface configuration for cryopumping in transition and viscous flow regimes," *Proc. Soc. Vac. Coaters Techcon* **39**, 13-18 (1996).
65. J. V. Ramsay, R. P. Netterfield and E. G. V. Mugridge: "Large-area uniform evaporated thin films," *Vacuum* **24(8)**, 337-340 (1974).
66. R. P. Netterfield: "Uniform evaporated coatings on rotating conical workholders," *J. Vac. Sci. Technology* **19**, 216-220 (1981).
67. J. C. Sellers: "The disappearing anode myth: strategies and solutions for reactive PVD from single magnetrons," *Surface and Coatings Technology* **94-95**, 184-188 (1997).
68. P. Sieck: "Effect of Anode Location in Deposition Profiles for Long Rotatable Magnetrons," *Proc. Soc. Vac. Coaters Techcon* **37**, 233-236 (1994).
69. S. Shiller, K. Goedicke, V. Kirchhoff and T. Kopte: "Pulsed Technology - a New Era of Magnetron Sputtering," *Proc. Soc. Vac. Coaters Techcon* **38**, 293-297 (1995).
70. V. Kirchhoff and T. Kopte: "High-Power Pulsed Magnetron Sputter Technology," *Proc. Soc. Vac. Coaters Techcon* **39**, 117-122 (1996).
71. V. Kirchhoff, T. Kopte and U. Hartung: "Antireflective Layers Produced by Pulsed Magnetron Sputter Technology," *Proc. Soc. Vac. Coaters Techcon* **39**, 242-247 (1996).
72. S. Shiller, V. Kirchhoff, K. Goedicke and P. Frach: "Advanced Possibilities for the Stationary Coating of Substrates by Means of Pulsed Magnetron Sputtering," *Proc. Soc. Vac. Coaters Techcon* **40**, 129-134 (1997).

73. S. Shiller, V. Kirchhoff, T. Kopte and M. Schulze: "Special Features of the Pulsed Magnetron Sputter Technology for Glass Coaters," *Proc. Soc. Vac. Coaters Techcon* **40**, 168-173 (1997).
74. S. Shiller, V. Kirchhoff , F. Milde, M. Fahland and N. Schiller: "Plused Plasma Activated Deposition of Plastic Films," *Proc. Soc. Vac. Coaters Techcon* **40**, 327-332 (1997).
75. G. Hoetzch, O. Zywitzki and H. Sahm: "Structure, Properties and Applications for PVD Al_2O_3 Layers - A Comparison of Deposition Technologies," *Proc. Soc. Vac. Coaters Techcon* **40**, 77-85 (1997).
76. J. Stumpfel, G. Beister, D. Schulze, M. Kammer and St. Rehn: "Reactive Dual Magnetron Sputtering of Oxides for Large Area Production of Optical Multilayers," *Proc. Soc. Vac. Coaters Techcon* **40**, 179-186 (1997).
77. T. Rettich and P. Wiedemuth: " New Application of Medium Frequency Sputtering for Large Area Coating," *Proc. Soc. Vac. Coaters Techcon* **41**, 182-186 (1998).
78. U. Heister, J. Krempel-Hesse, J. Szczyrbowski and G. Bräuer: " New Developments in the Field of MF-Sputtering with Dual Magnetron to Obtain Higher Productivity for Large Area Coatings," *Proc. Soc. Vac. Coaters Techcon* **41**, 187-192 (1998).
79. H. Schilling, J. Szczyrbowski, M. Ruske and W. Lenz: " New Layer System Family for Architectural Glass Based on Dual Twin-Magnetron Sputtered TiO^2," *Proc. Soc. Vac. Coaters Techcon* **41**, 165-173 (1998).
80. D. Beisenherz, G. Bräuer, J. Szczyrbowski and G. Teschner: "Mass Production of Anti-Reflex Film System," *Proc. Soc. Vac. Coaters Techcon* **41**, 266-269 (1998).
81. Y. Tachibana, H. Ohsaki, A. Hayashi, A. Mitsu and Y. Hayashi: "High Rate Sputter Deposition of TiO_2 from TiO_{2-x} Target," *Proc. Soc. Vac. Coaters Techcon* **42**, 286-289 (1999).
82. F. Milde, M. Dimer, P. Gantenbein, C. Hecht, D. Pavic, and D. Schulze: "Industrial Scale Manufacture of Solar Absorbent Multilayers by MF-Pulsed Plasma Technology," *Proc. Soc. Vac. Coaters Techcon* **42**, 163-168 (1999).
83. U. Heister, J. Bruch, T. Willms, C. Braatz, A. Kastner and G. Bräuer: "Recent Developments on Optical Coatings Sputtered by Dual Magnetron Using a Process Regulation System," *Proc. Soc. Vac. Coaters Techcon* **42**, 34-38 (1999).
84. R. Scholl, A. Belkind and Z. Zhao: "Anode Problems in Pulsed Power Reactive Sputtering of Dielectrics," *Proc. Soc. Vac. Coaters Techcon* **42**, 169-175 (1999).
85. D. Lishan, S. Onishi, D. Johnson, A. Lateef; "Stable High Rate Reactive Al_2O_3 Deposition," *Proc. Soc. Vac. Coaters Techcon* **43**, 321-326 (2000).
86. M. Fahland, V. Kirchoff, C. Charton and P. Karlsson: "Roll-to-Roll Deposition of Multilayer Optical Coatings onto Plastic Webs," *Proc. Soc. Vac. Coaters Techcon* **43**, 357-361 (2000).
87. J. W. Seeser, P. M. LeFebvre, B. P. Hichwa, J. P. Lehan, S. F. Rowlands, and T. H. Allen: "Metal-Mode Reactive Sputtering: A New Way to Make Thin Film Products," *Proc. Soc. Vac. Coaters Techcon* **35**, 229-235 (1992).
88. D. M. Mattox: "Ion Plating Fundamentals," *Vacuum Technology & Coating* **1(6)**, 22-28 (2000).
89. N. Boling, B. Wood and P. Morand: "A High Rate Reactive Sputter Process for Batch, In-line, or Roll Coaters," *Proc. Soc. Vac. Coaters Techcon* **38**, 286-289 (1995).

90. R. P. Netterfield and J. V. Ramsay: "Correction of Parabolic Errors in Fabry-Perot Interferometers," *Appl. Opt.* **13,** 2685-2688 (1974).
91. I. C. Stevenson and G. Sadkhin: "Optimum Location of the Evaporation Source: Experimental Verification," *Proc. Soc. Vac. Coaters Techcon* **22**, Paper O-12 (2001).
92. D. Gibson, J. Gardner, and G. Chester: "Automated Control for Production of High Precision Vacuum Deposited Optical Coatings," *Proc. Soc. Vac. Coaters Techcon* **43**, 17-22 (2000).

5

Materials and Process Know-How

5.1. INTRODUCTION

In the preceding chapters, we have discussed the functional and design concepts and techniques for optical thin film coatings and typical equipment which might be used in the realization of a physical coating on a substrate. We now address the additional factors of materials and processes needed to carry a design to its fruition. At the design stage, the index of refraction and dispersion are assumed values for the materials. This assumption is usually based on the best information available at the time of the design. The index includes the absorption (imaginary part of the index) which defines the spectral range over which the materials are usable. Guenther[266] provides a useful review of the physical and chemical aspects of coatings. The durability of the materials in various environments such as abrasion, humidity, salt fog, solvents, etc., should be considered in the choice of materials before even starting a design with those materials. Durability in the space environment may also be of concern and has been reported by Blue and Roberts[263]. The availability of the equipment and processes to satisfactorily deposit a given material is usually a significant consideration also. The thrust of this chapter, therefore, is to share as much as practical of our experience with various materials and processes and to refer the reader to additional information in the areas discussed.

5.1.0. Measuring Spectral Results in the Real World

When we leave the design world where almost anything is possible and enter the real world of a deposited optical film, we need ways to measure the reflectance and transmittance of our film as a function of wavelength. The usual way of doing this is with a spectrophotometer which covers the spectral region of interest. The first recording spectrophotometer developed by Cunningham and Hardy[162,163] was oriented toward color measurements of almost any type of sample, diffuse or specular, reflecting and/or transmitting. It functioned strictly in the visible spectrum, had an integrating sphere, and was photometrically very precise and accurate because of careful consideration given to these factors. The biggest market for spectrophotometers over the decades has become chemists, biologists, etc., and not the physics and optical coating industry. As a result, the instruments on the commercial market are not, for the most part, oriented toward the needs of optical coatings. High resolution and wavelength (or wavenumber) accuracy are more critical to chemists, etc., than the precise photometric values. On the other hand, we in the coating industry place our emphasis just the other way around. We are usually forced to make the best that we can of each situation. If a given spectrophotometer can be relied upon to be reproducible, then we can at least calibrate its measurements. There are occasional papers in the journals and sections in texts on the fine points of the measurement of optical coatings such as Arndt et al.[59], but seldom a review of "common" practice for the person new to the field. We will here attempt to help with that problem of "how to measure" before embarking on what to do with those measurements.

5.1.0.1. Spectrophotometers

An early spectrophotometer may have consisted of a light source, a monochromator set at a given wavelength, and a detector with output display (originally it was probably the eye and a notebook). One would note the display reading and then insert a sample of interest in the optical path to see how the reading changed. The ratio of the two would represent the transmittance of the sample at that one wavelength. This would represent a simple single beam (SB) spectrophotometer. The accuracy of the measurement would depend on several factors such as: did the source brightness or detector and display response change from the time of calibration with no sample until the time when the sample was measured; was the response of the detector and display linear with sample transmittance, etc. The Hardy spectrophotometer and most since that time have been double beam (DB) instruments to overcome drifts of source or detector systems. A DB system divides the beam from the monochromator into a reference and sample path which are alternately sampled by the same detector. This alternate sampling is usually at a rapid rate as compared to the change of wavelength in a scanning spectrophotometer. Any change in source or detector response affects both the reference and sample and is thereby canceled in effect. Most DB instruments further overcome the detector linearity issues by having an optical means or mechanical "comb" which reduces the reference signal until it matches the sample signal. At such a point, as the channels alternate (reference and sample), the modulation goes to zero. If they do not match exactly, the phase of the modulation indicates whether the reference needs to be increased or decreased to match the channels. The linearity of the system is then dependent on the linearity of the comb system or optical scheme. The Hardy spectrophotometer used polarization optics to accomplish this and was thus only dependent on the accurate control of the polarizer angle.

In the early 1960s, Fourier spectrometers (as reviewed by Strong[164]) became commercially available. All these except the one described by Willey[165] are basically SB instruments and subject to the drift and linearity problems. The advantage of these instruments over dispersive (prism or grating) instruments is the much higher signal to noise (S/N) that can be achieved, the high spectral resolution, and the speed at which a spectrum can be obtained. These factors are all related and can be traded off with one another. The basic advantage stems from the fact that the detector is looking at all of the spectrum for the whole measurement time. The signal is then integrated for that whole time along with the noise. This "multiplex" advantage in S/N over a conventional dispersive spectrum scanning instrument goes as the square root of the number of spectral elements resolved. Therefore, if 1024 spectral elements were to be resolved, the advantage would be 32 times in S/N. This has principally been beneficial in the infrared spectrum where systems tend to be S/N limited. Most commercial infrared instruments are now Fourier Transform IR (FTIR) spectrometers.

Because of the SB nature of the Fourier instruments, the linearity of the detector is usually the major limitation of photometric accuracy. The drift is not usually as great a problem if frequent calibrations are made. The signal levels on the detectors need to be kept low enough to be in a linear range for the detector. This is exacerbated by the fact that the Fourier transform of a spectrum has a large peak near "zero path difference" in the interferometer scan which is many orders of magnitude larger than the bulk of the signal that is developing the detailed resolution of the resulting spectrum. It is advisable to check the photometric reproducibility and accuracy of FTIRs by repeated 100% transmitted scans and samples of known transmittance. IR materials such as Ge, Si, ZnSe, ZnS, and NaCl can be used for transmittance level calibration. This is because their indices are well known and therefore the reflection losses and resulting transmittance in their non-absorbing regions are well known. Care should be taken that the reflected light is in fact "lost" and does not get back to the detector. This usually requires only an appropriate (small) tilt of the sample in the beam.

Willey[165] developed a true double beam Fourier spectrophotometer with an integrating sphere much like an IR equivalent to the original Hardy spectrophotometer. The S/N advantage of the Fourier system was needed to overcome the orders of magnitude signal loss of an integrating sphere. The sphere, on the other hand, made it practical to measure diffuse samples as easily as specular samples in both transmittance and reflectance. Many of the applications of this instrument were related to its ability to measure the diffuse IR reflectance and thereby emittance of any sample. The first model of the instrument was used to help develop the Space Shuttle's ceramic reentry tiles; later units played key roles in solar collector development and other thermal control coatings.

Different spectrophotometers have different f-number beams in their sample compartments and different minimum size points in their beams. Older instruments may have been f/10 while some newer ones may be f/4. This beam convergence/divergence may need to be taken into account where coatings are angle sensitive. For example, a 100 GHz DWDM filter will shift a significant portion of its bandwidth (0.4 nm) with a one degree tilt. This means that such a filter cannot be properly measured with an ordinary spectrophotometer because the beam divergence will distort the band shape significantly. A typical beam in a grating instrument is also a millimeter or more wide and about a centimeter high at its smallest point. Again, the DWDM filter of 1.5 mm square could not be measured properly in such a beam. A new family of instruments have appeared for the DWDM field such as Optical Spectrum Analyzers (OSAs). These are built around the tunable lasers, Gradient Index Lenses (GRINs), and fiber optics of the field that they serve. They usually measure in dB rather than %T. Although an OSA is very different in detail, in the final result it is a spectrophotometer.

Infrared instruments are subject to effects of the absorption of water vapor and CO_2 in the atmosphere. These are most noticeable at about 2.7, 3.2, and 6.2 μm for H_2O, and at 2.7, 4.3, and 15 μm for CO_2. If the atmosphere in the instrument changes from when the 100%T calibration was run to when the sample is run, there may be spectral artifacts at some of these wavelengths. If this is a problem, the instrument can be purged with dry nitrogen to eliminate the effects. Similar things can be said for regions of the ultraviolet (UV). The region from 4 to 200 nm in the UV is referred to as the "vacuum ultraviolet" because the atmosphere absorbs at those short wavelengths and instruments must be evacuated to be effective. We understand that a dry nitrogen purge might be usable for some cases in the UV.

5.1.0.2. Measuring Transmittance

The typical spectrophotometer is designed to measure transmittance. We will later discuss ways to measure reflectance, but this section deals with transmittance only. Figure 5.1a illustrates a typical sample beam in the sample compartment of a grating spectrophotometer as seen from above and Fig. 5.1b shows the beam as seen from the side. The slit of the monochromator is typically imaged in the sample compartment; often near the center as shown, but sometimes at one side of the compartment. The reference beam is usually identical and toward the back of the compartment. The slit image is much higher than it is wide. We will assume the light travels from left to right in these figures.

A normal practice would be to run a 100%T calibration with no sample in the beams and then a 0%T with an opaque sample in the sample beam while being careful not to interfere with the reference beam. The sample of interest is then placed in the sample beam and its transmittance spectrum scanned. Barring errors and instrumental non-linearity, the displayed result is the actual transmittance of the sample.

There are a few problems to be aware of and to avoid. Figure 5.1c shows what can happen if a sample has some wedge in it. The beam is deflected and all of the transmitted light may not reach the detector (or be vignetted). Different instruments have different sensitivities to wedge effects. It is useful to rotate the sample about the horizontal beam axis to see if that influences the readings; this

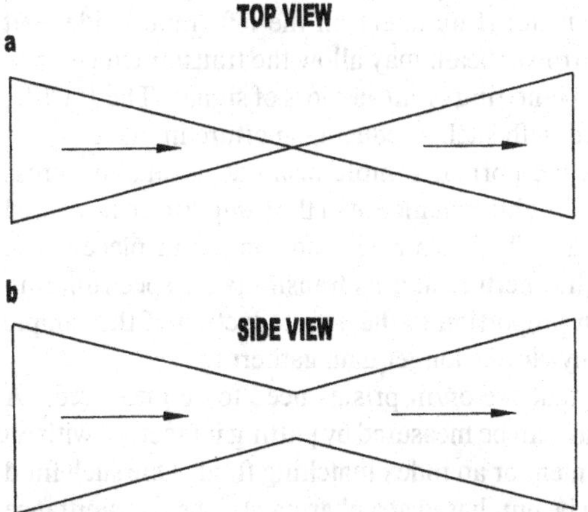

Fig. 5.1a,b. Typical sample beam in the sample compartment of a grating spectrophotometer as seen from above (a) and as seen from the side (b).

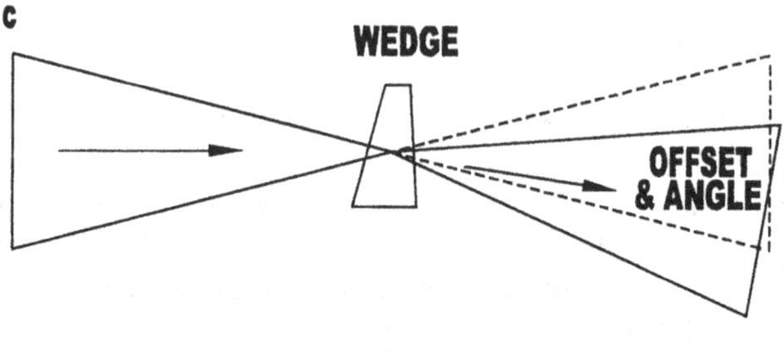

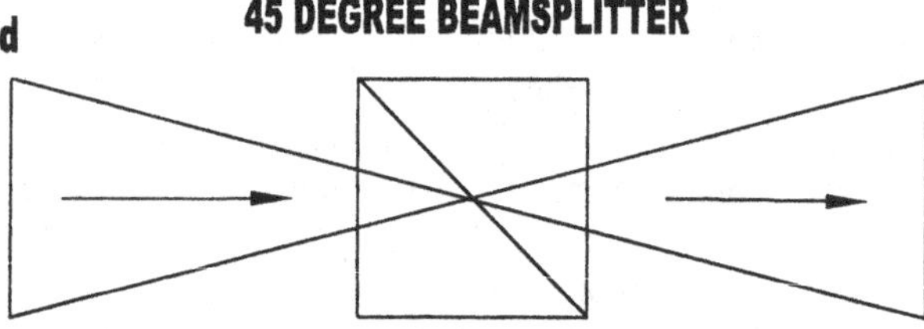

Fig. 5.1c,d. The effect of wedge in a sample (c), measuring a 45° beamsplitter (d).

may be an indication of wedge in the sample. Similar problems can occur with thick samples like plates and prisms. As seen in Fig. 5.1e, a tilted thick plate can offset the beam and cause some of it to not reach the detector. Figure 5.1f shows how a thick plate or prism also would cause a focus shift which might be a problem for some instruments. It may be possible to overcome some of these effects by reducing the beam diameter (f/number) on the left (input) side as it enters the compartment. This narrower beam may allow the transmitted beam to have some offset without causing vignetting or further loss of signal. The 100%T, of course, needs to be recalibrated with such a reducing aperture in place.

If a sample is smaller than the normal sample beam at its smallest cross section, one can insert an aperture slightly smaller than the sample to be measured at the sample position and run a 100%T scan with no sample in place. The sample can then be added over the aperture and its transmittance spectrum run. This will reduce the S/N ratio in proportion to the area reduction of the sample beam; but this can be overcome by slower/longer data gathering.

Coatings on beamsplitters that are on/in prisms need to be measured. A beamsplitter coated on a 45° prism can be measured by putting it together with an uncoated prism using uncured cement or an index matching fluid. One such fluid is "oil of wintergreen" which can be purchased at a pharmacy. The transmittance

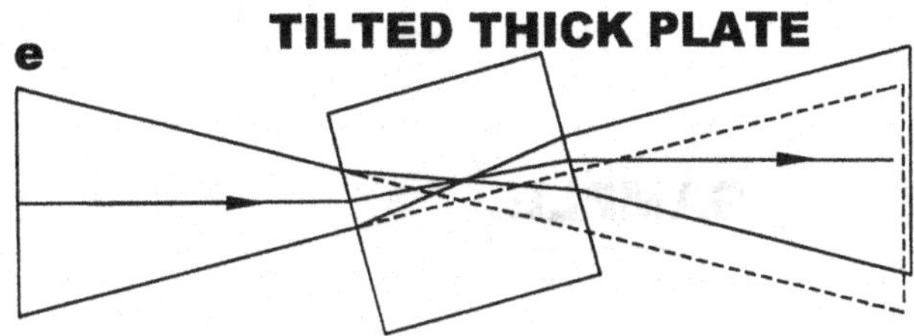

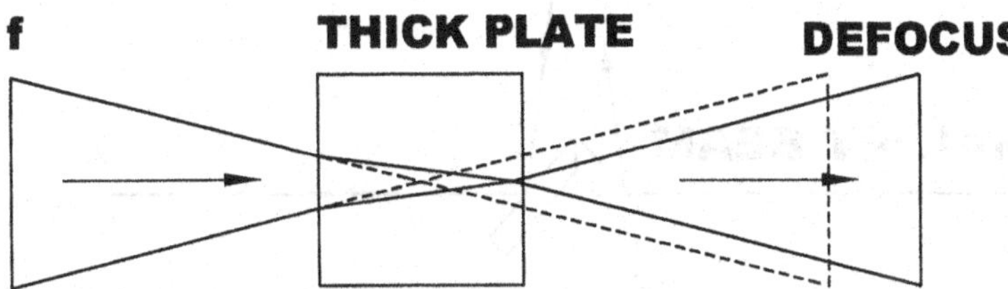

Fig. 5.1e,f. Beam offset (e) and beam defocus (f) caused by a thick sample plate.

of the assembly can then be measured as in Fig. 5.1d. The above-mentioned effects of offset and defocus need to be considered. Another problem is polarization. All spectrophotometers have some complex polarization effects in the beams. From earlier discussions, we know that coatings at anything but near normal incidence will have polarization effects. One practice is to measure the assembly once and then rotate it 90° about the horizontal beam axis and measure it again. The average of the two scans should represent the effect of random polarization on the coating. The transmittance of beamsplitters in parallel sided prisms at other than 45° can also be measured this way.

It is also possible to insert linear polarizers in ***both*** sample and reference beams (typically on the left where the beams enter the compartment) to only pass the desired polarization through the sample. There are also depolarizers which are crystalline in nature and change the state of polarization rapidly across the aperture of the beam so as to homogenize the polarization of the beams.

5.1.0.3. Measuring Reflectance

Since most spectrophotometers have been made for transmittance measurements, it is necessary to use attachments in the sample beam to accommodate the desired reflectance measurements. In some cases, the instrument manufacturers have

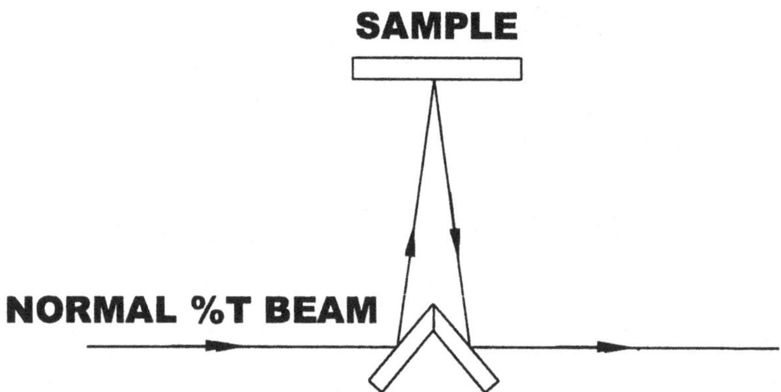

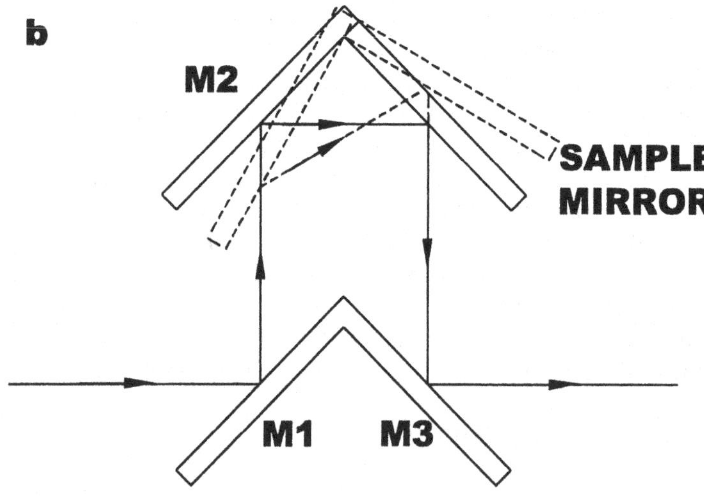

Fig. 5.2a,b. A near-normal reflectance attachment (a) and a variable angle reflection attachment (b).

attachments available; and in other cases there are a wide variety of attachments available from companies like Harrick[166] and Labsphere[167].

The simplest and first essential reflection attachment is illustrated in Fig. 5.2a. The normally horizontal transmittance beam is reflected upward by a mirror to the sample, and the beam reflected from the sample is reflected again into the optical path by another mirror. The angle of incidence on the sample mirror should be as small as practical to avoid the changes in reflectance due to angle; 8° to 12° is typical and satisfactory. This is probably the best fixture for measuring ARs and other low reflecting coatings.

There are several alternatives to calibrate this attachment. First, a 0%R line can be run with no sample in position. Then, a thin glass slide of known index can be placed in the sample position and the spectrum measured. If the glass is of index about 1.5 with no absorption in the region of interest, then the reflection should be about 8%. All subsequent samples can be compared with this calibration of the 8% reflectance. The glass slide can be verified by measuring its %T (which should be about 92%) if the absorption can be neglected. Another alternative is the have a sample of known index with a significant wedge between the bottom and top surfaces. The back surface should then be at such an angle that none of the light reflected from it gets back into the sample beam that reaches the detector. This wedge would then reflect about 4% depending on the exact index of the glass. A variant of this wedge technique is to have a coating sample on a thin plane parallel glass witness piece which is then oiled to a wedge with index matching fluid to create the same effect as the wedge alone. In these cases for calibration and coating measurement, it is only the first surface reflection that is measured. Yet another approach is to coarse grind the back side of a sample so that the light reflected from the back is scattered widely, and very little of the light from the back gets back into the beam that reaches the detector. The suitability of this approach will be more dependent on the f/number of the instrument. And finally, another scheme is to paint the back side of the flat witness with what amounts to an index matching fluid filled with carbon black (black paint). In this case, all the light reaching the back surface enters the fluid or black laquer (such as black Krylon™ spray paint) and is absorbed by the carbon black so that only the front surface reflection is measured.

This simple reflectance attachment can also be used for high reflectors if a standard of known high reflectance can be used to calibrate it. However, a "VW" attachment has advantages for high reflectors and can be self calibrating to give "absolute" reflectance values. This system is attributed to John Strong[168] and Nilsson[179] describes its use. Zwinkels et al.[178] describe its application in some detail and point out possible "pitfalls". Figure 5.2c and d show how it works. A 100%R calibration spectrum is run with the reference mirror in the position shown in Fig. 5.2c. A 0%R calibration is run with the reference mirror removed. The

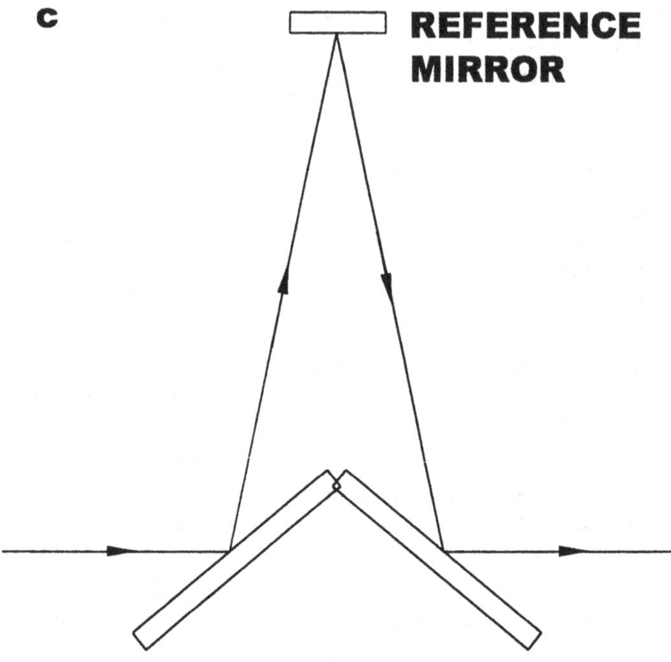

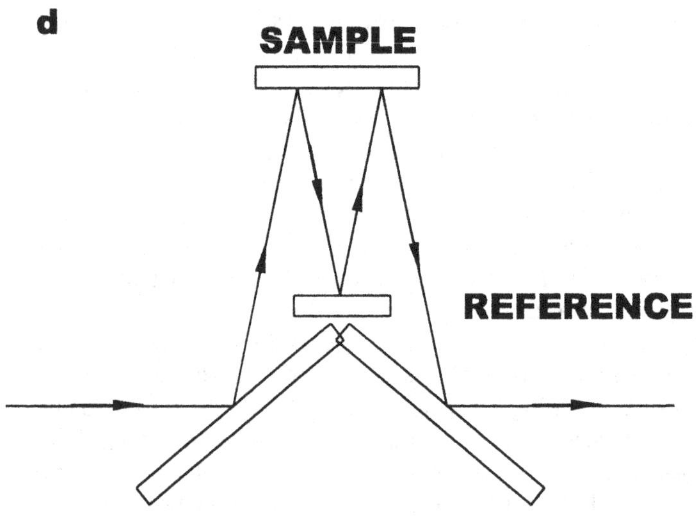

Fig. 5.2c,d. The "V-W" reflection attachment in the V configuration (c) and the W configuration (d).

reference mirror is then positioned as shown in Fig. 5.2d, and the sample mirror positioned as shown. The sample thus reflects the beam two times and all of the other reflections are the same between c and d. The resulting spectrum is then the square of the %R. This technique is very sensitive for high reflectors. The absoluteness of the measurement depends on how well the details of alignment, etc., are executed. If one analyzes the effects of errors of this system versus the simple reflectance attachment, it become apparent that the VW is not good for AR coatings but is good for high reflectors. The reverse is true for the simple reflectance attachment.

A variant of the VW has appeared which is called the "VN" attachment. Figure 5.2e adapted from Boudet et al.[177] illustrates one such scheme. The general concept is similar to the VW, but the beam reflects only once from the sample mirror and gives %R directly as a result. Here, both the M2 and M3 mirrors must be repositioned after the 100% calibration. The objections that we have heard from users of some of these attachments is that the alignment is "touchy" and not very stable, and further verification with a known sample is advisable. One former

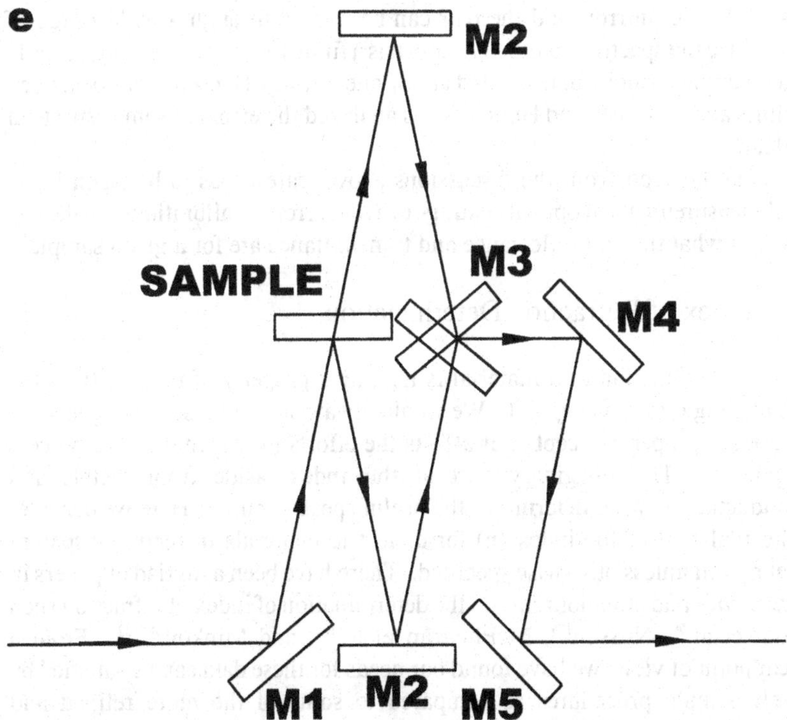

Fig. 5.2e. The "VN" reflection attachment, where M3 switches paths from V to N.

manufacturer of such attachments has commented: "For a limited time, we did offer a VN accessory at the request of one of the spectrometer manufacturers. We discovered that this accessory requires very precise alignment of both the accessory and the spectrometer, in addition to a perfectly symmetrical source and detector. Since these criteria are rarely met, the VN concept does not work as well as the VW design." The version illustrated here[177] is probably the most stable of those that we are aware of and is reported to operate at only 6° angle of incidence.

Another interesting absolute reflectance measuring scheme has been reported by Shaw and Blevin[169], but probably not suited to most coating laboratory's needs.

It should also be noted that all these attachments have a longer optical path than that of the normal transmittance compartment path. This means that there will be focus error effects unless optical power is added via lenses or curved mirrors to correct the focus errors.

There are special variants of the simple, VW, and VN attachments designed to work at 45° and sometimes other fixed angles of incidence. More general variable angle reflectance attachments are also available such as illustrated in Fig. 5.2b and also described by Zwinkels et al.[178] Here, the sample is mounted at right angles to the M2 mirror, and the pair can be rotated through a wide range of angles. Since the spectrophotometer beam has a finite f/number, there are actually a range of angles which are measured at any one setting. Hansen[170] has described these units and their use, and Hunter[171] has analyzed the errors of some aspects in great detail.

As can be seen from the discussions above, care needs to be taken in all spectral measurements of optical coatings to avoid errors. Calibrations are the key to knowing what the real reflectance and transmittance are for a given sample.

5.1.1. Index of Refraction Determination

The index of refraction of a material is its major property of interest from the optical coating effects viewpoint. We would always hope that scattering is not a major optical property except as it affects the efforts to reduce it in the process development. The imaginary part of the index, aside from metals and semiconductors, is what determines the useful spectral range. Here we consider only the real part of the index (n) for dielectric materials in their transparent spectral region unless otherwise specified. There have been a myriad of papers in *Applied Optics* and other journals on the determination of index of refraction such as Arndt et al.[59], Nilsson[179], McPhedran et al.[180], and Minkov[181,182]. From a practical point of view, we have found our needs for these data can be satisfied by relatively simple procedures as compared to some of the more refined and sophisticated ones published. The film design software[172] which we use also provides features to derive index values from spectral data which functions along the general ideas described below.

Our practice has been to deposit a layer of the high index material (TiO_2 in this example) of about one wave of optical thickness (four QWOTs or four turning points) on a substrate of known index as illustrated in the optical monitor trace in Fig. 5.3a. This case is 4 QWOTs at 800 nm monitoring wavelength. We then record the reflectance and transmittance spectrum over the region of interest. This produces a reflectance spectrum with a few peaks in the visible spectrum labeled "measured" in Fig. 5.3b. We pick off the wavelength and magnitude of the maxima and minima of reflectance and also at points every 20 nm from 500 to 800 nm and every 10 nm from 400 to 500nm (where the changes are more rapid). These points are entered as reflectance design goals in a thin film design computer program. We then enter the known substrate index (1.52) and an estimate of the coating index (2.2) and thickness (363.6 nm). We then ask the program to optimize the thickness and a dispersive index for a best fit to the input data points. We assume no absorption and use the simple "QUAD" formulation given in Eqn. 5.1. This function is in the FTG Software[172] that we use to define index as a function of wavelength (λ in nm) where A and B are constants. This gives us 2.073452 and 65419.87 for A and B in the index equation and 359.48 nm for thickness. Figure 5.3.c also shows the results of this estimate after this optimization as compared to the measured data. Note that the reflection loss from the back side of the substrate must be properly taken into account. Some care might be needed to be sure that the optical thickness is not in error by one half wave which would give peaks that would more or less line up with the data.

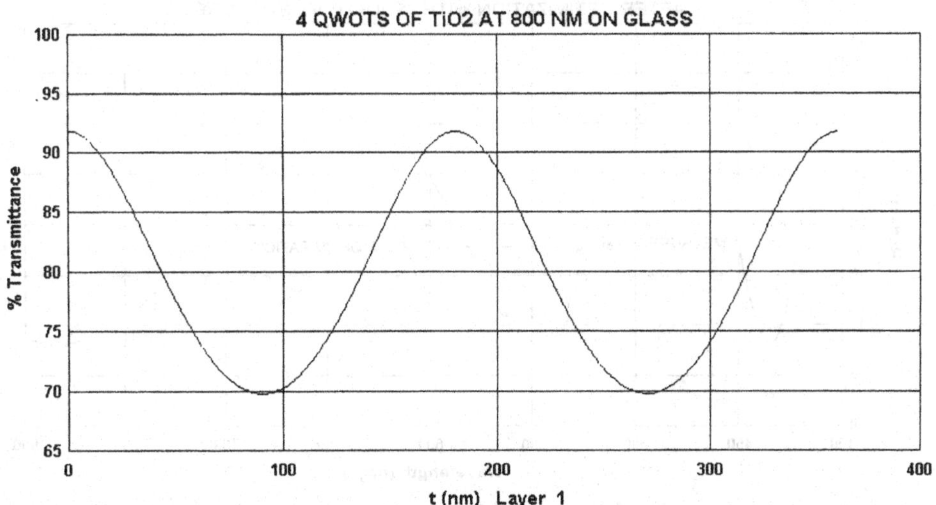

Fig. 5.3a. Simulated optical monitor trace when depositing four (4) QWOT's of TiO_2.

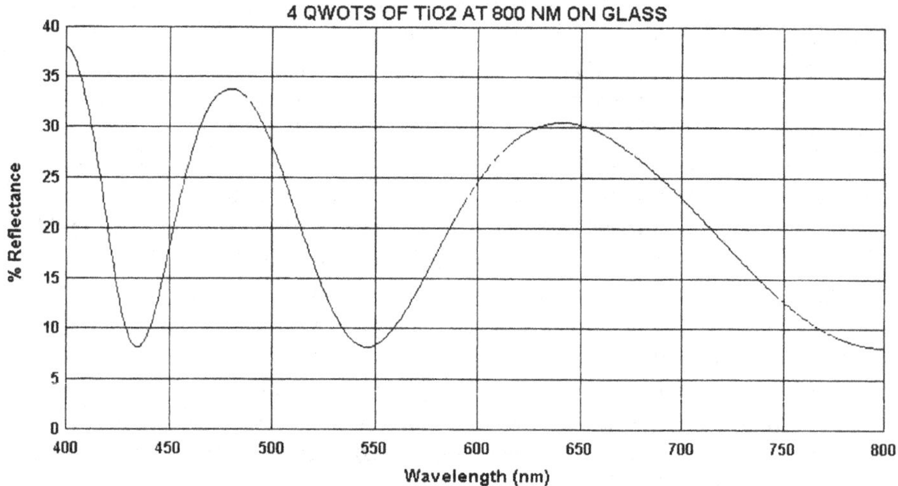

Fig. 5.3b. Simulated reflectance spectrum of four (4) QWOTs of TiO_2 as in Fig. 5.3a.

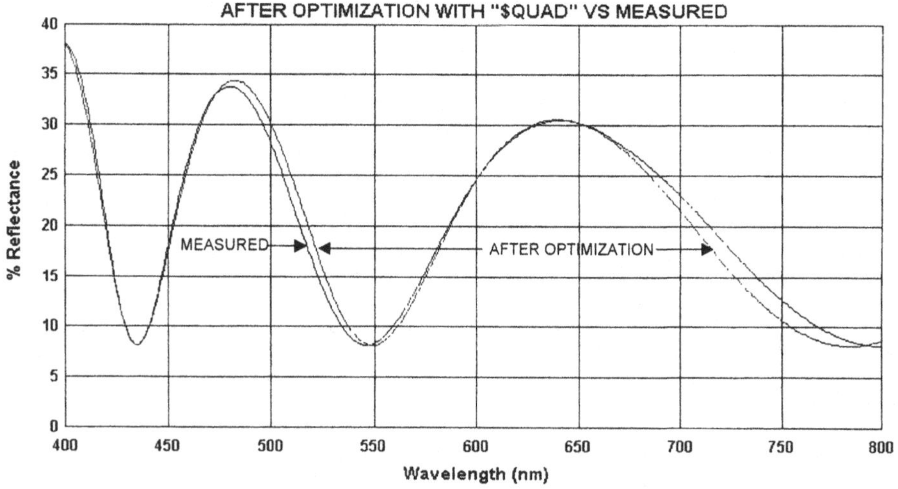

Fig. 5.3c. Comparison of measured reflectance data with best fit using the "QUAD" function in an optimization to find the sample's thickness and index with dispersion.

If any doubt exists, we try an estimated thickness about one half wave thicker and thinner to see if the optimized fit is better. This then has eliminated most of the likely ambiguity which could occur. The resulting values are almost as accurate as the run to run variations of index in many cases. However, when we use these values to compare the transmittance spectra, we do notice some indication of absorption in the peaks and valleys from 400 to 500 nm as seen in Fig. 5.3d.

Macleod[173] points out that the reflectance spectrum is much less sensitive to the effects of absorption than the transmittance spectrum. The minima in Fig. 5.3c are essentially at the reflectance of the bare substrate (absentee layers) but the maxima in Fig. 5.3d are lower at the 400-450 nm end than the bare substrate would be.

Our normal practice is to determine the k-values also or at least to be sure they are not significant. Therefore, we almost always also want to run the reflectance spectrum of the sample. As mentioned above, the same points are picked off the reflectance spectrum as for the transmittance spectrum. If the extrema of the reflectance and transmittance of the same sample are not at essentially at the same wavelengths, the measurements are suspect and need to be investigated. A given film or layer actually has only one physical thickness which applies for all wavelengths. In this discussion, we assume homogeneous layers.

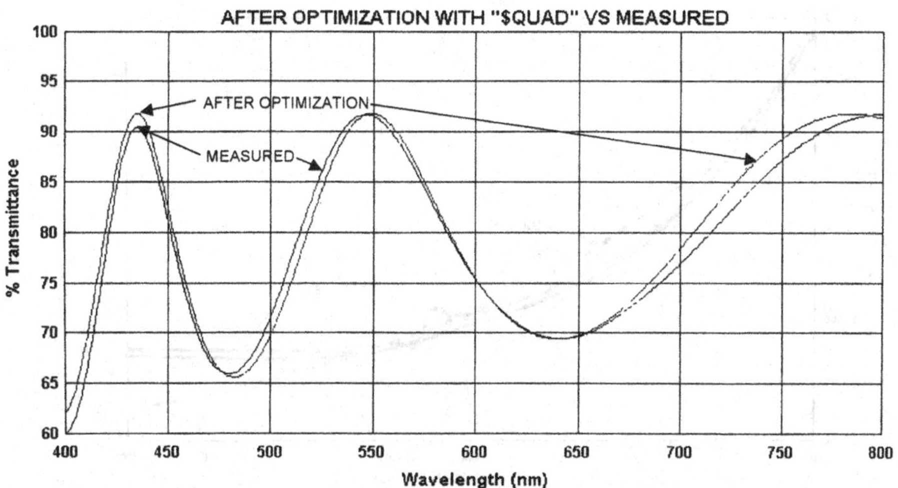

Fig. 5.3d. Comparison of measured transmittance data with best fit using the "QUAD" function in an optimization to find the sample's thickness and index with dispersion.

Macleod[173] has an extensive discussion of the effects of inhomogeneities on these index fitting attempts. It is important to arrive at the best practical estimate of that physical thickness because errors will cause the index values to be generally higher or lower in proportion to that error. We then start fitting the n and k values to the optimization targets from the long wavelength end in sections of small spectral width such as 20 nm or smaller. We use the thickness found using $QUAD as a fixed value for the rest of this process. In a case such as TiO_2, it is hoped that the long wavelength part of the spectrum will be essentially free from absorption so that it may be neglected. Then a copy of the file of targets is edited to include only those from 780 to 800 nm. The program is asked to optimize using $NK, which fits single values of n and k with no dispersion. This fitting will sometimes yield a small negative k-value. This is obviously not realistic and must be due to the error in our data and/or modeling scheme. In such a case, we could revert to $N, set the k to zero, and reoptimize. However, we will choose to collect these "piecewise" pairs of estimated n and k across the spectrum. Once we have them all, we plot the resulting n and k versus wavelength as in Figs. 5.4 and 5.5. In Fig. 5.4, we show the true values of the index, the $QUAD fit resulting from the measurements, and the piecewise fit of successive small spectral sections. The true k values are plotted in Fig. 5.5 along with the piecewise fit results. In both cases, the piecewise fit shows fluctuations which are not consistent with real materials. We are generally safe in assuming that a smooth fit to the results will best represent the real n and k. The true values shown give an indication of how

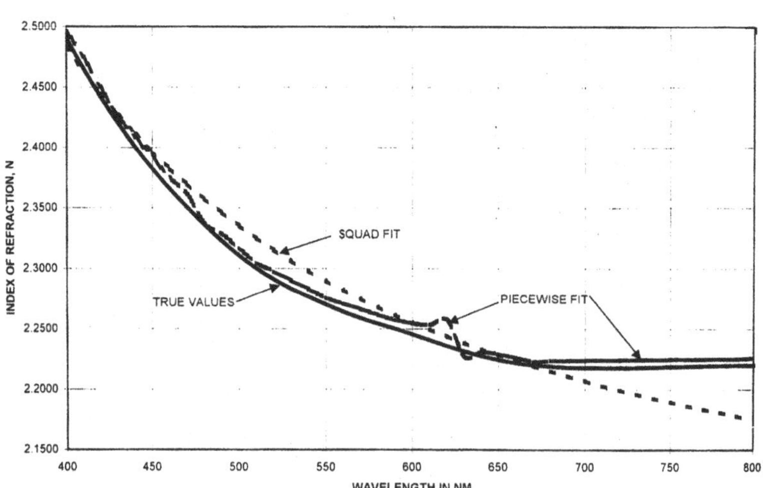

Fig. 5.4. Comparison of measured index data, best fit using the "QUAD" function, and results of a piecewise n and k fit.

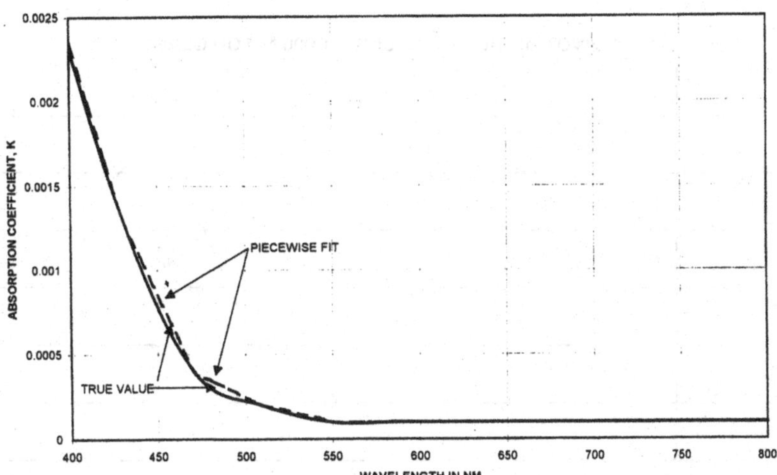

Fig. 5.5. Comparison of true absorption data with the results of the piecewise fit.

well the piecewise fit and $QUAD fit have worked. FilmStar[172] has implemented essentially this procedure which makes it much quicker and easier to accomplish; they have a variety of ways to approach a fit.

Once we know the index versus wavelength of the high index, we deposit a QWOT of this high index material on a witness piece first and then four QWOTs of the low index (SiO_2 in this case) on top. This will give an optical monitor curve like that shown in Fig. 5.6a and a reflectance spectrum like Fig. 5.6b. Otherwise, the contrast might be low for an index like 1.46 on a substrate of 1.52. Again, we pick off the spectral data points and use them as optimization targets. We now consider the index versus wavelength of the high index QWOT as known and fixed. Since the thickness of that first QWOT may not be exactly a QWOT, we use the thickness of the first and second layers in the optimization along with the index of the second layer (low index). We repeat the same sequence of optimization with dispersion for the low index layer. The results of this sequence are the index values of both the high and low index materials as a function of wavelength.

Dispersion can be fit to a variety of equations of varying complexity with one to five or more variables. One variable, the index, would imply no dispersion. Four variables in the right equation per the work of Herzberger[174] would probably describe the index of all optical glasses (without absorption) well from one end to the other of its range of good transmittance. A similar number of terms may be

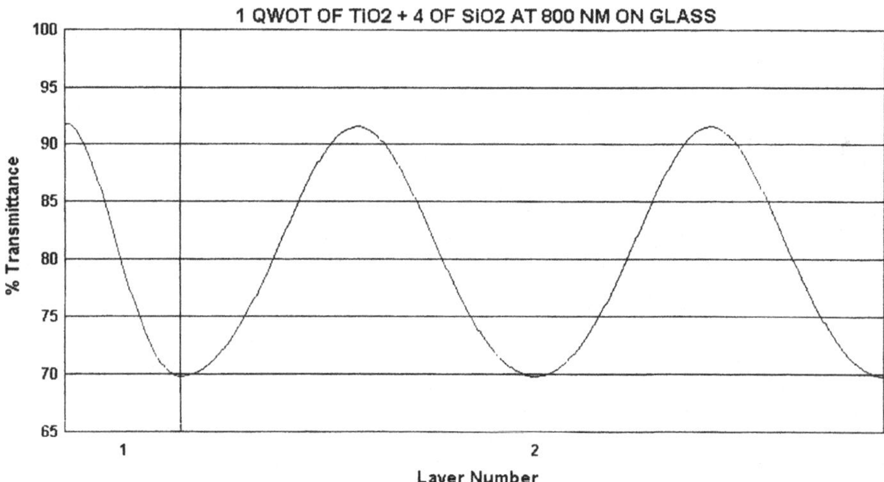

Fig. 5.6a. Simulated optical monitor trace when depositing one (1) QWOT of TiO_2 and then four (4) QWOTs of SiO_2.

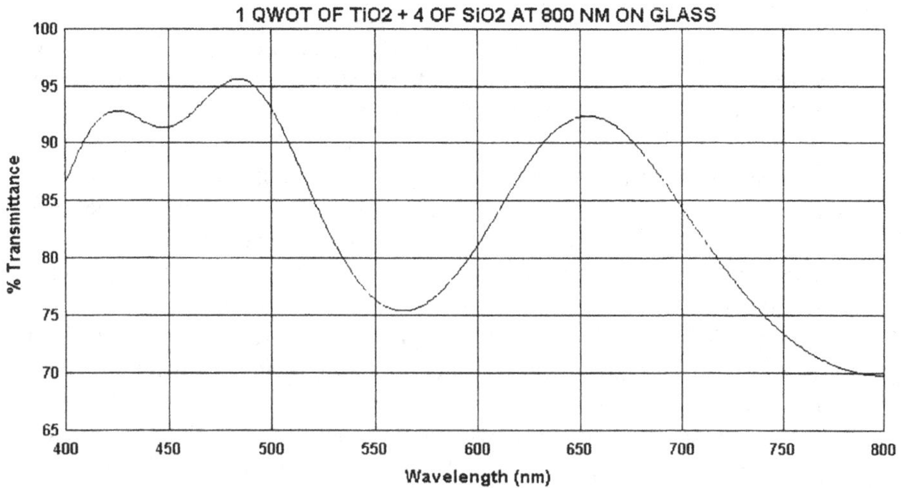

Fig. 5.6b. Simulated transmittance spectrum of the coating in Fig. 5.6a.

needed to describe the k-value also. We have found the simplest practical equation to be that the index is equal to a constant (A) plus a second coefficient (B) divided by the wavelength squared as shown in Eqn. 5.1.

The technique given above depends on the index of the coating being sufficiently different from the substrate to give meaningful results. In the case of TiO_2 or MgF_2 on crown glass, this is not a problem. In the case of SiO_2 on crown

$$n = A + \frac{B}{\lambda^2} \tag{5.1}$$

glass, there would be much less contrast from the maxima to the minima (indices of 1.46 on 1.52). That is why we overcome this by depositing a QWOT of TiO_2 (high index) on the substrate before the four (4) QWOTs of SiO_2. Note that depositing SiO_2 on a fused silica substrate should produce no interference changes if the indices were identical. This can sometimes be a useful test if one is trying to produce SiO_2 from SiO where the index of the film depends heavily on its degree of oxidation.

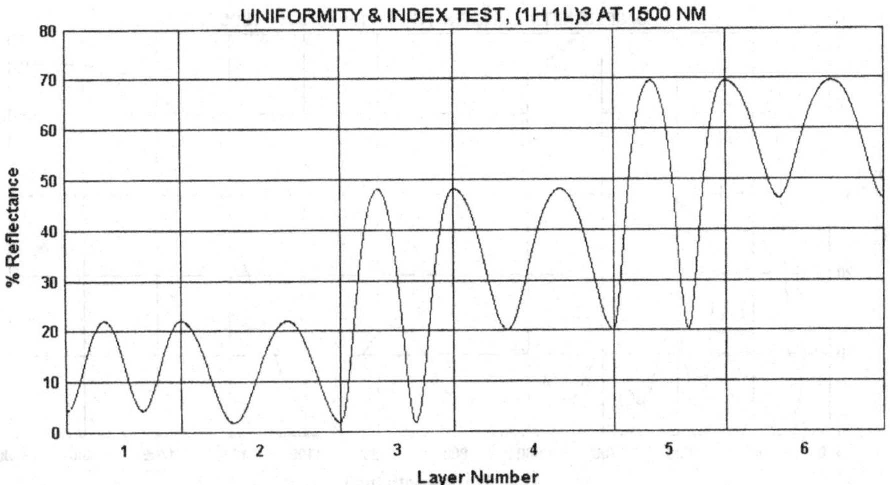

Fig. 5.7. Simulated optical monitor trace when depositing six alternating layers of high and low index which are each three (3) QWOTs thick at the monitoring wavelength.

There is another procedure that can give good information on the index values of both a high and low index material in a single test run. It is also a good test for uniformity over some area which has been sampled by witness pieces or different areas on one test piece. A great deal of information can be gleaned from this type of test data. The test coating is (1H 1L)3 at 1500 nm, in this case. Two or four pairs (or more) is an option here, but three pairs or a total of six (6) layers is probably the most practical and sensitive. Many box coaters are not equipped to monitor at 1500 nm, so we have shown in Fig. 5.7 how the monitor curves would look at 3/4 waves or 500 nm(20000 cm^{-1}) rather than the 1/4 wave at 1500 nm (6666.7 cm^{-1}). These layer terminations would be easy to monitor.

Figure 5.8 shows the perfect spectrum on a wavelength scale while Fig. 5.9 shows it on a wavenumber (cm^{-1}) or frequency scale, on which the spectrum is symmetric. Figure 5.10 illustrates that the top of the reflectance bands at 1/4 and 3/4 waves are sensitive to changes of index. Here we see the changes in reflectance with ± 0.01 changes in the difference between the high and low indices. This figure is also for exact and equal QWOTs of high and low index layers. If the high and low layer thicknesses are respectively 1% greater and 1% less than a QWOT (or the reverse), the shape of the spectrum changes significantly as seen in Fig. 5.11.

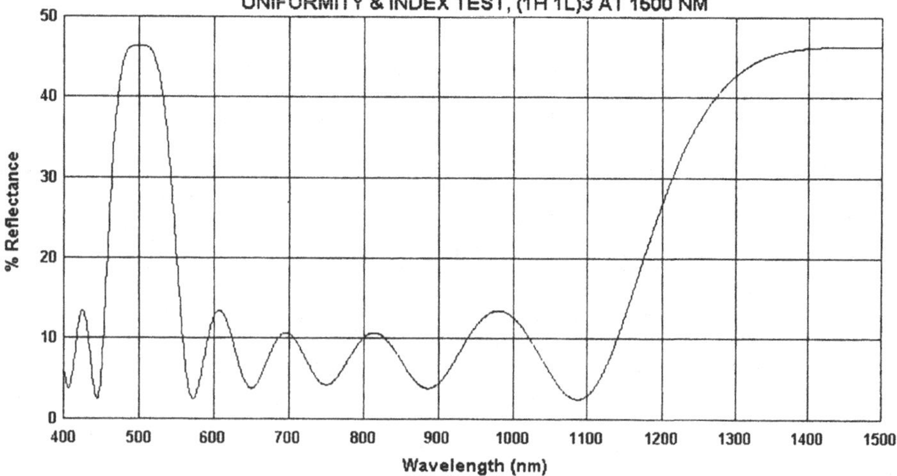

Fig. 5.8. Reflectance spectrum of the ideal design from Fig. 5.7 on a linear wavelength scale showing the one QWOT point at 1500 nm and the three QWOT point at 500 nm.

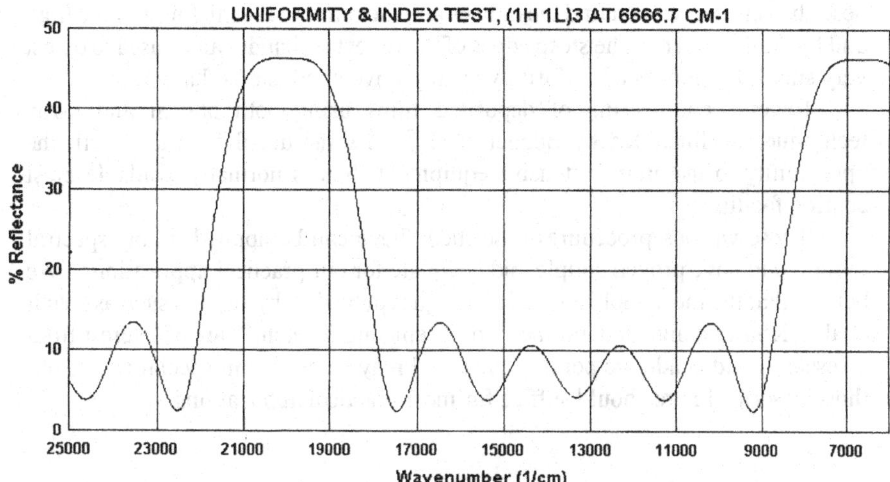

Fig. 5.9. Reflectance spectrum of the ideal design from Fig. 5.7 on a linear wavenumber scale showing the symmetric nature of such a presentation.

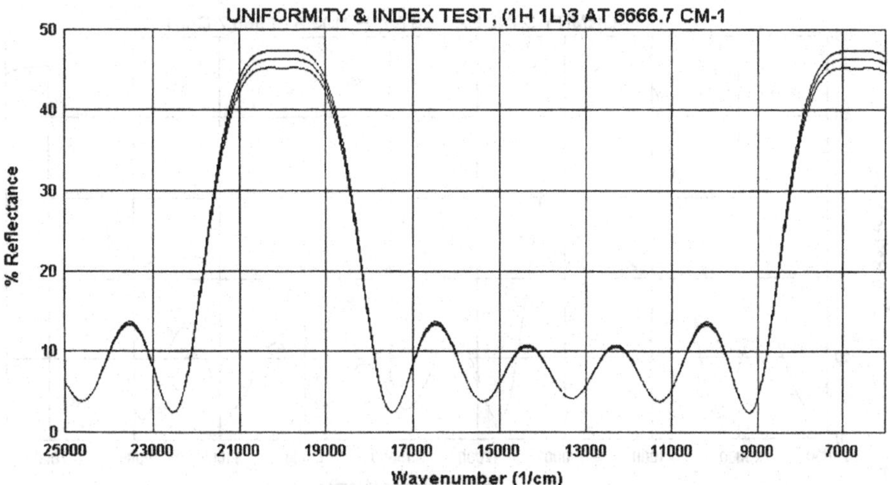

Fig. 5.10. The reflectance bands at 1/4 and 3/4 waves with ± 0.01 changes in the difference between the high and low indices.

The top of the reflector starts to slope and the undulations between the reflectors change their relative height. By the type of "reverse engineering" procedures described earlier, we can find the indices and the ratios of the thickness of the high and low index layers. The steep edges of the reflection bands can be used to detect very small differences of uniformity by any wavelength shifts that occur.

Reverse engineering of deposited films using both optical and x-ray techniques is illustrated by Boudet et al.[177] in some detail for those with the opportunity to use more extensive equipment than is normally available most coating facilities.

These various procedures described above can be applied in any spectral region, and have proved simple and adequate for our practical applications. We believe that the more sophisticated techniques published by authors such as Arndt et al.[59], Khawaja and Bouamrane[175], or Zheng and Kikuchi[176] are of interest from a research and academic point of view and may be needed in special cases, but those described here should suffice for most practical applications.

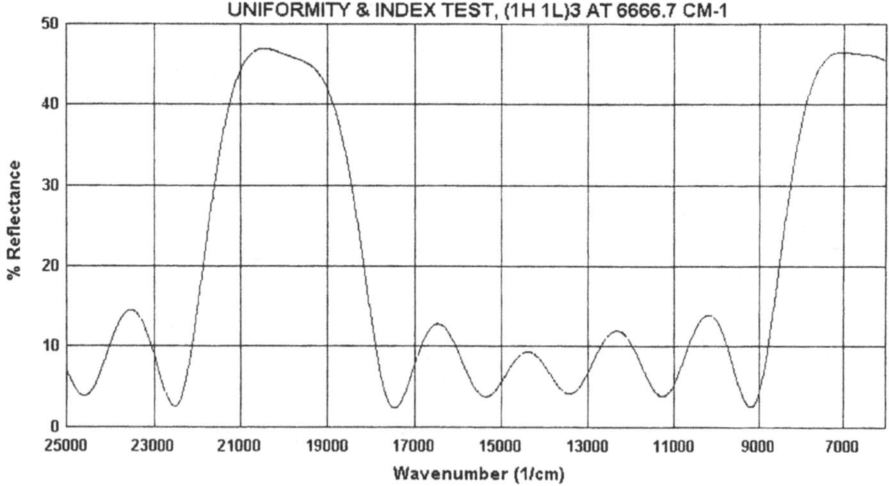

Fig. 5.11. The shape of the reflectance bands change with ± 0.01 QWOT changes in the thickness ratio between the high and low index layers.

5.2. PROCESS KNOW-HOW

There are innumerable processes by which we could deposit optical thin films. The list is a veritable "alphabet soup" of PVD, CVD, PECVD, RIP, IAD, ECR, ECR-PECVD, DLC, etc. The thrust of this book with respect to materials and processes is toward the currently more common practices in optical thin film production wherein our greatest personal experience lies. We here attempt to share as much "know-how" as practical from our experience and that of others. However, we will also give a brief overview of some of the other processes as we understand them which may become beneficial in the future for general or special applications.

Alfred Thelen[110] made an interesting observation that optical coating technology has three major phases of emphasis and seems to cycle through these phases. Figure 5.12 is after Thelen's presentation of the subject. The three phases are: *Process, Design,* and *Control.*

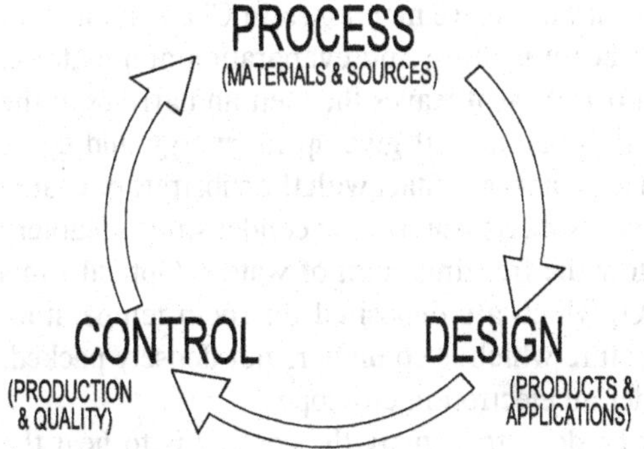

Fig. 5.12. Thelen's view of the historic cycles in optical coating technology.

PRIORITIES

ERA	PROCESS	DESIGN	CONTROL
1945	1	2	3
1960	3	1	2
1975	2	3	1
1990	1	2	3
2005	3	1	2

Fig. 5.13. Thelen's time table of the priorities in optical coating technology by era.

A new process with materials and sources is developed and then requires design work to exploit its unique new capabilities and features in products and applications. This then requires the control to be advanced as needed to support the optimum implementation in production and quality. Thelen conjectures that we have gone through one full cycle and are now (1990-2005) in a Process emphasis phase again as shown in Fig. 5.13. Our secondary emphasis now would be design, and control advancement would be at a low ebb because we have recently come through large advances in control technology. By 2005, if not sooner, design will be the major emphasis. On the other hand, it may be that we have entered an era wherein all three areas progress more or less in parallel. It appears to us that the design activity is no longer in response to new processes, but has progressed with a life of its own.

5.2.1. Film Growth Models and Observations

One overriding feature of processes that is a differentiating factor is the energy of the process at the film formation location. As we mentioned in Chap. 4, an atom evaporated from a boat or electron beam melting and evaporation of a material only has an energy on the order of 0.1 eV as it leaves the melt and arrives at the substrate. If the substrate is cold, the atom will give up its energy and come almost immediately to rest as soon as it makes contact with the substrate or coated surface. This growth is similar to atmospheric water vapor condensing as feathery frost on surfaces that are well below the freezing point of water. Optical films such as evaporated MgF_2 and TiO_2 which are deposited on room temperature substrates have a microscopic structure which is columnar, not densely packed, and looks somewhat like frost under an electron microscope.

The first thing which might be done to densify the coating is to heat the substrate so that the energy of the arriving atom does not get removed as quickly. The atom then has enough mobility to migrate and fall into a lower potential energy spot and fill a hollow before it is quenched and loses all of the energy it had upon arrival. When MgF_2 is deposited on substrates heated to 300°C , the films are much more dense and hard, and they are more like freezing rain. In the case of freezing rain, the liquid water flows on the surface somewhat before it loses its energy and solidifies.

Greene[445] discusses nucleation, film growth, and microstructural evolution. He states that: "The primary deposition variables which determine the nucleation and growth kinetics, microstructural evolution, and, hence, physical properties of films grown by physical vapor deposition are: the film material, the incident film flux, the kinetic energy of species incident at the film growth surface, the film growth temperature, the flux of contaminants, and the substrate material, surface cleanliness, crystallinity, and orientation...Note that the flux of contaminants which competes with the flux of film material for incorporation during deposition is strongly dependent upon the base pressure, pumping speed, and the design of the vacuum system while substrate surface cleanliness depends also upon pre-deposition processing."

Movchan and Demchishin[183] (M-D) introduced the first structure zone model (SZM) that showed three zones of film growth as a function of the substrate temperature (Ts) divided by the melting temperature (Tm) of the material being deposited. The Ts/Tm ratios for the three zones for metals are: 1) <0.3, 2) 0.3 to 0.45, and 3) >0.45. They conclude that the boundary between zones 1 and 2 is somewhat lower at <0.26 for oxides. The nature of the deposited film in zone 1 is like the feathery frost mentioned above with columnar growth separated by significant voids. Zone 2 has densely packed columns, whereas zone 3 has a polycrystalline structure.

Thornton[184], working just with sputtered metals, advanced the model from one dimension to two by studies adding the effects of argon gas pressure in the sputtering process and pointing out that the size of the columns and structure grows with thickness of the film. He identified a transition zone between 1 and 2 consisting of densely packed fibrous grains. The zone "boundaries" change as a function of argon pressure in sputtering as shown in Fig. 5.14 (after Thornton).

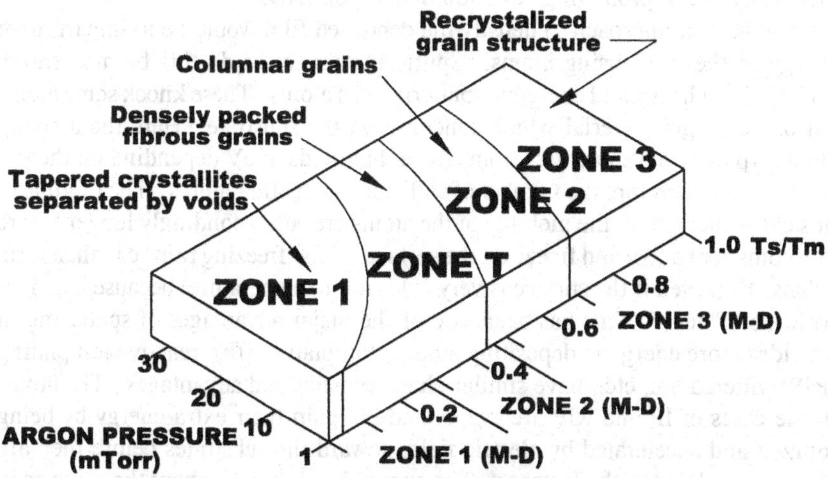

Fig. 5.14. Thornton's zone model of coating structure as a function of the ratio of substrate to melting temperatures and the argon pressure while sputtering.

He speaks of the low adatom mobility at temperature ratios <0.1 where the initially nucleated growth continues in the direction of available coating flux which is significantly affected by intergrain shading. Müller[185] confirmed this behavior by computer simulation in two dimensions along with the behavior in zones 2 and 3. Müller concluded that the transition from porous to more densely packed growth structure was caused by enhanced adatom mobility at increased substrate temperature. He further showed that high deposition rates interfere with the void filling process of adatom mobility and require higher temperatures to get the same density at high rates.

Adamik et al.[441] studied NdF_3 stratified with MgF_2 and also CaF_2. They show TEM, XRD, and AFM analyses of the films. They discuss the fact that additives and impurities can change the structural evolution per the M-D and/or Thornton Zone Models. The impurities can inhibit or promote the evolution of specific structures and shift the zone boundaries to higher or lower substrate temperatures. The grain size increased with CaF_2 stratification and decreased with MgF_2 stratification.

Müller[186] also used the simulation system to model postdeposition annealing of porous films. In this case, he studied gold. His model showed that even at room temperature the small voids would migrate and eventually coalesce with larger voids. He further states that: "In the case of weak bonding between film and substrate, interfacial voids are formed, which reduce the film's adhesion." His colleagues Netterfield and Martin[187] reported extensive related experimental work on gold films which showed that oxygen ion-assisted deposition (IAD) was much more effective in promoting adhesion than argon IAD.

A second approach to densify the deposited film would be to impart more energy to the condensing atoms. Sputtering processes do this by momentum transfer from heavy and energetic sputtering gas atoms. These knock some atoms out of the target material which condense on the substrate. Here the arriving atoms typically have energies from tens to hundreds of eV depending on the gas pressure and therefore the number of collisions with other atoms. The lifetime of this extra energy and the mobility of the atoms are correspondingly longer before the atoms cool down and freeze on the surface. This "freezing rain" can then form a dense film even if the surface is very cold (room temperature) because it arrives with more "heat." This has been one of the major advantages of sputtering, it provides more energetic depositing atoms. Ion plating (IP), reactive ion plating (RIP), filtered arc, etc., have similar characteristics and advantages. The atoms in the cases of IP and RIP are evaporated but gain their extra energy by being ionized and accelerated by electric fields toward the substrates before they are again neutralized with electrons. The energy levels can be about the same as in sputtering.

Guenther[188] reviewed the M-D and Thornton models and the history of the field, and he then proposed an extended model. He mentioned that many other factors probably influence the results such as: rate, thickness, annealing temperature, atom to ion arrival rate (or ion-to-molecule arrival rate), diffusivity coefficient (fast and slow surface kinetics regimes), etc. He discusses the various aspects extensively. For the purposes of this discussion, however, we will point out his proposal that there is a fourth zone (4) beyond 3 which he calls "vitreous-amorphous" zone as shown in Fig. 5.15. Here the ratio Ts/Tm >1.0 based on his experience with RLVIP. He talks of the activation energy rather than the substrate temperature, since the incoming material and ions are energetic. This implies the formation of a transient liquid monolayer, which is immediately quenched into a solid state on the "cold" substrate and preserves the amorphous structure of the liquid in the solidified form. Guenther[280] shows a graphic example of the columnar growth of TiO_2 and a confirming comparison between an actual micrograph and a 2D model. He also shows a micrograph of a film section of interlaced layers of TiO_2 only, but where alternate layers are E-beam evaporated and ion plated and thereby Zone 1, T, or 2 versus Zone 4. It is imaginable to the author that these conditions can be achieved by other processes such as IAD, APS, UBMS, etc.

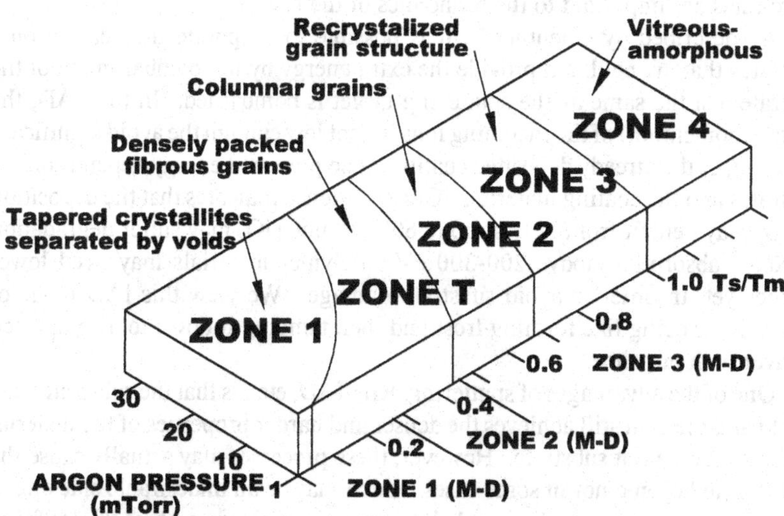

Fig. 5.15. Guenther's addition to the work of Thornton and of Movchan and Demchishin on the zonal observations as in Fig. 5.14.

The ion-to-molecule arrival rate is a significant factor in IAD. Netterfield et al.[57] discuss this in detail with respect to CeO_2 in particular, but the work has general applicability. They speak of the critical arrival rate (CAR) that is required to produce stable films, i.e., fully densified, maximum index, no humidity shift, etc. They report that excess above the CAR has some effect in reducing the index of CeO_2. They conjecture: "that either additional oxygen is being incorporated into the CeO_2 film or that closed isolated voided volumes are being created in the film." The reported effects were with 1200 eV oxygen ions; we wonder if this would be the case in the 100-300 eV regime.

Let us digress briefly to give an example of the calculation of the adatom to ion arrival rate. Consider SiO_2 at a deposition rate of 5 A/sec or 0.5 nm/sec or $5x10^{-8}$ cm./sec. SiO_2 has a density of 2.2 gm/cm^3 and a molecular weight of 60. Therefore, a cm^3 would contain 2.2/60 moles times (Avogadro's number) $6.02x10^{23}$ molecules/mole = $0.22x10^{23}$ molecules. Therefore, a 5 A/sec rate would have $0.22x10^{23}$ times $5x10^{-8}$ cm./sec = $11x10^{14}$ molecules/sec (molecule arrival rate). If the measured ion current on the depositing surface is $175x10^{-6}$ amperes (coulombs/sec), this times $0.625x10^{19}$ charges/coulomb, would give $10.95x10^{14}$ ions/sec (ion arrival rate). This implies a molecule-to-ion arrival rate of $11x10^{14}/10.95x10^{14}$ or about a 1:1 arrival rate.

It can be seen that a lower deposition rate is required if an IAD source is operated at maximum ion current and the CAR has not been satisfied. This is at the root of our drive for higher power ion sources for robust industrial processes where rates are important to the economics of the results.

As described by Guenther[188], it is possible to evaporate the coating onto substrates that are cold, but provide the extra energy by ion bombardment of the substrate just the same as the sputtering target is bombarded. In this IAD, the energy (momentum) of the incoming ions is kept low enough the avoid significant sputtering of the already deposited coating. The preferred energy depends on the substrates and the coating materials. Our experience indicates that the deposition of SiO_2 may benefit from IAD at up to 600 eV, but TiO_2 may suffer degradation by excess absorption above 200-300 eV. Polymer materials may need lower energies yet, in order to avoid substrate damage. We view this IAD mode of deposition as being like forming frost and then tamping it down to a compacted icy layer with our shoe.

One of the advantages of sputtering, RIP, IAD, etc., is that the substrates can be cold and the film still achieves the denser and harder properties of the material evaporation on a hot substrate. However, these processes may actually cause the substrates to become hot in some cases, which may be an undesirable side effect. Plastics, for example must be kept below some temperature such as 80-150°C to avoid melting and deterioration. Here IAD has been used beneficially, but too much ion energy can damage the plastic substrates. The extra energy provided by all of the processes mentioned except substrate heating impinge only on the surface. If the substrate is a good heat conductor, the effect of the added energy on the substrate temperature may be minimal. If the substrate is a good thermal insulator, the effective surface temperature may be higher with a steep temperature gradient as one moves into the substrate. This might be an asset or a liability, depending on the materials and circumstances. The surface heat might help the dense growth of the deposit, but it might be detrimental to some substrate surfaces such as polymers.

Many of the above analogies and comments may apply also to CVD processes. The typical CVD reactor for semiconductor layer growth has had the substrates heated to very high temperatures (over 500°C) and fully dense coatings have been produced. Plasma Enhanced CVD (PECVD) provides higher energy molecules at lower substrate temperatures and some ion plating effects may appear with accelerating electric fields, etc. The CVD processes may gain much of their high energy effects from the chemical reactions and interactions themselves. In general, these are more complex than the PVD processes and will be briefly touched upon in Sec. 5.7.

5.2.2. Chiral and Sculptured Coatings

The M-D zone 1 results discussed above are not always to be avoided, however. Motohiro and Taga[316] have provided an extensive report on their work with polarizing coatings which take advantage of the zone 1 columnar structure with voids in between. The depositions are at controlled angles to the surface normal of the substrate (which may also be appropriately rotated about its surface normal) to produce coatings that have circular (and other) polarizer functions like fixed liquid crystals. Hodgkinson et al.[315], Wu et al.[189], and Messier et al.[190] have done related work with structured coatings, and they show micrographs of an interesting variety of structures created this way. Scattering in such constructions is a serious consideration and has been studied extensively by Hodgkinson et al.[315,317]

Mbise et al.[273] studied angular selective window coatings by depositing at large angles to the surface normal. Such coatings might be used in automobile and architectural windows for energy control and visual comfort. They found chromium most effective, but titanium, aluminum, and chrome and aluminum oxides were also tried.

5.2.3. Stress in Coatings

Stress is a major issue in optical coatings, and it will appear often in the discussion of various materials. The literature points to the fact that the Zone 1 coatings described above tend to be tensile, while the Zone 3 and 4 coatings can be quite compressive. Either extreme of stress in a coating can cause problems with the adhesion of the film to the substrate and substrate warping. Ennos[11] reported extensive experimental work to measure stresses in the coated materials using the processes of that time (up to1965). Macleod[4] discusses (pp. 401 to 402) the reduction of tensile stress in porous coatings by water absorption. Smith[191] and Pulker[192] both have extensive discussions of the physics of stress in films and some of the problems associated with it. The Thornton[184] studies and Guenther's[188] review also point to the fact that Zone 1 films tend to be tensile and higher zones become increasingly compressive. This was certainly our earlier experience[120] with the high compressive stress in Zone 4 films. Wolfe[196] discusses using high angles of incidence and lower rate to pressure ratios to reduce stress in sputtered applications. These factors would tend to move the conditions into the lower zone numbers for less compressive stress. Gluck et al.[197] discuss the successful reduction of stress by using mixtures of a variety of materials in the infrared. Quesnel et al.[198] studied the stresses of IBS and evaporated YF_3 and LiF. The evaporated films were tensile and the IBS films were compressive, as might be expected from the foregoing discussions. They studied the influence of humidity, temperature, and aging on stress. Water absorption tends to add compressive stress, but is mostly reversible if the film is dried by evacuation, etc. Aging reduced the compressive stress in the LiF deposited by IBS, but it increased the compressive stress in the IBS YF_3. The thermal coefficient of expansion (TCE) of the film and substrate will also affect stress. The difference between them will change the resulting stress as a function of the deposition temperature. We once found that we had to reduce the deposition temperature for an SiO_2/TiO_2 stack from 300 to 270°C on KG3 glass to avoid such stress cracking.

Lechner et al.[199] reported on RLVIP coated Ta_2O_5. It appears that fully dense films had high compressive stress and some absorption after coating; but when annealed for 4 hours at 350°C, the stress and absorption decreased significantly and the thickness increased by over 1%. They attribute this to improved film stoichiometry.

Lucheta[200] gives an extensive analysis of coating stress using elasticity theory. His motivation centered around diamond and related coatings. He concluded that "thermally induced stresses alone will be much higher than normally encountered in coating work." Takashashi[201] published an important paper on how the coefficient of thermal expansion of the substrate is critical to the temperature stability of the center wavelength narrow bandpass filters and the stress relationships that exist.

There appears to be a relationship wherein coatings with higher compressive stress exhibit higher hardness. This would seem intuitive since a more densely packed material should be harder than a less dense material. This would point to the probable desirability of some compressive stress in a coating up to but short of the level which would cause other problems such as cracking and/or adhesion failure.

5.2.3.1. The Measurement of Stress in Coatings

The measurement of stress is most commonly done by deposition on thin bars (as in the case of Ennos[11]) or disks of known flatness and mechanical properties. The shape change due to the film stress is measured during and/or after the coating deposition. The stress can be calculated[11,198] in various ways from the measured bending of the substrate. Lechner et al.[199] give the following procedure for intrinsic stress measurements. "The residual intrinsic stress of the films is measured interferometrically (Newton fringes) by the deformation of coated circular glass discs (1.1mm thickness, 50mm diameter in their case, or silicon discs are often used). The radius of curvature can be determined using the fringes numbered m and j from the interference pattern with Eqn. 5.2:

$$r = \frac{d_m^2 - d_j^2}{4\lambda (m - j)} \tag{5.2}$$

From the different radius of curvature before and after coating, knowing additionally the Young's modulus (E_s), Poisson's ratio (ν_s) and the corresponding thickness t_s and t_f of substrate and film, the intrinsic stress σ_f can be calculated according to Eqn. 5.3:

$$\sigma_f = \frac{E_s}{6(1 - \nu_s)} \frac{t_s^2}{t_f} \left(\frac{1}{r_{f+s}} - \frac{1}{r_s}\right) \tag{5.3}$$

For such measurements flat substrates showing a concentric circular Newton fringe pattern are required. Since, however, only a few substrates have such an ideal finish, substrates with a slightly elliptical deformed fringe pattern can also be accepted. To obtain a mean film stress value, the mean bending radius of the sample is determined by 45° stepwise rotated fringe diameter measurements.

5.2.4. Laser Damage in Coatings

With the advent of lasers in the 1960s, a significant new field evolved in the study and attempts to understand what factors can cause and inhibit laser-induced damage to substrates and coatings by the concentrated energies in small volumes made possible by lasers. Laser "damage" is useful for cutting and/or melting materials, etc., but is harmful in other cases such as lens and window transmittance and mirror reflectance. Van Stryland et al.[312] found that laser-induced breakdown field increased as $V^{-1}\ t_p^{-1/4}$, where V is the focal volume and t_p is the laser pulse width. Arora and Dawar[208], who studied effects in bulk silicon, stated that the laser-induced damage threshold (LIDT) fluence (energy density) depends on wavelength, pulse width, number of pulses, and the surface and bulk quality of the material. In their case, they concluded that thermally induced stress and long-lived electron traps seem to be the most probable causes of laser damage in bulk silicon. In the case of coatings, all these and other factors may come into play, and the thresholds are typically lower than bulk materials by a significant amount. Contamination, inclusions, etc., cause significant adverse effects. We will first review some of the published conclusions with respect to LIDTs in bulk.

Bettis[216] showed that the intrinsic laser-induced breakdown thresholds of solid, liquids and gases correlate well with the expression $N^{2/3} / (n^2 - 1)$, where N is the atomic number and n is the refractive index of the material. He further states that there is general agreement that LIDT scales with pulse duration *t*, as $t^{1/4}$. He refers to the work of Soileau et al.[225] for the scaling with wavelength. Their results for NaCl and KCl indicate little wavelength (measured in eV) dispersion in the damage threshold as measured in electric field strength from 0.0 (DC) to nearly 2.0 eV (10.6 to 0.53 μm). The approximate bulk breakdown field in these materials from their data can calculated as: Field (V/cm) = $(1.6 + 0.425\ \text{eV}) \times 10^8$. O'Connell et al.[211] reported on the bulk laser damage properties of PMMA. They concluded that it has a significantly lower LIDT than silicate glasses, that the process in PMMA is cumulative, and there is some relaxation to previous damage resistance with time. They further conjecture that the damage is caused primarily by impurities, i.e., an extrinsic mechanism, which must be of submicron size. O'Connell[212] later reported on extensive work that supported the conclusion of the extrinsic light-absorbing defects as the primary cause of laser damage in PMMA. This paper also goes into detail on modeling damage in closed form equations for both bulk and surface damage thresholds. He assumes that "the material contains absorbing impurities and defects with deterministic damage thresholds, so that, if in the illuminated area or volume there is at least one defect whose damage threshold is equaled or exceeded by the local beam fluence (i.e., a damageable defect), damage occurs." Porteus and Seitel[213] showed the absolute "onset" intensity of laser damage to be the best way to measure and describe LIDT in that it is independent of the test laser and the spot size. They reported on this in detail and O'Connell[212] expanded upon that work. Ziegler and Schepler[215] did measurements on silver gallium selenide and concluded that energy density, not peak-power density, was the determining factor in LIDT. Dahmani et al.[250-252] have studied the relationship of stress to damage and crack propagation by lasers in fused-silica and borosilicate glass. They found that externally applied stress improves the LIDT by breaking hoop-stress symmetry.

When we look at coatings in particular, the roles of structure, voids, inclusions, etc., become even more dominant in LIDT work. Kozlowski[226] provides an extensive discussion of damage-resistant coatings and the various factors involved. Shaw-Klein et al.[217] put forth a model to predict LIDT in coatings, including those that have porous structure such as sol-gel coatings and the Zone 1, T, and 2 PVD morphologies discussed above. They confined their attention to damage due to bulk absorption at extrinsic inclusions and point defects. The general concept is that a defect absorbs heat from the laser, expands, and causes mechanical failure by the thermal stresses created. This would imply that films which are less dense due to voids would have less thermal conductivity and be more subject to damage. Sol-gels and less dense evaporated films, on the other hand, have shown better LIDTs than this would predict. They conjecture that the porous films allow the pressure stress to relax in a shorter distance than would be the case in a dense film, and therefore the porous film would be expected to have a better LIDT.

Trench et al.[218] address the nature and appearance of some of these defects. They studied HfO_2/SiO_2 films where the principle defects were nodules due to ejecta from the hafnia evaporation. They say that, "it has generally been accepted that the low damage threshold is associated with nodular defects in the coatings." They further indicate that defects with heights of 0.7 μm are most susceptible to laser damage, whereas the pits formed by the ejection of such a nodule do not increase laser damage susceptibility. In the laser conditioning described below, they propose that nodules are popped out by the lower fluence of conditioning to cause no further effects, or that the nodules are welded to the coating such that they cannot pop out. Stolz et al.[219] give more detail on these nodules and their minimization. Chow et al.[221] describe reducing nodules in hafnia. However, in all of these reports[218,219,221] it appears that the reduced nodule density does not improve the LIDT. It appears that they used activated oxygen but no IAD and no crucible liner. This leaves some question as to the relative impact of these nodules on the LIDT versus some other mechanism such as basic absorption in the film. Poulingue et al.[237] artificially generated defects with diamond and silica particles under controlled conditions to study and model such defects. They concluded that the nodule size is the critical parameter for LIDT in high reflectors and that defect density is not critical. For silica, the critical seed size is 0.5 to 1 μm which induces a critical nodule diameter of about 4 μm. Below this size, there is a sharp decrease in laser coupling (and thereby damage). Guenther[280] showed a 2D model of defect growth as might apply to these cases. He also showed ESCA depth profiles and binding energy plots of E-beam versus LVRIP hafnia layers.

Laux et al.[209] reported the comparison of two layer versus 14 layer AR coatings in the UV, coated by molecular-beam deposition, wherein the 14 layer coating had a higher LIDT. They conclude that this is due to the thin layers being more amorphous where the crystalline structure has not had an opportunity to develop. There are less interface defects, and therefore the losses are less. Gao and Wang[220] studied damage mechanisms for KrF laser coatings and concluded that large compressive stresses due to the laser fluence cause cracking in the films. Krajnovich et al.[222] did an extensive study of AR coatings on CaF2 for KrF lasers at 248 nm and found the following coating materials quite satisfactory: AlF_2, Al_2O_3, LaF_3, MgF_2, and SiO_2. Carniglia, et al.[459] had an interesting earlier paper on the successful use of a half wavelength optical thickness barrier layer on the substrate before the regular coating. They reported that the damage was due to thermal expansion of the material near the substrate-film boundary, and to be caused by polishing or cleaning residue rather than by absorption in the coating materials themselves.

Laser conditioning has evolved as a technique to improve the LIDT of a coating before it is subjected to maximum fluence levels. The surface is irradiated with a series of pulses lower than the threshold. In many cases, this can prepare a coating that would have initially damaged (at a given fluence level) so that it withstands an even higher level of fluence. Arenberg et al.[214] (in 1989) reported a significant discorrelation between laser tests of unconditioned and conditioned coatings, which became good correlation when the samples were either equally unconditioned or conditioned. This certainly pointed to a significant phenomenon worth investigating. Eva et al.[210] reported on doing this at 248 nm on LaF_3/MgF_2 coatings. They concluded that the conditioning effect was a function of the radiation dose rather than energy density or pulse rate, and further that the absorptance must originate from imperfections (contaminants, structural, and stoichiometric defects) in the layers which are reduced by conditioning in many cases. They also concluded that conditioning induces structural and stoichiometric changes in the film. As mentioned two paragraphs above, Trench, Stolz, Chow, et al. think that nodules are a major factor in LIDT, but their results leave this possibly inconclusive at this point.

The absorption of laser energy is one (if not the key) factor in LIDT. Photothermal deflection (PD) spectroscopy (PDS) has become one of the preferred methods to measure absorption. Commandré and Roche[227] provide an extensive review of PDS. Jackson et al.[223] have analyzed these sensitive and versatile methods in some detail, and Welsch and Ristau[224] have expanded upon that work and included an extensive list of references. There are many other papers which can be found including PDS in the work reported. The basic concept of the method is to illuminate the surface to be tested at normal incidence with a modulated laser pump beam while a probe laser beam (typically HeNe) is passed parallel to and very close to the surface. The probe beam is deflected by the mirage effect of gas above the surface being heated by the energy absorbed by the surface/coating from the pump beam. The deflection of the beam is measured and can be used to calculate the absorption of the surface and/or film.

To the extent that LIDT is a function of absorbed energy, the design of a coating can be used to minimize the electric field at critical levels in the coating such as at interfaces between layers. Interfaces generally appear to be regions of greatest extrinsic absorption due to defects, inclusions, etc. The problem of interface absorption has been successfully circumvented in some cases by "blending" depositions from one layer to the next. This is done by opening the shutter for the next material for a period where both high and low index materials are coevaporated producing a mixed layer of intermediate index. This makes little difference in the optical characteristics of most designs, but can also be accommodated in detail in the design process if necessary.

As we showed in the earlier chapter, the absorption of energy in a metal film depends mostly on the electric field strength in the film due to the illumination at a given wavelength. Koslowski[226] touches on this design technique of the "shifted electric field." Apfel[235] describes the principles and procedures for such designs in some detail. Diso et al.[236] show that a stepwise reflectivity profile in a coating has no adverse effect in the LIDT and that the difference from one section to the other can be explained on the basis of the electric-field intensity.

The subject of LIDT is by no means a closed book at this time (2002). The concerned party should avail themselves of the latest literature when appropriate. From a practical optical coating point of view, it still appears advisable to strive for very low absorption in coating films at the expected laser wavelengths and to produce films which are as defect-free as practical. The latter often depends on minimizing the particles which get onto the surface to be coated from when it is a fresh and pristine surface through the coating deposition process. Seeds for defects are known to come from dust and also deposition source "spitting" as discussed in the references above[218,219,221,226].

5.2.5. Rain Erosion of Coatings

In today's environment of high speed aircraft for the military, rain erosion has appeared as a significant problem. Fortunately, the effect on coated windows at a significant angle of attack to the rain drops (like 45°) is greatly reduced.

Windows and other surfaces which must be used at near normal incidence represent the major problem. This includes cases such as missile domes and protective/sealing windows on forward-looking IR (FLIR) systems.

Seward et al.[450] showed the effects of rain erosion damage on window and dome materials such as MgF_2, sapphire, spinel, and ZnS with and without "diamond-like carbon" (DLC) coatings. They also showed their single and multiple impact jet apparatuses to simulate rain impact. They discuss the fracture stress of these materials as a function of velocity and the various damage velocity thresholds. Adler et al.[452] showed micrographs of damage and compared testing with nylon beads and waterdrops. Adler and Mihora[451] report on analytic studies of raindrop interactions with surfaces and even carry the technology into the space environment with "hydrometeors" of frozen water. Knapp and Kimock[154] described DLC coatings on ZnS and ZnSe which used a Ge layer to promote good adhesion for films up to 2 μm thick. Fuller and Fisher[449] discuss multilayers with DLC by PECVD. They reported reasonable rain erosion resistance. Hassan and Propst[453] deposited DLC coatings on Ge and ZnS by PECVD. They reported good rain erosion and abrasion resistance. Blackwell et al.[454] elevated the performance to new levels with a combination of boron phosphide (BP) and DLC. The BP has an index of about 3.0, and therefore the DLC (at about 2.0) functions as an AR for Ge. A coating with 10 μm of BP overcoated with DLC is said to have rain erosion thresholds more than an order of magnitude better than DLC alone.

The thresholds for damage seem to be an issue of impact resistance of the combined structural properties of substrate and coating layers. Stress and adhesion in the coatings are important in addition to the other mechanical properties. There are many examples where the coating holds together but the substrate has cracked under the coating due to the transmitted forces of the impacts. Doubinina et al.[448] described an IR AR where this seemed to be the case. Most of the reports referenced above were concerned with the 8 to 12 μm spectral band. We have seen coatings for the 3 to 5 μm band which successfully used a 20 to 30 μm-thick layer of Al_2O_3 on the substrate with appropriate AR layers over and under this layer. It appears that the AR layers stand up reasonably well when there is a layer below them that does not flex too much. The fine points of rain erosion protection are moving rapidly. It is therefore advisable for the reader with specific interests in this area to check the latest SPIE journals and proceedings as good sources for what is being published in the unclassified literature.

5.3. MATERIALS

We saw in the first chapters on design that it is almost always advantageous to use the highest and lowest indices available for all dielectric coatings in the spectral regions of interest. It appears that this understanding was not apparent to most experts in the field prior to the last decade or two. It seems that a much broader selection of materials were used in common coatings in the 1970s. We conjecture that this may have been motivated by at least two things. One is that: before the common use of E-guns, there was a need for materials which could be evaporated from resistance sources and also have good adhesion while giving the needed optical properties. Also, the low energy of arrival of thermally evaporated atoms before the use of ion assisted deposition, made adhesion, packing density, and durability even more material dependent than they are now. The second motivation may have been that the design concepts were driven by attempting to have only multiples of integer quarter waves in a design. These were the easiest to analyze and monitor with the earlier technology. We showed in Chapter 1 that any index between the highest and lowest available can be simulated to any desired degree. Therefore, only the extremes should be needed. However, we will qualify that statement (pending further investigation) to say that: at non-normal incidence, the benefits of intermediate indices have been touted for years and there *may* be some reason for this in nature which has yet to be explained. We will proceed with the assumption that the two-material approach will best solve most practical problems.

In the visible and near infrared region, the highest and lowest indices would usually be TiO_2 and MgF_2 (indices on the order of 2.35 and 1.38 at 550 nm). We mentioned that we have most often compromised and used SiO_2 (1.46) for the low index when larger numbers of layers and thicknesses are needed due to process limitations. We have yet to develop a satisfactorily scatter-free MgF_2 deposition for these cases with TiO_2, although this may well be possible with the right process development. There are numerous comments in the literature that MgF_2 becomes excessively stressed (tensile) in thicker films with most of the present deposition techniques.

In the infrared from 8 to 12 μm, germanium (Ge @ 4.1) and thorium fluoride (ThF_4 @ 1.35) would be preferred choices for the high and low index extremes. However, the ThF_4 is radioactive and thereby the subject of many searches for a substitute low index material such as that by Sulzbach[1] and Lubezky et al.[2] A problem was posed to the coating designers to see how well an AR coating could be designed with Ge and ZnS as reported by Aguilera et al.[3] and discussed in Chapter 2. For these reasons and many others, the choice of materials is more often ultimately dictated by the physical properties which can be achieved by given processes than by index of refraction. This is why we will link materials and processes together in many cases.

We will deal with the subject of materials in two ways in this chapter. We will discuss some of the more common materials that we have worked with and then discuss some specific processes used with those materials. We will also mention some of the other materials with which we have not had extensive experience but that have been used in the past and or elsewhere. In this latter case, we will provide some references for the interested reader. In discussing specific material depositions using IAD, we will mention the type of ion/plasma source and its operating conditions. The more detailed discussion of some ion/plasma sources will be covered in Sec. 5.4.

5.3.1. Some Specific Materials

We will share our experience with various materials in this section. These approaches and results are not necessarily the best that have been or could be done, but represent our findings and understanding to date. Where we have extensive experience (not all of which may have been positive), we will say correspondingly more than where we have less experience. In the later case, we will provide some references. Ritter[9] provided a wealth of understanding and information on materials in use in 1976. Macleod[4] has an extensive list of materials, data, and references which may also be helpful to the reader.

5.3.1.1. Silicon Compounds

Silicon oxides have been most prevalent as coating materials, but silicon nitride and silicon oxynitrides are receiving increased interest.

Silicon Oxides, SiO to SiO_2

Silicon dioxide, silica, SiO_2, or "quartz" is one of the preferred low index materials for a variety of reasons. Its use in designs was seen extensively in the first chapters. Although it is of somewhat higher index than MgF_2 at 1.38 in the visible spectrum, it can often be deposited with less porosity and scattering. It is relatively durable and can have a good laser damage threshold[203]. With the advent of electron beam evaporation sources (E-guns) several decades ago, it became practical to evaporate fused silica. There are many papers written on its use; a few more recent ones are cited here. The dioxide is useful at 248 nm where Krajnovich et al.[222] give its index as 1.50. Scherer et al.[205] reported that all films were compressive and showed extensive analyses of how these stresses reduced with time. They found this "to be driven by a hydrolysis of SiO_2 strained bonds, probably followed by a network rearrangement responsible for stress relaxation."

Chow and Tsujimoto[203,204], and Harris[206] discuss using improved E-beam control to obtain more reproducible results with SiO_2. This study worked to optimize the sweep and other parameters to avoid "carve-in" (which most writers call "tunneling") and gain other desirable properties of the depositions.

We had found that the deposition of critical optical thin film stacks with silicon dioxide from an E-gun was severely limited by the stability of the evaporation pattern or angular distribution of the material which is also a motivation of the works reported above[203-206]. We had not obtained satisfactory results in some of the more demanding applications with either solid discs or granular SiO_2 starting material. The amount of material deposited on a central monitor chip or control crystal in a box coater did not have a reproducible ratio to that received at other positions in the chamber. Figure 5.16 illustrates the variable distribution of silicon dioxide evaporant from an E-gun as is commonly experienced in physical vapor deposition. The "cloud" may be broad or narrow and not necessarily normal to the general surface of the material if there is any tunneling. This is explained by the erratic melting/sublimation of silicon dioxide surfaces in both granular and solid disc forms of silica.

One hypothesis for this unstable behavior is illustrated in Fig. 5.17. The surface of the silica is locally heated with an E-gun by the electrons impinging from above in this figure, and the surface becomes white hot. There is some sublimation of the silica and apparently surface melting and evaporation, but no

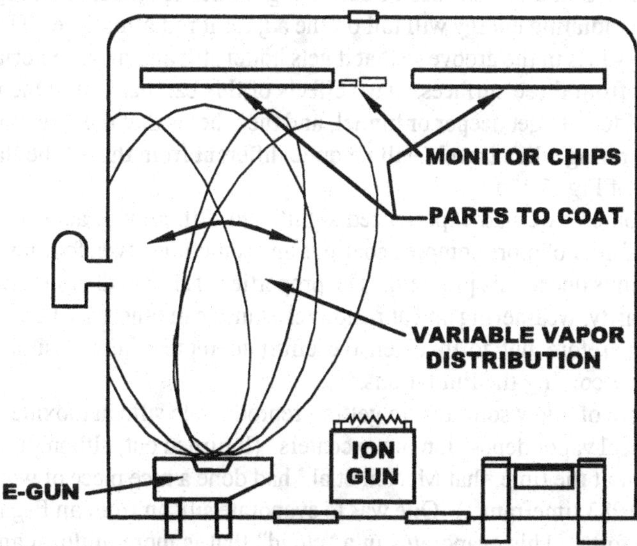

Fig. 5.16. The variable distribution of silicon dioxide evaporant from an E-gun as is commonly experienced in physical vapor deposition.

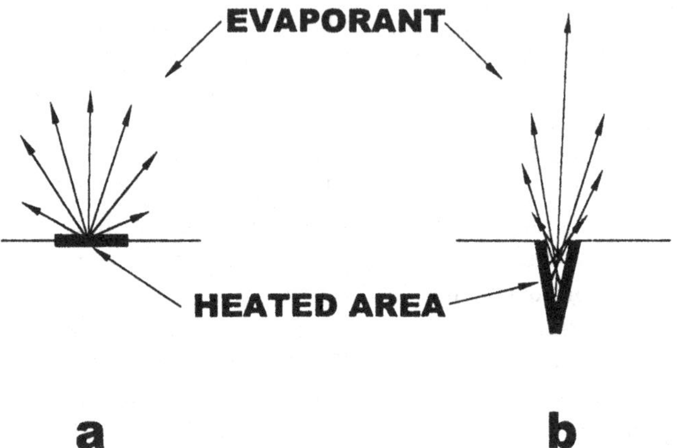

Fig. 5.17. The vapor distribution of silicon dioxide evaporant from a flat surface versus a groove or tunnel where the radiant heat is confined and causes extra heat and evaporation.

significant flow of molten material. A great deal of the impinging energy is dissipated in the heat radiated by the white hot surface. If this surface is primarily horizontal and upward facing, as in Fig. 5.17a, the energy will radiate into the relatively colder environment of the chamber and be lost from the silica. However, if a groove has formed in a flat surface or between granules of silica, as in Fig. 5.17b, much of the radiating energy will fall on the adjacent radiating face. This results in less energy loss in the groove so that it gets hotter still and more material will be evaporated from these surfaces. Two effects of this can be seen. One is that the groove will tend to get deeper or tunnel, and the other is that the direction in which the evaporating silica travels will be quite different from that of the flat horizontal surface of Fig. 5.17a.

In our box coaters, we had experienced significant difficulty in achieving satisfactory repeatability of more complex coatings and sometimes even four layer antireflection coatings due to this problem. The properties of silicon dioxide such as its low index, clarity, weathering and abrasion resistance can otherwise be very desirable. We therefore put forth extensive effort to retain these desirable properties while overcoming the limitations.

We were aware of a few solutions to getting reproducible silicon dioxide at high rates by physical vapor deposition in boxcoaters. (It turned out, although we were not aware of it at the time, that McNeil et al.[5] had done a nice piece of work in this area in the 1983 timeframe.) One was to evaporate silicon from an E-gun in a molten liquid form. This evaporates in a "cloud" that is more uniform and repeatable than the silica cloud. The silicon must then be properly oxidized to silicon dioxide as it is being deposited. This has been done by reactive ion plating[120] and some plasma enhanced chemical vapor deposition (PECVD) processes. We were not in a position to acquire such a specialized chamber, and therefore we chose another approach. The second scheme, which was chosen, is to evaporate silicon *monoxide* (SiO) and complete its oxidation with ion assistance. This could be retrofit into our existing chambers and processes. Silicon monoxide sublimes and is difficult to get satisfactory results using an E-gun, somewhat like silicon dioxide. However, it can be readily evaporated from resistance "boats" with a uniform and reproducible cloud. SiO has been a commonly used material for barrier layers on plastic packaging sheet material. The boats used are typically a maze-like structure to limit the "spitting" or flying hot "ashes" that tend to occur from an open boat when there are any small grains in the material loaded into the boat. After an "outgassing" process (under shutter) which drives off the "sparks," the vapor deposition can be very uniform and stable. We found the R. D. Mathis[460] Company's SO-36 resistive source to be satisfactory for our work which required relatively large amounts of material. To evaporate at about 1 nm/second at the center of a Balzers BAK760, we applied about 600 amps at 1.7-2.1 volts (depending on how well the electrical contacts had been made). Our goal was to obtain good silicon dioxide at a rate of 1 nm/second. This cannot normally be achieved by simply evaporating silicon monoxide in an oxygen background, even at elevated temperatures. If the pressure were high, the films would be porous and weak. If the pressure were low, the films would be absorbing and of high index. Without any additional oxygen, SiO films are of about 1.9 index and brown in color. However, in the NIR at or beyond 1000 nm, SiO has little or no absorption and has been successfully used with silicon as the high index material for NIR coatings. An adequate supply of energetic ions or neutrals of oxygen are needed to obtain SiO_2 at the rates and properties desired. Kaufman type ion sources can be used to a certain extent at lower rates by combinations of oxygen and argon, but the nature of the Kaufman gun's filament and grids makes its use with oxygen in a high power mode unfavorable. Some gridless ion guns such as the DynaVac PS1500[461], on the other hand, work quite well using oxygen as the ionized gas.

The general configuration of the chamber used is shown in Fig. 5.18. The ion source was aimed at the calotte in an orientation for best uniformity. The aim point was approximately midway between the points in the calotte which were directly over the silicon monoxide and titania sources. These sources had uniformity masks at about 3 cm below the calotte and directly over the sources. The titania that was used as the high index material was evaporated from an E-gun diagonally opposite the silicon monoxide source. The operational details of the ion/plasma source will be covered in Sec. 5.4.

We have used this deposition arrangement and technique extensively for the production of stacks ranging from 40 to 90 layers of silicon dioxide and titania.

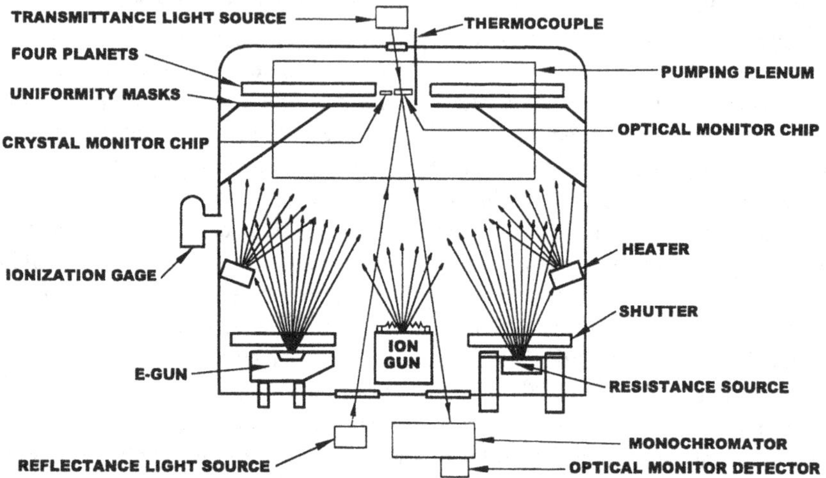

Fig. 5.18. The general configuration of the chamber used to produce SiO_2 from SiO (and also HfO_2 from Hf metal).

For E-beam evaporation of silica from granules or discs without IAD, we typically might find a 15 to 20 nm spectral shift with change of water content (humidity) from wet to dry. With IAD from SiO, shifts of less than 1 nm and even zero have been achieved. The deposition conditions of such a run were: 225°C, 1 nm/second for both materials, 30 SCCM of oxygen through the ion gun, 300 drive volts, 5 drive amps, 17 amps for neutralizer, and the chamber pressure was mostly in the range of 1.0×10^{-4} torr. The choice of temperature was based on the fact that the coatings showed cracking due, presumably, to differential thermal expansion when deposited on substrates at higher temperatures. These films also passed the adhesion and severe abrasion tests of MIL-C-675. The resulting index of the silica could be made anything from 1.45 to values higher than 1.50, depending on the deposition conditions. We found that the conditions which gave little or no humidity shift with this type of coating were not necessarily compatible with some laser damage requirements. However, we think that this might be overcome with further process optimization. Additional oxygen background pressure was required to reduce the laser damage and give an index of about 1.46, but the humidity shift partially returned. More ion current at a low ion voltage is probably required to get a combination of high laser damage threshold and low humidity shift. The ion current available limits the rate at which material can be deposited and properly oxidized and densified. Therefore, the highest ion currents available are usually the most desirable.

We have generally been able to achieve good and repeatable uniformity over a full chamber diameter calotte (1.2 m) by the use of masks. However, the ion gun parameters must be stable or the uniformity can be seen to have some changes. A change in chamber pressure will change the mean free path of the ions and neutrals and thereby change the effects of the uniformity masks. The ion beam has a distribution which is more concentrated on the axis of the beam. It appears that if an adequate amount of ions reach the less bombarded areas, the excess ions in the central areas are not detrimental. We have had some indication that this is less true when the drive volts are in the higher (500 V) regions. The hypothesis is that some ion sputtering or etching may be occurring at the higher drive volts and thereby the thicknesses of the layers are slightly reduced at the center of the beam impingement area.

We have found the oxygen-ion-assisted deposition of silicon monoxide to be a solution to the reproducibility problems of silicon dioxide films. We have been able to deposit films at acceptably high rates in production environments.

In the 2000-2001 timeframe, many new coating systems moved toward production of DWDM filters using DIBS to deposit SiO_2 from fused silica as the low index material.

Silicon Nitride

Early work with silicon nitride as an AR for Ge, Si, and GaAs by rf-diode sputtering was reported by Laff[285]. Eisenstein and Stulz[286] did related work with ARs for semiconductor injection lasers. Laird and Wolfe[290] state that Si_3N_4 is extremely hard, durable, and corrosion resistant. Martin et al.[282] reactively sputtered SiO_2/Si_3N_4 multilayers and found very good temperature and humidity stability. Netterfield et al.[283] report the index of Si_3N_4 at 633 nm as high as 2.1 and the useful spectral range as 250 to 9000 nm. They evaporated Si from an E-beam source and added the nitrogen by IAD from a Kaufman source and show the 633 nm index at 1.92 for their process. Lambrinos et al.[284] give details of their DIBS process to produce Si_3N_4. They show n and k values as a function of process parameters. The k seems to climb from 600 nm to shorter wavelengths. Bovard et al.[287] deposited Si_3N_4 using LVRIP. Their extensive study showed an index at 400 nm of nearly 2.1 and the absorption only starts to increase at about 400 nm to large values at 300 nm. They mention the absorption edge at 260 nm and ~7000 nm as the onset of the Si-N bond absorption band.

Silicon nitride has appeared extensively in the large area coating field in the past decade. This has be aided by the application of C-Mag™ sputtering[289] of silicon to Si_3N_4 and SiO_2. The application to low emissivity coatings was described by Laird and Wolfe[290]. They report that the films are extremely hard, durable, and corrosion-resistant. An advantage of using silicon nitride is that the silver does not need an additional protective barrier layer to prevent oxidation during subsequent layer deposition. A drawback of using it is the high degree of stress that tends to reduce the adhesion. They reduced the stress by using dual cathodes and a modified magnetic configuration. Thin NiCr barrier layers were also used to increase the adhesion between the nitride and the silver. Further application to low emissivity coatings was described by Nadel.[291]

Lipin and Machevski[292] reported on flexibility of using biased reactive rf-sputtering of silicon nitride. Krempel-Hesse[293] described the application of Si_3N_4 for AR and production of photovoltaic solar cells. The need for large coated areas at low cost makes sputtered Si_3N_4 appear attractive. Tsai et al.[129] investigated the plasma-enhanced chemical vapor deposition (PECVD) of silicon nitride and silicon to make multilayer coatings.

We have been associated with work reactively sputtering Si_3N_4 for its optical and tribological properties. Films as hard as 20 GPa were produced. This is comparable to Al_2O_3 (Sapphire) in the bulk.

5.3.1.2. Titanium Oxides, TiO to TiO_2

Titanium dioxide (TiO_2) has been our favored high index material for the visible and near infrared spectrum because of its high index and relative robustness. Figure 5.19 shows the transmittance of hafnia, yttria, zirconia, niobia, and titania

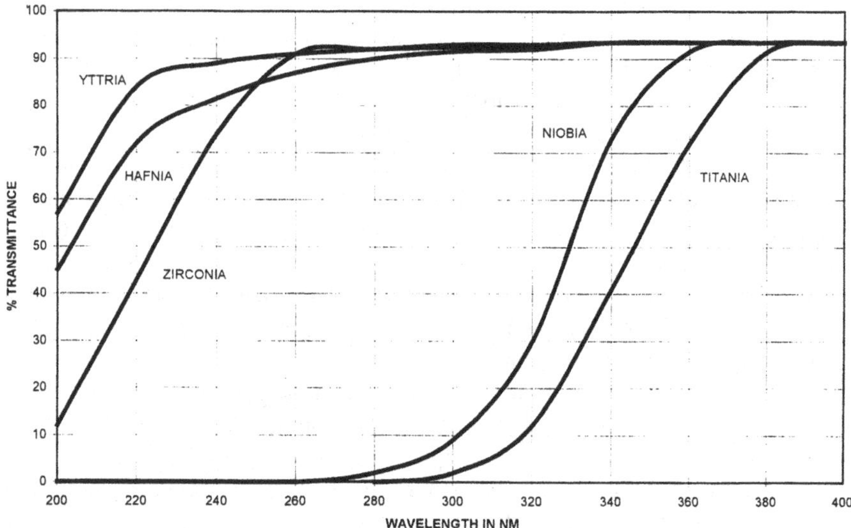

Fig. 5.19. Transmittance of hafnia, yttria, zirconia, niobia, and titania films in the ultraviolet.

films in the ultraviolet. Niobia has the most similar index and transmittance. TiO_2 appears historically to be less popular than other materials with some practitioners because of difficulties in obtaining a stable result. Bennett et *multi* al.[228] published the results of many TiO_2 films produced by a great variety of techniques and laboratories. That report shows the range of results which can be obtained by different processes. We believe the root of the problems that cause difficulties for some laboratories/facilities can be seen in the extensive work of Pulker et al[6]. They started with various materials including TiO_2, Ti_3O_5, Ti_2O_3, TiO, and pure Ti. These represent oxygen-to-titanium atom ratios for the starting materials of: 2.0, 1.67, 1.5, 1.0, and 0.0 respectively. They found that the 1.67 material was stable and gave consistent layers of repeatable index of 2.205 at about 550 nm. However, the first layer of 2.0 material gave an index of about 2.06 followed by layers of nearly 2.20 index. Figure 5.20 is adapted from their paper. The 1.5 material gave and index of 2.27 for the first layer and then settled in to nearly 2.205 for subsequent layers. The 1.00 material, on the other hand, required several layers to drop from an index of 2.38 toward 2.2. All of their results were stated to be free from absorption. One criterion for almost any TiO_2 deposition is to have negligible absorption in the working spectral region. This will place a lower limit on the background oxygen pressure and some constraints on temperature and deposition rate. Granqvist[368] also discusses titanium oxide films in his handbook on electrochromic materials. His review is valuable as an additional data source and another point of view.

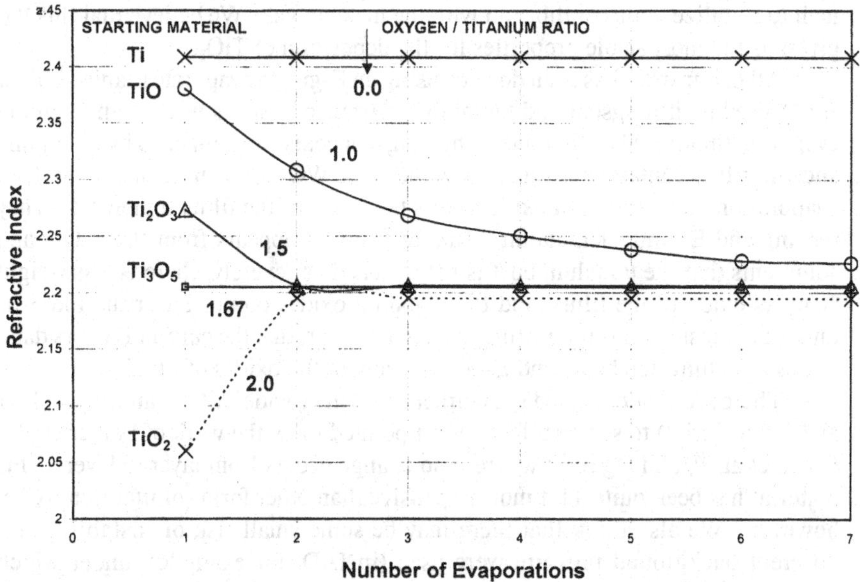

Fig. 5.20. Index of refraction of TiO_X films as a function of the number of evaporations for various starting materials; taken from the report of Pulker et al[6].

Our interpretation of their results is that the material being evaporated in the crucible or boat reaches an equilibrium with the background or makeup oxygen in the chamber after some period of being melted or some number of layers of evaporation. The TiO_2 (2.0 material) is releasing excess oxygen until it comes to equilibrium, while the TiO (1.0 material) must take up oxygen from the background to reach equilibrium. This seems to be consistent with the concept of "congruent" vaporization as studied by Chiao et al.[274] It is also consistent with the findings of Ogura.[281] It can be seen that this latter process is slower as evidenced by the data of Pulker et al.[6] where it takes many layers for TiO to approach a stable index. As a result, our philosophy and practice has been to attempt to achieve the most stable process practical by melting in new tablets of TiO_2 starting material in an E-gun under its shutter with the background pressure of oxygen which is to be used during the actual process. As the premelt is started, the power to the gun is raised and the material begins to outgas. The power is raised until the pressure is somewhere below that which would cause E-gun arcing (we used 5×10^{-4} torr). The molten material in the crucible (usually with a molybdenum liner to reduce heat loss) is kept at a temperature where little material is actually evaporating, but gas is being driven off, as evidenced by the chamber pressure gauge. After a few minutes, the chamber pressure drops and the power can be increased to keep the material melting and outgassing. When the *whole* charge in the crucible is well melted and the chamber pressure has come down to the oxygen makeup gas pressure setting to be used in production, a few minutes of additional soaking is allowed at production deposition power and background oxygen pressure for the melt to stabilize at its equilibrium with the makeup gas. We believe that this has given us the most stable properties for the deposition of TiO_2.

All of our work has been done by using an E-gun to evaporate titania. Pulker et al.[6] used both tungsten and tantalum resistance boats as well as an E-gun to evaporate titania. They found that the tungsten boats were attacked by the titania and the films contained up to 70 weight % of WO_3 after many layers of TiO_2 evaporation. The tungsten oxide lowers the index of the films somewhat. They recommend E-guns or tantalum boats instead. It appears from their data and comments that the tantalum boat is not attacked as severely, there is less weight % of its oxide in the films, and the tantalum oxides do not lower the index as much as tungsten. It is interesting to note, however, that the current Balzers data[7] shows only tungsten boats and E-guns for any of the oxides of titanium.

There are vendors in today's market who can provide Ti_3O_5[281] and Ti_2O_3 (1.67 and 1.5 material) to support the concept pointed to by the work of Pulker et al.[6], Chiao et al.[274], and Ogura[281] wherein no change occurs from layer to layer. This material has been quite a bit more expensive than other forms of titanium oxide, however. We also think that there may be some small risk of instability if a different background pressure were used (in IAD, for example); under which circumstance a different atomic ratio might be the more stable value. For this reason, we prefer to use the least expensive uncontaminated material available and "melt it in" as described above.

Another approach (which we do not agree with) was taken some years ago by engineers at Leybold Heraeus. They made a 19-pocket E-gun. At the beginning of each run they placed one or two pellets of new material into each pocket. Each new layer used another pocket with fresh material. In our view, this is like trying to balance on an acrobat's beam for stability rather than standing flat on the floor!

The resulting index of refraction and freedom from absorption is a significant function of the background pressure of oxygen and temperature as shown by Pulker et al[6]. Higher temperatures for the substrates give higher index, for example, 2.63 at 500 nm for 400°C. However, higher temperatures are usually less desirable for other reasons and IAD has become a popular way to approach the higher indices at low or even room temperature. It is usually necessary to provide enough O_2 to avoid absorption which reduces transmittance, but it may also be necessary to reduce absorption to increase the laser damage threshold.

Starting from the metal gives a significantly different index (2.41) from the 2.205 that the various oxides converge upon with increasing numbers of depositions from the same material charge. Pulker et al.[6] allude to this being a case where the Ti deposits in a dense layer with a crystal structure which then oxidizes to the rutile crystal form of TiO2. Bulk rutile is the most dense form of titania. All forms tend to this stable phase when heated above about 400°C. At low deposition temperatures without energetic processes like sputtering or IAD, the oxide starting materials tend to give amorphous phases and/or the anatase crystalline form as the temperature is increased. These are lower index in the bulk form than rutile, and thin films may be lower still due to packing density per the Thornton Zones 1 and T described above. Jang et al.[270] studied E-beam evaporated TiO_2 with various oxygen flow rates for crystalline structure changes, and they analyzed the films with various new tools such as atomic force microscopy (AFM). At a constant deposition temperature of 250°C, they concluded that increased oxygen flow rate increased the rutile content versus the anatase. Guenther[280] shows a graphic example of the columnar growth of TiO_2 and a confirming comparison between an actual micrograph and a 2D model. He also shows a micrograph of a film section of interlaced layers of TiO_2 only, but where alternate layers are E-beam evaporated and ion plated and thereby layers in Zone 1, T, or 2 versus layers in Zone 4.

There have been interesting applications made of the chemically active properties of titania. There have been many studies of the use of TiO_2 as a photocatalyst to treat air, water, etc., through the photolysis of organics and toxic gases[238-240] Street light glazing, etc., have been coated with TiO_2 to breakdown organic road and smog deposits. The concept is that UV light activates the photocatalytic action of TiO_2 to break down organic deposits that then evaporate or wash away. The action is also proported[241-243] to make the surfaces "superhydrophilic" so that water droplets lie very flat on the surface with near zero contact angles. The combination of these effects is highly desirable for automotive mirrors, windows and also architectural glazing. There are ongoing discussions of the relative potency of the various crystalline phases of TiO_2 with respect to these effects. Anatase seems to be the most effective, but it is not yet clear that it is exclusively so. The production of a specific phase is a function of substrate temperature, other energy added to the deposition (such as IAD), etc.[244-245] Chen et al.[275] concluded that rutile formation in a coating was enhanced by substrates with an Al_2O_3 content, while it was reduced by a Na_2O content. The determination of which phases exist in a coating are best done by Raman spectroscopy, X-ray diffraction (XDS), and IR analysis.[246-249] Such coatings are already found on the external rear-view mirrors of some production automobiles.

Before we deal with the use of IAD for TiO_2, let us first examine the valuable results provided by Pulker et al.[6] We have taken the data published there and applied the tools of the Design of Experiments (DOE) field described in the next chapter to develop a formula and graphics of how the index of TiO_2 might be expected to vary with oxygen pressure, substrate temperature, and deposition rate. We took the data from their Tables I and II and performed a least squares fit to first and second order equations for a plane. The second order effects were found

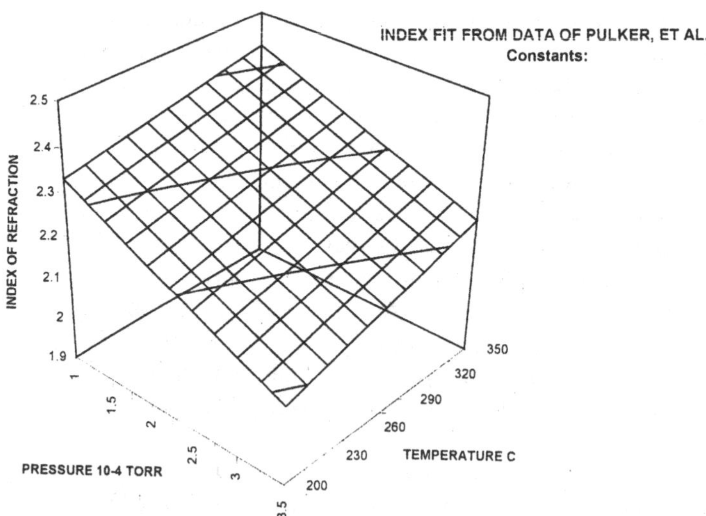

Fig. 5.21. The index of refraction as a function of temperature and pressure from the data of Pulker et al.[6]

to be negligible to within the experimental error. Figure 5.21 shows the index of refraction from their data as a function of temperature and pressure. It can be seen that the highest index occurs at the highest temperature and lowest pressure and that there are fairly linear relationships of the variables.

Table 5.1 is the data input of index versus pressure and temperature, and Table 5.2 shows the results from the multiple regression analysis of this data by the DOE software. This was used with other input to derive Eqn. 5.4 for the index of refraction of the titania film (n_t) as a function of the variables. Using Eqn. 5.4, we can predict what the index of refraction at 530 nm will be as a function of pressure, temperature, and rate. The pressure (P) is entered in units of 10^{-4} torr, the temperature (T) in degrees Celsius, and the rate (R) in Å/second.

$$n_t = 2.28 - .119P + .518x10^{-3}T + .11x10^{-3}PT + .0107R \qquad (5.4)$$

The data from Pulker et al.[6] in Tables I and II are not consistent with the data in their Table III for the index of refraction as a function of deposition rate. We have derived the slope of the rate portion of the equation from Table III, but the offset has been determined from the data in Tables I and II. It is not clear as to the source of the inconsistency of Table III; perhaps there is a typographic error in the temperature or pressure conditions reported by Pulker et al.[6] for that set of experiments.

Factor	A	B	Results
EXPERIM.	PRES	TEMP	REFRACT.
Row #	TORRx10-4	CELSIUS	INDEX
1	1.5	200	2.28
2	1.5	250	2.32
3	1.5	300	2.36
4	1.5	350	2.38
5	1	200	2.33
6	1.5	200	2.28
7	2	200	2.23
8	2.5	200	2.17
9	3	200	2.12
10	3.5	200	2.1
11	1	300	2.38

Factor	A	B	Results
EXPERIM.	PRES	TEMP	REFRACT.
Row #	TORRx10-4	CELSIUS	INDEX
12	1.5	300	2.36
13	2	300	2.33
14	2.5	300	2.26
15	3	300	2.22
16	3.5	300	2.18
17	1	350	2.41
18	1.5	350	2.38
19	2	350	2.36
20	2.5	350	2.29
21	3	350	2.25
22	3.5	350	2.22

Table 5.1. Design of Experiments input for the analysis of TiO_2 index of refraction data from Pulker et al.[6] used to generate Eqn. 5.4 and Fig. 5.21.

Y-hat Model					
Factor	Name	Coeff	P(2 Tail)	Tol	Active
Const		2.32301	**0.0000**		
A	ESSURE 1	-0.11928	**0.0000**	0.046	X
B	PERATU	0.000518	**0.0002**	0.131	X
AB		0.00011	**0.0361**	0.035	X
Rsq	0.9853				
Adj Rsq	0.9828				
Std Error	0.0116				
F	401.2133				
Sig F	0.0000				

Factor	Name	Low	High	Exper
A	ESSURE 1	1	3.5	2.5
B	PERATU	200	350	300
Prediction				
	Y-hat		**2.262440739**	

Table 5.2. Design of Experiments analysis results for TiO_2 index of refraction data from Pulker et al.[6] used to generate Eqn. 5.4 and Fig. 5.21.

Now that we have seen the way that the index of TiO_2 can vary with deposition conditions, it becomes easy to imagine why some practitioners have been less than enthusiastic about the material. From the partial derivatives of Eqn. 5.1, it can be seen that a 0.01 variation in the index of TiO_2 would result from a change of any of the following deposition parameters: 0.1×10^{-4} torr of oxygen pressure, 17°C in substrate temperature, or 1Å/second of deposition rate. The pressure seems to be the most sensitive and may be the most difficult to maintain. The material would have to be well outgassed so that no pressure bursts occur. It would also be desirable to be sure that some makeup oxygen was needed to achieve the desired background (such as 1×10^{-4} torr) so that a steady pressure could be maintained by a gas pressure controller. Melt conditions in the crucible would need to be fairly steady to keep the rate constant also. These conditions can be achieved with today's technology and some care and understanding. We therefore have found TiO_2 to be our preferred high index material in the visible and near IR spectrum.

The use of IAD with TiO_2 can generally give the same benefits as with SiO_2 such as reduced/eliminated humidity shift, higher density, etc. Some time ago, we developed a TiO_2/SiO_2 process using Argon IAD by a Kaufman gridded ion source with an oxygen background in the chamber. We were able to get rid of the humidity shift in a laser blocking filter with very tight edge requirements. We found that the SiO_2 (evaporated from an E-gun) could accommodate 600 eV bombardment without ill effects. However, the TiO_2 would have absorption if the ions (from the Kaufman ion source) were at much more than 200 eV. We used 600 eV for the silica layers and 200 eV for the titania layers. It was found that the silica layer after a titania layer could only use 200 eV for about the first 1/10th of the layer or there would be absorption. Our interpretation of this was that the higher energy (600 eV) ions would cause absorption producing damage to the titania and even penetrate some layer thickness of silica before the SiO_2 was thick enough to protect the TiO_2. We concluded that IAD for TiO_2 should be in the 200 eV range from a gridded source and 333 eV or less from a gridless source where the mean energies are estimated to be about 60% of the drive voltage. This has been consistent with our subsequent experience.

Most of our more recent TiO_2/SiO_2 processes have used 300 eV drive voltage as the goal for gridless sources for both types of material. This avoids changing drive voltage with each layer, and it has been found satisfactory for both types of material. The drive voltage is then chosen as high as the TiO_2 would allow, and the deposition rates of the material can be chosen to be as high as allowed by full densification with no absorption.

Once the TiO_2/SiO_2 process parameters have been optimized for a given chamber with or without IAD, any design for the appropriate spectral region can be adapted to that chamber by proper number of layers and thicknesses. If the process chosen was at an elevated temperature, the influence of different substrates might have to be accounted for. The optimum process might be done at near ambient temperature, at high deposition rates (1 nm/second), and would normally be one with little or no humidity shift or absorption and scattering. It might also be optimized for the best laser damage threshold. We have yet to simultaneously achieve all these goals, however. We would expect to use such a TiO_2/SiO_2 process for the 400-2000 nm spectral region and perhaps to 3000 nm or beyond. The materials have good environmental durability and are not expensive or toxic.

An interesting study was reported by Hsu and Lee[457] using single and dual-ion-beam sputtering. Although their processes were relatively slow and mostly used post baking, the study shows various properties and interactions that are useful for insight.

5.3.1.3. Magnesium Compounds

Magnesium Fluoride

Magnesium fluoride (MgF_2, index 1.32-1.39) has been used successfully for over half a century. It has been the most common optical coating on glass as a QWOT layer antireflection coating. It can be hard and relatively insoluble. It transmits well from about 120 nm in the vacuum UV out to about 7 μm the mid-infrared.

Krajnovich et al.[222] give its index at 248 nm as 1.43. Olsen and McBride[8] showed that a 2.75 mm thick single crystal of MgF_2 was clear from at least 200 nm out to about 6 μm and then started to increase in absorption toward longer wavelengths. At 10 μm, the transmittance fell to about 2%. A thin layer can be used as a top protective layer on coatings in the 8 to 12 μm region, although it has too much absorption for a thick layer in that region.

The hardness, durability, and density of MgF_2 without IAD is a strong function of the substrate temperature during deposition. At room temperature, the films can usually be rubbed off with your finger, have high humidity shifts, an index of about 1.32 in vacuum, and a packing density of 0.82 according to an extensive study by Ritter[9]. With a 300°C deposition temperature, Ritter[9] shows that the packing density is about 0.98 and the index nearly 1.39. The 300°C films will pass an eraser rub test and have low humidity shift. The variation in index and density are nearly linear between room temperature and 300°C.

MgF_2 is normally evaporated from a resistance boat source of molybdenum or tantalum where it melts to form a puddle and evaporates. The use of an E-gun with this and other fluorides seems to have some problems. Granular MgF_2 from an E-gun can easily cause "explosive" scattering of the material if the E-bean puts too much power on the material. It is conjectured that the rush of the fluoride vapors from evaporated granules below the surface of the granule pile "blows" the solid granules above like dust. It is therefore necessary to use great care and slowly bring up the power on the E-gun. We found the charge in the E-gun often becomes almost black. It is our opinion that a boat is the best approach unless there are extenuating circumstances forcing the use of an E-gun.

Magnesium fluoride films tend to have a columnar structure as shown by Guenther[10], high tensile mechanical stress as shown by Ennos[11], and scattering. As mentioned earlier, we have found this scattering a limiting factor in using MgF_2 for very broadband AR coatings with maybe six (6) layers of MgF_2. The mechanical stress can in principle be reduced by admixture of a second material of higher cation radius. Thielsch et al.[279] state that BaF_2 is such a material which also has a vapor pressure close to MgF_2. They investigated the thermal evaporation of various mixture ratios and substrate temperatures. The pure BaF_2 has an index of about 1.48 or 0.1 higher than MgF_2, so the index of the mixture rises in proportion to the BaF_2 content. They gathered useful data, but the results achieved were not the hoped-for degree of stress reduction on hot substrates.

The use of IAD to improve MgF_2 films has been investigated by several groups.[12-14,327] The bottom line is that films with the robustness of those deposited on 300°C substrates can often be deposited at ambient temperature with IAD. This is a valuable addition to many coating applications. Martin et al.[13] showed that the additional absorption in the UV between 120 and 180 nm was probably due to MgO. It would seem that this is of no concern for those working in the visible spectrum. The report of Kennemore and Gibson[12] points to the likelihood that energies lower than the 125-150 eV that they used may be optimal for MgF_2. The use of IAD for MgF_2 on plastics is almost mandatory to get reasonable adhesion and hardness. Our own experience points to the fact that MgF_2 cannot be "hit too hard" with ions, i.e., the eV must be low to avoid absorption due to "damage." This damage is conjectured by several authors such as Tsou and Ho[320] to be preferential sputtering of the fluorine from the film and disassociation losses of fluorine at the evaporation from the boat. The films are less than stoichiometric with the Mg:F ratio perhaps as low as 1:1.86[12] with oxygen filling most of the gap as MgO. Martin et al.[14] reported a somewhat different result with a 1:1.99 ratio. They used oxygen ions rather than argon and state that an ion-scattering spectroscopy (ISS) study "indicates that for films deposited by either evaporation or IAD the stoichiometry and surface concentration of oxygen is the same after exposure to atmosphere." It is a bit counterintuitive that oxygen is a good gas for the IAD of MgF_2! Is it possible that the lower atomic mass of the oxygen causes less differential sputtering than the argon?

Laux and Richter[319] and Laux et al.[209] used molecular-beam deposition at near room temperature for MgF_2 and reported on the packing density measurement technique. Their films had a packing density of about 85% of the bulk value.

Thomas[276] made porous coatings of MgF_2 from colloidal suspensions. These had indices as low as 1.2 because of the porosity and were therefore excellent AR coatings on low index substrates such as fused silica. They also had good laser damage resistance.

The photolithographic industry for semiconductor technology moves continually to shorter wavelengths. In recent years the emphasis has been moving from 355 nm to 193 nm. MgF_2 is a suitable material at these wavelengths. Scattering becomes ever more critical as the wavelength gets shorter. Quesnel et al.[277] addressed this problem in making MgF_2/LaF_3 multilayers by IBS. They found a need to replenish the lost fluorine atoms in the process by adding diluted fluorine gas. They achieved lower roughness and scattering than by conventional evaporation techniques. Ristau et al.[278] compared the results of IBS with both boat and E-beam PVD for MgF_2. They found the index at 248 nm to be 1.40-1.41 for the IBS films and higher by about .03 for the PVD films. It is conjectured that this is due to MgO and other contaminants in the PVD films. The IBS film showed an index of about 1.387 at 500 nm. Coating stresses were compressive for IBS and tensile for PVD.

MgF_2 has been and will continue to be an attractive and well used low index material. It appears, however, that there is still an opportunity for study and improvement of it with respect to stress and scattering.

Magnesium Oxide

Magnesia (MgO), as shown by Balzers,[7] must be evaporated by an electron beam source and it sublimes. The material seems to be hard, durable, and have attractive UV transmittance. In 5.5 mm thick bulk samples, it was shown by Oppenheim and Goldman[60] to transmit well to 6 μm in the IR and have some transmittance out to 9 μm. Therefore, in thin films it might be useful out to at least 9000 nm. Roessler and Walker[61] measured the bulk properties in the UV and reported no absorption down to 190 nm with indices of 1.86 at 250 nm and 2.06 at 190 nm. They showed a k-value of 0.1 at 166 nm with an n-value of 2.65. This points to the potential for MgO as a UV coating material.

Apfel[62] reported MgO/MgF_2 stacks with good transmittance from 220 to 400 nm, but that the number of layers might be limited to about sixty due to film stresses. Bradford et al.[63] investigated thin films of MgO from 220 nm to 8 μm and found no losses in that region. They obtained indices of 1.70 at 500 nm on ambient temperature substrates and 1.74 on 300°C substrates. Their report implies that these higher temperature depositions are virtually bulk-like in density and index, and that they show no water vapor absorption at 3 μm. The one problem mentioned is that exposed surfaces of MgO develop a hazy, bluish scattering surface due to interactions with atmospheric CO_2. This was not found to be a problem with inner layers of MgO. It was successfully used as the first layer of a classical MHL-index 3-layer AR coating with MgO/CeO_2/MgF_2. We speculate as to whether Apfel's[62] work with MgO and MgF_2 might be beneficially extended.

Krannig[328] used planar magetron sputtering of magnesium oxide film to be used as a protective dielectric coating for plasma display panels. Hartwig et al.[329] mention the use of MgO as a vapor barrier layer in web coating. Vedam and Kim[330] used MgO as an example of an inhomogeneous film for their analysis technique. The material has also been used as a thin adhesion layer in IR coatings.

5.3.1.4. Germanium

Germanium is the preferred high index material at index 4.0 or greater from 2 μm to at least 14 μm. As we saw in Chap. 1, Aguilera et al.[3] used Ge as the high index material to design high performance broad band ARs for Ge. It is a fairly robust material and well behaved during deposition. Its high index means that fewer layers can be used than with lower index materials to get a given reflectance. Balzers' data[7] indicates that it can be evaporated from tungsten and graphite boats, but our only experience has been using an E-gun. The material melts at about 937°C and forms a liquid in an E-gun and evaporates easily at about 1400°C. It was shown by de Sande[15] that Ge has less than full packing density when evaporated by an E-gun, but that IAD or laser-deposition gives near bulk density.

Oh[16] reported on significant gains using IAD on three layer ARs of ThF_4, Ge, and ThF_4 like that shown in Sec. 1.3.5 and five layer versions with two Ge layers. He found the best results using IAD only on the germanium layers and reported increased strength, packing density, and reduced columnar microstructure.

Rafla-Yuan et al.[318] studied Ge films with ellipsometric spectrometry in the spectral range 0.3 to 1.7 μm. This region, of course, has significant absorption and is not often considered in the case of germanium. However, the semiconductor nature of the material here can make it of interest in some specific applications, and the observations of the structure and density are of more general interest. They compared monocrystalline wafers of Ge with films evaporated by E-beam on substrates at 130°C. They found the films to be amorphous and inhomogeneous for these process conditions. The film index at 500 nm was higher (5.0) than that of the wafer (4.25). They attribute that to the evaporated film being in a non-equilibrium amorphous state, which resulted in a higher film density.

We have produced various 8-12 μm band filters with dozens of layers of ThF_4/Ge on Ge substrates. There was a great deal of residual stress in these thick coatings but they did pass the environmental tests (most of the time!). We also found[17] that there was significant change in absorption if the chamber temperature was too high. We backed down from 300 to about 200°C and the problem went away. We later became aware through Guenther[266] of the work of Evangelisti et al.[18] where they found that Ge has a transition from amorphous to crystalline in a narrow range from 240 to 280°C. There is also some earlier work by Dettmer et al.[262] and Adamsky[265] who studied the crystallinity of Ge and report even lower transition temperatures.

On one occasion, the Ge/ThF_4 multilayer filter production started to have excessive absorption. It took us about a week of tests to discover that the thermocouple which controlled the chamber temperature had been moved about 10 cm in the chamber cleaning process. As a result, the parts were exceeding the Evangelisti temperature limit even though the thermocouple was at the required temperature. When it was properly repositioned, the process returned to producing good parts.

5.3.1.5. Thorium Fluoride

Thorium fluoride (ThF_4) is one of the best low index materials for the region from 260 nm to beyond 12000 nm. However, it does have some radioactivity which we will discuss below. The index ranges from about 1.52 in the visible down to about 1.38 in the 10000 nm region and approaches 1.60 at its short wave limit as reported by Heitmann and Ritter.[19] They describe how they determined that thorium oxyfluoride ($ThOF_2$) actually deposits ThF_4 only because ThO_2 has a much higher melting/evaporation temperature and stays behind in the boat. ThF_4 evaporates at a somewhat lower temperature than MgF_2 from molybdenum or tantalum boats. We have generally used boats with reentrant covers to avoid flying sparks of fine particles of ThF_4. The films formed seem to be even more robust than MgF_2. ThF_4 would probably be the preferred material in the visible except for its somewhat higher index and its radioactivity. In fact, it appears to have been preferred in a past era before concerns about radioactivity. In the infrared region, it still has much to recommend it because no fully comparable material has been reported for the 8000-12000 nm region. Harrington et al.[306] concluded that ThF_4 was one of the lowest absorbing materials at the HF/DF chemical laser wavelengths (2.8-µm and 2.3-µm). As we mentioned in Sec. 5.3.1.4 on Ge, Thomas Oh[16] produced coatings with ThF_4 that were more robust than those done with other materials.

Al-Jumaily et al.[20] have described their results successfully using IAD with 300 eV argon ions at ambient temperature. Their results point to the fact that ThF_4 behaves much like MgF_2 in the sense of packing density and humidity shifts. Enough IAD can yield almost fully densified films which show almost no H_2O absorption at the 3 micron water band in the IR spectrum. This implies that a low spectral shift with humidity and greater general robustness is to be expected.

The issue of the radioactivity of ThF_4 is an interesting one. It appears to us that there has been an overreaction to the hazard. It is certainly true that proper precautions and understanding are needed, but there seems to be a certain amount of unfounded fear. Otten[21] gives a very good report of what is generally required for safe and legal use of ThF_4. He tells that the only time that the radiation monitor went off in a properly operated facility was when some operator returned to work after being treated by a doctor with radioactive iodine. Admittedly we have seen and heard of cases where ThF_4 was not treated with the proper respect in the past. A related sad case was reported once on the TV show "60 Minutes" wherein patients in the late 1940s were injected with thorium compounds for X-ray analysis. This was a grave mistake resulting from ignorance at that early stage in the knowledge of radioactivity and its effects on living tissue. On the other hand, we have heard it said that one could salt their food with ThF_4 and it would do no harm because it would pass through the system before any real harm was done. It is our understanding that the greatest hazard is if ThF_4 dust or flakes enter the lungs. Since there is no mechanism to eliminate it, it would stay there forever and continue to emit its low level of radiation to the detriment of its host.

Otten[21] describes everything that was necessary for a proper ThF_4 coating operation. Our conclusion is that all of the health hazards of ThF_4 can readily be protected against except one: the paperwork may kill you! It is our understanding that the bureaucratic problems may be even worse in some European countries.

5.3.1.6. Zinc Sulfide

Zinc sulfide (ZnS) is known for its advantages and notorious for its disadvantages as an optical coating material. It has a high index (2.35), a broad transmittance range (400 to 13 μm), ordinarily has compressive stress, good environmental durability, and can be evaporated rapidly from resistance sources. The index properties of the clear bulk version of the material, Cleartran™, have been reported by Debenham[272]. ZnS is also used in electroluminescent displays which we will not address here other than to refer the interested reader to a related paper.[271] It is known as a "dirty" material and, linked with this, it has a low sticking coefficient to substrates at elevated temperatures. Because it did not require an E-gun and has a high index, it has been used for decades in both the visible and infrared spectrum. We have had only limited personal experience with Zn, but we will describe our interpretation of what we read in the literature.

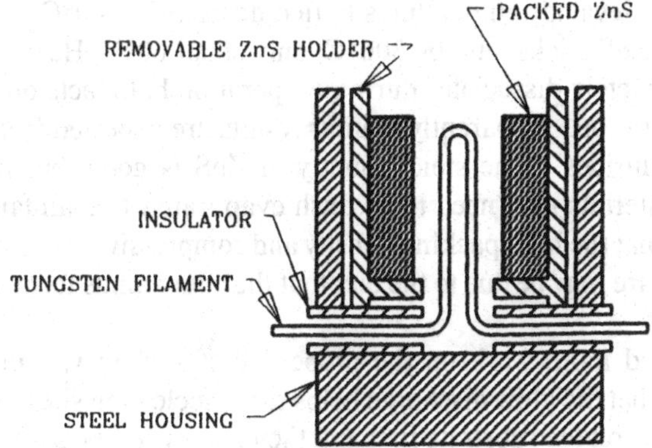

Fig. 5.22. The "howitzer" source which is attributed to Francis Turner et al. at Bausch and Lomb (in 1958).

Cox et al.[22] reported good optical and physical properties for ZnS as a reflector in the vacuum ultraviolet. Cox and Hass[23] reported on its use in the infrared. They indicated that the best films were deposited on 150°C substrates after glow discharge cleaning. They used the "howitzer" source shown in Fig. 5.22 which they attribute to Francis Turner et al. at Bausch and Lomb (in 1958). They said that it allowed "high deposition rates without noticeable decomposition." That was 10 nm/second with no noticeable absorption in the visible. It had to be well degassed as it was brought up to evaporation temperature.

Perry[24] used ZnS and ThF_4 to make He-Ne laser mirrors in the early 1960s. He stated that the mirrors were quite durable and could withstand careful cleaning with acetone, distilled water, and dry nitrogen to restore their original quality. He documented an interesting observation that using a glow discharge *created* surface contamination by small particles. He said that it appeared that the glow discharge sputtered particles from the chamber walls and fixtures.

Ammann and Wintemute[25] studied 1.08-μm laser damage to ZnS films. They deposited the ZnS at 6 nm/second from a molybdenum boat onto room temperature substrates with no glow discharge.

Netterfield[26] investigated the inhomogeneity of the index of ZnS with film thickness and found it lower in index (2.327) at the substrate and asymptotically approaching about 2.35 after 300 nm thickness. He deposited ZnS at about 0.5 nm/second on substrates at ambient temperature from an electron beam gun. Wu[27] came to similar conclusions on the inhomogeneity and also found a thin surface layer of lower index.

Ritter[28] investigated the "sticking coefficient" versus temperature that had been observed from earlier times. He graphed his results comparing the ratio of film thickness deposited at elevated temperatures to that deposited at 40°C. At 150°C, only 90% as much ZnS sticks, and by 300°C, only about 60%. He cites evidence that the ZnS completely dissociates during evaporation but reacts on a surface during film condensation. Apparently similar results are expected from sulfides, selenides, and tellurides. The stoichiometry of ZnS is good, but he mentions that the other materials may need to be flash evaporated to maintain stoichiometry. Ritter says that the high packing density and compressive stress of ZnS even at room temperature may be due to the fact that the compounds form at the substrate surface.

Hunter et al.[29] reported more work on the properties of ZnS in vacuum ultraviolet. They point out that the glow discharge helps form nucleation sites for the ZnS which must nucleate on the substrate first and then react with the sulfur. Without the glow, very irregular condensation is reported. This is in interesting contrast to the problems reported by Perry.[24] Hunter et al.[29] deposited at room temperature from a "howitzer" source. They say that tantalum, molybdenum, or tungsten boats can be used, but that the howitzer heats the ZnS by radiation. They describe the construction and use of the howitzer in some detail. Our own experience with ZnS was using a tantalum or molybdenum "pepper-pot" type boat with no direct path from the material load to the outside. This minimizes the flying ashes which would tend to cause "spatter" on the substrates as in SiO sublimation.

Hass et al.[30] reported that UV radiation has a major effect on ZnS. They found that a 15-20 nm thick layer of ZnS had completely converted to ZnO as a result of UV exposure in air!

Bangert and Pfefferkorn[31] described experiments with the type of crystal structure of ZnS after evaporation from boats and an electron beam gun. They concluded that E-gun films should be more stable.

Himel et al.[32] deposited ZnS on liquid-nitrogen-cooled substrates and found that the films actually had tensile stress rather than the usual compressive stress. In cooperation with Ruffner et al.[33] they carried the work further. They observed that minimum stress and optical losses occured at about -50°C. Their work using IAD did not result in improvements, but suggested that lower energy ions and other than argon gas might give better results in the future.

Rudisill et al.[34] described evaporating ZnS from a vitreous carbon crucible with an overhead pancake tungsten filament. Macleod[4] on (page 365 in the Second Edition) mentions this technique for ZnS and SiO. It is much less prone to spitting because the surface is heated by radiation as in the howitzer, and the hottest part is the evaporating surface.

It can be seen that there are a lot of things to consider in the use of Zn. It also appears that the understanding has improved over the decades, but that there is not necessarily a consistent explanation for all of the behaviors of ZnS even now. The reason that we have had little experience with it is a conscious avoidance of its use. This is because of its "dirty" nature, i.e., it does not all condense on the first surface that it encounters. It tends to get behind masks and baffles and require extra cleaning that is not usually needed with most other materials. It also seems to have a high vapor pressure at temperatures like the 300°C that may be used for TiO_2 and SiO_2 coatings. As a result, coating operations tend to use separate chambers for ZnS processes and visible hard coating processes. After using Zn, if it is necessary to go back to 300°C processes, the residual Zn on the walls and fixtures may sublime and/or migrate onto the substrates and make a non-stick layer before the intended coating is deposited. We have had to do a thorough chamber cleaning and then bake out the chamber at the hottest temperature that it will take for a several hours to get rid of the ill effects of the Zn.

The merits and faults of ZnS need to be carefully weighed in any application. In almost any case it is a complex judgement call as to whether to use it or not.

5.3.1.7. Zinc Selenide

Much that has been said about ZnS can also be said about zinc selenide (ZnSe). It seems to have a higher sticking coefficient as a function of temperature than ZnS, but similar behavior. It is therefore also somewhat of a dirty material. We have evaporated it from a boat onto substrates at 250°C in layer stacks with Ge without difficulty. Bulk ZnSe has a somewhat higher index than ZnS, about 2.6 at 500 nm, and less absorption in the visible and beyond 12 μm out to 14 μm. It is not as hard in the bulk as ZnS. Harrington et al.[306] concluded that ZnSe was one of the lowest absorbing materials at the HF/DF chemical laser wavelengths (2.8-μm and 2.3-μm).

Rudisill[34] reports depositing ZnSe on 150°C substrates from an alumina crucible heated in a tantalum boat. Gluck et al.[35] evaporated ZnSe onto 150°C substrates from tantalum boats. They investigated mixed films of Si/ZnSe and also ZnSe/LaF_3 in attempts to get materials for long-wavelength IR gradient-refractive-index optical filters. We also have made gradient coatings successfully with ZnSe and Ge.

Rizzo et al.[294] used RF magnetron sputtering for ZnSe and were able to control the stress from tensile to compressive at near bulk densities.

5.3.1.8. Lead Telluride

Lead telluride, PbTe, is one of the highest index IR materials. Palik[82] reports that the bulk material has a k-value under 0.1 from 3.8 μm to 25 μm and an index of 6.02 at 4 μm. It may be transparent as a thin film material from shorter than 3.8 μm to nearly 40 μm. Thin film reports imply an index of about 5.1 to 5.5 in the IR. We have no personal experience with it, but refer the reader to Macleod[4] (page 395) for a summary of its properties. He recommends tantalum boats and Balzers[7] lists molybdenum also. Balzers lists the index as 5.6. The material sublimes, and it is recommended not to heat it any more than necessary as it may lose long wavelength transparency due to increase free-carriers. Material purity seems to be an issue for the same reason. Substrate temperatures up to 250°C are said to be beneficial. Health precautions are necessary, see Sec. 5.3.1.31.

Seeley et al.[36,37] have applied PbTe extensively out to 40 μm with good results. They[37] also have a discussion of other materials used to work in the infrared beyond the common 14 μm edge of the atmospheric window. Zhang et al.[260,261] report in detail on the material and its index at a wide range of environmental temperatures.

5.3.1.9. Hafnium Compounds

Hafnium Oxide

Hafnia, or hafnium dioxide (HfO_2), has become relatively popular in recent times. The properties of the bulk material have been reported by Wood et al.[207] along with yttria stabilized cubic hafnia, and cubic zirconia. There apparently was little done with it until the availability of electron beam sources. Ritter[9] shows data that imply its usefulness down to about 220 nm in the UV, and Baumeister and Arnon[38] state that its high transparency extends down to 235 nm. The latter reports indices at about 500 nm which vary around 2.0 for deposition on 250°C substrates, depending on the deposition conditions. Smith and Baumeister[40] deposited HfO_2 and found the mean index between prebaked and postbaked at 550 nm to be 2.007. The n and k at 250 nm were 2.214 and .018. This work was before the widespread use of IAD, and shows potential for improvement to a stable index between 2.05 and 2.1 with oxygen IAD. Cox and Hass[39] show the benefit of HfO_2 instead of SiO_2 as a protective overcoat on aluminum for the 8-12 μm region. Smith and Baumeister[40] reported some later work with hafnia and other materials that showed similar results. Kruschwitz and Pawlewicz[192] gave the optical and durability properties of hafnia and other materials. Our own experience showed it to be a preferred high index material in the UV as compared to zirconia. Figure 5.19 shows the transmittance of hafnia, yttria, zirconia, niobia, and titania films in the UV. The yttria transmits a bit further into the UV, but it has an index of 1.72 in the visible as compared to hafnia's 1.93.

There have been several articles pointing to the tunneling effects of HfO_2 and attempts to overcome the problems including Chow et al.[203,204] and Harris[206]. We found a more satisfactory approach, when an ion source is available. The deposition was treated somewhat like that of converting SiO to SiO_2 as in Sec. 5.3.1.1. The hafnium metal was evaporated from a vitreous carbon liner in an E-gun pocket. The metal was totally molten and therefore evaporated evenly without any tunneling or "spitting". A tungsten liner was tried first, but it was attacked by the molten metal. The liner provides some thermal barrier so that the metal can remain liquid, similar to the TiO_2 process that we prefer. The oxygen to form HfO_2 was provided by a DynaVac PS1500[461] high power ion/plasma source at 1 KW of drive power (about 10A at 100V). This supported a 1 nm/sec rate in a 1.4 m box coater. The films were clear at 350 nm with an index of 2.08+i0 and had the following values at shorter wavelengths: 300 nm, 2.11+i.001; 260 nm, 2.2+i.01; 240 nm, 2.27+i.025; 2.25 nm, 2.17+i.039.

Stolz et al.[219] and Chow et al. [221] describe depositing hafnia from the metal. It appears that they used activated oxygen but no IAD and that crucible liners were not used. Jensen et al.[331] describe their optimization of IAD parameters for low humidity shift hafnia.

We have also used hafnia as a glue layer of 5-10 nm thick between Ge and ThF_4 as a stress reduction/adhesion promoting means.

Hafnium Fluoride

Traylor-Kruschwitz and Pawlewicz[192] deposited HfF_4 at 250°C. They found an index of about 1.56 in the visible and near IR spectrum and 1.46 in the 8 to 12 μm band. It was clear in the near IR but the k went from 0.010 to 0.070 in the 9 to 12 μm range. Their films were environmentally satisfactory except for the 24 hour humidity followed by the slow tape pull test.

5.3.1.10. Niobium and Neodymium Compounds

Some years ago we experimented briefly with Nb_2O_5 which behaved for us somewhat like Ta_2O_5 and has optical properties more like TiO_2. Wolfe[308] sputtered niobium oxide for large scale applications and found it has an index of refraction of 2.28. He states that the deposition rate of NbO_x is about 2.5 times greater than TiO_x in his process, and that this material also gives the lowest amount of compressive intrinsic stress of all reactively sputtered oxides. Figure 5.19 shows the transmittance of hafnia, yttria, zirconia, niobia, and titania films in the UV.

Similarly, we looked at neodymium oxide (Nd_2O_3) and fluoride (NdF_3) on which Hass et al.[41] reported. Smith and Baumeister[40] deposited NdF_3 and found the mean index between prebaked and postbaked at 550 nm to be 1.617. McNally et al.[193] list the material as a commonly used substitute for thorium fluoride, but that it exhibits stress cracks in thick layers. Traylor-Kruschwitz and Pawlewicz[192] deposited NdF_3 at 250°C. They found an index of about 1.7 in the visible and near IR spectrum and 1.6 in the 8 to 12 μm band. It was clear in the near IR but the k went from 0.010 to 0.025 in the 10 to 12 μm range.

Laux and Richter[319] and Laux et al.[209] used molecular-beam deposition at near room temperature for NdF_3 and reported on the packing density measurement technique. Their films had a packing density of about 85% of the bulk value. Adamik et al.[441] studied NdF_3 stratified with MgF_2 and also CaF_2. They show TEM, XRD, and AFM analyses of the films. They discuss the fact that additives and impurities can change the structural evolution per the M-D and/or Thornton Zone Models. The impurities can inhibit or promote the evolution of specific structures and shift the zone boundaries to higher or lower substrate temperatures. The grain size increased with CaF_2 stratification and decreased with MgF_2 stratification.

It does look as if these materials and praseodymium oxide (Pr_6O_{11}) might be worthy of further investigation for possible improved durability over the more common materials.

5.3.1.11. Yttrium Compounds

Yttrium Oxide

Yttria (Y_2O_3) has become a material of some interest. It requires an electron beam to evaporate it. Hass et al.[41] reported briefly on it, and that it showed a variation with film thickness (inhomogeneity). Smith and Baumeister[40] deposited Y_2O_3 and found the mean index between prebaked and postbaked at 550 nm to be 1.789. The n and k at 250 nm were 1.914 and .005. Cox and Hass[39] showed it to be very favorable as a protective coating for aluminum, particularly with respect to high angles of incidence in the infrared from 8-12 µm. Wickersheim and Lefever[42] reported the transmittance of the bulk material in 1.89 mm thickness to go out to beyond 8 µm. Heitmann[43] had data on thin films of yttria that transmitted well from 300 nm to 14 µm. Lubezky et al.[44] used yttria with success to protect silver mirrors which passed eraser rub, cellophane tape test, and 24 hour humidity exposure. Phillips et al.[332] investigated the use of yttria in barrier layers by web coaters. Figure 5.19 shows the transmittance of hafnia, yttria, zirconia, niobia, and titania films in the UV. The yttria transmits a bit further into the UV, but it has an index of 1.72 in the visible as compared to hafnia's 1.93.

Yttrium Fluoride

Hass et al.[41] included YF_3 in an early survey on rare earth materials. McNally, et al.[193] report that films up to a QWOT for 2.6 µm could be prepared, with low stress that were transparent out to 14 µm with a refractive index of 1.36 at 10.6 µm. They report no water absorption bands at 3 and 6 µm if the films are evaporated at high temperature. The disadvantages are possible adhesion problems on ZnSe and that it does not pass MIL-C-675 for severe abrasion resistance. Mendes et al.[195] reported on extensive investigations of YF_3 with and without CaF_2 doping which were followed by a report of further work by Jacobson et al.[194] The coatings were deposited with IAD using argon only at a maximum of 300 eV and varying ion fluxes. Their initial conclusions were that moderate ion flux and substrate temperatures gave the best results including low tensile stress and good transmittance in the UV, visible, and IR spectral regions. The indices from 300 to 1100 nm ranged from 1.45 to 1.54 depending on the deposition conditions. The coatings passed tape tests after humidity. Their later results confirmed these findings and that the deposition temperatures needed to be below 180°C. Quesnel et al.[198] compared YF_3 which was E-beam evaporated at 150°C and ion beam sputtered at 60°C. It is interesting to see that their micrographs of both results look amorphous. They also provide extensive analyses of structure, stress, mechanical properties and aging of YF_3 and LiF.

5.3.1.12. Zirconium Dioxide

Zirconium dioxide (ZrO_2) or zirconia is described by Macleod[4] as hard and tough, but inhomogeneous. Klocek[310] gives data on the bulk material with index 2.166 at 550 nm for cubic zirconia. Figure 5.19 shows the transmittance of hafnia, yttria, zirconia, niobia, and titania films in the UV. The yttria transmits a bit further into the UV, but it has an index of 1.72 in the visible as compared to zirconia's 1.86. Balzers[7] indicates that ZrO_2 must be evaporated from an electron beam source, but that ZrO (monoxide) can be evaporated from tungsten boats. This sounds a bit like SiO and SiO_2. The spectral range for thin films covers at least 340 to 12000 nm with an index of about 2.05 at 500 nm. However, Fig. 2 from Ritter[9] implies that ZrO_2 might be used down to less than 250 nm in some cases. Smith and Baumeister[40] deposited ZrO_2 and found the mean index between prebaked and postbaked at 550 nm to be 2.037. The n and k at 250 nm were 2.307 and .010. Clapham et al.[45] used zirconia in polarization devices for its specific index (2.05). They found that the films sometimes needed to be baked to get rid of absorption. Coleman[46] was not successful in sputtering zirconia from the oxide. However, he produced sputtered films of ZrO_2 by reactive sputtering of the metal in an argon-oxygen mixture and found the index at 1060 nm to be 2.2. This is consistent with the probable bulk value of about 2.25 at 500 nm as implied by the report of Wood et al.[311,47] Ritter[9] reported densities of 0.67 of bulk for depositions at 30°C and only 0.82 at 300 °C. These results further imply that zirconia as a thin film material has usually lacked full densification and should benefit by the proper application of IAD to increase its index to the bulk value and overcome inhomogeneity. This was confirmed by Cho and Hwangbo[202], wherein they could even change the sign of the inhomogeneity using the proper parameters in IAD. Stetter et al.[48] discussed the inhomogeneity in detail and described the properties of "Substance 1" which is a mixture of ZrO_2 and $ZrTiO_4$. The results of Smith and Baumeister[40] also appear to suffer from lack of densification as evidenced by the typical index of 2.05 at 500 nm.

Martin et al.[49] used 600 eV argon IAD on ZrO_2 and found increased absorption presumably due to preferential sputtering loss of the oxygen. When they added an oxygen backfill of 3.8×10^{-5} torr, the films showed lower losses and had an index of 2.15 at 550 nm. This reduced the humidity shift to negligible amounts, implying near full densification. Sainty et al.[50] were successful in protecting aluminum and silver films with ZrO_2 deposited with 700 eV argon only IAD on room temperature substrates. Klinger and Carniglia[51] showed that zirconia films deposited at 300°C are crystallographically inhomogeneous which accounts from some of the inhomogeneity noted by many observers over the years. They found that optical thicknesses of less than a QWOT at 600 nm of deposition was cubic phase, and after that the film was "only" the monoclinic phase (which has a lower index of refraction). They posed several then unanswered questions resulting from their findings. Martin and Netterfield[309] studied the deposition of zirconia *in situ* with ion scattering spectroscopy and concluded that complete substrate coverage occurs with 0.3 nm films. Dobrowolski et al.[52] deposited zirconia reactively (without IAD) from the metal with higher (2.14) than normal index of refraction and lower absorption. They report on their process in great detail.

We conjecture that the use of only oxygen for IAD at appropriate ion energy might result in even higher index, more homogeneous films than have been reported to date. The question then would be whether this ZrO_2 would then have more desirable properties than our favored TiO_2 processes.

5.3.1.13. Tantalum Pentoxide

Tantala (Ta_2O_5) can be used much like TiO_2 except the index is somewhat lower (1.9 to 2.0) and it tends to need baking more frequently to get rid of absorption. Smith and Baumeister[40] deposited Ta_2O_5 and found the mean index between prebaked and postbaked at 550 nm to be 2.138. The n and k at 300 nm were 2.311 and .0085. Balzers[7] shows that it can be evaporated from tungsten boats, but our experience has all been to evaporate from an electron beam source with an oxygen background of $1\text{-}2\times10^{-4}$ torr. Milam et al.[53] studied tantala with silica coatings extensively in search of better laser damage thresholds. They also found that baking at 400°C in air made major changes in absorption and residual stress in the coatings. Demiryont et al.[54] reported on ion-beam sputtering of tantalum oxide. Their indices ranged from 1.92 to 2.18 for non-absorbing films at 550 nm, and the absorption set in at wavelengths shorter than 365 nm and the cutoff wavelength was 252 nm.

Lechner et al.[199] reported RLVIP coated Ta_2O_5. It appears that fully dense films had high compressive stress and some absorption after coating; but when annealed for 4 hours at 350°C, the stress and absorption decreased significantly and the thickness increased by over 1%. They attribute this to improved film stoichiometry.

Ogura[281] found that a starting mixture of Ta_2O_5 and 7 wt% Ta metal gave the most stable results. He describes the oxygen outgassing that occurs with simple Ta_2O_5 from an E-gun similar to the case with TiO_2. In this case, the Ta metal in the mixture reacts with the liberated oxygen to maintain an oxygen depleted melt in the crucible. He found film stresses to be compressive in many cases. We have never attempted outgassing Ta_2O_5 in the crucible as we prefer for TiO_2, but we know of no reason why it would not be as satisfactory and obviate the need for more complex and/or expensive mixtures.

In the 2000-2001 time period, many new coating systems moving toward production of DWDM filters were using DIBS to deposit Ta_2O_5 reactively from titanium metal targets. Their low index material was SiO_2 from fused silica targets also by DIBS. Takashashi[201] published and important paper on how the coefficient of thermal expansion of the substrate is critical to the temperature stability of the center wavelength narrow bandpass filters and the stress relationships that exist. Zöller et al.[455,456] reported on temperature stabilization. The choice of substrate material also appears to be critical to gain the minimum sensitivity of such filters to temperature shift. Faber et al.[458] touches on this substrate choice and other aspects of DWDM filter design and production. If a different high index material were used, we would expect a different substrate material to be optimum for low temperature shift.

5.3.1.14. Aluminum Compounds

Aluminum Oxide

Alumina (Al_2O_3) is a commonly used intermediate index material at about 1.6-1.7. Ritter[9] reports that it has 100% packing density at ambient and elevated temperatures and a transparent band from 200 to 7 μm. Eriksson et al.[55] show the infrared transmittance to approach 10 μm and the index at 500 nm to be 1.60. Krajnovich et al.[222] give its index at 248 nm as 1.72. It is evaporated from an electron beam source. This is done with or without some background oxygen in the chamber, each case might be examined to determine the need or benefit of the oxygen. As is common with other materials, increased substrate temperature leads to higher index films, which seems to be at odds with the Ritter[9] report of 100% density at ambient temperature.

Ross et al.[295] did extensive work to produce thick hard wear layers of alumina on polymer substrates in an IAD PVD process. They achieved good coatings of 4 μm thickness. They concluded that "Some type of ion bombardment is required to produce crack-free Al_2O_3 coatings that are thick enough to offer enhanced wear and appearance retention." "That bombardment may be by ion-assist, plasma-assist, or by the self-biasing which occurs during sputtering." Kennedy et al.[296] applied this work to solar reflectors. They tested designs with up to 5.2 μm of alumina as a protective overcoat on silver. Their results are promising toward the goal of a reflector which will function well for at least 10 years in the outdoor environment.

Thielsch et al.[442] investigated the use of Al_2O_3 by plasma IAD with SiO_2 for UV applications. They found high LIDTs and environmentally stable optical characteristics.

We have had some private conversations indicating that the filtered arc process has be successfully used to produce alumina films with bulk (corundum) density and hardness (22 Gpa). Fietzke et al.[297] reported on alumina deposited by dual pulsed magnetron sputtering for tribological applications. They achieved full density and hardness only at 700°C; therefore, the optical applications of that process may be limited at full hardness.

Aluminum Nitride

Martin et al.[298] give the details of depositing AlN by reactive IAD of E-beam evaporated aluminum metal. They report the index to be up to 2.10 at 633 nm, and the films were highly transparent over the region of 275-800 nm. They found the stresses could be quite high, and ion energies and arrival rates needed to be carefully controlled to minimize these effects. The films are known to be quite hard and have merit as tribological coatings. Kusaka et al.[299] used planar magnetron sputtering to produce AlN. They found that residual stresses could be changed from compressive through tensile by the nitrogen gas pressure of the process.

Garzino-Demo and Lama[300] mention its application to automotive mirrors because it is transparent, very hard, and very resistant to pollution and wear. Verhoeven, and Steenbergen[301] mention the use of AlN in optical media. There are also many papers on aluminum oxynitrides such as the application by Holmes and Biricik[302] to graded-interface and rugate filter technology.

Aluminum Fluoride

Aluminum fluoride (AlF_3) is mentioned by Ritter[28] as subliming from molybdenum crucibles, having transparency from 195 nm to 10 μm and an index of 1.38, and a low packing density. Krajnovich et al.[222] give the index as 1.36 at 248 nm. Cerac[66] claims that it has been used for eximer laser mirrors, is non-hygroscopic, has no absorption from 250 to 1000 nm, and has excellent abrasion resistance. McNally et al.[193] report that films deposited at high temperatures (250-300°C) do not show water absorption or stress cracks for a QWOT at 6 μm. They find a refractive index of 1.36 at 230 nm and 1.28-1.30 at 3 μm and a long-wave cut-off at 7 μm. The films are very humidity sensitive with poor environmental resistance. AlF_3 is a component of Cryolite, $NaAlF_4$, which has been used for years, but is not known for durability when unprotected. Ritter[9] discusses the behavior of Cryolite in some detail during evaporation.

5.3.1.15. Cerium Compounds

Cerium Oxide

Cerium dioxide (CeO_2) was used by Cox and Hass[23] as a single layer AR on germanium for the 2 to 5 μm region. They evaporated it from heavy tungsten boats on 200°C substrates and got an index of about 2.2. They found a water absorption band at about 3 μm which is consistent with later findings of porosity. Hass et al.[56] reported further on the details of CeO_2 and found it to be very durable.

The index was a strong function of substrate temperature and reached about 2.45 at 500 nm on a 300°C substrate. The material absorbs at wavelengths shorter than 400 nm. The outgassing before evaporation is critical to repeatable high index values. Macleod[4] (page 391) describes the material as awkward to handle due to inhomogeneities. He reports that molybdenum boats cannot be used because of reactions which produce a white powdery coating. Smith and Baumeister[40] deposited CeO_2 and found the mean index between prebaked and postbaked at 550 nm to be 1.999. Ritter[28] reports that the material is usable to 12 μm.

Like many of the materials discussed above, CeO_2 suffers from lack of packing density when evaporated by historic techniques. Hodgkinson et al.[315] show scanning electron micrographs of films deposited at 300°C which exhibit a very columnar structure. Even at 300°C, CeO_2 is below the Thornton Zone 1 boundary because its melting point is about 2600°C. Therefore it is not surprising that the structure is quite columnar. However, to circumvent this columnar structure and lack of density, Netterfield et al.[57] had great success using oxygen IAD at about 300 eV at ambient temperature to get low absorption films of index up to 2.49 at 550 nm. Without IAD, they report films with a packing density of about 0.55 and large moisture shifts. It appears that a fully densified film may have an index greater than 2.45 at 500 nm. Inoue et al.[269] analyzed the surface morphology of epitaxial CeO_2 with respect to crystallinity and added further understanding of the films. It could be that IAD will make CeO_2 much more popular than it had been without it.

Cerium Fluoride

Cerium fluoride (CeF_3) was investigated by Hass et al.[41] who used a heavy tungsten boat. They found an index of 1.63 at 500 nm and very satisfactory mechanical and chemical durability. They noted absorption maxima at 234 and 248 nm and negligible absorption at longer than 300 nm. Cox et al.[67] used CeF_3 in extensive tests of three layer ARs. Fujiwara[68] evaporated mixtures of CeF_3 and CeO_2 from molybdenum boats and could get any index desired between 1.60 and 2.13 with reasonable reproducibility. He reported satisfactory chemical and mechanical durability. Macleod[4](page 394) mentions obtaining very stable films of index 1.63 at 550 nm by evaporation from a tungsten boat. Smith and Baumeister[40] deposited CeF_3 and found the mean index between prebaked and postbaked at 550 nm to be 1.619. Ritter[28] (page 33) comments for some reason that the stability of CeF_3 is not very satisfying. Traylor-Kruschwitz and Pawlewicz[192] deposited CeF_3 at 250°C. They found an index of about 1.64 in the visible and near IR spectrum and 1.57 in the 8 to 12 μm band. It was clear in the near IR but the k went from 0.010 to 0.032 in the 9 to 12 μm range. Their films passes environmental tests of moderate abrasion, tape, and adhesion even after 24 hour humidity.

5.3.1.16. Scandium Oxide

Scandia (Sc_2O_3) or scandium oxide was discussed by Heitmann[43] as similar to Yttria. He reported the value for bulk material to be 1.964. His work showed 1.87 and 1.92, depending upon annealing, at 550 nm for evaporated films. He also showed good transmittance for films out through the 8 to 12 μm window in the atmosphere. Rainer et al.[58] investigated the material extensively for high power UV laser applications. They found high laser damage resistance in use at 248 nm. Arndt, et multi al.[59] measured the index of a batch of "identical" samples of Sc_2O_3 prepared by electron beam evaporation on 150°C substrates. The index was found to be 1.86 ± 0.01 at 550 nm as a mean of the measurements by seven different institutions with different approaches to the measurement. We are not aware of any studies using oxygen IAD with Sc_2O_3, but suspect that it might give a positive result. Note that this material has been found to be expensive, see Sec. 5.3.1.32.

5.3.1.17. Zinc Oxide

Zinc oxide (ZnO) is most recognized as an element of UV screening sun cream. It has the property of absorbing wavelengths shorter than about 400 nm and thereby blocks harmful ultraviolet radiation to the skin. This can also be of interest in protecting polymers which may suffer chain scission by an energetic photon or even become crosslinked and therefore more brittle by radicals produced by such a photon. Ye and Tang[64] state that the absorption edge is at 385 nm. They used magnetron sputtering with indium doping to get inexpensive conductive coatings. Fabricius et al.[65] also sputtered ZnO, but to make UV detectors. In Sec. 5.3.1.6, we mentioned that Hass et al.[30] reported that UV radiation has a major effect on ZnS. They found that a 15-20 nm thick layer of ZnS had completely converted to ZnO as a result of UV exposure in air.

Aluminum-doped ZnO (ZAO) has become widely used as a transparent conductive oxide (TCO) coating. Yoshimura et at.[333] DC magnetron sputtered c-axis crystallographically oriented ZAO (C-ZAO) and achieved resistivity as low as 1000 μΩcm. Ogawa et al.[335] used DC magnetron sputtering from sintered ZAO targets and Zn/Al targets reactively. Their ZAO films were compared favorably with ITO for resistivity, abrasion resistance, and transparency in the visible spectrum. Laird and Wolfe[290] compare ZnO and silver for low emissivity applications and protective layers for them such as Si_3N_4. Nadel[291] reviews low emissivity applications and the use of ZnO therein. May et al.[334] compare the results of processing with DC and medium frequency (MF) sputtering in a production coater. Aluminum-doped Zinc oxide films were deposited by reactive sputtering from ceramic oxide targets in the dual magnetron arrangement with 10 Ω sheet resistance and maximal transmittance of 88.5 % at substrate temperatures <150°C. Szyszka[337] deposited reactively MF sputtered ZAO from metallic targets. Films prepared by this technique exhibit low resistivity of 300 μΩcm at 200°C substrate temperature and 480 μΩcm at 100°C substrate temperature (film thickness of 500 nm).

5.3.1.18. Lead Fluoride

According to Ritter[28] (page 34), lead fluoride (PbF_2) is used as a high index material in the UV, 1.998 at 300 nm. He says that it would be reduced by contact with hot tantalum, tungsten, or molybdenum boats, so that Pt or ceramic boats are needed. Ennos[11] showed that it has a relatively low stress which starts as compressive and becomes increasingly more tensile with thickness, but this is independent of the deposition rate. He deposited from a platinum boat. Sankur[69] investigated the evaporation of PbF_2 with lasers by had some problems with higher extinction coefficients. Destro et al.[70] used an aluminum oxide coated tungsten crucible to evaporate lead fluoride for their property analyses and obtained an index of 1.775 to 1.78 at 500 nm. Horowitz and Mendes[71] used a tungsten boat for their investigations with PbF_2. Harrington et al.[306] concluded that PbF_2 was one of the lowest absorbing materials at the HF/DF chemical laser wavelengths (2.8-μm and 2.3-μm). Traylor-Kruschwitz and Pawlewicz[192] deposited PbF_2 at 250°C. They found an index of about 1.46 in the visible and near IR spectrum and 1.3 in the 8 to 12 μm band. It was clear in the near IR but the k went from 0.010 to 0.030 in the 8 to 12 μm range. Their films passed moderate abrasion and tape, but not 24 hour humidity followed by slow tape pull.

5.3.1.19. Calcium Fluoride

Calcium fluoride (CaF_2) is mentioned by Heavens[72] as being evaporated at pressures above 10^{-4} torr to achieve an index of about 1.23 to 1.28 as an approach to the ideal index of 1.22 for a SLAR on crown glass. However, he says that the resulting film is not very satisfactory. Ritter[9] reports that the packing density of CaF_2 deposited at room temperature is only about 0.57. This is consistent with the bulk index given by Ennos[11] as 1.434, and implies the normal film is full of voids and might be expected to be weak and prone to shifts with change in humidity. Ennos[11] show that the high initial tensile stress decreases with thickness of the film. Heavens and Smith[73] noted that increased film thickness results in considerable visible scatter, and they saw evidence of strong inhomogeneity wherein the films were less dense with increasing thickness. Zukic et al.[74] did extensive work with materials for the vacuum UV. They showed that CaF_2 had an extinction coefficient less than 0.1 for wavelengths longer than 130 nm and an index of about 1.40 at 200 nm when deposited on 250°C substrates. They used an electron beam source at 10 kV. Balzers[7] shows that CaF_2 can be evaporated from tungsten, tantalum, or molybdenum boats and that it sublimes. Traylor-Kruschwitz and Pawlewicz[192] deposited CaF_2 at 250°C. They found an index of about 1.39 in the visible and near IR spectrum and 1.32 in the 8 to 12 μm band. It was clear in the near IR but the k went from 0.010 to 0.022 in the 10 to 12 μm range. Their films did not pass the slow tape pull test after 24 hour exposure to high humidity. CaF_2 will transmit out to beyond 12 μm in the IR. Its lack of full densification seems to be the basis of its present limited usefulness. Gluck et al.[259] produced films at room temperature with IAD and laser evaporation having near bulk density, improved durability, and a near bulk index of ~1.44. This points to the fact that future improvements in IAD of fluorides may make this and other fluoride materials more useful.

Thomas[276] made porous coatings of CaF_2 from colloidal suspensions. These had indices as low as 1.2 because of the porosity and were therefore excellent AR coatings on low index substrates such as fused silica. They also had good laser damage resistance.

5.3.1.20. Barium Fluoride

Barium fluoride (BaF_2) has similar physical characteristics to CaF_2. Malitson[75] measure the bulk properties from 265 nm to 10.346 μm which ranged in index from 1.512 to 1.396 with an index of 1.477 at 500 nm. Laufer et al.[76] showed that the bulk material would transmit down to 135 nm in the VUV, but that the edge moved to longer wavelengths at elevated temperatures. Gibbs and Butterfield[77] measured the absorption of thin films at 10.6 μm and showed the needlelike crystals in the deposited film. Kemeny[78] deposited BaF_2 on room temperature substrates and found packing densities of about 0.66 for low deposition rates. He found a nearly linear increase of density with increasing rates up to 0.83 for a 20 nm/second rate. The index values were lower than the bulk values by a factor consistent with the packing density. At 5 nm/s, the index at 1000 nm was 1.30 as compared to 1.47 for the bulk. Zukic et al.[74] found an index of about 1.70 at 200 nm and an extinction coefficient of less than 0.1 out to about 135 nm. The physical limitations of BaF_2 again seem to result from lack of full densification.

5.3.1.21. Ytterbium Fluoride

Traylor-Kruschwitz and Pawlewicz[192] deposited YbF_3 at 250°C along with many other materials and reported comparative results. They found an index of about 1.52 in the visible and near IR spectrum and 1.48 in the 8 to 12 μm band. It was clear in the near IR but the k went from 0.010 to 0.055 in the 8 to 12 μm range. They report that the fundamental absorption became increasingly large at wavelengths less than 0.6 μm and that their film showed a considerable amount of water band absorption compared with other fluorides. This might imply that the film was not fully dense. The film passed moderate abrasion, adhesion tests with fast tape pull, and slow pull adhesion even after 24 hour humidity. McNally et al.[193] list the material as worthy of further investigation. Jacobson et al.[194] showed some results with YbF_3 wherein IAD had been used. They concluded that high deposition temperatures (>180°C) were detrimental. Their data correlated with that of Ref. 192 for index but showed no UV-visible absorption. This seems to imply that the conditions used by Jacobson, et al.[194] might give better results with respect to absorption.

5.3.1.22. Lanthanum Compounds

Hass et al.[41] reported in 1959 on the properties of the oxide and fluoride of lanthanum with indices of about1.9 and 1.6 in the visible spectrum. Smith and Baumeister[40] deposited La_2O_3 and found the mean index between prebaked and postbaked at 550 nm to be 1.868. The n and k at 250 nm were 2.012 and .002. LaF_3 had a mean index of 1.602 at 550 nm and 1.650 at 250 nm with a k of .001.

Lanthanum fluoride is of interest as a UV material. The bulk values were measured by Wirick[229] showing an index of 1.65 at 258 nm. Krajnovich et al.[222] give the index of LaF_3 films at 248 nm as 1.59. Targrove et al.[230] reported in some detail on LaF_3 using IAD with argon and oxygen. They show that the stoichiometry tends to be fluorine deficiency due to preferential sputtering of the lighter F atoms from the films. The addition of oxygen can fill the voids but at the expense of intrinsic absorption at short wavelengths due to the lanthanum oxide. This is similar to the results of Martin et al.[13,14] with MgF_2. Traylor-Kruschwitz and Pawlewicz[192] also include LaF_3 in their studies and reported the index to be 1.58 at 600 nm. McNally et al.[193] surveyed materials as ThF_4 replacements and commented that LaF_3 might otherwise be good but that it has high stress which would make it not suitable for quarter wave optical coatings in the IR. Kolbe et al.[231] reported further IAD and IBS work on LaF_3 for UV applications to 200 nm.

Harrington et al.[306] mentioned that although LaF_3 was not one of the lowest absorbing materials at the HF/DF chemical laser wavelengths (2.8-μm and 2.3-μm) that they tested, it might be improved by reactive-atmosphere process treatment.[307]

Other lanthanum compounds have been investigated for various possible applications. Friz et al.[232] report on Merck's "Substance H4" which is a lanthanum/titanium oxide mixture of index 2.1 at 550 nm and a transmittance range of 360 to 7000 nm. Koenig and Friz[233] also report on lanthanum oxide

(La_2O_3) and aluminum oxide mixtures for mid-index ranges (1.7-2.0) in the visible and UV. These are referred to as M2 and M3, and the transmit well to 250 nm in the UV. Friz et al.[234] then reported on using plasma IAD in the deposition of these three mixed materials. They found these films more absorptive than conventionally evaporated films, but conjecture that improvements could be made by process adjustments.

The photolithographic industry supporting semiconductor technology moves continually to shorter wavelengths. In recent years the emphasis has been moving from 355 nm to 193 nm. LaF_3 is a suitable material at these wavelengths. Scattering becomes ever more critical as the wavelength gets shorter. Quesnel et al.[277] addressed this problem in making MgF_2/LaF_3 multilayers by IBS. They found a need to replenish the lost fluorine atoms in the process by adding diluted fluorine gas. They achieved lower roughness and scattering than by conventional evaporation techniques.

5.3.1.23. Rhodium

Coulter et al.[313] reported extensive work on the properties of rhodium. They evaporated it from an E-gum at room temperature and 300°C. They found the reflectance at 546 nm to be 73.8 and 78.1% respectively. They state that the substrate temperature seems to be the only parameter that affects *n* and *k*, not rate or pressures up to 2×10^{-5} torr. The reflectance is higher in the red and infrared, probably giving a slightly yellow appearance to mirrors of rhodium. They found that about 2-3 nm of Nichrome was needed on the substrate for good adherence, but it had no detrimental effects to the coating. They report from others that sputtered films had not needed a glue layer of Nichrome. They also report good results with two reflection enhancing layers of SiO_2/TiO_2 of $SiO_2/$ CeO_2 to gain over 91% visible reflectance. The bare rhodium coating is described as extremely hard and chemically very durable. Both bare and enhanced version withstand one hour in boiling 5% salt water and ten hours in 10% NaOH and also 10% HCl solutions. The plain Rh mirrors lost 10% reflection when heated to 400°C (probably due to formation of Rh_3O_4 on the surface) whereas the overcoated versions showed no change in reflectance.

Rhodium was used as a test coating to compare various techniques to determine *n* and *k* by Arndt et multi al.[59]. They found significantly different values than Coulter et al.[313] had found. The work of Arndt et al. was extended by Aspnes and Craighead[314] to further compare techniques and attempt to resolve the discrepancies found. They concluded that the differences were due to microstructure and that Rh oxides and voids were included in the films measured. The films were not opaque, so we expect the Coulter et al. results to be more useful for opaque films.

5.3.1.24. Chromium

Chromium metal is sometimes used for a beamsplitter and often used as a "glue" layer to promote adhesion. Holland[79] discusses its use in both applications. Glue layers might range from 2 to 50 nm, but he reported 30 nm as the useful value to promote adhesion under aluminum mirror layers.. Smithson et al.[336] studied the growth of gold films on polyimide (PI) substrates with and without a chrome underlayer. Micrographs are shown of the size of islands of nucleation after 4 nm of gold are deposited. The islands are large on the bare PI and smaller on a chrome treated surface. They chose 2.5 nm of Cr as a base layer. The Cr causes the domain size of the gold to be smaller (which results in lower conductivity of a thin layer). This work shows how the substrate and deposited material interact to determine structure.

Because Cr as an optical layer is reflective (~60%) in the visible spectrum and environmentally durable, it can make a useful first surface mirror. As seen in earlier chapters, it is valuable in metal dielectric combinations to design both neutral density filters and more complex films where its absorption is used to advantage.

Granular material can be evaporated from tungsten boats or larger pieces from an electron beam source. The material sublimes, but a surface oxide can inhibit evaporation/sublimation. Chromium electroplated tungsten filaments can be used, but usually need to be thoroughly outgassed under a shutter before deposition to avoid contamination. Chang et al.[355] discuss the evaporation of Cr from E-guns in both pellet and rod form. For applications requiring large quantities of chromium, the rod feed remains stable in terms of subliming surface and therefore the stability of the deposition uniformity for the life of the rod. Baker and Iacovazzi[80] patented a gold mirror combination using Cr as a glue layer. We know of at least one group which used Cr as a glue layer on plastic eyeglasses. Henderson and Weaver[81] evaporated Cr chips from a spiral triple-stranded tungsten wire basket. Their electron micrographs showed a continuous film structure for 10 nm and greater thicknesses, and they reported on the optical properties. Another successful application as a glue layer for gold on plastic required a minimum of 5 nm of Cr. An extensive table of n and k is found in Palik.[83] Ennos[11] reported Cr as having the highest tensile stress which he encountered.

5.3.1.25. Aluminum

Aluminum is one of the most commonly evaporated (and sputtered) mirror coatings for both decorative and technical coatings. It is easily sputtered and evaporated from an electron beam source. The most extensively used technique is probably still to evaporate aluminum canes or pieces of aluminum wire from tungsten filaments as described in Chapter 6. It has the best reflectance of the common metals in the UV, but is not as high as Cu, Ag, and Au in the infrared. Ennos[11] showed that Al initially had a high tensile stress which reduced to a small compressive stress at an opaque thickness and which reduced further after deposition. Apfel[84] (see Sec. 1.5) gives a useful triangle diagram of the optical properties of aluminum and other metals versus thickness.

Funk et al.[303] studied the influence of residual gas (water vapor) and substrate temperature on evaporated aluminum films. Although their work may have been primarily oriented toward semiconductor industry applications, it offers valuable insight for optical applications. They investigated substrate temperatures from 60 to 300°C. The other pertinent parameter of their work was the ratio of impacting aluminum particles to water vapor molecules, "normalized vapor deposition rate." There is also a discussion of the regions delineated by the ratios of the absolute substrate temperature to the melting temperature of the metal as relates to the crystal size which results. This looks very much like the M-D or Thornton structures. They state that higher substrate temperatures and normalized rate yield larger grains. They also say that most evaporations in the industry take place between 30 and 60% of the melting temperature. The water vapor in the chamber is broken down into hydrogen, oxygen, and/or hydroxyl groups by the fast electrons from the E-beam and by catalytic cracking on the fresh aluminum surface. The oxygen is efficiently gettered by the aluminum leaving hydrogen as the dominant residual gas in the chamber, as shown by RGA. They show that the reflectance of the films at 800 nm decreases with increasing temperature. However, at the highest normalized rates, the decrease is much less up to 250°C. The roughness of the films and formation of hillocks are shown and seem to relate to the reflectance. Our experience with films appearing to be "hazy" when deposited at over 150°C may be due to this. They state that there are no hillocks at very high normalized rates where the crystallites grow primarily with <111> orientation. This is all consistent with the empirical findings of Hass et al.[304,305] many decades ago. They also concluded that the best reflectance was achieved by the best vacuum and highest deposition rates.

In recent times, Larruquert et al.[347-351] have studied the deposited properties of aluminum in detail for applications in the far ultraviolet (FUV). Aluminum is the most favorable material for this region, but it is highly degraded by oxidation. As a result, it must be maintained in a vacuum. However, it is of interest for space-based astronomical observations where some vacuum exists. They have examined the 77-120 nm region in detail including the n and k values for that region. Aznárez et al.[352] have described the instrumentation which they used for this work. Low earth orbit is know to have a high atomic oxygen environment. Therefore they studied and demonstrated[353] the ability to recoat a surface as many as ten times to extend the predicted life in a low earth orbit by an order of magnitude. Larruquert and Keski-Kuha[354] developed a coating of Al/MgF2/Mo which optimized the reflectivity at 83.4 nm line of O^+ and rejected 121.6 nm Lyman α hydrogen line.

We discuss aluminum (Al) in Chapter 6 as a simple illustration of the application of Design of Experiments methodology to process development.

5.3.1.26. Silver

Holland[79] showed that silver (Ag), like aluminum, gives better reflectance if deposited as rapidly as possible and on substrates that are not hot. He attributes this to greater agglomeration at high temperatures and slow rates which in turn cause greater absorption. Silver does not wet a tungsten filament but tends to form droplets with high surface tension. It can be evaporated from a tight spiral of stranded tungsten which does not let the silver droplets fall through. Silver is easily sputtered and can be evaporated from an electron beam source. Jackson and Rao[85] wrapped a few turns of thin platinum wire around a V-shaped tungsten filament before wrapping the silver wire around that. They state that Ag will wet the Pt but not the tungsten. Huebner et al.[86] evaporated the Ag from a tantalum boat and measured the optical constants. Canfield and Hass[87] used tungsten boats for their VUV work. Bennett and Ashley[88] evaporated silver from a tantalum boat for their ultrahigh vacuum investigations. Hass et al.[89] used a tungsten boat for silver in their development of a silver coating protected with Al_2O_3 and SiO. We have applied the very high rate sputtering of silver at low gas pressures described by Radzimski and Posadowski[341]. We found it to be a very satisfactory technique. Sargent et al.[116] describe the sputtering and properties of silver in the metal mode process and the application to a NBP filter with alumina. Lubezky et al.[44] protected the Ag with yttria for better IR performance.

The adhesion of silver to different materials is a significant issue. Burger and Gerenser[338] reported an extensive study of the chemical binding of silver to polymers. They state that: "Metal atoms will diffuse along a polymer surface until a chemically reactive site is encountered. If no strong chemical interactions are possible between the metal and polymer, then metal-metal bonding will occur and lead to cluster growth." They found that oxygen glow discharge (OGD) significantly increased the number of nucleation sites and thereby adhesion to polymers. The OGD has an undesirable side effect of breaking bonds in the polymer backbone and weakening the polymer's mechanical strength. They conclude that: "The promotion of adhesion by plasma treatment relies on striking an appropriate balance between the process of creating desirable nucleation sites and the unwanted side effect of mechanically weakening a polymer." McClure et al.[346] addressed the adhesive weakening effect on polymers by an amorphization process that they describe in detail. They state that, "the treatment consists of melting the surface, followed by rapid cooling, with no detectable change in the chemical composition of the films. Since it is a thermal process, a wide range of energy sources could be used to produce the amorphous surface. Short pulse flashlamps are promising candidates for scale-up to industrial levels." Grace et al.[339] extended this surface activation investigation on silver-PET films and found that nitrogen glow discharge gave even better environmental durability. Shi et al.[340] studied the application of nitrogen plasma treatment of acrylic-based polymer.

Kennedy et al.[296] used about 50 nm of copper "back protective layer" under the silver reflecting layer as a sacrificial layer to protect the silver. Pellicori[90] investigated the corrosion and scattering of protected silver films. He suspected an electrochemical reaction between the Ag and the inconel or Cr binder (glue) layer, and suggested the use of an oxide such as Al_2O_3 or SiO_x instead. Song et al.[91] found, on the other hand, that an underlayer of copper and reflectance-boosting dielectric overlayers provided better durability to silver. They presented evidence that some of the copper migrates up the grain boundaries to the surface and performs a sacrificial role in cathodically protecting the silver. Garzino-Demo and Lama[300] speak of how "In the past automotive mirrors have been produced by wet chemical deposition, which first coated the back surface of the glass with a very thin tin layer to improve the adhesion onto the glass surface of a subsequent silver layer followed by a copper layer, to prevent silver tarnishing, and a paint layer to protect against physical damage and chemical attack." Bussjager and Macleod[322] expanded on the work of Song et al.[91] by adding a 1 nm layer of copper over the silver. This did not influence the reflectance of the silver, but it did reduce the deterioration of silver due to water. However, the resistance of the surface to sulfur attack was decreased. Bennett et al.[93,94] investigated the scattering and microstructure of silver films. Hwangbo et al.[95] reported on the use of IAD on Ag and Al films. Vergöhl et al.[267] optimized the parameters for magnetron sputtering of silver. Wolfe[196] describes the use of silver between NiCr plus silicon nitride layers for a durable solar control stack. Laird and Wolfe[290] extended this work further the next year. Wolfe[342] also developed durable anti-static AR coatings for video displays using silver and the processes developed earlier. Szczyrbowski et al.[343] developed low emissivity coatings based on silver that can withstand the 500°C heat of glass tempering. Treichel et al.[344] investigated the details of a "blocker" layer used on top of the silver to protect it from attack by the aggressive sputtering of the next oxide layer on top if it. Such blockers have typically been thin layers of Ti of NiCr as in Wolfe's work.

It is necessary to have representative values for the n- and k-values of silver to design multilayers using it. Palik[82] gives the composite optical constants resulting from various investigators. Apfel[84] (see Sec. 1.5) has a useful triangle diagram illustrating that silver has high reflection with very low absorption in the visible and IR. Ordal et al.[92] developed tables of values for silver and many other metals.

Fan and Bachner[326] analyzed the application of $TiO_2/Ag/TiO_2$ and tin-doped indium oxide coatings for solar energy collectors. Pracchia and Simon[323] studied the influence of dielectric materials from a design aspect in silver based heat mirrors of the dielectric-metal-dielectric (DMD) type. They found that lower indices gave higher cutoff wavelengths but the transition becomes more gradual. Eisenhammer et al.[324] further studied the design issues and concluded that the optimal heat mirror for their solar applications was: $TiO_2/Ag/TiO_2/Ag/Y_2O_3$. Lee et al.[325] worked with the actual deposition of $TiO_2/Ag/TiO_2$ and $TiO_2/Ag/TiO_2/Ag/TiO_2$ mirrors with IAD. They found an unexpected effect when the silver of the second (and fourth) layer is deposited on the titania layer. There appeared to be an anomalous layer of dielectric-like material before the expected silver effects occurred in the optical monitoring. They modeled this and concluded that the anomalous layer was a mixed layer of TiO_2 and silver with a thickness of 2.56 nm and an equivalent complex index of 2.015-*i*0.016. They also found that the five-layer design had better heat rejection characteristics and was more durable in a humid environment.

Nahrstedt et al.[321] describe the electroless silver process which requires no vacuum so that large astronomical mirrors can be coated without removal from their mountings (a great reduction in cost and risk). They also compare the durability of other mirrors in the tarnishing environment of high sulfur content. Other mirrors include bare aluminum and protected silver.

5.3.1.27. Gold

Gold (Au) has the highest reflectance of known materials in the infrared from 1000 nm to longer wavelengths. Being a noble metal, it has great chemical durability. It has little scratch resistance because of its malleability. Balzers[7] lists it as being evaporated from tungsten or boron nitride boats or by electron beam sources. Jackson and Rao[85] state that gold does not wet tungsten, and they used the same platinum wire wrap mentioned in the description of silver. However, Advena et al.[96] used a tungsten filament without mention of any wetting problem. Bennett and Ashley[88] used molybdenum boats for Au because gold quickly alloys with tantalum, making it unusable. Otherwise, they used the same approach for both silver and gold. Gold typically has low adhesion to glass surfaces. A glue layer of chromium is commonly used. Smithson et al.[336] studied the growth of gold films on polyimide (PI) substrates with and without a chrome underlayer. Micrographs are shown of the size of islands of nucleation after 4 nm of gold are deposited. The islands are large on the bare PI and smaller on a chrome treated surface. They chose 2.5 nm of Cr as a base layer. The Cr causes the domain size of the gold to be smaller (which results in lower conductivity of a thin layer). This work shows how the substrate and deposited material interact to determine structure. Martin et al.[97,264] used oxygen IAD to greatly improved adhesion of gold to glass. They found that Ar IAD was only of small benefit, but oxygen IAD gave a hundredfold improvement in adhesion. The IAD needs to be discontinued after opacity is reached and the film finished without it. The oxygen doping reduces the reflectance of the film.

LeBlanc and Parent[345] give extensive details of the sputtering process for gold onto compact discs.

Apfel[84] (see Sec. 1.5) shows the behavior of the reflectance, transmittance, and absorptance of gold versus the thickness in a triangle diagram. An extensive summary of the optical properties versus wavelength is found in Palik.[82]

Interesting interactions of "inert" gold are worth noting. Sato et al.[98] showed that solar selective surfaces with high absorption in the visible and high reflection in the IR can be made by heat treating aluminum coated with gold. They describe the diffusion of aluminum into the gold to produce an alloy $AuAl_2$. Lee and Lue[99] describe related processes with gold on silicon. Gold blacks are somewhat like soot produced by evaporating gold in a high pressure (many torr) environment of an inert gas like helium or nitrogen. Small particles (~80 μm) of gold aggregate into structures with very low packing density as shown by O'Neill et al.[100] Zaeschmar and Nedoluha[101] discuss the theory. McKenzie[102] shows that oxygen used in the process will cause tungsten oxide to be formed from the filament and change the properties of the deposit. Advena et al.[96] give details of the deposition process and resulting films.

5.3.1.28. Indium-Tin Oxide

Indium-Tin Oxide (ITO) or In_3O_5-SnO_2 has the advantage of relatively good electrical conductivity while having relatively good visual transmittance. Such films have many applications and have come into great demand for data display panels, resistance-heated windows for defrosting, electrochromic windows and mirrors, etc. These include architectural glazing for solar selective windows and controllable transmittance windows. Holland[79] gives the history of ITO and the probable mechanism of its functionality. Vossen[103] provided an additional review of ITO in his chapter on transparent conducting coatings in general. Cormia et al.[372] provided a summary of the history of ITO coating from a polymer web or roll-to-roll coating perspective. There has been a great deal of activity over the recent decades in these areas.

Hamberg and Granqvist[401] evaluated reactively E-beam deposited ITO from 0.25 to 50 μm. They state that the data is explained in detail from a theory encompassing scattering of free electrons by ionized impurities. Mayr[373] provided an in-depth discussion of the atomic/molecular mechanisms of ITO's properties. He showed the sheet resistance and visible transmittance versus oxygen flow rate in the deposition process, and also showed how the specific resistance passes through a minimum with increasing oxygen flow rate. Dobrowolski et al.[106] evaporated Merck Substance A with oxygen IAD onto glass substrates at temperatures from ambient to 150°C. Process parameters and results are reported. They achieved transparent, conducting ITO films at 150°C or less. Lee et al.[387] evaporated ITO by E-beam with IAD and reported the process parameters and resulting characteristics. Gilo et al.[398] deposited ITO by E-beam with and without IAD. They found the IAD films still conductive when overcoated with MgF_2 but not oxides. The non-IAD films with MgF_2 overcoat were not conductive. Traces of indium were found in the MgF_2, so they speculate that the indium diffused through the fluoride. Czukor et al.[376] reactively sputtered In:Sn (90:10) alloy on polymer substrates in the web coater environment. They investigated a range of process parameters and were able to consistently produce resistivity of <5 Ωμ and k at 400 nm <0.050 on PET polymer. Gilbert et al.[377] later performed a related study of alloy and oxide targets to that of Czukor et al.[376] They confirmed the earlier conjecture that the oxide targets seemed to provide a more stable process. The stresses appeared to be lower for the films produced by alloy targets. Yoshimura et al.[382] showed the change of film properties as a function of the density of sintered ITO targets; the densest being the most beneficial. Lewis et al.[383] show extensive scanning electron micrographs of ITO targets produced by various processes and compare their performance. J. L. Grieser[378] compared DC sputtered ITO with ion beam assisted sputtering and concluded that IBAD did *not* improve the results. Gibbons et al.[380] compared the production merits of a rotating ceramic target with conventional planar targets of ITO. They state that, "The cylindrical target has about 10 times the useful target inventory of a planar target," and that, "the period of useful production time can be increased by a factor of about five." May et al.[334] compare the results of processing with DC and medium frequency (MF) sputtering from oxide targets in a production coater. They demonstrate the advantages of the dual magnetron (MF) mode with high dynamic deposition rates. Treece et al.[381] used an interesting methodology to optimize the deposition parameters of ITO on PET. Their principal variable was the water vapor in the system, and they found an optimum point for that.

Methods of monitoring and control of ITO deposition were studied by Patel et al.[389] They applied a residual gas analyzer (RGA), a plasma emission monitor (PEM), and an optical gas controller (OGC). They concluded that all these contributed to enhancing the control of the process. Patel et al.[390] discussed further progress in closed loop control using RGA and PEM and measurements of their results. Patel[392] also discusses the application of this technology to ITO deposited on polymeric webs. Gibbons et al.[391] describe the successful use of similar control systems that are applicable to ITO depositions.

Huang et al.[379] subjected ITO coated soda-lime glass to the tempering process of heating to 650°C and then rapid air quenching. They found that the heating process caused a diffusion of alkali ions from the glass into the ITO. This in turn poisons the conductivity of the ITO and thereby also changes the optical properties. It turned out that 60 nm of SiO_2 on the glass before the ITO was an optimum compromise between blocking the ion diffusion and causing cracks in the coating. Tsai et al.[288] carried this further and found that codeposited SiO_2-TiO_2 formed an even more effective barrier to the effects of the Na atom diffusion/migration at higher temperatures. They also showed that annealing all of the ITO films that they produced in air for one hour to 300 or even 500°C decreased the sheet resistance. This they attributed to the ITO changing from amorphous to crystalline.

Saif et al.[375] studied the addition of aluminum doping to ITO to enhance thermal and chemical stability. They concluded that: "Al doping improved the thermal stability of ITO films, especially the higher resistance films. Al doping slightly improves the environmental stability. The concentration of Al doping in ITO film is very important, as excessive concentrations of Al may cause degradation of ITO."

Hamberg and Granqvist[105] modeled the radiative properties of heavily n-doped semiconductors in general and of ITO in particular. This allows them to compute the luminous, solar, and thermal properties. They state that the model is also applicable to WO_3 films. The *n* and *k* optical constants of ITO and indium oxide were reported by Bright[108] over the 350 to 1200 nm spectral range as a function of deposition conditions. Baouchi[384] reported an atomic force microscopy (AFM) study of ITO by reactive and non-reactive sputtering. The latter were smoother, but LCD makers (to whom roughness is important) vary on which is better for their particular processes. Lehan[385] used the ratio of specular to total transmission (forward scatter) to measure the grain size (roughness) of ITO films, and he found good correlation with scanning electron microscopic (SEM) measurements. Zhang et al.[386] measured and reported the *n* and *k* of ITO films deposited by DC magnetron sputtering with post deposition annealing. They concluded that, "The resistivity correlates very strongly with the integral of the *k* spectrum from 600 to 1100 nm." Henry et multi al.[388] studied the gas barrier performance and microstructure (by SEM) of ITO and Al_2O_3. They found ITO superior due to smoother films with less defects.

Hamberg and Granqvist[374] studied transparent heat-reflecting mirrors of ITO with a MgF_2 top layer. The MgF_2 not only improves the transmittance and reflectance, but can be used to improve the perceived color. They were able to achieve quite satisfactory "color free" results in both transmittance and reflectance.

Gilbert et al.[362] discussed the application of ITO to switchable windows. Wolfe[342] also developed durable anti-static AR coatings for video displays using ITO as a key layer. Kuhlmann[393] describes the production application of ITO as a high index and shielding layer in an AR on CRTs. Laird et al.[394] similarly describe a conductive AR for glass and plastic substrates. Gibbons et al.[395] describe depositing ITO for display applications. They give *n* and *k* values for their ITO and the relationship between the k and resistivity of the films. Wolfe[308] gave further details on AR coatings with ITO. Wang and Lee[396] studied what can be done to design AR coatings which incorporate ITO for EMI shielding. They considered the classical AR with two QWOTs of ITO as the high index layer and a low index layer on the outside, but also other solutions with the ITO on the outside. Blacker et al.[397] describe their development of an anti-static AR for eyeglasses using ITO on plastic substrates with both sputter and E-beam deposition.

ITO with 0% Tin

Indium oxide (In_2O_3) can be used as a transparent conductive coating, and it is reported to have good properties for the IR in the 3-5 μm region. Lubezky et al.[399] reported on the deposition of the oxide by thermal evaporation of the metal from an alumina crucible and using activated oxygen of the material. Today, this might best be done with IAD. They achieved good sheet resistance and the layers withstood adhesion, humidity, and moderate abrasion. Adurodija et al.[400] investigated the effects of substrate temperature on pure In_2O_3 and ITO films. The lowest resistivity was found from the pure material at <100°C. Our limited experience with In_2O_3 is that the process can be very sensitive to temperature and oxygen pressure. This is perhaps to be expected since we understand that the most important factor influencing the resistivity and transparency of the films is the deviation from the 2:3 stoichiometry.

5.3.1.29. Electrochromic Materials, Tungsten Oxide, Etc.

The ability to change the optical performance of a coated surface electrically has gotten a lot of attention because of its potential usefulness in many applications. Castellion[358] filed a patent in 1972 on an electrochromic mirror which "rapidly" (15 seconds) changed reflectivity. He mentions tungsten oxide and molybdenum oxide in the claims. There have been innumerable patents since then on various EC devices, chemistries, and schemes. Vanadium compounds and particularly tungsten oxide are the subject of many investigations such as Mathew et al.[104] for their electrically controllable properties. Nishide and Mizukami[161] showed that fully oxidized tungsten is clear in the visible and has an index of refraction of about 2.0 or higher. Demiryont and Nietering[357] investigated the details of evaporating WO_3 reactively and conventionally from the starting forms of the metal wire or the oxide powder. The melting points of W and WO_3 are given as 3410 and 1473°C respectively, which makes the reactive evaporation technique attractive. They found the material to be generally amorphous and have a highest index of 2.11 at 550 nm. They also report the n and k values under various conditions. They observed the better EC properties in films that were amorphous, porous, and oxygen deficient.

Andersson et al.[127] used a lithium perchlorate doped polymer laminate in a "Smart Window" construction. This report has extensive details and references on the EC work to that date (1988). Demiryont[359] describes a design technique for EC windows. Caskey et al.[356] discuss the application of thin film electrochromic (EC) devices to automotive rear view mirrors. They report that: "tungsten trioxide can be changed to hydrogen tungsten bronze ($H{:}WO_3$) if hydrogen is driven into the oxide and energy is supplied to complete the reduction reaction. This all happens in the solid state. The bronze constitutes an absorption center for light. When the voltage is reversed, the hydrogen is removed from the tungsten bronze, thereby oxidizing the film to its transparent state." They give the n and k values of their WO_3 layers in the bleached and colored states. They say that "these devices operate with a combination of hydrogen and hydroxyl ions within the solid state stack. The films deposited must permit incorporation of water vapor, and must support the hydrolysis required to free ions for transport during electrochromic coloration and bleaching." The films are porous and therefore one must deal with the tensile stresses. Byker[360] patented an EC light modulator, as might be used in an automotive rear view mirror, utilizing certain organic nitrogen-containing compounds and having an advantage of being self-erasing.

Lippens et al.[361] described roll coating of EC half-cells of ITO and WO_3. They used cooler drum temperatures to get the desired film porosity. Gilbert et al.[362] discussed the application of ITO, ZnO, and thin metal layers such as Ag to switchable windows. Mathew et al.[364] showed tests of an EC system which darkened in approximately 10 seconds and cleared an order of magnitude faster. Sapers et al.[365] demonstrated the performance of potential EC windows that have many of the necessary attributes for architectural applications. WO_3 and NiO were the electrochromic and counter electrochromic layers, respectively. Lampert[363] reviewed the history (as of 1995) and the economics of EC devices, and he conjectured concerning the future of the field. He stated that: "The great challenge is to make these technologies, through advanced deposition and fabrication techniques, economical in the 100-250 \$/m^2 price range." Lampert[366,367] again (1999 and 2000) reviewed the state-of-the-art of switchable glazing. The field has now been named "Chromogenics." He reviews a very broad range of applications being pursued wherein the most mature seem to be automotive and architectural.

The most conspicuous and commercialized of these are automotive rear view mirrors.

Granqvist[368] has provided an extensive handbook of information and results in the field up through its publication in 1995. He mentions that WO_3 sublimes (from Ta, Mo, or W boats, crucibles, or E-gun) and that the vapor is molecular in nature, the dominant species being trimeric W_3O_9. The resistivity of the films drops from 10^9 Ω cm when deposited on room temperature substrates to 10^{-2} Ω cm when deposited on 500°C substrates. The bulk oxide is about 10^{-5} Ω cm. Reactive evaporation with oxygen can produce stoichiometric films or anything between that and $WO_{2.5}$ which results from no oxygen with the evaporation. Hydrogen can be incorporated in the film by the evaporation conditions or later due to exposure to atmospheric water vapor. Many other details are reported in the handbook on the structure of the films. Granqvist also touches on TiO_2 as an EC material.

Larsson et al.[371] reported on the application of EC coatings for emittance control on a nanosatellite for when it moves in and out of the earth's shadow and solar radiation. They used reactive DC magnetron sputtering for the deposition of WO_3 and ITO.

There have also recently been some reports[369,370] of WO_3 and titanium oxides being deposited as gas sensors for H_2S, NO, NO_2, and NH_3.

5.3.1.30. Cubic Boron Nitride

Cubic boron nitride (cBN) is second only to diamond in hardness but is not found in nature. It was first synthesized by Wentdorf[402] by high temperature and pressure similar to the conditions for diamond in 1957. It has been made commercially as an abrasive by heat and pressure in the presence of a catalyst.[403] As such, the crystal sizes have been no greater than 0.5 mm. More details on the bulk crystals can be found in Mishima[405] and Yoo et al[404]. It has been of interest as a potential machine tool coating because it does not readily react with ferrous metals as does diamond. The semiconductor industry has been interested because of its wide bandgap, good thermal conductivity, and that it can be doped both p- and n-type.

Cubic boron nitride has potential as a hard optical coating for the visible and possibly the near infrared. Its index of refraction is about 2.1 and its hardness is on the order of 50 GPa or greater[406], where diamond is about 80 GPa and sapphire is about 21 GPa. The primary residual problem at this time is that the films have very high compressive stress when deposited under conditions practical for most potential applications. As of the year 2001, there does not appear to be any successful commercial application of the material due to these process difficulties. However, a study of the research on cBN over the past decade may also aid in the understanding of other materials and process.

It has been found that thin films of cBN can only be deposited with the assistance of appropriate fluxes of ions in a proper energy range. The nucleation conditions seem to be most critical, after which the growth parameters may be somewhat less stringent. Most of the conceivable process approaches have been tried and reported such as CVD, PECVD, IAD-PVD, sputtering with IAD, dual ion beam sputtering, unbalanced magnetron sputtering, pulsed laser deposition, etc.

Most of the literature points to a mixed initial layer on the substrate (usually silicon in most reports) which initially deposits in an apparently amorphous form (aBN). All of the literature further points to the growth process of cBN starting from a layer of graphite-like hexagonal boron nitride (hBN) which may also be (or contain) rhombohedral (rBN)[407] or a disordered turbostratic (tBN) phase of boron nitride. As the layer builds in thickness, some form of hBN is identified, primarily by its infrared transmission or reflection spectrum at ~780 and ~1360 cm^{-1}. Without the appropriate conditions of ion bombardment, the film continues to grow as hBN up to any arbitrary thickness. The right ion bombardment appears to cause the hBN to form as tBN with an orientation perpendicular to the substrate[408]. With the proper ion conditions, cBN starts to nucleate and grow on top of the h/tBN layer which is of the order of 10 nm thick, which again is best identified by the IR spectrum of cBN near 1055-1060 cm^{-1}. This peak may be shifted as far as 1100 cm^{-1} by stress and other effects[408, 409] in the films. Essentially all processes to date result in cBN films which are highly stressed. This stress builds with thickness and thereby limits the thickness which can remain adhered to the substrate. Delamination is sometimes even found to be because the Si substrate yielded internally and some Si stayed with the parting BN layer. Litvinov et al.[410] report that thick adherent films deposited at ~1000°C cause the Si substrate to relieve some stress by internal plastic deformation.

Although some of the work reported may have been motivated by achieving single crystal growth, the reported results seem to all be nanocrystalline with sizes of the order of 5-50 nm. Although substrates of diamond give superior results[411], it is not apparent that any crystalline substrate is necessary since the amorphous and hBN precursor layers do not appear to be epitaxial even on single crystal Si substrates. It appears from the literature[412, 413] as though the ion bombardment causes the hBN to orient in a tBN and/or rBN phase on the substrate such that the lattice spacing matches 2:3 with the lattice spacing of the cBN (on a nanocrystalline basis). Yamada-Takamura et al.[407] state that they have clear evidence of rBN in the transition from tBN to cBN. The rBN is said to have a "diffusionless" transition from rBN to cBN, and therefore cBN should be easier to achieve. One conjectural interpretation of this is that the atoms maintain the same relative positions in a lattice, but that the angles of the bonds between them change. The interfaces between the t/rBN and the cBN, as seen under high resolution transmission electron microscopy (HRTEM) and normal TEM with selective viewing of the crystal phase by diffraction masking, do not appear to be planar or epitaxial but random and rough on the scale of the nanocrystals.

There have been several models put forth to explain how the energetic ions cause the cBN phase to nucleate. There is evidence that once the nucleation occurs, much less ion energy/momentum is needed to have the cBN continue to form. The four models discussed in the review by Mirkarimi, et al.[408] are: sputter, thermal-spike, stress, and subplantation. The sputter model is based on experience with some other materials such as diamond where the undesired graphitic phase is preferentially sputtered or etched away and leaves only the cBN to continue its growth. It has been reasonably well demonstrated that this is not the case with hBN and cBN because there is very little differential in their sputter rates. However, it has been reported[414] that hydrogen and methane do selectively etch hBN. The other three models are still in contention as to which or what combination of them is the best model for understanding the cBN nucleation and growth. Ulrich et al.(44) state that thermal-spike cannot be the model for their ion plating process, they talk more in terms of stress and subplantation. However, Franke et al.[416] state that, from their evidence, the thermal-spike is favored.

Our current interpretation of these models is as follows. The thermal-spike model is that the incoming ion/atom rapidly gives up its energy to some of the atoms in the bombarded film matrix. This pulse creates an increase in local temperature and pressure which allows a transition in crystal phase as the energy is rapidly quenched. The subplantation model as described by Lifshitz et al.[417] is the shallow surface implantation of impinging species. These densify the film, fill voids, displace atoms, and contribute to the energy and rearrangements of atoms. The implantation of a new atom in a matrix would seem to increase the local "pressure" at its final resting place. This leads to the concepts of the stress model which has sometimes spoken of interstitial atoms and dislocations creating local stress which causes the tBN to convert to cBN. It would seem that there must be some truth to each of these three models, and it just remains to sort out the details of a combination of effects. All of this is helpful to guide the process developer. The better understanding of the physics should minimize the efforts need to achieve a practical application of such coatings. However, a workable process will ultimately depend on the macroscopic application of an empirical process.

The literature points to a minimum substrate temperature of about 100°C and a minimum ion energy of about 125eV[418,419] (maybe lower in some processes[420]) for the cBN conversion from hBN. There is further a minimum flux as measured in the arrival ratio of ions to depositing atoms that is necessary for nucleation. However, Johansson[421] reports that momentum transfer is not fundamental to the growth of cBN. The upper limit on such flux is where the depositing atoms are resputtered and leave no gain in film thickness. It is fairly clear that high ion energy atoms/ions and high temperature are not needed to grow cBN. However, the relief of stress is reported by both high temperature annealing at 600-1000°C[410, 422, 423] and high energy atom/ion bombardment[424]. Some work has been done with dopants to reduce stress. In general, they appear to also reduce hardness[425] and can even inhibit the formation of cBN[426]. Under the right conditions a doping of silicon seems to reduce stress without significant loss of hardness[415]. There is some justification to conjecture whether, once the cBN has been nucleated and growing, the ion energy/momentum[427] and temperature both can be reduced so that the stress in the growing film is minimal and needs no dopant or further annealing.

5.3.1.31. Toxicity of Coating Materials

When working with any coating material, it is important to be aware of any associated hazards such as toxicity. We have already touched on the radioactivity of ThF_4 and what is necessary for its safe application. Clearly no coating material should be allowed to enter the body as a solid (dust), liquid, or gas. Dust of any of the materials has no way to exit the lungs once it has entered. We will list those materials which we have discussed that are listed as significantly toxic by Plunkett,[109] but the reader is cautioned to make all the necessary investigations of current information on any material which they use. One major source for such information are the Material Safety Data Sheets (MSDS) which are provided for each material.

Selenium salts are listed by Plunkett[109] as highly toxic, but he states that the element is probably harmless. Selenium is taken for the health by some! Coating rooms using ZnSe should have good ventilation to remove the H_2Se gas that is obvious from its smell when a chamber is opened after coating with ZnSe. This gas is also generated when ZnSe is attacked by an acid, even vinegar.

Lead telluride has been said to be toxic and cumulative. Although this is not obvious from Plunkett,[109] PbTe should be investigated and approached with caution. Our advice is: read the MSDSs and treat all coating materials as if they had flu-germs on them.

5.3.1.32. Relative Cost of Coating Materials

Sometimes the cost of the coating material is no issue if it is the only material that will do the job. However, it is well to keep material cost in mind if any quantity production is planned. We have surveyed the cost of vacuum deposition grade materials from a material supplier's catalog. There are wide variations in cost from one form and purity to another among the same basic material. We have tried to use a representative average for ranking purposes. The reader is advised to examine the current market and more than one supplier for any materials to be seriously considered in a particular application.

Among the expensive materials, we were surprised to find Sc_2O_3 was several times the cost of gold. Germanium was about one sixth, silver and ZnSe were about one tenth, and Ta_2O_5, HfO_2, AlF_3, Y_2O_3, and ThF_4 were about one sixteenth the cost of gold.

The medium-priced materials were: PbTe, CeF_3, SiO, Ti_2O_3, BaF_2, Nb_2O_5, TiO, ZnS, Nd_2O_3, MgO, ZrO_2, and aluminum in approximate decreasing order of cost. Aluminum might bear some further examination as it should be relatively inexpensive even in deposition grade.

The less expensive materials in order of approximately decreasing cost were: ZnO, Cr, Al_2O_3, CeO_2, CaF_2, SiO_2, PbF_2, TiO_2, MgF_2, and indium-tin oxide (ITO).

Note that TiO_2 is, at worst, half the cost of the other oxides of titanium. This is why we prefer it as the starting material as described in Sec. 5.3.1.2. Also, we think that Ta_2O_3 is unattractive at more than three times the cost of TiO_2. Scandia has no appeal for us because of cost, unless it was the only material to solve the problem at hand.

5.4. ION SOURCES

As we have already seen in the discussions of some materials, ion-assisted deposition (IAD)[5,12,13,14,16,20,49,50,54,57,106] has become well established as a means of improving optical coating performance and reducing cost. Some performance improvements include durability, stability with humidity and temperature, and better stoichiometry. Cost reductions result from low temperature deposition and increased rates for reactive processes. Most of the early process development mentioned above was performed using gridded Kaufman-type sources. These are characterized as commonly being used with a relatively collimated beam of ions with a well defined and narrow distribution of ion energies (electron volts, eV). Cuomo et al.[111] give an extensive description of the various gridded sources and others. The grids are commonly made of graphite to minimize sputtering of the grids and do not stand up well to the use of oxygen in the source. The internal filaments are subject to relatively high gas pressures and are also vulnerable to oxygen. For these reasons, we have worked more extensively with the broad beam sources which work better with oxygen under high power usage.

The rate at which a deposition can occur and produce the desired properties with IAD is limited by the ion beam power. Twice the power would allow twice the rate. We speak here of beam power as the eV of the ions multiplied by the ion current of the beam. For most materials, as we discussed in Sec. 5.3.1.2 on TiO_2, there is an upper limit on the eV beyond which the coating properties are degraded due to ion "damage." Investigations of commercially available End-Hall and Cold Cathode sources have shown that the former tends to have a lower than desired eV capability and the latter tends to be too high. For some years, the commercially available embodiments of both sources exhibited operational stability problems due to source component changes with use and power supply instabilities. The eV characteristics of both are affected by gas flow and chamber pressure. The excess pressures needed for voltage control and high power operation can degrade the performance of the deposited films. We will discuss the concepts and techniques used to overcome these limitations and increase the power and stability by several times.

Our original motivation for this work was to obtain as powerful a source of oxygen ions with neutralizing electrons (plasma source) as possible and as economically as practical for the reactive deposition of SiO_2 from SiO as mentioned above. However, the results are applicable to other ions such as nitrogen, argon, etc., and other deposition materials. We reported[112] on the development of processes which convert the SiO to SiO_2 during deposition by the use of ion-assisted deposition. The additional oxygen must be supplied in a sufficiently energetic process to provide the material conversion during the SiO deposition on the surface to be coated. The deposition rate (1-2 nm per second), the uniformity, and the repeatability of the processes must also be adequate for the production of the coating at an economical rate. These experiments have been done with commercially available End-Hall and Cold Cathode ion sources and a new plasma source. A comparison of the results and behavior observed with each type of source are discussed below. Our experience to date has included four types of ion sources: the Ion Tech gridded Kaufman source, the Denton Cold Cathode CC-102R, the Commonwealth Scientific Mark II End-Hall source, and the DynaVac PS1500[461] high power plasma source of our own design. We will also discuss the Denton CC-105 recently reported by Morton[462]. In all cases, our process speed was limited by the rate at which we could deposit the materials and obtain the desired properties. This rate is limited in turn by the ion current density which can be provided. We therefore want to operate the ion sources at the highest beam power practical. We also have learned from the literature and our own observations that there is an upper limit on the eV of the ions. Excess eV will cause damage to the deposited materials and therefore to the optical properties, particularly absorption. The power density of the ions cannot practically be increased by added voltage beyond some point such as about 300 eV for TiO_2, although SiO_2 might tolerate 600 eV. Our principal requirements for the SiO to SiO_2 conversion were processes which ran for many hours. It is our understanding (and experience) that the filaments and grids of the Kaufman type source could not be expected to survive the full power oxygen operation required for long periods, and therefore we made no attempt to use these sources in this case. The Cold Cathode source was said to "prefer" operation with oxygen over argon because of sputtering effects.

5.4.1. Cold Cathode Source

Our first efforts were with the CC-102R Cold Cathode source. Dobrowolski et al.[106] used this source for the IAD of ITO. The End-Hall source has some similar characteristics to the Cold Cathode and we also developed it into a somewhat satisfactory solution to our requirements. In both cases, we operated at the maximum power capability of the sources consistent with stable and long term operation.

The cross section of the Cold Cathode Source on which we have previously reported[112] is seen in Fig. 5.23. A molybdenum anode is surrounded by an aluminum cavity which is in turn surrounded by a ring of permanent magnets which produce a magnetic field. The gas is admitted to the source chamber through six (6) small orifices below the anode ring by a gas flow controller. The source is water cooled. It can be biased to offset the voltage of the whole system with respect to ground, but we have no experience with this; we have only operated at zero bias volts. There is a tungsten neutralizer over the aperture which we typically operated at its maximum current of 20 amps. The neutralizer voltage ranged from 12 volts upward with age. The filament sometimes lasted more than 10 hours and was easy and inexpensive to replace, so that it was not as big a problem as in a Kaufman gridded type source. The power supply had three

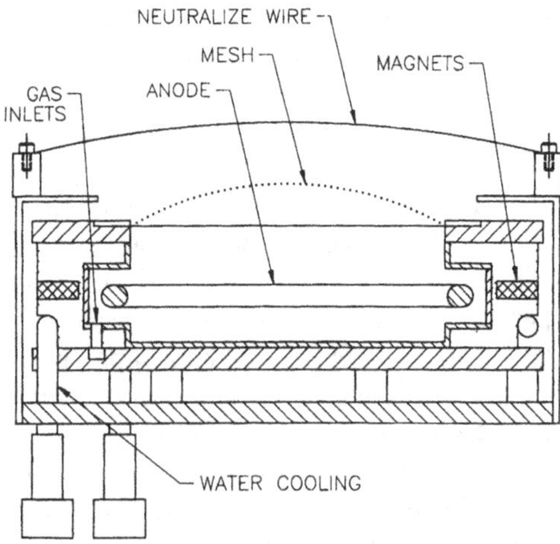

Fig. 5.23. The cross section of the Denton Cold Cathode Source.

controls: neutralizer amps, bias volts, and drive amps. There were displays of drive volts and beam amps. We set the neutralizer to maximum current, the bias to zero (but not off), and controlled only the drive volts through the drive current and the gas flow.

The behavior of the Cold Cathode Source is a strong function of the gas pressure, flow conditions through the cavity, the neutralizer emission, and the "cleanliness" or surface conditions inside the source cavity. We assume that the drive voltage is somewhat linearly related to the actual output beam voltage and similarly that the drive amps is related to the beam amps. The drive voltage is primarily a function of the drive amps applied and the gas flow through the source. There is also a significant effect of source cleanliness by which we mean the state of deposits which build up in the cavity with use. We found it necessary to clean the source after every usage of about eight hours.

If any of these four types of sources is operated without enough neutralization, one observes sparks on the substrates and fixturing, and arcing indications on the source power supply. These arcs/sparks usually cause damage to the coating and the substrate. On the other hand, we have seen no ill effects from excess neutralization. We observe that, if the pressure in the cavity of the Cold Cathode source drops too low, the drive voltage increases and at some point the source will arc. This seems to disrupt the steady operation of the ion beam conditions, and therefore we avoid this condition if at all possible. Because neutralization is essential, all of these sources should be considered plasma sources, and the Denton source is *not* a cold cathode source in practice!

Our early experiments with the largest standard aperture supplied with the "Cold Cathode" source pointed to the desirability of keeping the drive volts below 400 for minimum absorption in the films and above 300 for maximum densifying effect (beam power). We prefer to operate at 300 V or less. We could seldom approach the lower limit of 300 with the large aperture because the gas flow required for that with the typical source conditions and the resulting chamber pressures would be too high to allow the production of the desired robust films.

The use of this source with a reduced aperture increases the pressure inside the cavity while decreasing the chamber pressure for a given gas flow. In this configuration in a 1100 mm chamber, the operating conditions are typically: 20-40 SCCM oxygen flow, 1 amp drive current, 250-300 drive volts (250-300 drive watts), and resulting chamber pressure of $0.7\text{-}2.2\times10^{-4}$ torr. This has been reported to produce results which are equal to or better than the larger aperture results.

When operating at the high powers which we used, if the neutralizer were turned off, one could observe the molybdenum anode to be glowing a dull red. We accidentally operated the source without water cooling at one time and that partially demagnetized the magnets of the source. The result of the lower magnetic field is to increase the drive voltage under otherwise constant conditions. Since this is very undesirable, operation without water cooling is to be assiduously avoided. The magnet sets are expensive to replace. We have also experimented with a higher current power supply and seemed to have overheated and damaged the magnets at about 2.0 amps of drive current even with water cooling. The standard power supplies provided to us by the manufacturer did not stand up to continuous operation at full power. There was an internal meltdown of one of the PC boards after extended use on at least six occasions with three separate power supplies.

We have used this deposition arrangement and technique extensively for the development of stacks ranging from 40 to 90 layers of SiO_2 and TiO_2. Insignificant spectral shift with changing humidity can be achieved using this IAD system. Without IAD, we typically may see a 15 to 20 nm spectral shift. The deposition conditions in a 760 mm chamber of such a run using the large source aperture might be: 225°C, 1 nm per second for both materials, 48 SCCM of oxygen through the ion source, 450-500 drive volts, 1.0 drive amps, 20 amps for the neutralizer, and the chamber pressure would be mostly in the range of 1.0 to 1.5×10^{-4} torr. The SiO layers getter more than the TiO_2 and therefore the higher chamber pressures are associated with the TiO_2. These films also pass the adhesion and severe abrasion tests of MIL-C-675. The resulting index of the SiO_2 is about 1.50. We have found that the conditions which give little or no humidity shift with this type of coating are not necessarily compatible with some laser damage requirements. Additional oxygen is required to reduce the laser damage and give an index of about 1.46, but the humidity shift will partially return, perhaps 6-8 nm. There may also be some correlation between the excessively high (450-500) drive volts and the laser damage threshold.

The ion source parameters must be reasonably stable or the uniformity has been seen to have some changes. The ion beam has a distribution which is clearly more concentrated on the axis of the beam. It appears that if an adequate amount of ions reach the less bombarded areas, the excess ions in the more bombarded areas are not detrimental. We have had some indication that this is less true when the drive volts are in the high (500 V) region. The hypothesis is that some ion etching may be occurring at the higher drive volts and thereby the thicknesses of the layers are slightly reduced at the center of the beam impingement area.

5.4.2. End-Hall Source

The End-Hall source has many similarities to the Cold Cathode source described above and some differences. Figure 5.24 illustrates the general configuration. Kaufman and Robinson[113] and Cuomo, Rossnagel, and Kaufman[111] describe these sources in some detail, but we will also give a brief description here for the convenience of the reader. The gas to be ionized is admitted to the throat of the anode at a controlled flow rate. Electrons from the DC or AC current-heated cathode bombard the gas. A high voltage is applied to the anode and ionization of the gas occurs. The magnetic field in the anode region is primarily axial which enhances the effectivity of the electrons. The ions are accelerated upward from the anode. Electrons from the cathode also serve as the neutralization for the ion beam.

It can be seen that the cathode is similar to the neutralizer of the Cold Cathode source. The magnetic fields and gas feeds are similar also. The anode configurations are different in detail, but similar in function. The major difference seems to be that the gas is confined to a smaller space where ionization occurs in the End-Hall source described here. This allows lower ion voltages at lower gas flow rates and therefore lower total chamber pressures.

The source which we have used had limits of: 5.0 amps of drive current, 175 anode or drive volts, 50 SCCM gas flow, 25 amps of cathode or neutralizer current. We had chosen: 4.0 amps drive current, 120 anode volts, and 19.0 amps of starting cathode current because the controller was unstable at the high limits of voltage, current, and gas flow. The controller/power supply was automated to start the discharge and then control the beam to preset values. A starting anode voltage and gas flow are preset, we used 130 volts and 20 SCCM. Upon starting, the controller heated the cathode, stabilized the gas flow to the preset value, and increased the anode voltage until the discharge started. After a number of seconds,

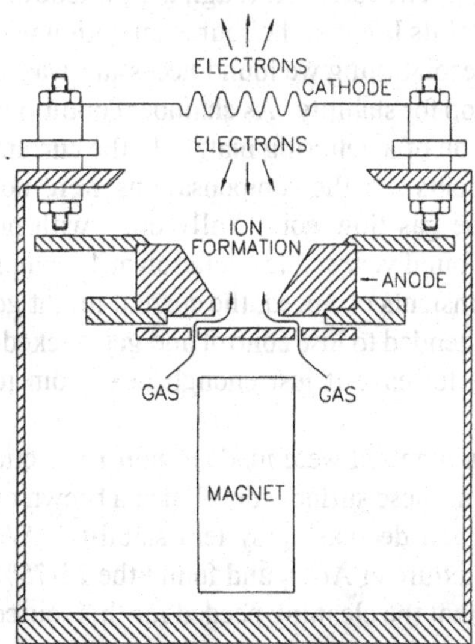

Fig. 5.24. The general configuration of End-Hall Source which has many functional similarities to the Cold Cathode Source.

the controller adjusted the system to the chosen running values set on the knobs of the controller. In principle, the controller then maintained the ion beam at constant voltage and current. The voltage was controlled by the gas flow in a servo loop, the drive current was maintained by a current control loop, and the neutralizer had a third loop which attempted to maintain neutralizer current.

We initially had great difficulty in achieving stable operation of this system. At first, we were told to be concerned about cleanliness of the source-head components, gas leaks in the interface of the head to its mounting plate, and coating on the cables in the chamber. We did find it necessary to shield the ion source from the electron gun cabling near the baseplate of the chamber to overcome arcing, noise, and interactions. However, our other concerns above turned out to be inconsequential. As long as all of the electrical connections were good and no major coating buildups occurred on insulators leading to the head, we had no difficulties with the head.

The controller was another story. Since our need was to get the maximum ion current density practical at an acceptable voltage from the source, we operated at the limits of the system capability. The three control loops seemed to fight each other. We were only able the achieve stable operation after making two changes to the control system. First we removed the neutralizer/cathode power control from the controller and replaced it by a manual variable AC transformer control. We set the cathode current to about 20 amps before starting the source and left the variable transformer set for the rest of the day. This provided satisfactory results and the longest cathode life (about 10 hours). The current through the cathode did drop to the 15 amps range toward the end of its life, but the source operation was stable until the cathode burned out. The second thing we found necessary was to properly compensate the current control loop for stability. As chamber conditions changed such as pressure during evaporation of a gettering material, the current and gas flow controls experienced transients. If the compensations were not correct, this lead to oscillations and/or the gas flow going fully open with no control. With the changes mentioned, we found very stable operation at 4.0 amps and 120 volts (480 drive watts). When transients occurred, the current might go near 5.0 amps. If it exceeded 5.0 amps, it tended to lose control and get "locked-up." We therefore operated at 4.0 amps to leave it just enough headroom to accommodate normal process transients.

The anode parts and gas distributor plate used were made of non-magnetic stainless steel. When operated with oxygen, these surfaces developed a brown or orange color and we seem to have observed decreasing system stability. We experimented with 50/50% and 25/75% mixture of Ar/O_2 and found the 25/75% most stable. We were very pleased to find that the cleaning needed for this source when we used the 25/75% mix was very easy. When setting up for a new run and before installing a new cathode, we scrubbed the cone of the anode and the exposed spot on the gas distributor plate with ScotchBright™ and removed the dust produced with a vacuum cleaner. There has been some later indication that the argon may be unnecessary in some cases and operation with pure oxygen may be satisfactory. The cathodes are a consumable item which we replaced before every long run. It is important to be sure that good electrical contact is made, particularly with the spring fingers of the push-on connectors. This seems to be hypersensitive if the controller has not been modified as we had done to remove the neutralizer control loop.

We found the End-Hall source with the control system as we had modified it to be quite stable, reproducible, and easy to maintain. The beam gave good results and uniformity over a 1150 mm diameter calotte when aimed at the 70% radius. The densification and oxidation of the SiO/SiO_2 seemed better than the Cold Cathode source under the same conditions. The effects of gettering were not as apparent here since the End-Hall source seemed to achieve higher beam current with less gas flow. The make-up oxygen supplied to keep constant chamber pressure was probably greater with the End-Hall source.

Our previous work with gridded Kaufman sources showed that TiO_2 tended to be damaged to the point of some absorption by ions in excess of 200 eV while SiO_2 was not adversely affected by 600 eV. The mean ion eV of the End-Hall source has been estimated[113] to be 60% of the anode voltage and by similarity the Cold Cathode and PS1500 sources may be about 60% also. This would lead to the estimation that the End-Hall source was providing about 72 eV ions and the Cold Cathode about 200+ eV. If the drive currents for the two sources can be compared at 4.0 amps and 1.0 amps respectively, this implies a relative maximum ion power of 72x4 to 200x1, or a 288:200 power ratio between the End-Hall and Cold Cathode sources. This is consistent with our observations in that the End-Hall seems somewhat stronger, but not overwhelmingly so. It appears that the End-Hall source is challenged to operate at a higher anode voltage for more beam power up to the point of film damage (200ev/.60) while the Cold Cathode source is challenged to operate at a lower anode voltage (333 V) to avoid damage. Both sources have proved usable but have significant room for improvement as provided by the manufacturers.

5.4.3. PS1500 Plasma/Ion Source

Our general needs would be best met by a system which operated stably at as much below 300 drive volts (V_D) as practical in an oxygen or argon background pressure of 1×10^{-4} torr and at the highest drive current (A_D) practical. The experience with the above sources and their limitations provided the basis for further developments toward higher power. Figure 5.25 shows the source. We will describe the measured performance of a production unit at 1500 drive watts.

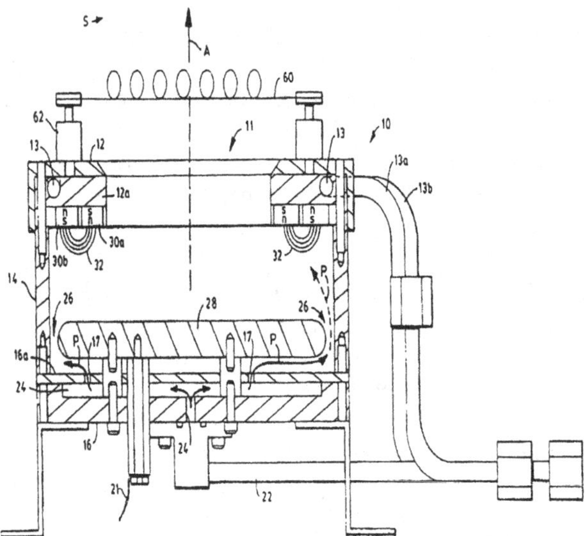

Fig. 5.25. The general configuration of the PS1500 source.

There was also a prototype unit which was operated continuously at 3000 drive watts (W_D). These are as compared to the 300 W_D of the Cold Cathode or 480 W_D of the End-Hall. Unless otherwise noted, the test data are all given for operation with a gas mixture of 75/25% O_2/Ar at an A_D of 10 amps and the neutralizer current (A_N) was set just above that needed to avoid arcing and sparking.

With all of these ion sources, electrons must be supplied by a neutralizer to combine with the ions and avoid a positive charge buildup on the substrates. Too little neutralization is indicated by arcing in the chamber. Sparks will be seen occasionally on substrates and holders and other points in the chamber. It is our practice to use about 0.5 amps additional neutralizer current than is needed to just eliminate arcing and sparking. In the case of the PS1500, neutralizer current usually ranges from 10-15 amps. All of the testing reported here has been done under these conditions. The A_N versus drive current and gas flow is shown in Fig. 5.26 for a 440 mm length of 0.5 mm diameter tungsten wire coiled to fit the connections above the source. The required A_N is only a slight function of gas flow and only moderate function of drive current. At low A_D, it appears that most of the neutralizer power is used to raise its temperature to where thermal electrons are emitted. Only small increases of A_N are needed to neutralize an order of magnitude more ions at 10 A_D. The A_N required for neutralization increases with drive amps because more ions are produced.

The voltage versus gas flow is primarily a function of the design of the ion source and the pumping speed of the chamber. Figure 5.27 shows the drive

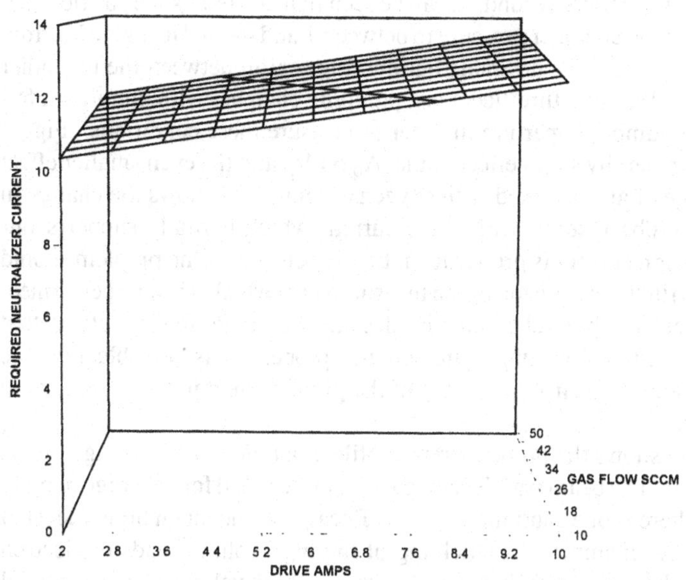

Fig. 5.26 The neutralizer current versus drive amps required by the new plasma source for no substrate sparking for a 440 mm length of 0.5 mm diameter tungsten wire at 30 SCCM.

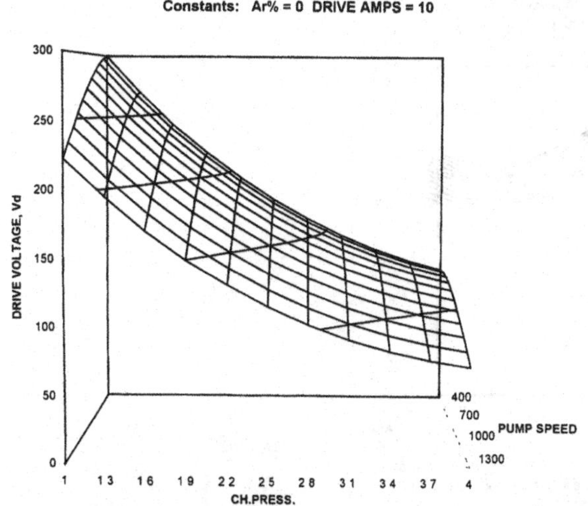

Fig. 5.27 The drive voltage versus chamber pressure and system pumping speed for the PS1500 plasma source.

voltage characteristics of the PS1500 with oxygen in a chambers with pumping speeds of 400 to 1400 liters/second. It can be seen that at 1400 l/s the gas flow can be used to control the chamber pressure to between 1 and 4×10^{-4} torr give V_D from 73 to 223 volts. This is the desirable but missing region between the two other gridless sources. We have throttled a fast pumping chamber with aluminum foil over part of the pumping aperture in order to measure these conditions. Figure 5.28 shows the generally small effect of the A_D on V_D and the even smaller effect of the percentage of argon mixed with oxygen. Figure 5.29 shows the change in the biased ion probe reading with drive current which partially supports our assumption that ion current is proportional to drive current. One problem should be pointed out with the simple ion probe measurements which we have performed. The probe responds only to charged particles and would not register atoms that became neutralized before impinging on the probe. It is possible that the departure from a rectilinear relationship of the probe current to the drive current is due to this effect.

Figure 5.30 shows the probe current profile as a function of the angle to the axis of the source. The beam was developed to be quite broad for coverage of more surface area. There is one caution, however. Because of the much higher level of plasma in a given chamber, the shielding of the high voltage leads of electron beam guns must be very carefully done. An exposed high voltage lead or terminal at many kilovolts will attract positive ions which will cause arcing in the E-gun. This is similar to the situation in a reactive ion plating system. We have done this shielding in two different ways. One system had false floor of sheet metal shields

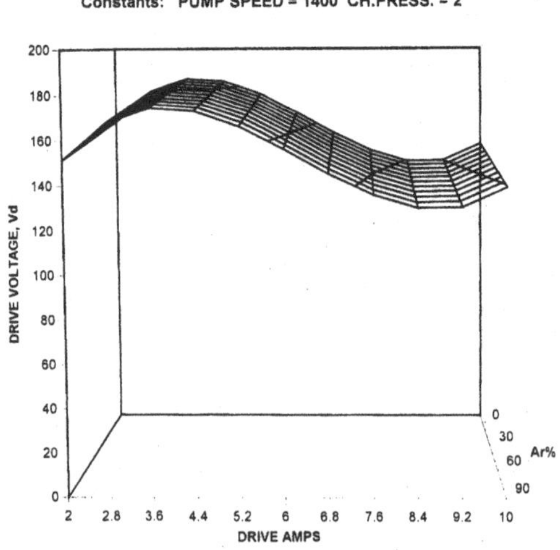

Fig. 5.28 Drive voltage as a function of drive amps and percentage of argon to oxygen in the gas flow for the PS1500 plasma source.

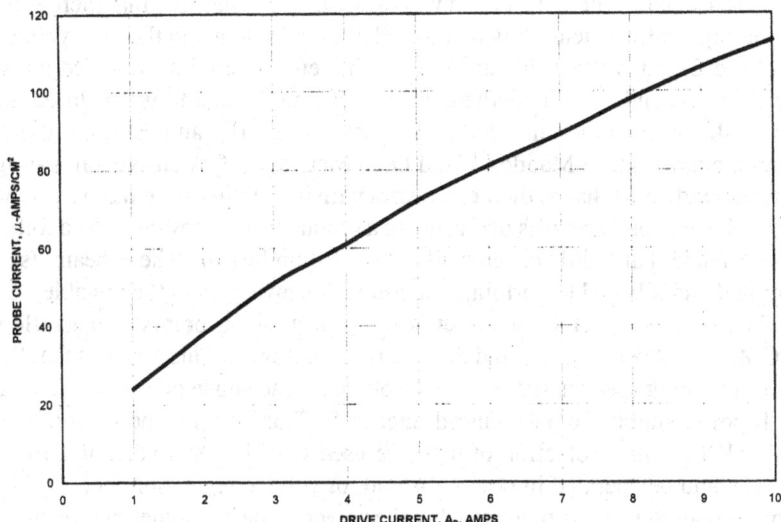

Fig. 5.29 The change in the biased ion probe reading with drive current. This partially supports our assumption that ion current is nearly proportional to drive current.

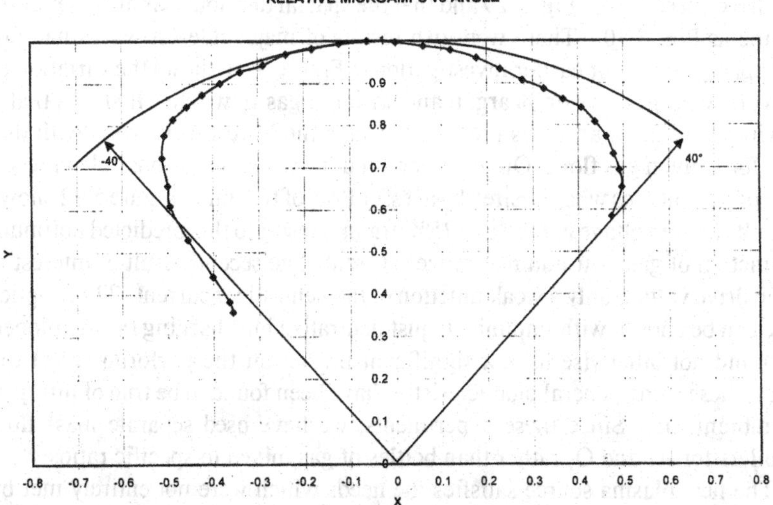

Fig. 5.30 Ion probe current profile as a function of the angle to the axis of the source.

above the base plate where the E-gun was located. We made sure that there were no major gaps in the shielding which would allow ions to get to the high voltage leads of the E-gun without the ions having first encountered a grounded metal surface. In a second case where there were no base plate shields, we enclosed the E-gun with heavy aluminum foil except at the hearth and E-beam itself. Conversations with Don Maddox[118] and Leon McCrary (of extensive ion plating experience) indicated that in their early work with ion plating systems an extreme case of this same problem was overcome in a unique way. They describe a foiled baseplate shield that tightly covered all but the E-gun hearth. The E-beam itself was ignited and allowed to perforate the foil for its own path to the crucible.

The variables available on the new source are: gas type (or mixture), gas flow in SCCM, neutralizer current, and drive current or power. This assumes that the chamber pumping speed is fixed. As the source for the anode power, we used an MDX II power supply from Advanced Energy[446]. This unit had the capability to supply 15KW from a selection of taps;, we used tap #4 at 800 volts and 18.75 amps. It could be operated in power, current, or voltage regulation modes. The current regulation mode is preferred, but the power mode can sometimes be more stable in certain regions of operation. The results of interest are: drive voltage (ion eV), ion current, ion distribution in space, and stability. MDX-5 units have also been used with their 10 amp maximum on drive current; this has been satisfactory since it allows 10 amps at 150 drive volts when working with the 1.5 KW power limitation of the PS1500.

For the purposes of this particular investigation, we focused on the drive voltage result and we assumed that the ion current was approximately proportional to the drive current as in Fig. 5.29 and that the spacial distributions are essentially the same as Fig. 5.30. There were two results of major interest which had not been apparent from our earlier investigations. Figure 5.31 shows the variation of drive volts with gas mix or % argon and with the gas flow rate. It shows that a mix with about 33% argon is predicted to give the lowest drive volts with this source for a given gas flow. On the other hand, it would still be possible to reach higher drive voltages when desired by a lesser flow of this mix. Figure 5.32 shows the results Drive voltage at the mix (25% argon) closest to this predicted optimum as a function of gas flow and neutralizer current. The second result of interest is that the drive volts is only a weak function of the neutralizer current. This implies the A_N can be chosen with impunity to just neutralize ion charging (as mentioned above) and not otherwise have a significant impact on the performance of the source. These same general characteristics have been found to be true of nitrogen and ambient air. Since these experiments, we have used separate mass flow controllers for Ar and O_2 rather than bottles of gas mixed to specific ratios.

The new plasma source satisfies the needs which were not entirely met by previous sources, but those sources might be adequate for some less stringent

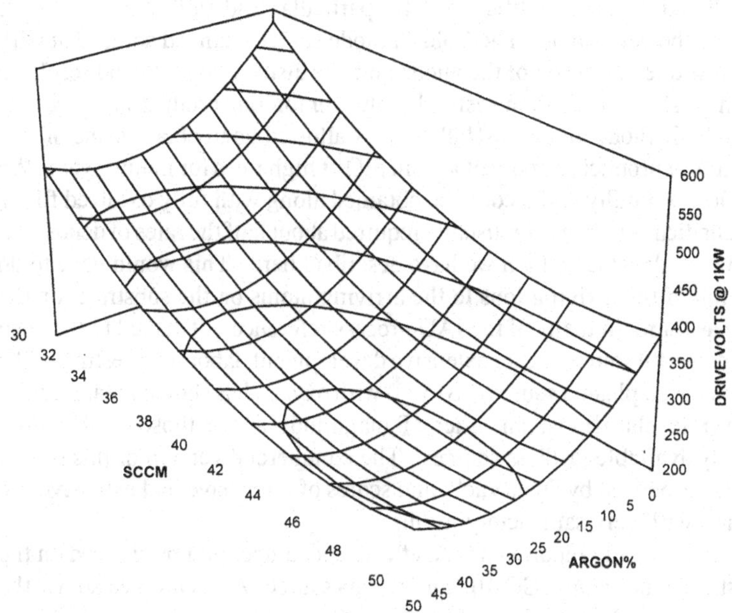

Fig. 5.31 Drive voltage as a function of percentage of argon in the oxygen gas mix and the gas flow rate.

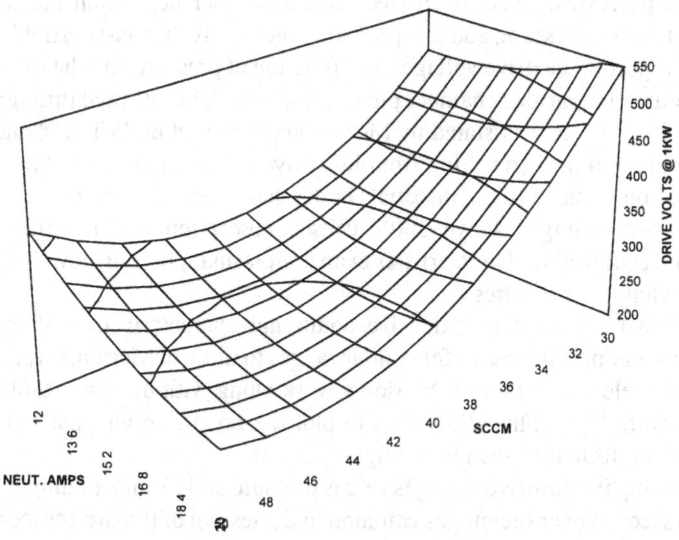

Fig. 5.32 Drive voltage as a function of gas flow rate and neutralizer current using a 25% argon gas mixture at one (1) kilowatt of drive power.

requirements. After some modifications, the particular End-Hall source which we used proved to be very stable. The Cold Cathode source changed somewhat with operating time due to erosion of the anode and gas distributor plate and resulting deposits; and it showed less than desirable behavior for long production processes. Both the Cold Cathode and End-Hall sources allowed us to obtain some of the desired results of producing good SiO_2 from SiO at high rates for many layers. We found that low humidity shifts could be obtained along with fully oxidized films. This full densification, however, usually required about half the rates of deposition or twice the ion densities which we have described here. This would have to do with the ratio of the arriving ions to the arriving atoms on the substrate, or the arrival rate as found in many of the IAD process references. This led to the need for a more powerful source. After extensive development, laboratory testing of the resulting improved plasma source (PS1500) with oxygen has shown great stability at ion power levels almost an order of magnitude above those of the other commercially available gridless sources. The commercial version of this source has now been produced by DynaVac[114] and scores of units have had extensive use in production with very satisfactory results.

Morton[463] recently reported on the effects of chamber pumping speed on the characteristics of the Denton CC-105 ion/plasma source. As mentioned above, the principal variables affecting the performance of some of the common ion/plasma sources used in optical coatings are pumping speeds, gas flows, and drive currents. The first two parameters result in a given chamber pressure which is important to most deposition processes. The mean free path and other deposition factors depend on the chamber pressure, and low pressure is generally the most desirable. Morton's work reported the drive voltages as a function of pressure and the drive current to the source for various chamber pumping speeds. The gas used through the source was oxygen. It was stated by Morton and reported by Willey[463] and others that low ion voltages (resulting from low drive voltages) are desirable to avoid disassociation damage to the materials being deposited. Willey found for example that drive voltages above 300V caused absorption in TiO_2 films. Therefore, the most desirable characteristics of an ion/plasma source are low drive voltages at low chamber pressures.

We had also recently done extensive additional characterization of the DynaVac PS1500 ion/plasma source for similar ranges to those of Morton's work. We processed the relevant data from Morton's work along with our own results using DOE software.[464,465] This allowed us to plot the results in an "apples to apples" comparison format as seen in the figures below.

When we compare the drive voltages versus pressure and chamber pumping speed (in liters/second) over the ranges common to the testing of the two sources, we produce Figs. 5.33 and 5.34. The common ranges are: 1 to 4×10^{-4} torr pressure, 400-1400 l/sec pumping speed, up to 3.5 drive amps, and using pure oxygen for gas. High pumping speed has the advantage of reducing the drive voltage and chamber pressure. Morton's work extended to a pumping speed of 2350 l/sec, but our test work with the PS1500 to date has only been done in a chamber with up to 1400 l/sec pumping speed. The comparison was limited to this, and to 3.5 drive amps because that was the limit of Morton's work. The PS1500 will normally operate at drive currents up to 10 drive amps and/or drive power limits of 1500 watts continuously. Figure 5.35 also shows the CC-105 over the more extended pumping speed range to 2350 l/sec. reported by Morton. Figure 5.36 shows the PS1500 when operated at its extended range of 10 amps of drive current in a 1400 l/sec chamber.

With both sources, there is not a strong influence of the drive current on the drive voltage output. Figures 5.37 and 5.38 show the comparison over the common range of testing from 1 to 4 drive amps and 400-1400 l/sec. Figure 5.39 shows the CC-105 over the full range of measured pumping speeds to 2350 l/sec. Figure 5.40 shows the PS1500 over its full range of drive current to 10 amps.

We measured the integrated ion current output of the PS1500 at 11% of the drive current. We were unable to measure any additional neutrals which might be coming to the surface, since the ion probe used would only register charged particles. At 10 drive amps and 57 cm from the source, we measured 160 microamps per square cm. The breadth of the beam output is seen in Fig. 5.30. This is broader and flatter than most ion sources and is well suited to large "box coaters" as used in the optical coating industry.

It can be seen from Figs. 5.34 and 5.35 that the PS1500 achieves lower drive voltages at lower pressures for the same pumping speeds. Greater pumping speeds

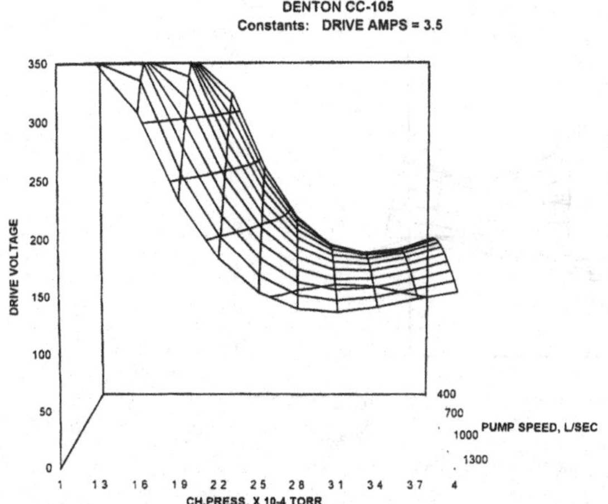

Fig. 5.33 Drive voltage versus pressure and chamber pumping speed for the Denton CC-105 cold cathode source over the common range of testing.

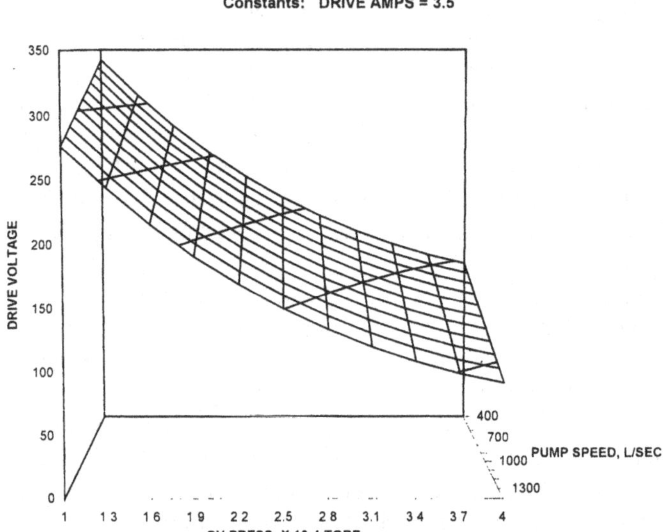

Fig. 5.34 Drive voltage versus pressure and chamber pumping speed for the DynaVac PS1500 plasma/ion source over the common range of testing.

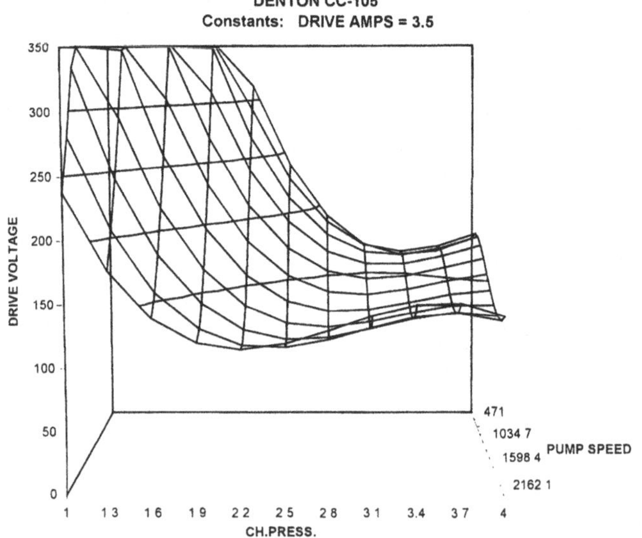

Fig. 5.35 Drive voltage versus pressure and chamber pumping speed for the Denton CC-105 cold cathode source over the full range of testing reported.

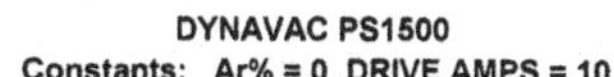

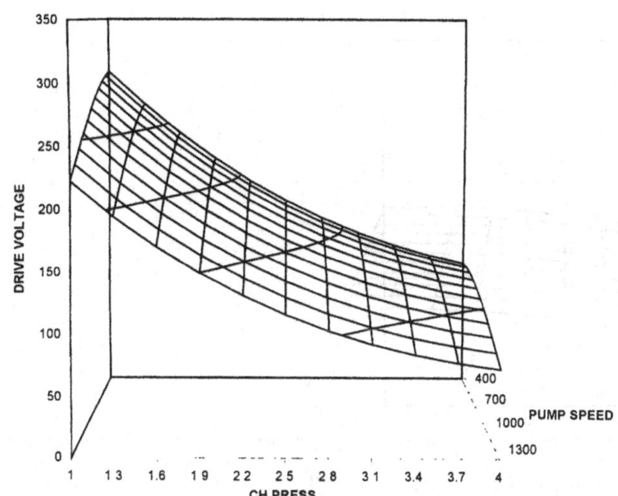

Fig. 5.36 Drive voltage versus pressure and chamber pumping speed for the DynaVac PS1500 at the maximum 10 amps drive current for the MDX-5 power supply.

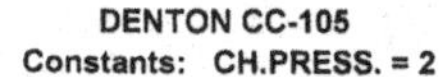

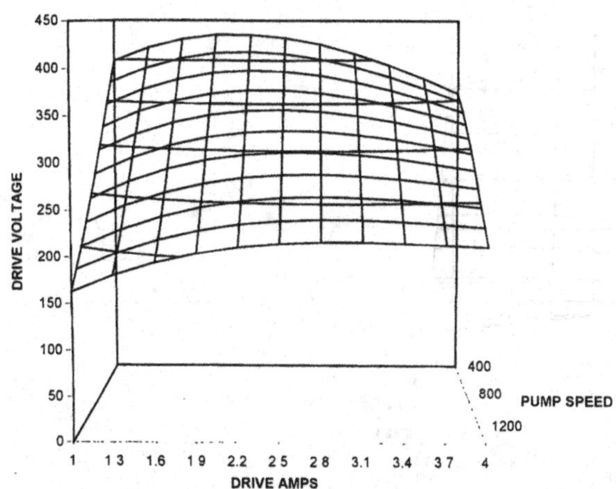

Fig. 5.37 Drive voltage as a function of drive current and chamber pumping speed for the Denton CC-105 over the common range of testing.

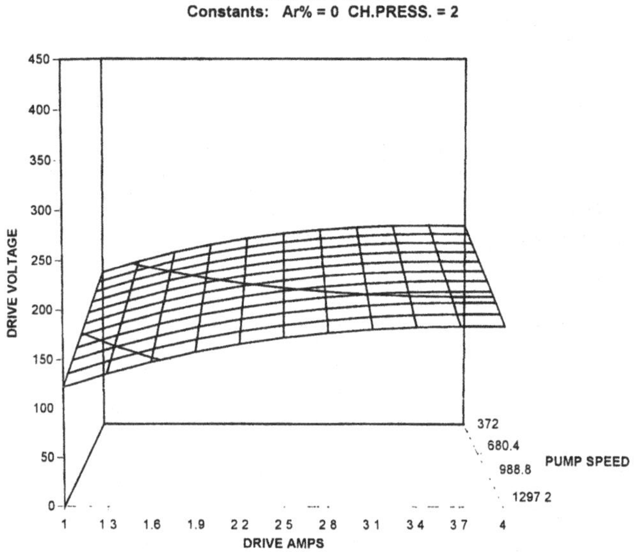

Fig. 5.38 Drive voltage as a function of drive current and chamber pumping speed for the DynaVac PS1500 over the common range of testing.

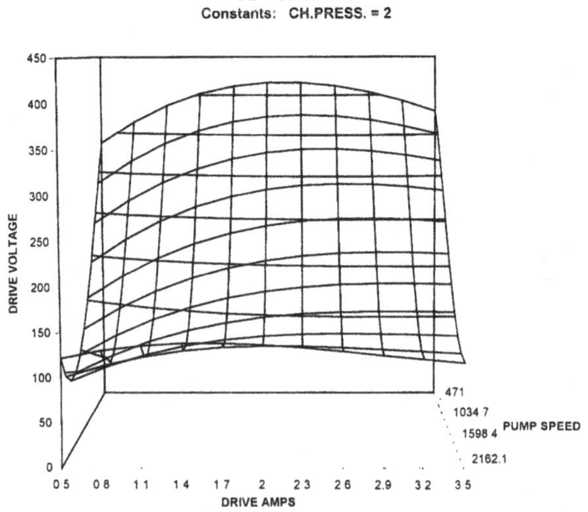

Fig. 5.39 Drive voltage as a function of drive current and chamber pumping speed for the Denton CC-105 over the full range of testing reported.

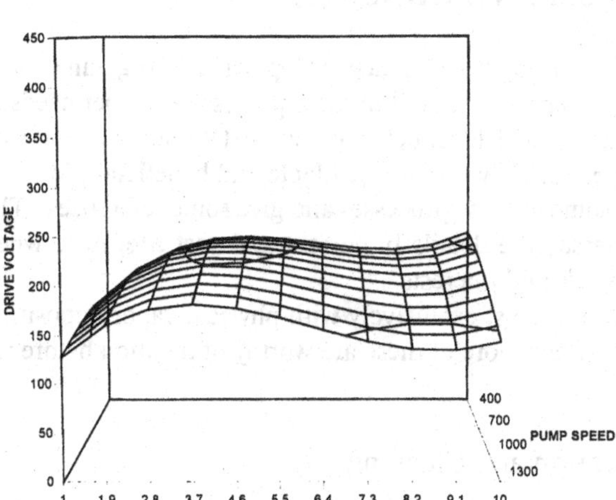

Fig. 5.40 Drive voltage as a function of drive current and chamber pumping speed for the DynaVac PS1500 over the full range of testing at the maximum 10 amps drive current for the MDX-5 power supply.

benefit both sources proportionately. The PS1500 achieves less than 100V at 3.5A with the 1400 l/sec system. At 300V, the pressure is $<1.0 \times 10^{-4}$ torr for the PS1500. The high drive currents of the PS1500 (10A) and thereby high ion currents (1.1 amps) provide the ability to deposit films at proportionately higher rates than lesser powered sources for the same ion to adatom arrival ratios.

From Figs. 5.33 to 5.40 and Morton's report[463], it is apparent that high pumping speeds are preferred for most ion-assisted deposition (IAD) processes because of the lower drive voltages and lower operating pressures.

Niederwald et al.[447] have also compared three different ion/plasma sources in similar processes. They compared the Cold Cathode, End-Hall, and APS systems. They reported the highest refractive indices with the End-Hall and the smoothest films from the APS.

Ion-assisted deposition has proven itself as a means of producing superior properties in almost all oxide films. Fluoride films, on the other hand, have not yet been as successfully treated by IAD. We see this as the challenge for the near future where there are the greatest opportunities for improvement. It appears that the techniques used to date tend to liberate fluorine and leave metal atoms behind without making up the lost fluorine atoms, except perhaps by replacing them with oxygen (as seen in the case of MgF_2[14] discussed in Sec. 5.3.1.3). The unfluorinated (or unoxidized) metal atoms then cause absorption. If the fluorine atoms can be replaced or never lost in the first place, IAD should show the same improvements for fluorides as it has with oxides.

5.5. OTHER PROCESSES TO CONSIDER

There are an enormous variety of ways to deposit optical coatings and more variations being developed every day. The thin films process developer needs to be circumspect and keep current with these developments to take advantage of new opportunities and equipment as they become available and beneficial. We can only touch briefly here on some of these processes and give some references. The reader is encouraged to pursue the details in current and past literature when appropriate to their own needs and interests.

We dealt in some detail almost exclusively with physical vapor deposition (PVD) processes thus far. A few more of these are worthy of mention before we go on beyond PVD.

5.5.0. Surface Preparation and Cleaning

Before we go into other deposition processes, let us briefly address the preparation of substrates for coating. The removal of foreign materials from a surface is usually essential to allow the deposited films to adhere well to the surface. Cleaning is the principal part of this, but surfaces may also by made more active to promote adhesion by chemical or plasma processes. Guenther[253] reported studies of the influence of surface layers "naturally" formed on the substrates of glasses sensitive to acid or alkali where they acted as the first layer of the coating and changed the reflectance. We have seen cases where a visible "splotchiness" appears on some such coated surfaces after coating. In that case, it was necessary to coat the lenses of those types of glass soon (hours) after polishing to avoid the atmospheric effects on the surface chemistry, or it was possible to very carefully store the surfaces in an inert environment until coating.

This author is not personally experienced in a broad variety of cleaning processes, but refers the reader to an extensive discussion of the subject by Mattox.[254] There are also more than one hundred references on "cleaning" among the Society of Vacuum Coaters Technical Conference Proceedings. The advent of IAD and other processes has somewhat supplanted the earlier practice of "Glow Discharge" cleaning just before deposition. Guenther (and also Perry[24]) used to conjecture that glow discharge "cleaning" might do more harm than good in some cases because it might sputter debris and contaminants from the chamber walls and onto the substrates. The directionality of many ion/plasma sources can reduce this concern and provide an ion/plasma activated cleaning action which is primarily substrate-specific. Such cleaning with oxygen ions would seem to be able to remove hydrocarbons by oxidation and sputtering.

There are a few additional papers in the literature with helpful cleaning techniques. Coated mirrors with a limited protective overcoat or none can easily become scratched or "sleeked" by any kind of wiping. Some time ago McDaniel[255] described how to paint on a collodion solution with a camel hair brush and then peel off the dust and collodion after it gels. He points out that this can be done without removing the mirror from its instrument in many cases, thus avoiding the need for realignment of the optical system.

In the latter part of the 1970s it was found that ultrasonic cleaning can cause surface deterioration. House et al.[256] in the United States reported a reduction in the LIDT in some cases where ultrasonic cleaning was used. McLauchlan and Gibbs[257] in Australia reported similar results including micrographs of the surface modifications. Brenci et al.[258] from Italy confirmed the same conclusions, all in the same year (1977). The conclusion from this is that: if LIDT is of concern, ultrasonic cleaning should be used carefully and sparingly if at all.

5.5.1. Physical Vapor Deposition

We discussed sputtering in Chapter 4. This has had its limitations when depositing oxides. If metal oxides are to be reactively sputtered from the metal, the tendency of the cathode to form an oxide layer and change its sputtering characteristics has been a major difficulty. Optical Coating Laboratory, Inc.[115,116] developed its MetaMode™ process to overcome this limitation. They move the parts quickly from the environment of a metal sputtering source to the environment of an oxidizing plasma source where the thin metal layer is converted to the metal oxide. Similar things can be done with nitrides. The typical configuration is drumlike wherein the parts rotate many times by the alternating sources and build the desired thickness from many very thin sublayers of the same oxidized material.

Pulker[117] has a brief review of Reactive Ion Plating (RIP) in which he first references the work of Mattox and Kominiak[118] in 1972. Much of the early application of the RIP process seems to have been outside of the optical industry in what is referred to as the Ion Vapor Deposition (IVD) of aluminum of "Ivadizing." This field is exemplified by the work of Nevill.[119] In the past decade, there has been expanded interest in this area by the optical industry which was facilitated by the introduction of the Balzers BAP800 RIP system. Guenther et al.[120,121] described the system, process, and some applications to optics. This system generates ionized atoms of the deposition materials (along with other species) which are attracted to the negatively charged substrates with energies of the order of 50 eV. This is much higher than the non-IAD evaporation energies and usually of a higher power density (due to higher current) than most IAD processes. Fully densified films are often achieved with a vitreous rather than crystalline or dendritic structure. Some difficulties were encountered due to a tendency to high compressive stress which is often produced.

An interesting process that has recently become of interest is the vapor deposition of polymer materials for optical purposes. Shaw and Langlois[122,123] and Martin et al.[124] describe evaporating a monomer in a vacuum and curing it to a polymer with UV or an electron beam. Their work has been in the web coating environment, and hundreds of layers have been deposited with control adequate for dichroic film applications. The work to date has been very successful for web coatings of polymer-aluminum-polymer as barrier layers for food and other packaging applications. Yializis[125] illustrates that one of the benefits here comes from the monomer flowing and covering blemishes that would have otherwise caused pinholes in the aluminum film which reduce its barrier properties. The blemish-covering properties are helpful on decorative parts, and the protective nature of the polymer over metal coatings of aluminum and silver has been demonstrated.

It may be that polymer thin films will accrue greater interest in the future for optical applications. Gittleman and Arie[126] sputtered Teflon between two layers of aluminum to make a write-once optical storage disk. Andersson et al.[127] used a lithium perchlorate doped polymer laminate in a "Smart Window" construction. Biederman[268] describes sputtering polymers as coatings.

5.5.2. Dip, Spin, and Spray Coatings

Pulker[117] (pp. 96-114) has an extensive discussion of immersion- or dip-coating. Here the substrate is dipped into a solution which forms a film of uniform optical thickness as it is properly withdrawn from the solution. The layer may be baked or fired to finalize its form and composition. Multiple layers can be applied in this way. The Schott Company in Germany pioneered this work many decades ago and it is currently used for a broadband AR coating on picture framing glass with very good results. It can also be used for architectural windows. The films can be made at least as durable as evaporated films. It is not out of the question to use such processes on optical instrument windows and lenses, but curved lenses might cause uniformity problems and some glasses might deteriorate at the post processing temperatures. Ulrich and Weber[128] used these processes to deposit films for light-guides.

Thomas[276] made porous coatings of CaF_2 from colloidal suspensions. These had indices as low as 1.2 because of the porosity and were therefore excellent AR coatings on low index substrates such as fused silica. They also had good laser damage resistance. The preparation was by spinning. The thickness was adjusted by rotation speed of the substrate and/or suspension concentration.

5.5.3. Chemical Vapor Deposition

Most of the processes we have discussed above are PVD wherein materials are vaporized and condensed on the substrates with a composition having the stoichiometry essentially the same as the starting material. The vaporization might be from melting and evaporation or from sputtering. Reactive evaporation starts to depart from this definition of PVD. This is particularly true if the starting material is a metal which is oxidized during the deposition such as starting with Ti of Si and producing TiO_2 or SiO_2. Reactive evaporation then leads us into Chemical Vapor Deposition (CVD). Pulker[117] (p. 117) gives the definition: "Basically CVD is a material synthesis in which constituents of the vapor phase react chemically to form thin solid films as a solid-phase reaction product which condenses on the substrate." In essence, reactive evaporation is just this; it is just that one of the components came to be in the vapor phase by heating or sputtering. In the common case, however, CVD is done with constituents which are normally in the gaseous phase. Pulker[117] has an extensive discussion of CVD if the reader is interested in more detail and references, we will only make a few general comments here. CVD has been used extensively in the semiconductor industry over the last few decades. Most of the reactions require higher substrate temperatures than normal for PVD, such as 500-1000°C. To date, these processes are not applicable to organic substrates such as polymers. Many of the semiconductor processes have dealt with dangerous gases and difficult environmental protection issues. Carlsson[443] discusses CVD in more detail for those needing additional information.

5.5.4. Plasma-Enhanced CVD

Plasma-Enhanced CVD (PECVD) or Plasma-Activated CVD (PACVD) has become a very active field in the last decade. It allows CVD reactions to occur at much lower temperatures than ordinary CVD such as 300°C and even less. The plasmas had typically been generated in earlier times by the usual DC glow discharge but in more recent times by RF or microwave. Martinu and Poitras[466] have given an extensive review of this technology and its applications. Sherman[444] also provides a broad review of PECVD. Pulker[117] (p. 135) points out that the lower pressures of PACVD give the benefit of large diffusion constants. He mentions the deposition of SiO_2, TiO_2, and other optical materials by this process. He shows that the deposition of Si_3N_4 by atmospheric pressure CVD at 900°C with low pressure PECVD at 300°C give virtually the same properties. Tsai et al.[129] used RF PECVD at about 250°C to produce 15-layer LWP filters in the near IR from a-Si and a-SiN_x films.

The application of PECVD to barrier coatings has been variously reported over the past five years.[130-136] SiO_2 has been produced from tetramethyldisiloxane (TMDSO) and from hexamethyldisiloxane (HMDSO). Exploratory work has been done with Al_2O_3 from trimethylaluminum (TMA) by Knoll and Theil.[136]

The use of PECVD for ophthalmic applications has also been reported.[137-139] Tetraethoxysilane(TEOS) has been used as the precursor. A plasma source and its application to PECVD was described be Johnson.[140] Bertella et al.[141] used PECVD for a hard carbon wear layer on magnetic disks. Izu et al.[142] coated silicon solar cells using RF PECVD.

Electron Cyclotron Resonance (ECR) has added another means of generating a plasma. Hwang et al.[143] has used ECR-PECVD to produce diamond-like carbon films on polycarbonate material. Sauvageau et al.[144] deposited SiO_2 films using ECR-PECVD for superconducting applications and Bourget and Lane[145] deposited SiC over large areas using this process.

The benefits such as lower deposition temperatures and more complete reactions gained by adding energy to the depositing/reacting atoms and molecules are apparent from all of the foregoing discussions. Photoinitiation is a commonly used method to activate polymerization. Bonds in a photoinitiator additive are broken by an energetic photon, usually in the UV. These dangling bonds are then highly reactive and promote electron exchanges which cause polymerization. Another form of enhanced CVD has been done using photons. Demiryont et al.[146] have performed laser-assisted CVD (L-CVD) of aluminum oxynitrides using TMA and a laser at 193 nm.

5.5.5. Plasma Polymerization

There is essentially a continuum of processes which includes the activated CVD processes discussed above and plasma polymerization. Pulker[117] (p. 270) points out that plasma polymerization had been performed by the middle of last century. The introduction of acetylene, ethylene, styrene, benzene, etc., into a glow discharge can be made to deposit a solid polymer. This does not differ significantly from PECVD in that the plasma generates radicals and ions which can then react with each other. If only inorganic materials are involved, the products might not be polymers, but Pulker points out that polymers can also be produced from only an inorganic mixture of CO, H_2, and N_2. Because of the complexity of the process mechanisms, the number of variables, and the number of reactions which could take place, plasma polymerization is not fully characterized and understood. Most processes seem to have been arrived at by experimentation. In effect, one is dealing with plasma/vapor phase polymer chemistry, whereas most polymer chemistry has been done in solution without extra electrical energy input. It appears to be a vast opportunity for further theoretical and experimental research and development.

Tien et al.[147] used plasma polymerization to produce films for integrated optics. Hollahan et al.[148] made protective and AR films on alkali metal halide crystals. Wydeven and Kubacki[149] made an AR for polymethylmethacrylate (PMMA). Hora and Wohlrab[150] describe the application of plasma polymerization to ophthalmic hard coatings on plastic lenses and Wielonski[151] reports on the application to plastic aircraft window hard coatings. In the next section, we will touch briefly on the very active related field of diamond-like carbon hard coatings.

5.5.6. Hard Carbon Coatings

Diamond-like hard carbon (DLC) coatings have gone from a laboratory process in the early 1980s to such widespread use as to be found in many if not most supermarkets as protection on the bar-code scanner window surfaces. The benefits of DLC include hardness, low friction, and chemical resistance.

Dischler et al.[152] , Bubebżer et al.[433,434], and Koidl et al.[435] reported the deposition of amorphous carbon at 50 nm/minute from benzene in an RF discharge in 1982-1984. Enke[153] gave additional detail in his work using similar techniques. He had showed us surfaces coated on germanium which would not be scratched by a razor blade. Knapp and Kimock[154] applied DLC for the supermarket scanners using methane gas to produce a deposition through a Kaufman type ion (plasma) source instead of the previously used RF sputtering-like processes. They also applied the coating to ZnS and ZnSe. Russell[155] gave a review of many aspects of diamond technology and its marketplace and how DLC fits into the picture. Daniels[156] described the successful application of DLC on polycarbonate plastic by an RF planar diode process. He showed some of the stress-related problems and that a film of about 3 microns gave very good abrasion resistance on 1/8" polycarbonate. Hughes and Kimock[157] showed the further commercialization of DLC to glass and polycarbonate using Kaufman type sources. Bartella et al.[141] showed the application of DLC as a thin protective overcoat on magnetic media using sputtering technology of carbon plus C_2H_2 gas for higher deposition rates. Lavine[158] describes a related system to deposit a 20-30 nm layer of DLC on hard disk media. Hwang et al.[143] deposited DLC on polycarbonate with ECR-PECVD at 50°C from CH_4 and C_2H_6 at rates of about 0.2 nm/second and reported on their extensive parameter investigations. Leonhardt et al.[159] mentioned the deposition of DLC using a laser arc method. Shabalin et al.[429,430] deposited DLC on hard discs at 1-2.5 nm/second from butane and ethylene with an ion source. Their source operated from 750 to 1500 discharge volts. Hieke et al.[431] describe the use of medium frequency discharge for PECVD rather than RF. They start with acetylene or methane and report film properties equal to RF processes except that the films are more compliant and smoother. The process is expected to be more economical and more easily scaled to large sizes.

J. Robertson[438] described the deposition mechanism on an atomic scale. He speaks of the sp^3 bonding arising, in 50-500 eV ion beam processes, C ions entering subsurface atomic sites and producing quenched-in density increases. The optimum energy is about 100eV because the ions must have enough energy to penetrate the surface, but not so much as to cause annealing out of the density increase. He also describes the more complex case of a-C:H. Martinu[439] reviews the deposition and testing of hard coatings on plastics. Scaglione et al.[440] studied the use of DIBS to deposit DLC. They report a peak hardness of over 61 GPa as a function of momentum transfer.

Tribological applications of DLC are widespread. Metal-containing DLC (Me-DLC) have appeared in the literature. Schaefer et al.[432] describe some of these applications for the manufacturing/fabrication industry.

Raiche and Jeffries[160] analyzed diamond deposition using laser-induced fluorescence. Although this dealt with the dc arcjet deposition of more high temperature processes, it adds some information for better understanding of these processes. Martin et al.[437] analyzed the hardness and structure of DLC by arc evaporation and found the hardness to be 13000 Hv.

Until recent decades, the properties of diamond had been considered achievable only by great heat and pressure. The process of using plasma activated atoms/molecules has made it practical to attain diamond-like properties at essentially room temperature. We have touched on many of the approaches which have been taken to accomplish this, and many more will probably develop in the future. The deposition of actual diamond, as opposed to DLC, has been the subject of a whole realm of work done in the past decade which we have not even touched on except in the last reference.[160] We have even heard of small quantities of real diamond being produced by the artful use of an acetylene torch. It has been amazing in both DLC and true diamond to see how processes become simpler and easier with each contribution to the progress of understanding, methodology, and equipment.

Chardonnet et al.[428] report on the successful application of microwave PECVD produced diamond films of 1-2 μm thickness in Brewster angle polarizer stacks for 10.6 μm lasers. They achieved 1000:1 extinction ratios and expect that even better ratios are not difficult to achieve.

There is a book[436] reviewing the whole subject and a journal devoted to the subject: *Diamond and Related Materials* published by Elsevier Science Publishers in Amsterdam.

5.6. SUMMARY

The synergism of one technologist building upon the foundation of another has moved the achievements in the field of materials and processes (and many others) much more rapidly than could be expected from the efforts of one investigator or laboratory. The increased communication and accessibility of information in recent times has accelerated the progress. It can be seen from many of the discussions of this chapter that the development of energetic deposition processes has made a major impact on durability and performance of what can be produced in optical thin films. In the sense of Thelen's[110] view as seen in Figs. 5.12 and 5.13, we certainly have experienced major work and progress in the area of processes in the 1990s.

5.7. REFERENCES

1. F. C. Sulzbach: "Infrared Antireflection Coatings Without Thorium Fluoride," *Proc. Soc. Vac. Coaters* **36**, 102-108 (1993).
2. I. Zubezky, E. Ceren, Z. Taubenfeld, and H. Zipin: "Efficient and durable AR coatings for Ge in the 8-11.5 μm band using synthesized refractive indices by evaporation of homogeneous mixtures," *Appl. Opt.* **22**, 1828-1831 (1983).
3. J. A. Aguilera, J. Aguilera, P. Baumeister, A. Bloom, D.Coursen, J. A. Dobrowolski, F. Goldstein, D. E. Gustafson, and R. A. Kemp: "Antireflection coatings for germanium IR optics: a comparison of numerical design methods," *Appl. Opt.* **27**, 2832-2840 (1988).
4. H. A. Macleod: *Thin Film Optical Filters,* 2nd Ed. (MacMillan, New York, 1986) "Appendix Characteristics of thin-film materials."
5. J. R. McNeil, A. C. Barron, S. R. Wilson, and W. C. Herrnann, Jr.: "Ion-assisted deposition of optical films: low energy vs high energy bombardment," *Appl. Opt.* **23**, 552-559 (1984).
6. H. K. Pulker, G. Paesold, and E. Ritter: "Refractive indices of TiO_2 films produced by reactive evaporation of various titanium-oxygen phases," *Appl. Opt.* **15**, 2986-2991 (1976).
7. *Coating Materials and Selector Guide* (slide rule) from Balzers, 8 Sagamore Park Rd., Hudson, NH 03051.
8. A. L. Olsen and W. R. McBride: "Transmittance of Single-Crystal Magnesium Fluoride and IRTRAN-1 in the 0.2- to 15-μ Range," *J. Opt. Soc. Am.* **53**, 1003-1005 (1963).
9. E. Ritter: "Optical film materials and their applications," *Appl. Opt.* **15**, 2318-2327 (1976).
10. K. H. Guenther: "Growth structures in a thick vapor deposited MgF_2 multiple layer coating," *Appl. Opt.* **26**, 188-190 (1987).
11. A. E. Ennos: "Stresses Developed in Optical Coatings," *Appl. Opt.* **5**, 51-61 (1966).
12. C. M. Kennemore and U. J. Gibson: "Ion beam processing for coating MgF_2 onto ambient temperature substrates," *Appl. Opt.* **23**, 3608-3611 (1987).
13. P. J. Martin and R. P. Netterfield: "Ion-assisted deposition of magnesium fluoride films on substrates at ambient temperatures," *Appl. Opt.* **24**, 1732-1733 (1985).
14. P. J. Martin, W. G. Sainty, R. P. Netterfield, D. R. McKenzie, D. J. H. Cockayne, S. H. Sie, O. R. Wood, and H. G. Craighead: "Influence of ion assistance on the optical properties of MgF_2," *Appl. Opt.* **26**, 1235-1239 (1987).

15. J. C. G. de Sande, C. N. Afonso, J. L. Escudero, R. Serna, F. Catalina, and E. Bernabeu: "Optical properties of laser-deposited a-Ge films: a comparison with sputtered and E-beam-deposited films," *Appl. Opt.* **31**, 6133-6138 (1992).
16. T. I. Oh: "Broadband AR coatings on germanium substrates using ion-assisted deposition," *Appl. Opt.* **27**, 4255-4259 (1988).
17. R. R. Willey: "Antireflection coatings for germanium without zinc," *Proc. SPIE* **1168**, 205-211 (1989).
18. F. Evangelisti, M. Garozzo, and G. Conte: "Structure of vapor deposited Ge films as a function of substrate temperature," *J. Appl. Phys.* **53**, 7390 (1982).
19. W. Heitmann and E. Ritter: "Production and Properties of Vacuum Evaporated Films of Thorium Fluoride," *Appl. Opt.* **7**, 307-309 (1968).
20. G. A. Al-Jumaily, L. A. Yazlovitsky, T. A. Mooney, and A. Smajkiewicz: "Optical properties of ThF_4 films deposited using ion-assisted deposition," *Appl. Opt.* **26**, 3752-3753 (1987).
21. M. J. Otten: "The use of thorium fluoride in the regulatory environment," *Proc. Soc. Vac. Coaters* **32**, 220-224 (1989).
22. J. T. Cox, J. E. Waylonis, and W. R. Hunter: "Optical Properties of Zinc Sulfide in the Vacuum Ultraviolet," *J. Opt. Soc. Am.* **49**, 807-810 (1959).
23. J. T. Cox and G. Hass: "Antireflection Coatings for Germanium and Silicon in the Infrared," *J. Opt. Soc. Am.* **48**, 677-680 (1958).
24. D. L. Perry: "Low-Loss Multilayer Dielectric Mirrors," *Appl. Opt.* **4**, 987-991 (1965).
25. E. O. Ammann and J. D. Wintemute: "Damage to ZnS thin films from 1.08-μm laser radiation," *J. Opt. Soc. Am.* **63**, 965-971 (1973).
26. R. P. Netterfield: "Refractive indices of zinc sulfide and cryolite in multilayer stacks," *Appl. Opt.* **15**, 1969-1973 (1969).
27. Qihong Wu: "Inhomogeneity of the refractive index of ZnS film," *Appl. Opt.* **26**, 3753-3754 (1987).
28. E. Ritter: "Dielectric Film Materials for Optical Applications," *Physic of Thin Films* **8**, 1-49 Academic Press (1975).
29. W. R. Hunter, D. W. Angel, and G. Hass: "Optical properties of evaporated films of ZnS in the vacuum ultraviolet from 160 to 2000 A," *J. Opt. Soc. Am.* **68**, 1319-1322 (1978).
30. G. Hass, J. B. Heaney, W. R. Hunter, and D. W. Angel: "Effect of UV irradiation on evaporated ZnS films," *Appl. Opt.* **15**, 2480-2481 (1980).
31. H. Bangert and H. Pfefferkorn: "Condensation and stability of ZnS thin films on glass substrates," *Appl. Opt.* **19**, 3878 (1980).
32. M. D. Himel, J. A. Ruffner, and U. J. Gibson: "Stress modification and reduced waveguide losses in ZnS thin films," *Appl. Opt.* **27**, 2810-2811 (1988).
33. J. A. Ruffner, M. D. Himel, V. Mizrahi, G. I. Stegeman, and U. J. Gibson: "Effects of low substrate temperature and ion assisted deposition on composition, optical properties, and stress of ZnS thin films," *Appl. Opt.* **28**, 5209-5214 (1989).
34. J. E. Rudisill, M. Braunstein, and A. I. Braunstein: "Optical Coatings for High Energy ZnSe Laser Windows," *Appl. Opt.* **13**, 2075-2080 (1974).
35. N. S. Gluck, D. B. Taber, J. P. Heuer, R. L. Hall, and W. J. Gunning: "Properties of mixed composition Si/ZnSe and ZnSe/LaF_3 infrared optical thin films," *Appl. Opt.* **31**, 6127-6132 (1992).

36. J. S. Seeley, R. Hunneman, and A. Whatley: "Infrared multilayer interference filter manufacture: supposed longwave limit," *Appl. Opt.* **18**, 3368-3369 (1979).
37. J. S. Seeley, R. Hunneman, and A. Whatley: "Far infrared filters for the Galileo-Jupiter and other missions," *Appl. Opt.* **20**, 31-39 (1981).
38. P. Baumeister and O. Arnon: "Use of hafnium dioxide in multilayer dielectric reflector for the near uv," *Appl. Opt.* **16**, 439-444 (1977).
39. J. T. Cox and G. Hass: "Protected Al mirrors with high reflectance in the 8-12 μm region from normal to high angles of incidence," *Appl. Opt.* **17**, 2125-2126 (1978).
40. D. Smith and P. Baumeister: "Refractive index of some oxide and fluoride coating materials," *Appl. Opt.* **18**, 111-115 (1979).
41. G. Hass, J. B. Ramsey, and R. Thun: "Optical Properties of Various Evaporated Rare Earth Oxides and Fluorides," *J. Opt. Soc. Am.* **49**, 116-120 (1959).
42. K. A. Wickersheim and R. A. Lefever: "Infrared Transmittance of Crystalline Yttrium Oxide and Related Compounds ," *J. Opt. Soc. Am.* **51**, 1147-1148 (1961).
43. W. Heitmann: "Reactively Evaporated Films of Scanty and Yttria," *Appl. Opt.* **12**, 394-397 (1973).
44. I. Lubezky, E. Ceren, and Z. Klein: "Silver mirrors protected with yttria for the 0.5-14 μm region," *Appl. Opt.* **19**, 1895 (1980).
45. P. B. Clapham, M. J. Downs, and R. J. King: "Some Applications of Thin Films to Polarization Devices," *Appl. Opt.* **8**, 1965-1974 (1969).
46. W. J. Coleman: "Evolution of Optical Thin Films by Sputtering," *Appl. Opt.* **13**, 946-951 (1974).
47. D. L. Wood, K. Nassau, and T. Y. Kometani: "Refractive index of Y_2O_3 stabilized cubic zirconia: variation with composition and wavelength," *Appl. Opt.* **29**, 2485-2488 (1980).
48. F. Stetter, R. Esselborn, N. Harder, M. Friz, and P. Tolles: "New materials for optical thin films," *Appl. Opt.* **15**, 2315-2317 (1976).
49. P. J. Martin, H. A. Macleod, R. P. Netterfield, C. G. Pacey, and W. G. Sainty: "Ion-beam-assisted deposition of thin films," *Appl. Opt.* 22, 178-184 (1983).
50. W. G. Sainty, R. P. Netterfield, and P. J. Martin: "Protective dielectric coatings produced by ion-assisted deposition," *Appl. Opt.* **23**, 1116-1119 (1984).
51. R. E. Klinger and C. K. Carniglia: "Optical and crystalline inhomogeneity in evaporated zirconia films," *Appl. Opt.* **24**, 3184-3187 (1985).
52. J. A. Dobrowolski, P. D. Grant, R. Simpson, and A. J. Waldorf: "Investigation of the evaporation process conditions on the optical constants of zirconia films," *Appl. Opt.* **28**, 3997-4005 (1989).
53. D. Milam, W. H. Lowdermilk, F. Rainer, J. E. Swain, C. K. Carniglia, and T. T. Hart: "Influence of deposition parameters on laser-damage threshold of silica-tantala AR coatings," *Appl. Opt.* **21**, 3689-3694 (1982).
54. H. Demiryont, J. R, Sites, and K. Geib: "Effects of oxygen content on the optical properties of tantalum oxide films deposited by ion-beam sputtering," *Appl. Opt.* **24**, 490-495 (1985).
55. T. S. Eriksson, A. Hjortsberg, G. A. Niklasson, and C. G. Granqvist: "Infrared optical properties of evaporated alumina films," *Appl. Opt.* **20**, 2742-2746 (1981).
56. G. Hass, J. B. Ramsey, and R. Thun: "Optical Properties and Structure of Cerium Dioxide Films," *J. Opt. Soc. Am.* **48**, 324-327 (1958).

57. R. P Netterfield, W. G. Sainty, P. J. Martin, and S. H. Sie: "Properties of CeO_2 thin films prepared by oxygen-ion-assisted deposition," *Appl. Opt.* **24**, 2267-2272 (1985).
58. F. Rainer, W. H. Lowdermilk, D. Milam, T. T. Hart, T. L. Lichtenstein, and C. K. Carniglia: "Scandium oxide coatings for high-power UV laser applications," *Appl. Opt.* **21**, 3685-3688 (1982).
59. D. P. Arndt, R. M. A. Azzam, J. M. Bennett, J. P. Borgogno, C. K. Carniglia, W. E. Case, J. A. Dobrowolski, U. J. Gibson, T. T. Hart, F. C. Ho, V. A. Hodgkin, W. P. Klapp, H. A. Macleod, E. Pelletier, M. K. Purvis, D. M. Quinn, D. H. Strome, R. Swenson, P. A. Temple, and T. F. Thonn: "Multiple determination of the optical constants of thin-film coating materials," *Appl. Opt.* **23**, 3571-3596 (1984).
60. U. P. Oppenheim and A. Goldman: "Infrared Spectral Transmittance of MgO and BaF_2 Crystals between 27° and 1000°C," *J. Opt. Soc. Am.* **54**, 127-128 (1964).
61. D. M. Roessler and W. C. Walker: "Optical Constants of Magnesium Oxide and Lithium Fluoride in the Far Ultraviolet," *J. Opt. Soc. Am.* **57**, 835-836 (1967).
62. J. H. Apfel: "Multilayer Interference Coatings for the Near Ultraviolet," *J. Opt. Soc. Am.* **56**, 553A (1966).
63. A. P. Bradford, G. Hass, M. McFarland: "Optical Properties of Evaporated Magnesium Oxide Films in the 0.22-8-μ Wavelength Region," *Appl. Opt.* **11**, 2242-2244 (1972).
64. Zhi-Zheng Ye and Jin-Fa Tang: "Transparent conducting indium doped ZnO films by dc reactive S-gun magnetron sputtering," *Appl. Opt.* **28**, 2817-2819 (1989).
65. H. Fabricius, T. Skettrup, and P. Bisgaard: "Ultraviolet detectors in thin sputtered ZnO films," *Appl. Opt.* **25**, 2764-2767 (1986).
66. Product data sheet. Cerac, Inc., P. O. Box 1178, Milwaukee, WI 53201 (1991).
67. J. T. Cox and G. Hass and A. Thelen: "Tripple-Layer Antireflection Coatings on Glass for the Visible and Near Infrared," *J. Opt. Soc. Am.* **52**, 965-969 (1962).
68. S. Fujiwara: "Refractive Indices of Evaporated Cerium Dioxide-Cerium Fluoride Films," *J. Opt. Soc. Am.* **53**, 880 (1963).
69. H. Sankur: "Properties of thin PbF_2 films deposited by cw and pulsed laser assisted evaporation," *Appl. Opt.* **25**, 1962-1965 (1986).
70. M. A. F. Destro, R. A. Stempniak, and A. J. Damiao: "Optical and crystalline properties of PbF_2 thin films," *Appl. Opt.* **32**, 7106-7109 (1993).
71. F. Horowitz and S. B. Mendes: "Envelope and waveguide methods: a comparative study of PbF_2 and CeO_2 birefringent films," *Appl. Opt.* **33**, 2659-2663 (1994).
72. O. S. Heavens: *Optical Properties of Thin Solid Films,* (Dover, New York, 1965).
73. O. S. Heavens and S. D. Smith: "Dielectric Thin Films," *J. Opt. Soc. Am.* **47**, 469-472 (1957).
74. M. Zukic, D. G. Torr, J. F. Spann, and M. R. Torr: "Vacuum ultraviolet thin films. 1: Optical constants of BaF_2, CaF_2, LaF_3, MgF_2, Al_2O_3, HfO_2, SiO_2, thin films," *Appl. Opt.* **29**, 4284-4292 (1990).
75. I. H. Malitson: "Refractive Properties of Barium Fluoride," *J. Opt. Soc. Am.* **54**, 628-632 (1964).
76. A. H. Laufer, J. A. Pirog, and J. R. McNesby: "Effect of Temperature on the Vacuum Ultraviolet Transmittance of Lithium Fluoride, Calcium Fluoride, Barium Fluoride, and Sapphire," *J. Opt. Soc. Am.* **55**, 64-66 (1965).
77. W. E. K. Gibbs and A. W. Butterfield: "Absorption of thin film materials at 10.6 μm," *Appl. Opt.* **14**, 3043-3046 (1975).

78. P. C. Kemeny: "Refractive index of thin films of barium fluoride," *Appl. Opt.* **21**, 2052-2054 (1982).
79. L. Holland: *Vacuum Deposition of Thin Films,* (Chapman & Hall, London, 1966).
80. M. L. Baker and R. A. Iacovazzi: "Aluminum, aluminum oxide, chromium, gold mirror," *Appl. Opt.* **24**, 920 (1985); U. S. patent no. 4,475,794.
81. G. Henderson and C. Weaver: "Optical Properties of Evaporated Films of Chromium and Copper," *J. Opt. Soc. Am.* **56**, 1551-1559 (1966).
82. E. D. Palik: *Handbook of Optical Constants of Solids,* (Academic Press, Boston, 1985).
83. E. D. Palik: *Handbook of Optical Constants of Solids II,* 374-387 (Academic Press, Boston, 1991).
84. J. H. Apfel: "Triangular coordinate graphical presentation of the optical performance of a semi-transparent metal film," *Appl. Opt.* **29**, 4272-4275 (1990).
85. D. A. Jackson and K. N. Rao: "Resolving Power in the Near Infrared of the Fabry- Perot Interferometer with Gold and with Silver Coatings," *J. Opt. Soc. Am.* **53**, 558-567 (1963).
86. R. H. Huebner, E. T. Arakawa, R. A. MacRae, and R. N. Hamm: "Optical Constants of Vacuum-Evaporated Silver Films," *J. Opt. Soc. Am.* **54**, 1434-1437 (1964).
87. L. R. Canfield and G. Hass: "Reflectance and Optical Constants of Evaporated Copper and Silver in the Vacuum Ultraviolet from 1000 to 2000Å," *J. Opt. Soc. Am.* **55**, 61-64 (1965).
88. J. M. Bennett and E. J. Ashley: "Infrared Reflectance and Emittance of Silver and Gold Evaporated in Ultrahigh Vacuum," *Appl. Opt.* **4**, 221-224 (1965).
89. G. Hass, J. B. Heaney, H. Herzig, J. F. Osantowski, and J. J. Triolo: "Reflectance and durability of Ag mirrors coated with thin layers of Al_2O_3 plus reactively deposited silicon oxide," *Appl. Opt.* **14**, 2639-2644 (1975).
90. S. F. Pellicori: "Scattering defects in silver mirror coatings," *Appl. Opt.* **19**, 3096-3098 (1980).
91. Dar-Yuan Song, R. W. Sprague, H. A. Macleod, and M. R. Jacobson: "Progress in the development of a durable silver-based high-reflectance coating for astronomical telescopes," *Appl. Opt.* **24**, 1164-1170 (1985).
92. M. A. Ordal, L. L. Long, R. J. Bell, R. R. Bell, R. W. Alexander, Jr., and C. A. Ward: "Optical properties of the metals Al, Co, Cu, Au, Fe, Pb, Ni, Pd, Pt, Ag, Ti and W in the infrared and far infrared," *Appl. Opt.* **22**, 1099-1119 (1983).
93. J. M. Bennett, H. H. Hurt, J. P. Rahn, J. M. Elson, K. H. Guenther, M. Rosigni, and F. Varnier: "Relation between optical scattering, microstructure and topography of thin silver films. 1: Optical scattering and topography," *Appl. Opt.* **24**, 2701-2711 (1985).
94. H. H. Hurt and J. M. Bennett: "Relation between optical scattering, microstructure and topography of thin silver films. 2: Microstructure," *Appl. Opt.* **24**, 2712-2720 (1985).
95. Chang Kwon Hwangbo, L. J. Lingg, J. P. Lehan, H. A. Macleod, J. L. Makous, and Sang Yeol Kim: "Ion assisted deposition of thermally evaporated Ag and Al films," *Appl. Opt.* **28**, 2769-2778 (1989).
96. D. J. Advena, V. T. Bly, and J. T. Cox: "Deposition and characterization of far-infrared absorbing gold black films," *Appl. Opt.* **32**, 1136-1144 (1993).
97. P. J. Martin, W. G. Sainty, and R. P. Netterfield: "Enhanced gold film bonding by ion-assisted deposition," *Appl. Opt.* **23**, 2668-2669 (1984).

98. T. Sato, K. Y. Szeto, and G. D. Scott: "Optical properties of aggregated gold on aluminum," *Appl. Opt.* **18**, 3119-3122 (1979).
99. Kung Lee and Juh Tzeng Lue: "Annealing effect on the optical properties of gold films deposited on silicon substrates," *Appl. Opt.* **27**, 1210-1213 (1988).
100. P. O'Neill. C. Doland, and A. Ignatiev: "Structural composition and optical properties of solar blacks: gold black," *Appl. Opt.* **16**, 2822-2826 (1977).
101. G. Zaeschmar and A. Nedoluha: "Theory of the Optical Properties of Gold Blacks," *J. Opt. Soc. Am.* **62**, 348-352 (1972).
102. D. R. McKenzie: "Selective nature of gold-black deposits," *J. Opt. Soc. Am.* **66**, 249-253 (1976).
103. J. L. Vossen: "Transparent Conducting Coatings," *Physics of Thin Films* **9**, 1-71 (1977).
104. J. G. H. Mathew, B. P. Hichwa, N. A. O'Brian, V. P. Raksha, S. Sullivan, and L. S. Wang: "Study of Vacuum Deposited, Thin Film Transmissive Electrochromic Devices," *Proc. Soc. Vac. Coaters* **38**, 194-197 (1995).
105. I. Hamberg and C. G. Granqvist: "Transparent and infrared-reflecting indium-tin-oxide films: quantitative modeling of the optical properties," *Appl. Opt.* **24**, 1815-1819 (1985).
106. J. A. Dobrowolski, F. C. Ho, D. Menagh, R. Simpson, and A. Waldorf: "Transparent, conducting indium tin oxide films formed on low or medium temperature substrates by ion-assisted deposition," *Appl. Opt.* **26**, 5204-5210 (1987).
107. O. Marcovitch, Z. Klein, and I. Lubezky: "Transparent conductive indium oxide film deposited on low temperature substrates by activated reactive evaporation," *Appl. Opt.* **16**, 2792-2795 (1989).
108. C. Bright: "Optical Constants of Evaporated and Sputtered Transparent Conductive Oxides," *Proc. Soc. Vac. Coaters* **36**, 63-67 (1993).
109. E. R. Plunkett: *Handbook of Industrial Toxicology*, (Chemical Publishing, New York, 1976).
110. A. Thelen: Private communication; to be published later.
111. J. J. Cuomo, S. M. Rossnagel, and H. R. Kaufman, *Handbook of Ion Beam Processing Technology*, (Noyes Publications, Park Ridge, NJ, 1989.)
112. R. R. Willey, "Achieving Improved Optical Thin Film Control and Uniformity of Silicon Dioxide by Using Ion Assisted Deposition," *Proc. Soc. Vac. Coaters* **36**, 75-81 (1993).
113. H. R. Kaufman and R. S. Robinson, *Operation of Broad-Beam Sources*, Sec. VI, (Commonwealth Scientific Corp., Alexandria, VA, 1984.)
114. Vacuum Technology Associates., dba DynaVac, 110 Industrial Park Road, Hingham, MA 02043.
115. J. W. Seeser, P. M. LeFebvre, B. W. Hichwa, J. P. Lehan, S. F. Rowlands, and T. H. Allen: "Metal-Mode Reactive Sputtering: A New Way to Make Thin Film Products," *Proc. Soc. Vac. Coaters* **35**, 229-235 (1992).
116. R. B. Sargent, T. H. Allen, and K. Takano: "Metal Mode Reactive Sputtered Metal-Dielectric Coatings," *Proc. Soc. Vac. Coaters* **36**, 68-74 (1993).
117. H. K. Pulker: *Coatings on Glass*, 248-250 & 269-271 (Elsevier, Amsterdam, 1984).
118. D. M. Mattox and G. J. Kominiak: *J. Vac. Sci. Technol.*, **9**, 528 (1972).

119. B. T. Nevill: "Diverse Applications of IVD Aluminum," *Proc. Soc. Vac. Coaters* **36**, 379-384 (1993).
120. K. H. Guenther, C. W. Fellows, and R. R. Willey: "Reactive Ion Plating–A Novel Deposition Technique for Improved Optical Coatings," *Proc. Soc. Vac. Coaters* **31**, 185-192 (1988).
121. K. H. Guenther, I. Penny, and R. R. Willey: "Corrosion-resistant front surface aluminum mirror coatings," *Optical Engineering,* **32,** 547-552 (1993).
122. D. G. Shaw and M. C. Langlois: "A New High Speed Process for Vapor Depositing Acrylate Thin Films: An Update," *Proc. Soc. Vac. Coaters* **36**, 348-352 (1993).
123. D. G. Shaw and M. C. Langlois: "Use of Vapor Deposited Acrylate Coatings to Improve the Barrier Properties of Metallized Film," *Proc. Soc. Vac. Coaters* **37**, 240-247 (1994).
124. P. M. Martin, J. D. Affinito, M. E. Gross, C. A. Coronado, W. D. Bennett, and D. C. Stewart: "Multilayer Coatings on Flexible Substrates," *Proc. Soc. Vac. Coaters* **38**, 163-167 (1995).
125. A. Yializis: "High Oxygen Barrier Polypropylene Films Using Transparent Acrylate-Al_2O_3 and Opaque Al-Acrylate Coatings," *Proc. Soc. Vac. Coaters* **38**, 95-105 (1995).
126. J. I. Gittleman and Y. Arie: "High-performance Al:polymer:AL trilayer optical disk," *Appl. Opt.* **23**, 3946-3949 (1984).
127. A. M. Andersson, C. G. Granqvist, and J. R. Stevens: "Electrochromic Li_xWO_3/polymer laminate/$Li_xV_2O_5$ device: toward an all-solid-state smart window," *Appl. Opt.* **28**, 3295-3302 (1989).
128. R. Ulrich and H. P. Weber: "Solution-Deposited Thin Films as Passive and Active Light-Guides," *Appl. Opt.* **11**, 428-434 (1972).
129. R-Y. Tsai, L-C. Kuo, and F. C. Ho: "Amorphous silicon and amorphous silicon nitride films prepared by a plasma-enhanced chemical vapor deposition process as optical coating materials," *Appl. Opt.* **32**, 5561-5566 (1993).
130. J. T. Felts: "Transparent Gas Barrier Technologies," *Proc. Soc. Vac. Coaters* **33**, 184-193 (1990).
131. R. J. Nelson: "Scale-Up of Plasma Deposited SiO_x Gas Diffusion Barrier Coatings," *Proc. Soc. Vac. Coaters* **35**, 75-79 (1992).
132. M. Izu, B. Dotter, S. R. Ovshinsky, and W. Hasegawa: "High Performance Clear Coat™ Barrier Film," *Proc. Soc. Vac. Coaters* **36**, 333-340 (1993).
133. J. E. Klemberg-Sapieha, L. Martinu, O. M. Kuttel, and M. Wertheimer: "Transparent Gas Barrier Coatings Produced by Dual-Frequency PECVD," *Proc. Soc. Vac. Coaters* **36**, 445-449 (1993).
134. E. Finson and J. Felts: "Transparent SiO_2 Barrier Coatings: Conversion and Production Status," *Proc. Soc. Vac. Coaters* **37**, 139-149 (1994).
135. R. B. Heil: "Mechanical Properties of PECVD Silicon-Oxide Based Barrier Films on PET," *Proc. Soc. Vac. Coaters* **38**, 33-39 (1995).
136. R. W. Knoll and J. A. Theil: "Effects of Process Parameters on PECVD Silicon Oxide and Aluminum Oxide Barrier Films," *Proc. Soc. Vac. Coaters* **38**, 425-431 (1995).
137. C. Misiano, E. Simonetti, P. Lagana, S. Menichella, G. Parone, and G. Taglioni: "Reactive Sputtering Deposition and PE-CVD for Ophthalmic Applications," *Proc. Soc. Vac. Coaters* **36**, 57-62 (1993).

138. S. Menichella, C. Misiano, E. Simonetti, L. DeCarlo, and M. Carrabino: "PE-CVD Hardening and Matching Coating for Ophthalmic Plastic Lenses," *Proc. Soc. Vac. Coaters* **37**, 37-40 (1994).
139. M. Fliedner, S. Beisswenger, R. Glötzelmann, K. Matl, and A. Zöller: "Plasma Ion Assisted Coating of Ophthalmic Optics," *Proc. Soc. Vac. Coaters* **38**, 237-241 (1995).
140. W. L. Johnson: "Comparison of Electrostatic Shielded and Unshielded Inductive Coupled Plasma Sources," *Proc. Soc. Vac. Coaters* **36**, 478-484 (1993).
141. J. Bartella, B. Cord, M. Geisler, and H. Zahel: "PECVD Carbon Overcoat for Magnetic Media," *Proc. Soc. Vac. Coaters* **37**, 299-306 (1994).
142. M. Izu, X. Deng, H. C. Ovshinsky, and S. R. Ovshinsky: "Roll-to-Roll RF PECVD for a-Si Solar Cell Manufacturing," *Proc. Soc. Vac. Coaters* **38**, 120-124 (1995).
143. Y-T. Hwang, J. C. S. Chu, and B. S. Mercer: "Deposition of Diamond-Like Carbon Films on Polycarbonate by ECR-PECVD Method," *Proc. Soc. Vac. Coaters* **37**, 363-368 (1994).
144. J. E. Sauvageau, C. J. Borroughs, M. W. Cromar, and J. A. Koch: "Optimization of ECR-Based PECVD Oxide Films for Superconducting Integrated Circuit Fabrication," *Proc. Soc. Vac. Coaters* **37**, 383-388 (1994).
145. L. Bourget and B. Lane: "Large Area ECR Processing," *Proc. Soc. Vac. Coaters* **37**, 369-374 (1994).
146. H. Demiryont, L. R. Thompson, and G. J. Collins: "Optical properties of aluminum oxynitrides deposited by laser-assisted CVD," *Appl. Opt.* **25**, 1311-1318 (1986).
147. P. K. Tien, G. Smolinsky, and R. J. Martin: "Thin Organosilicon Films for Integrated Optics ," *Appl. Opt.* **11**, 637-642 (1972).
148. J. R. Hollahan, T. Wydeven, and C. C. Johnson: "Combination Moisture Resistant and Antireflection Plasma Polymerized Thin Films for Optical Coatings," *Appl. Opt.* **13**, (1974).
149. T. Wydeven and R. Kubacki: "Antireflection coating prepared by plasma polymerization of perfluorobutene-2," *Appl. Opt.* **15**, 132-136 (1976).
150. R. Hora and C. Wohlrab: "Plasma Polymerization: A New Technology for Functional Coatings on Plastics," *Proc. Soc. Vac. Coaters* **36**, 51-56 (1993).
151. R. Wielonski: "Meter Square Plastic Windows Coated by Plasma Polymerization," *Proc. Soc. Vac. Coaters* **35**, 339-344 (1992).
152. B. Dischler, A. Bubenzer, G. Brandt, and P. Koidl: "Properties and Optical Applications of Hard Carbon Coatings," (A) *J. Opt. Soc. Am.* **72**, 1745 (1982).
153. K. Enke: "Hard carbon layers for wear protection and antireflection purposes of infrared devices," *Appl. Opt.* **24**, 508-512 (1985).
154. B. J. Knapp and F. M. Kimock: "Abrasion Resistant Diamond-Like Carbon Films for Optical Applications," *Proc. Soc. Vac. Coaters* **35**, 174-179 (1992).
155. C. J. Russell: "Diamond: From Sunglasses to Supercolliders," *Proc. Soc. Vac. Coaters* **35**, 254-259 (1992).
156. B. K. Daniels: "Diamond-Like-Carbon for Plastics," *Proc. Soc. Vac. Coaters* **35**, 260-265 (1992).
157. T. J. Hughes and F. M. Kimock: "Ion Beam Deposited Diamond-Like Carbon Coatings: Characteristics and Commercial Applications in Optics," *Proc. Soc. Vac. Coaters* **36**, 139-145 (1996).

158. D. Lavine: "A New Sputtering System for Hard Disk Media," *Proc. Soc. Vac. Coaters* **37**, 307-311 (1994).
159. G. Leonhardt, M. Falz, M. Kuhn, T. Lunow, E. Lopez, R. Wilberg, and H.-J. Scheibe: "Low Temperature PVD for Hard Coating of Dielectric Materials," *Proc. Soc. Vac. Coaters* **38**, 317-321 (1995).
160. G. A. Raiche and J. B. Jeffries: "Laser-induced fluorescence temperature measurements in a dc arcjet used for diamond deposition," *Appl. Opt.* **32**, 4629-4635 (1993).
161. T. Nishide and F. Mizukami: "Refractive indices of the tungsten oxide films prepared by sol-gel and sputtering processes," *Opt. Eng.* **34**, 3329-3333 (1995).
162. F. W. Cunningham and A. C. Hardy: "A recording photoelectric spectrophotometer," *J. Opt. Soc. Am.* **16**, 119 (1928).
163. A. C. Hardy: "A recording photoelectric color analyzer," *J. Opt. Soc. Am.* **18**, 96 (1929).
164. F. C. Strong III: "How the Fourier Transform Infrared Spectrophotometer Works," *J. Chem. Ed.* **56**, 681-684 (1979).
165. R. R. Willey: "Fourier Transform Infrared Spectrophotometer for Transmittance and Diffuse Reflectance Measurements," *Appl. Spectroscopy* **30**, 593-601 (1976).
166. Harrick Scientific Corp., 88 Broadway, Ossining, NY 10562-0997.
167. Labsphere, Inc., Shaker St., North Sutton, NH 03260-0070.
168. J. Strong: *Procedures in Experimental Physics*, 376, (Prentice-Hall, Englewood Cliffs, NJ, 1942).
169. J. E. Shaw and W. R. Blevin: "Instrument for the Absolute Measurement of Direct Spectral Reflectances at Normal Incidence," *J. Opt. Soc. Am.* **54**, 334-336 (1964).
170. W. N. Hansen: "Optical characterization of thin films: Theory," *J. Opt. Soc. Am.* **63**, 793-799 (1973).
171. W. R. Hunter: "Errors in using the Reflectance vs Angle of Incidence Method for Measuring Optical Constants," *J. Opt. Soc. Am.* **55**, 1197-1203 (1965).
172. FilmStar™ Design from FTG Software Associates, P. O. Box 579, Princeton, New Jersey 08542.
173. A. H. Macleod and C. Clark: "Problems of the Model in n and k Extraction," *13th Annual Vacuum Web Coating Conference*, Tucson, AR, p. 78-90, October 1999, Bakish Materials Corporation, New Jersey.
174. M. Herzberger: *Modern Geometrical Optics*, 121, (Interscience Publishers, New York, 1958).
175. E. E. Khawaja and F. Bouamrane: "Determination of the optical constants (n, k) of thin dielectric films," *Appl. Opt.* **32**, 1168-1172 (1993).
176. Y. Zheng and K. Kikuchi: 'Analytical method of determining optical constants of a weakly absorbing thin film," *Appl. Opt.* **36**, 6325-6328 (1997).
177. T. Boudet, M. Berger, O. Lartigue, and B. Hirren: "Optical and x-ray characterization applied to multilayer reverse engineering," *Appl. Opt.* **37**, 2175-2181 (1998).
178. J. C. Zwinkels, M. Noël, and C. X. Dodd: "Procedures and standards for accurate spectrophotometric measurments of specular reflectance," *Appl. Opt.* **33**, 7933-7944 (1994).
179. P. O. Nilsson: "Determination of Optical Constants from Intensity Measurements at Normal Incidence," *Appl. Opt.* **7**, 435-442 (1968).

180. R. C. McPhedran, L. C. Botten, D. R. McKenzie, and R. P. Netterfield: "Unambiguous determination of optical constants of absorbing films by reflectance and transmittance," *Appl. Opt.* **23**, 1197-1205 (1984).
181. D. Minkov: "Computation of the optical constants of a thin dielectric layer on a transmitting substrate from the reflection spectrum at inclined incidence of light," *J. Opt. Soc. Am. A* **8**, 306-310 (1991).
182. D. Minkov: "Singularity of the solution when using spectrum envelopes for the computation of the optical constants of a thin dielectric layer," *Optik* **90**, 80-84 (1992).
183. B. A. Movchan and A. E. Denchishin: *Fiz. Met. Metalloved.* **28**, 653 (1969).
184. J. A. Thornton: "Influence of apparatus geometry and deposition conditions on the structure and topography of thick sputtered coatings," *J. Vac. Sci. Technol.* **11**, 666-671 (1974).
185. K.-H. Müller: "Dependence of thin-film microstructure on deposition rate by means of a computer simulation," *J. Appl. Phys.* **58**, 2573-2576 (1985).
186. K.-H. Müller: "A computer model for postdeposition annealing of porous thin films," *J. Vac. Sci. Technol. A* **3**, 2089-2092 (1985).
187. R. P. Netterfield and P. J. Martin: "Nucleation and growth studies of gold films prepared by evaporation and ion-assisted deposition," *Appl. Surf. Sci.* **25**, 265-278 (1986).
188. K. H. Guenther: "Revisiting structure zone models for thin film growth," *Proc. SPIE* **1324**, 2-12 (1990).
189. Q. Wu, I. J. Hodgkinson, and A. Lakhtakia: " Circular polarization filters made of chiral sculptured thin films: experimental and simulation results," *Opt. Eng.* **39**, 1863-1868 (2000).
190. R. Messier, V. C. Venugopal, and P. D. Sunal: "Origin and evolution of sculptured thin films," *J. Vac. Sci. Technol., A* **18**, 1538 (2000).
191. D. L. Smith: *Thin Film Deposition*, 185-197 (McGraw-Hill, New York, 1995).
192. H. K. Pulker: "Mechanical Properties of Optical Thin Films," *Thin Films for Optical Systems*, 455-473 (Marcel Dekker, New York, 1995).
192. J. D. Traylor Kruschwitz and Pawlewicz: "Optical and durability properties of infrared transmitting thin films," *Appl. Opt.* **36**, 2157-2159 (1997).
193. S. M. McNally, S. D. Feiman, M. Friz, A. R. Wilson, and B. Kinsman: "Survey of Available Potential Replacements for Thorium Fluoride," *Proc. Soc. Vac. Coaters* **35**, 169-173 (1992).
194. M. R. Jacobson, S. F. Feiman, Z. Tianji, and H. A. Macleod: "Optical Properties of Undoped & CaF_2-Doped Yttrium Fluoride and Ytterbium Fluoride Coatings Fabricated by Conventional and Ion-Assisted Deposition," *Proc. Soc. Vac. Coaters* **37**, 248-253 (1994).
195. S. Mendes, M. R. Jacobson, S. F. Feiman, Z. Tianji, A. Ogloza, and H. A. Macleod: "Optical, Mechanical, and Microstructural Properties of YF_3 Coatings Fabricated by Conventional and Ion-Assisted Deposition," *Proc. Soc. Vac. Coaters* **36**, 109-118 (1993).
196. J. Wolfe: "Control of Stress-Corrosion Cracking in Thin Film Designs," *Proc. Soc. Vac. Coaters* **36**, 498-503 (1993).

197. N. S. Gluck, H. Sankur, and D. Taber: "Mixed Composition Infrared Optical Thin Films: Properties and Applications," *Proc. Soc. Vac. Coaters* **36**, 119-126 (1993).
198. E. Quesnel, B. Rolland, V. Muffato, D. Labroche, and J. Y. Robic: "The Ion Beam Sputtering: A Good Way to Improve the Mechanical Properties of Fluoride Coatings," *Proc. Soc. Vac. Coaters* **40**, 293-298 (1997).
199. W. Lechner, G. N. Strauss and H. K. Pulker: "Correlation Between Optical and Mechanical Properties of Ion Plated Ta_2O_5 Films," *Proc. Soc. Vac. Coaters* **41**, 287-299 (1998).
200. R. A. Lucheta: "Thermally induced interfacial stresses in a thin film on an infinite substrate," *Appl. Opt.* **30**, 2252-2256 (1991).
201. H. Takashashi: "Temperature stability of thin-film narrow-bandpass filters produced by ion-assisted deposition," *Appl. Opt.* **34**, 667-675 (1995).
202. H. J. Cho and C. K. Hwangbo: "Optical inhomogeneity and microstructure of ZrO_2 thin films prepared by ion-assisted deposition," *Appl. Opt.* **35**, 5545-5552 (1996).
203. R. Chow and N. Tsujimoto: "Silicon dioxide and hafnium dioxide evaporation characteristics from a high-frequency sweep E-beam system," *Appl. Opt.* **35**, 5095-5101 (1996).
204. R. Chow, P. L. Tassano, and N. Tsujimoto: "Oxide Vapor Distribution from a High-Frequency Sweep E-Beam System," *Proc. Soc. Vac. Coaters* **38**, 248-253 (1995).
205. K. Scherer, L. Nouvelot, P. Lacan, and R. Bosmans: "Optical and mechanical characteristics of evaporated SiO_2 layers. Long term evolution," *Appl. Opt.* **35**, 5067-5072(1996).
206. P. Harris: "Taking the lead in electron-beam deposition," *Vacuum & Thin film*, 26-29 February 1999.
207. D. L. Wood, K. Nassau, T. Y. Kometani, and D. L. Nash: "Optical properties of cubic hafnia stabilized with yttria," *Appl. Opt.* **29**, 604-607 (1990).
208. K Arora and A. L. Dawar: "Laser induced damage studies in silicon and silicon-based photodetrctors," *Appl. Opt.* **35**, 7061-7065 (1996).
209. S. Laux, K. Mann, B. Granitza, U. Kaiser, and W. Richter: "Antireflection coatings for UV radiation obtained by molecular-beam deposition," *Appl. Opt.* **35**, 6216-6218 (1996).
210. E. Eva, K. Mann, B. Anton, R. Henking, D. Ristau, P. Weissbrodt, D. Mademann, L. Raupach, and E. Hacker: "Laser conditioning of LaF_3/MgF_2 dielectric coatings at 248 nm," *Appl. Opt.* **35**, 5613-5619 (1996).
211. R. M. O'Connell, T. F. Deaton, and T. T. Saito: "Single- and multiple-shot laser-damage properties of commercial grade PMMA," *Appl. Opt.* **23**, 683-688 (1984).
212. R. M. O'Connell: "Onset threshold analysis of defect-driven surface and bulk laser damage," *Appl. Opt.* **31**, 4143-4153 (1992).
213. J. O. Porteus and S. C. Seitel: "Absolute onset of optical surface damage using distributed defect ensembles," *Appl. Opt.* **23**, 3797-3805 (1984).
214. J. W. Arenberg, M. E. Frink, D. W. Mordaunt, G. Lee, S. C. Seitel, and E. A. Teppo: "Correlating laser damage tests," *Appl. Opt.* **28**, 123-126 (1989).
215. B. C. Ziegler and K. L. Schepler: "Transmission and damage threshold measurements in $AgGaSe_2$ at 2.1 μm," *Appl. Opt.* **30**, 5077-5080 (1991).
216. J. R. Bettis: "Correlation among the laser-induced breakdown thresholds in solids, liquids, and gases," *Appl. Opt.* **31**, 3448-3452 (1992).

217. L. J. Shaw-Klein, S. J. Burns, and S. D. Jacobs: "Model for laser damage dependence on thin-film morphology," *Appl. Opt.* **32**, 3925-3929 (1993).
218. R. J. Trench, R. Chow, and M. R. Kozlowski: "What those Defects in Optical Coatings Really Look Like," *Proc. Soc. Vac. Coaters* **37**, 63-68 (1994).
219. C. J. Stolz, L. M. Sheehan, M. K. von Gunten, R. P. Bevis, and D. J. Smith: "The advantages of evaporation of Hafnium in a reactive environment to manufacture high damage threshold multilayer coatings by electron-beam deposition," *Proc. SPIE* **3738**, 318-324 (1999).
220. H. L. Gao and N. Y. Wang: "Possible damage mechanism of the dielectric coatings for a KrF laser," *Appl. Opt.* **32**, 7084-7088 (1993).
221. R. Chow, S. Falabella, G. E. Loomis, F. Rainer, C. J. Stolz, and M. R. Kozlowski: "Reactive evaporation of low-defect density hafnia," *Appl. Opt.* **32**, 5567-5574 (1993).
222. D. J. Krajnovich, M. Kulkarni, W. Leung, A. C. Tam, A. Spoon, and B. York: "Testing the durability of a single-crystal calcium fluoride with and without antireflection coatings for use with high-power KrF excimer lasers," *Appl. Opt.* **31**, 6062-6075(1992).
223. W. B. Jackson, N. M. Amer, A. C. Boccara, and D. Fournier: "Photothermal deflection spectroscopy and detection," *Appl. Opt.* **20**, 1333-1344 (1981).
224. E. Welsch and D. Ristau: "Photothermal measurements in optical thin films," *Appl. Opt.* **34**, 7239-7253 (1995).
225. M. J. Soileau, M. Bass, and E. W. Van Stryland: "Frequency dependance of breakdown fields in single crystal NaCl and KCl," *Natl. Bur. Stand. Spec. Publ.* **541**, 309-317(1978).
226. M. R. Kozlowski: "Damage-Resistant Laser Coatings," *Thin Films for Optical Systems*, ed. F. R. Flory, 521-549 (Marcel Dekker, Inc., New York, 1995).
227. M. Commandré and P. Roche: "Characterization of Absorption by Photothermal Deflection," *Thin Films for Optical Systems*, ed. F. R. Flory, 329-365 (Marcel Dekker, Inc., New York, 1995).
228. J. M. Bennett, E. Pelletier, G. Albrand, J. P. Borgogno, B. Lazarides, C. K. Carniglia, R. A. Schmell, T. H. Allen, T. Tuttle-Hart, K. H. Guenther, and A. Saxer: "Comparison of the properties of titanium dioxide films prepared by various techniques," *Appl. Opt.* **28**, 3303-3317 (1989).
229. M. P. Wirick: "The Near Ultraviolet Optical Constants of Lanthanum Fluoride," *Appl. Opt.* **5**, 1966-1967 (1966).
230. J. D. Targrove, J. P. Lehan, L. J. Lingg, H. A. Macleod, J. A. Leavitt, and L. C. McIntyre, Jr.: "Ion-assisted deposition of lanthanum fluoride thin films," *Appl. Opt.* **26**, 3733-3737 (1987).
231. J. Kolbe, H. Schink, and D. Ristau: "Optical Losses of Fluoride Coatings for UV/VUV Applications Deposited by Reactive IAD and IBS Processes," *Proc. Soc. Vac. Coaters* **36**, 44-50 (1993).
232. M. Friz, F. Koenig, and S. Feiman: "New Materials for Production of Optical Coatings," *Proc. Soc. Vac. Coaters* **35**, 143-149 (1992).
233. F. Koenig and M. Friz: "Development of Medium and High Refractive Index Coating Materials for the Visible and UV Spectral Range," *Proc. Soc. Vac. Coaters* **37**, 118-121 (1994).

234. M. Friz, U. B. Schallenberg, and S. Laux: "Plasma Ion Assisted Deposition of Medium and High Refractive Index Thin Films," *Proc. Soc. Vac. Coaters* **40**, 280-292 (1997).
235. J. H. Apfel: "Optical coating design with reduced electric field intensity," *Appl. Opt.* **16**, 1880-1885 (1977).
236. D. Diso, M. R. Perrone, A. Piegari, M. L. Protopapa, and S. Scaglione: "Single shot laser damage in ultraviolet mirrors with a stepwise reflectivity profile," *J. Vac. Sci. Technol. A* **18**, 477-484 (2000).
237. M. Poulingue, J. Dijon, B. Rafin, H. Leplan, and M. Ignat: "Generation of defects with diamond and silica particles inside High Reflection Coatings: influence on the laser damage threshold," *Proc. SPIE* **3738**, 325-336 (1999).
238. J. Sheng, J. Karasawa, and T. Fukami: "Thickness dependence of photocatalytic activity of anatase film by magnetron sputtering," *J. Mater. Sci. Letters* **16**, 1709-1711 (1997).
239. H. Wang, T. Wang, and P. Xu: "Effects of substrate temperature on the microstructure and photocatalytic reactivity of TiO_2 films," *J. Mater. Sci.; Mater. In Electronics* **9**, 327-330 (1998).
240. T. Wang, H. Wang, P. Xu, X. Xuechu, Y. Liu, and S. Chao: "The effect of properties of semiconductor oxide thin films on photocatalytic decomposition of dyeing waste water," *Thin Solid Films* **334**, 103-108 (1998).
241. T. Komatsu and M. Nakamura: "Anti-fog Element," *United States Patent* **5,854,708** (1998).
242. T. Komatsu and M. Nakamura: "Anti-fog Element," *European Patent Application* **EP 0 820 967 A1** (1998).
243. M. Hayakawa, E. Kojima, K. Norimoto, M. Machida, A. Kitamura, T. Watanabe, M.Chikuni, A. Fujishima, and K. Hashimoto: "Method of photocatalytically making the surface of a base material ultrahydrophilic, base material having ultrahydrophilic and photocatalytic surface, and process for producing said material," *European Patent Application* **EP 0 816 466 A1** (1998).
244. M. H. Suhail, G. M. Rao, and S. Mohan: "DC reactive magnetron sputtering of titanium-structural and optical characterization of TiO_2 films," *J. Appl. Phys.* **71**, 1421- 1427 (1992).
245. J. Szczyrbowski, G. Bräuer, M. Ruske, G. Teschner, and A. Zmelty: "Properties of TiO_2- Layers Prepared by Medium Frequency and DC Reactive Sputtering," *Proc. Soc. Vac. Coaters* **40**, 237-242 (1997).
246. H. Tang, K. Prasad, R. Sanjinès, P. E. Schmid, and F. Lévy: "Electrical and optical properties of TiO_2 anatase films," *J. Appl. Phys.* **75**, 2042-2047 (1994).
247. W. T. Pawlewicz, G. J. Exarhos, and W. E. Conaway: "Structural characterization of TiO_2 optical coatings by Raman spectroscopy," *Appl. Opt.* **22**, 1837-1840 (1983).
248. R. J. Gonzalez: "Raman, Infrared, X-ray, and EELS Studies of Nanophase Titania," *PhD. Dissertation to Virginia Polytechnic Institute* (1996).
249. M. Stamate, G. I. Rusu, I. Vascan: "I.R. absorption of TiO_2 thin films," *Proc. SPIE* **3405**, 951-954 (1998).

250. F. Dahmani, A. W. Schmid, J. C. Lambropoulos, and S. Burns: "Dependence of birefringence and residual stress near laser-induced cracks in fused silica on laser fluence and on laser-pulse number," *Appl. Opt.* **37**, 7772-7784 (1998).
251. F. Dahmani, S. J. Burns, J. C. Lambropoulos, S. Papernov, and A. W. Schmid: "Arresting ultraviolet-laser damage in fused silica," *Optics Letters* **24**, 516-518 (1999).
252. F. Dahmani, J. C. Lambropoulos, A. W. Schmid, S. Papernov, and S. J. Burns: "Crack arrest and stress dependence of laser-induced surface damage in fused-silica and borosilicate glass," *Appl. Opt.* **38**, 6892-6903 (1999).
253. K. H. Guenther: "Influence of the Substrate Surface on the Performance of Optical Coatings," *Thin Solid Films* **77**, 239-245 (1981).
254. D. M. Mattox: *Handbook of Physical Vapor Deposition (PVD) Processing*, 636-701 (Noyes Pubs., Westwood, NJ, 1997).
255. J. B. McDaniel: "Collodion Technique of Mirror Cleaning," *Appl. Opt.* **3**, 152-153 (1964).
256. R. A. House II, J. R. Bettis, and A. H. Guenther: "Untrasonic cleaning and laser surface damage threshold," *Appl. Opt.* **16**, 1130-1131 (1977).
257. A. D. McLauchlan and W. E. Gibbs: "Deterioration in optical materials as a result of ultrasonic cleaning," *Appl. Opt.* **16**, 554-546 (1977).
258. M. Brenci, P. F. Checcacci, R. Falciai, and A. M. Scheggi: "Deterioration in optical materials as a result of ultrasonic cleaning: comment," *Appl. Opt.* **16**, 3084 (1977).
259. N. S. Gluck, H. Sankur, and W. J. Gunning: "Ion-assisted deposition of CaF_2 thin films at low temperatures," *J. Vac. Sci. Technol. A* **7**, 2983-2987 (1989).
260. K. Zhang, J. S. Seeley, R. Hunneman, and G. J. Hawkins: "Optical and semi-conductor properties of lead teluride coatings," *Proc. SPIE* **1125**, 45-48 (1989).
261. K. Q. Zhang, J. S. Seeley, R. Hunneman, and G. J. Hawkins: "Optical and semi-conductor properties of lead teluride coatings," *Proc. SPIE* **1112**, 393-402 (1989).
262. K. Dettmer, H. Goebel, and F. R. Kessler: "Different Proportions of a-Ge and c-Ge in Dependence on the Substrate Temperature of Evaporated Films Determined by Refractive Index Analysis," *Phys. Stat. Sol.* **27**, 393-401 (1975).
263. M. D. Blue and D. W. Roberts: "Effects of space exposure on optical filters," *Appl. Opt.* **31**, 5299-5304 (1992).
264. P. J. Martin, W. G. Sainty, and R. P. Netterfield: "Oxygen-ion-assisted deposition of thin gold films," *Vacuum* **35**, 621-624 (1985).
265. R. F. Adamsky: "Effect of Deposition Parameters on the Crystallinity of Evaporated Germanium Films," *J. Appl. Phys.* **40**, 4301-4305 (1969).
266. K. H. Guenther: "Physical and chemical aspects in the application of thin films on optical elements," *Appl. Opt.* **23**, 3612-3632 (1984).
267. M. Vergöhl, N. Malkomes, B. Szyszka, F. Neumannm and T. Matthée: "Optimization of the reflectivity of magnetron sputter deposited silver films," *J. Vac. Sci. Technol. A* **18**, 1632-1637 (2000).
268. H. Biederman: "Organic films prepared by polymer sputtering," *J. Vac. Sci. Technol. A* **18**, 1642-1648 (2000).
269. T. Inoue, T. Nakamura, S. Nihei, S. Kamata, and N. Sakamoto: "Surface morphology analysis in correlation with crystallinity of CeO_2 (110) layers on Si(100) substrates," *J. Vac. Sci. Technol. A* **18**, 1613-1618 (2000).

270. H. J. Jang, S. W. Whangbo, H. B. Kim, K. Y. Im, Y. S. Lee, I. W. Lyo, and C. N. Whang: "Titanium oxide films on Si(100) deposited by electron-beam evaporation at 250°C," *J. Vac. Sci. Technol. A* **18**, 917-921 (2000).
271. A. N. Krasnov, P. G. Hofstra, and M. T. McCullough: "Parameters of vacuum deposition of ZnS:Mn active layer for electroluminescent displays," *J. Vac. Sci. Technol. A* **18**, 671-675 (2000).
272. M. Debenham: "Refractive indices of zinc sulfide in the 0.405-13 μm wavelength range," *Appl. Opt.* **23**, 2238-2239 (1984).
273. G. W. Mbise, D. LeBallac, G. A. Niklasson, and C. G. Granqvist: "Angular selective window coatings," *Proc. SPIE* **2255**, 182-192 (1994).
274. S.-C. Chiao, B. G. Bovard, and H. A. Macleod: "Repeatability of the composition of titanium oxide films produced by evaporation of Ti_2O_3," *Appl. Opt.* **37**, 5284-5289 (1998).
275. J.-S. Chen, S. Chao, J.-S. Kao, G.-R. Lai, and W.-H. Wang: "Substrate-dependent optical absorption characteristics of titanium dioxide thin films," *Appl. Opt.* **36**, 4403-4408 (1997).
276. I. M. Thomas: "Porous fluoride antireflective coatings," *Appl. Opt.* **27**, 3356-3358 (1988).
277. E. Quesnel, A. Petit dit Dariel, A. Duparré, J. Ferré-Borrull, and J. Steinert: "DUV Scattering and Morphology of Ion Beam Sputtered Fluoride Coatings," *SPIE* **3738**, 410- 416 (1999).
278. D. Ristau, W. Arens, S. Bosch, A. Duparré, E. Masetti, D. Jacob, G. Kiriakidis, F. Peiró, E. Quesnel, and A. Tikhonravov: "UV-Optical and Microstructural Properties of MgF_2-Coatings Deposited by IBS and PVD Processes," *SPIE* **3738**, 436-445 (1999).
279. R. Thielsch, M. Pommies, J. Heber, N. Kaiser, and J. Ullmann: "Structural and mechanical properties of evaporated pure and mixed MgF2-BaF2 thin films," *SPIE* **3738**, 539-548 (1999).
280. K. H. Guenther: "Optical thin films deposited by energetic particle processes," *SPIE* **1782**, 344-355 (1992).
281. S. Ogura: "Stable starting materials of tantalum pentoxide and titanium dioxide," *SPIE* **1782**, 335-343 (1992).
282. P. M. Martin, W. T. Pawlewicz, D. Coult, and J. Jones: "Observation of exceptional temperature humidity stability in multilayer filter coatings," *Appl. Opt.* **23**, 1307-1308 (1984).
283. R. P. Netterfield, P. J. Martin, and W. G. Sainty: "Synthesis of silicon nitride and silicon oxide films by ion-assisted deposition," *Appl. Opt.* **25**, 3808-3809 (1986).
284. M. F. Lambrinos, R, Valizadeh, and J. S. Colligon: "Effects of bombardment on optical properties during deposition of silicon nitride by reactive ion-beam sputtering," *Appl. Opt.* **35**, 3620-3626 (1996).
285. R. A. Laff: "Silicon Nitride as an Antireflection Coating for Semiconductor Optics," *Appl. Opt.* **10**, 968-969 (1971).
286. G. Eisenstein and L. W. Stolz: "High quality antireflection coatings on laser facets sputtered silicon nitride," *Appl. Opt.* **23**, 161-164 (1984).
287. B. G. Bovard, J. Ramm, R. Hora, and F. Hanselmann: "Silicon nitride thin films by low voltage reactive ion plating," *Appl. Opt.* **28**, 4436-4441 (1989).

288. R.-Y. Tsai, F.-C. Ho, and M.-Y. Hua: "Annealing effects on the properties of indium tin oxide films coated on soda lime glasses with a barrier layer of TiO_2-SiO_2 composite films," *Opt. Eng.* **36**, 2335-2340 (1997).
289. A. Belkind, J. Felts, and M. McBride: "Sputtering and co-sputtering of coatings using a C-Mag™ rotatable cylindrical cathode," *Proc. Soc. Vac. Coaters* **34**, 235-239 (1991).
290. R. E. Laird and J. D. Wolfe: "The Evolution of Durable, Silver Based, Low Emissivity Films Deposited by D. C. Magnetron Sputtering (ZnO to Si_3N_4)," *Proc. Soc. Vac. Coaters* **37**, 428-431 (1994).
291. S. Nadel: "Advanced Low-Emissivity Glazings," *Proc. Soc. Vac. Coaters* **39**, 157-163 (1996).
292. Y. Lipin, and E. Machevski: "Protective Properties of Silicon Nitride, Deposited by Bias Reactive Sputtering," *Proc. Soc. Vac. Coaters* **42**, 133-135 (1999).
293. J. Krempel-Hesse, R. Preu, H. Lautenschlager, and R. Lüdemann; "Sputtering of Silicon Nitride for Use in Crystalline Silicon Solar Cell Technology," *Proc. Soc. Vac. Coaters* **43**, 97-99 (2000).
294. A. Rizzo, L. Caneve, S. Scaglione, and M. A. Tagliente: "Structural and optical properties of zinc selenide thin films deposited by RF magnetron sputtering," *SPIE* **3738**, 40-47 (1999).
295. J. S. Ross, R. A. Hallman, D. J. Kester and J. D. Wisnosky: "Thick Hard Wear Layers for Plastic Using Vacuum Web Coating," *Proc. Soc. Vac. Coaters* **38**, 81-87 (1995).
296. C. E. Kennedy, R. V. Smilgys, D. A. Kirkpatrick, and J. S. Ross: "Optical performance and durability of solar reflectors protected by an alumina coating," *Thin Solid Films* **304**, 304-309 (1997).
297. F. Fietzke, K. Goedicke, and W. Hempel: "The deposition of hard crystalline Al_2O_3 layers by means of bipolar pulsed magnetron sputtering," *Surf. Coat. Technol.* **86-87**, 657-663 (1996).
298. P. Martin, R. Netterfield, T. Kinder, and A. Bendavid: "Optical properties and stress of ion-assisted aluminum nitride thin films," *Appl. Opt.* **31**, 6734-6740 (1992).
299. K. Kusaka, T. Hanabusa, and K. Tominaga: "Effect of a plasma protection net on residual stress in AlN films deposited by a magnetron sputtering system," *Thin Solid Films* **290-291**, 260-263 (1996).
300. G.A. Garzino-Demo and F.L. Lama: "Front Surface Wear and Corrosion Resistant Metallic and/or Dielectric Mirrors for Automotive Applications," *Proc. Soc. Vac. Coaters* **39**, 270-275 (1996).
301. J. Verhoeven, and C. Steenbergen: "Production of Optical Media and Thin Film Technologies," *Proc. Soc. Vac. Coaters* **38**, 380-385 (1995).
302. S.J. Holmes and V.W. Biricik: "Reactive Ion Beam Sputter Deposition of Graded-Interface Optical Thin Films of Aluminum Oxynitride and Silicon Oxynitride," *Proc. Soc. Vac. Coaters* **36**, 146-150 (1993).
303. K. Funk, M. A. Bösch, and T. Müller: "Influence of Residual Gas and Temperature on the Crystallization of Evaporated Aluminum Films," *Balzers Special Report* **BB 800 018 DE** (8909), Balzers AG, FL-9496 Balzers.
304. G. Hass, W. R. Hunter, and R. Tousey: "Reflectance of Evaporated Aluminum in the Vacuum Ultraviolet," *J. Opt. Soc. Am.* **46**, 1009-1012 (1956).

305. G. Hass, W. R. Hunter, and R. Tousey: "Influence of Purity, Substrate Temperature, and Aging Conditions on the Extreme Ultraviolet Reflectance of Evaporated Aluminum," *J. Opt. Soc. Am.* **47**, 1070-1073 (1957).
306. J. A. Harrington, J. E. Rudisill, M. Braunstein: "Thin film 2.8-μm and 2.3-μm absorption in single-layer films," *Appl. Opt.* **17**, 2798-2801 (1978).
307. R. C. Pastor and A. C. Pastor: *Mater. Res. Bull.* **10**, 117 (1975).
308. J.D. Wolfe: "DC-Sputtered Anti-Reflection Coatings for Plastic Webs Based on TiO_x and NbO," *Proc. Soc. Vac. Coaters* **40**, 377-381 (1997).
309. P. J. Martin and R. P. Netterfield: "Ion Scattering Spectroscopy Studies of Zirconium Dioxide Thin Films Pepared *In Situ*." *Surface and Interface Analysis* **10**, 13-16 (1987).
310. P. Klocek, ed.: *Handbook of Infrared Optical Materials*, 420-424 (Marcel Dekker, Inc., New York, 1991).
311. D. L. Wood, and K. Nassau: "Refractive index of cubic zirconia stabilized with yttria," *Appl. Opt.* **21**, 2978-2981 (1982).
312. E. W. Van Stryland, M. J. Soileau, A. L. Smirl, and W. E. Williams: "Pulse width and focal volume dependence of laser-induced breakdown," Phys. Rev. B 23, 2144-2151 (1981).
313. J. K. Coulter, G. Hass, and J. B. Ramsey, Jr.: "Optical constants and reflectance and transmittance of evaporated rhodium films in the visible," *J. Opt. Soc. Am.* **63**, 1149-1153 (1973).
314. D. E. Aspnes and H. G. Craighead: "Multiple determination of the optical constants of thin-film coating materials: a Rh sequel," *Appl. Opt.* **25**, 1299-1310 (1986).
315. I. Hodgkinson, S. Cloughley, Q. H. Wu, and S. Kassam: "Anisotropic scatter patterns and anomalous birefringence of obliquely deposited cerium oxide films," *Appl. Opt.* **35**, 5563-5568 (1996).
316. T. Motohiro and Y. Taga: "Thin film retardation plate by oblique deposition," *Appl. Opt.* **28**, 2466-2482 (1989).
317. I. Hodgkinson, P. I. Bowmar, and Q. H. Wu: "Scatter from tilted-columnar birefringent thin films: observation and measurement of anisotropic scatter distributions," *Appl. Opt.* **34**, 163-168 (1995).
318. H. Rafla-Yuan, J. D. Rancourt, and M. J. Cumbo: "Ellipsometric study of thermally evaporated germanium thin film," *Appl. Opt.* **36**, 6360-6363 (1997).
319. S. Laux and W. Richter: "Packing-density calculation of thin fluoride films from infrared transmission spectra," *Appl. Opt.* **35**, 97-101 (1996).
320. Y. Tsuo and F. C. Ho: "Optical properties of hafnia and coevaporated hafnia: magnesium fluoride thin films," *Appl. Opt.* **35**, 5091-5094 (1996).
321. D. Nahrstedt, T. Glesne, J. McNally, J. Kenemuth, and B. Magrath: " Electroless silver as an optical coating," *Appl. Opt.* **35**, 3680-3686 (1996).
322. R. J. Bussjager and H. A. Macleod: "Using surface plasmon resonances to test the durability of silver-copper films," *Appl. Opt.* **35**, 5044-5047 (1996).
323. J. A. Pracchia and J. M. Simon: "Transparent heat mirrors: influence of the materials on the optical characteristics," *Appl. Opt.* **20**, 251-258 (1981).
324. T. Eisenhammer, M. Lazarov, M. Leutbecker, U. Schöffel, and R. Sizmann: "Optimization of interference filters with genetic algorithms applied to silver-based heat mirrors," *Appl. Opt.* **32**, 6310-6315 (1993).

325. C.-C. Lee, S.-H. Chen, and C.-C. Jaing: "Optical monitoring of silver-based transparent heat mirrors," *Appl. Opt.* **35**, 5698-5703 (1996).
326. J. C. C. Fan and F. J. Bachner: "Transparent heat mirrors for solar-energy applications," *Appl. Opt.* **15**, 1012-1017 (1976).
327. D.A. Baldwin, J.B. Ramsey and J.E. Miles: "MgF_2 Optical Films: Ion-Beam-Assisted Deposition of Magnesium Fluoride in a Conventional Electron Beam Evaporator and the Resulting Film Properties," *Proc. Soc. Vac. Coaters* **40**, 243-247 (1997).
328. F. Krannig: "Determining Film Density from Index of Refraction Measurements," *Proc. Soc. Vac. Coaters* **43**, 7-10 (2000).
329. E. Hartwig, J. Meinel, T. Krug and G. Steiniger: "State-of-the-Art and Economy of Aluminum and Non-Aluminum Barrier Coatings," *Proc. Soc. Vac. Coaters* **35**, 121-127 (1992).
330. K. Vedam and S. Y. Kim: "Simultaneous determination of refractive index, its dispersion and depth-profile of magnesium oxide thin film by spectroscopic ellipsometry," *Appl. Opt.* **28**, 2691-2694 (1989).
331. T.R. Jensen, R.L. Johnson, Jr., J. Ballou, W. Prohaska and S.E. Morin: "Environmentally Stable UV Raman Edge Filters," *Proc. Soc. Vac. Coaters* **43**, 239-243 (2000).
332. R.W. Phillips, T. Markantes and C. LeGallee: "Evaporated Dielectric Colorless Films on PET and OPP Exhibiting High Barriers Toward Moisture and Oxygen," *Proc. Soc. Vac. Coaters* **36**, 293-301 (1993).
333. R. Yoshimura, N. Ogawa, K. Kuma, K. Yamamoto and T. Mouri: "Deposition of ZnO:Al Transparent Conductive Thin Films by DC Magnetron Sputtering," *Proc. Soc. Vac. Coaters* **35**, 362-364 (1992).
334. C. May, J. Strümpfel and D. Schulze: "Magnetron Sputtering of ITO and ZnO Films for Large Area Glass Coating," *Proc. Soc. Vac. Coaters* **43**, 137-142 (2000).
335. N. Ogawa, T. Iwamoto, T. Mouri, T. Minami, H. Satoh and S. Takata: "Transparent Conductive Thin Films of ZnO:Al by DC Magnetron Sputtering," *Proc. Soc. Vac. Coaters* **37**, 41-45 (1994).
336. R. L. W. Smithson, D. J. McClure, and D. F. Evans: "Nucleation and Growth of Gold Films on Bare and Modified Polymer Surfaces," *Proc. Soc. Vac. Coaters* **38**, 72-79 (1995).
337. B. Szyszka: "Properties of TCO-Films Prepared by Reactive Magnetron Sputtering," *Proc. Soc. Vac. Coaters* **43**, 187-192 (2000).
338. R. W. Burger and L. J. Gerenser: "Understanding the Formation and Properties of Metal/Polymer Interfaces via Spectroscopic Studies of Chemical Bonding," *Proc. Soc. Vac. Coaters* **34**, 162-168 (1991).
339. J. M. Grace, D.R. Freeman, R. Corts and W. Kosel: "Scale-Up of a Nitrogen Glow-Discharge Process for Silver -PET Adhesion," *Proc. Soc. Vac. Coaters* **39**, 436-440 (1996).
340. M. K. Shi, G. L. Graff, J. D. Affinito, M. E. Gross, G. Dunham, P. Mounier and M. Hall: "Plasma Surface Modification of Acrylic-Based Polymer Multilayer for Enhanced Ag Adhesion," *Proc. Soc. Vac. Coaters* **42**, 307-310 (1999).
341. Z. J. Radzimski and W. M. Posadowski: "Self-Sputtering with DC Magnetron Source: Target Material Consideration," *Proc. Soc. Vac. Coaters* **37**, 389-394 (1994).
342. J. Wolfe: "Anti-static Anti-reflection Coatings Using Various Metal Layers," *Proc. Soc. Vac. Coaters* **38**, 272-275 (1995).

343. J. Szczyrbowski, G. Bräuer, M. Ruske, G. Teschner, and A. Zmelty: "Temperable Low Emissivity Coating Based on Twin Magnetron Sputtered TiO_2 and Si_3N_4 Layers," *Proc. Soc. Vac. Coaters* **42**, 141-146 (1999).
344. O. Treichel, V. Kirchhoff and G. Bräuer: "The Influence of the Barrier Layer on the Mechanical Properties of IR-Reflecting (low-E) Multilayer Systems on Glass," *Proc. Soc. Vac. Coaters* **43**, 121-126 (2000).
345. A. LeBlanc and D. Parent: "Practical Solutions for Gold Sputtering of Recordable Compact Discs," *Proc. Soc. Vac. Coaters* **38**, 353-358 (1995).
346. D.J. McClure, D.S. Dunn and A.J. Ouderkirk: "Adhesion Promotion Technique for Coatings on PET, PEN, and PI," *Proc. Soc. Vac. Coaters* **43**, 342-346 (2000).
347. J. I. Larruquert, J. A. Méndez, and J. A. Aznárez: "Empiracal relations among scattering, roughness, parameters, and thickness of aluminum films," *Appl. Opt.* **32**, 6341-6346 (1993).
348. J. I. Larruquert, J. A. Méndez, and J. A. Aznárez: "Far-UV reflectance of UHV-prepared Al films and its degradation after exposure to O_2," *Appl. Opt.* **33**, 3518-3522 (1994).
349. J. I. Larruquert, J. A. Méndez, and J. A. Aznárez: "Far-ultraviolet reflectance measurements and optical constants of unoxidized aluminum films," *Appl. Opt.* **34**, 4892-4899 (1995).
350. J. I. Larruquert, J. A. Méndez, and J. A. Aznárez: "Optical constants of aluminum films in the extreme ultraviolet, 82-77 nm," *Appl. Opt.* **35**, 5692-5697 (1996).
351. J. I. Larruquert, J. A. Méndez, and J. A. Aznárez: "Degradation of far-ultraviolet reflectance of aluminum films exposed to atomic oxygen, In-orbit coating application," *Optics Communications* **124**, 208-215 (1996).
352. J. A. Aznárez, J. I. Larruquert, and J. A. Méndez: "Far-ultraviolet absolute reflectometer for optical determination of ultrahigh vacuum prepared thin films," *Rev. Sci. Instrum.* **67**, 497-502 (1996).
353. J. I. Larruquert, J. A. Méndez, and J. A. Aznárez: "Life prolongation of far ultraviolet reflecting aluminum coatings by periodic recoating of the oxidized surface," *Optics Communications* **135**, 60-64 (1997).
354. J. I. Larruquert and R. A. M. Keski-Kuha: "Multilayer coatings for narrowband imaging in the extreme ultraviolet," *SPIE* **3114**, 608-616 (1997).
355. P. Chang, W. K. Halnan and R. J. Hill: "Improvements in Equipment for the Evaporation of Subliming Materials in Reactive Gas Conditions," *Proc. Soc. Vac. Coaters* **42**, 122-128 (1999).
356. G. Caskey, D. Roberts, M. Hansen, D. Betz and M. Catlin: "Optical Considerations in Thin Film Electrochromic Devices," *Proc. Soc. Vac. Coaters* **35**, 345-350 (1992).
357. H. Demiryont, K. E. Nietering: "Tungsten oxide films by reactive and conventional evaporation techniques," *Appl. Opt.* **28**, 1494-1500 (1989).
358. G. A. Castellion: "Electrochromic (EC) mirror which rapidly changes reflectivity," *Assigned to American Cyanamid Co.* Pat. No. **3,807,832** (30 Apr. 1974).
359. H. Demiryont: "Quasi-symmetric electrochromic device for light modulation," *Appl. Opt.* **31**, 1250-254 (1992).
360. H. K. Byker: "Single-compartment, self-erasing, solution-phase electrochromic devices, solutions for use therein, and uses thereof," *Assigned to Gentex Corporation*

Pat. No. **4,902,108** (20 Feb. 1990).

361. P. Lippens, N. V. Bekaert S. A., Belgium; and P. Verheyen: "Electrochromic Half-Cells on Organic Substrate Produced by Subsequent Deposition of ITO and WO_3 in a Roll Coater Using Optical Plasma Emission Monitoring," *Proc. Soc. Vac. Coaters* **37**, 254-259 (1994).
362. L. R. Gilbert, S. P. Maki and D. J. McClure: "Applications of Vacuum Deposited Transparent Conductors in Switchable Windows," *Proc. Soc. Vac. Coaters* **38**, 111-119 (1995).
363. C. M. Lampert: "Switchable Glazing: Science and Technology of Smart Windows," *Proc. Soc. Vac. Coaters* **38**, 189-193 (1995).
364. J. G. H. Mathew, B. P. Hichwa, N. A. O'Brien, V. P. Raksha, S. Sullivan and L. S. Wang: "Study of Vacuum Deposited, Thin Film Transmissive Electrochromic Devices," *Proc. Soc. Vac. Coaters* **38**, 194-197 (1995).
365. S. P. Sapers, M. J. Cumbo, R. B. Sargent, V. P. Raksha, L. S. Wang, R. B. Lahaderne and B. P. Hichwa: "Monolithic Solid State Electrochromic Coatings for Window Applications," *Proc. Soc. Vac. Coaters* **39**, 248-255 (1996).
366. C. M. Lampert: "The State-of-the-Art of Switchable Glazing and Related Electronic Products," *Proc. Soc. Vac. Coaters* **42**, 197-203 (1999).
367. C. M. Lampert: "Smart Switchable Materials for the New Millennium—Windows and Displays," *Proc. Soc. Vac. Coaters* **43**, 165-170 (2000).
368. C. G. Granqvist: *Handbook of Inorganic Electrochromic Materials*, (Elsevier, Amsterdam, 1995).
369. V. Guidi, E. Comini, M. Ferroni, G. Martinelli, and G. Sberveglieri: "Study on nanosized TiO/WO_3 thin films achieved by radio frequency sputtering," *J. Vac. Sci. Technol. A* **18**, 509-514 (2000).
370. S. Santucci, C. Cantalini, M. Crivellari, L. Ottaviano, and M. Passacantando: "X-ray photoemission and scanning tunneling spectroscopy study on the thermal stability of WO3 thin films," *J. Vac. Sci. Technol. A* **18**, 1077-1082 (2000).
371. A.-L. Larsson, G. Niklasson, and L. Stenmark: "Thin film coatings with variable emittance," *SPIE* **3738**, 486-492 (1999).
372. R. L. Cormia, J. B. Fenn Jr., H. Memarian, and G. Ringer: "Roll-to-Roll Coating of Indium Tin Oxide—A Status Report," *Proc. Soc. Vac. Coaters* **41**, 452-457 (1998).
373. M. Mayr, "High Vacuum Sputter Roll Coating: A New Large Scale Manufacturing Technology for Transparent Conductive ITO Layers," *Proc. Soc. Vac. Coaters* **29**, 77-94 (1986).
374. I. Hamberg and C. G. Granqvist: "Color properties of transparent heat-reflecting MgF_2- coated indium-tin-oxide films," *Appl. Opt.* **22**, 609-614 (1983).
375. M. Saif, A. Wrzesinski, H. Patel, H. Memarian, and R. Ward: "Aluminum Doped Indium Tin Oxide (ITO) Films Prepared by DC Magnetron Sputtering," *Proc. Soc. Vac. Coaters* **40**, 230-236 (1997).
376. J. Czukor, W. Kittler, P. Maschwitz and I. Ritchie: "The Effects of Process Conditions on the Quality and Deposition Rate of Sputtered ITO Coatings," *Proc. Soc. Vac. Coaters* **34**, 190-195 (1991).
377. L. R. Gilbert, S. P. Maki and D. J. McClure: "Comparison of ITO Sputtering Process from Ceramic and Alloy Targets onto Room Temperature PET Substrates," *Proc. Soc. Vac. Coaters* **36**, 236-241 (1993).

378. J. L. Grieser: "Comparison of ITO Films Produced by Standard D.C. Sputtering and Ion Beam Assisted Sputtering," *Proc. Soc. Vac. Coaters* **38**, 155-162 (1995).
379. D. Huang, F. C. Ho, and R. R. Parsons: "Effects of surface compression strengthening on properties of indium tin oxide films deposited on automobile glass," *Appl. Opt.* **35**, 5080-5084 (1996).
380. K. P. Gibbons, C. K. Carniglia, and R. E. Laird: "Sputtering of ITO from a Rotating Ceramic Target," *Proc. Soc. Vac. Coaters* **41**, 159-164 (1998).
381. R. E. Treece, I. Eisgruber, R. Hollingsworth, J. Engel and P. Bhat: "Applying Combinatorial Chemistry Methodology to Optimizing the Growth of Indium Tin Oxide on Polymeric Substrates," *Proc. Soc. Vac. Coaters* **43**, 171-176 (2000).
382. R. Yoshimura, N. Ogawa, T. Iwamoto, and T. Mouri: "Studies on Characteristics of ITO Target Materials," *Proc. Soc. Vac. Coaters* **35**, 80-83 (1992).
383. B. G. Lewis, R. Mohanty and D. C. Paine: "Structure and Performance of ITO Sputtering Targets," *Proc. Soc. Vac. Coaters* **37**, 432-439 (1994).
384. A. W. Baouchi: "Atomic Force Microscopy Study of ITO Coatings Deposited on Glass by DC Magnetron Sputtering," *Proc. Soc. Vac. Coaters* **39**, 151-156 (1996).
385. J. P. Lehan: "Determinationof grain size in indium tin oxide films from transmission measurements," *Appl. Opt.* **35**, 5048-5051 (1996).
386. K. Zhang, A.R. Forouhi, and D. V. Likhachev: "Accurate and Rapid Determination of Thickness, n and k Spectra, and Resistivity of ITO Films," *Proc. Soc. Vac. Coaters* **42**, 255-260 (1999).
387. C.-C. Lee, S.-C. Shiau, and Y. Yang: "The Characteristics of Indium Tin Oxide (ITO) Film Prepared by Ion-Assisted Deposition," *Proc. Soc. Vac. Coaters* **42**, 261-264 (1999).
388. B. M. Henry, A. G. Erlat, C. R. M. Grovenor, G. A. D. Briggs, Y. Tsukahara, T. Miyamoto, N. Noguchi and T. Niijima: "Microstructural and Gas Barrier Properties of Transparent Aluminium Oxide and Indium Tin Oxide Films," *Proc. Soc. Vac. Coaters* **43**, 373-378 (2000).
389. H. Patel, M. Saif, and H. Memarian: "Methods of Monitoring & Control of Reactive ITO Deposition Process on Flexible Substrates with DC Sputtering," *Proc. Soc. Vac. Coaters* **39**, 441-445 (1996).
390. H. Patel, M. Saif and H. Memarian: "Progress in Monitoring and Control of Reactively Sputtered ITO Using Optical Emission and Mass Spectroscopy," *Proc. Soc. Vac. Coaters* **40**, 20-23 (1997).
391. K. P. Gibbons, R. E. Laird, and C. K. Carniglia: "High-Rate Sputter Deposition of TiO_2 Using Closed-Loop Feedback Control," *Proc. Soc. Vac. Coaters* **41**, 178-181 (1998).
392. H. Patel: "DC Magnetron Sputter Deposition of ITO on Polymeric Webs Using Plasma Emission for Process Monitoring and Control," *Proc. Soc. Vac. Coaters* **40**, 333-337 (1997).
393. B. E. Kuhlmann: "Optical Thin Film Coatings Deposited Directly on Cathode Ray Tubes," *Proc. Soc. Vac. Coaters* **39**, 264-269 (1996).
394. R. E. Laird, J. D. Wolfe and C. K. Carniglia: "Durable Conductive Anti-Reflection Coatings for Glass and Plastic Substrates," *Proc. Soc. Vac. Coaters* **39**, 361-365 (1996).

395. K. P. Gibbons, C. K. Carniglia, R. E. Laird, R. Newcomb, J. D. Wolfe and S. W. T. Westra: "ITO Coatings for Display Applications," *Proc. Soc. Vac. Coaters* **40**, 216-220 (1997).
396. R. Wang and C.-C. Lee: "Design of Antireflection Coating Using Indium Tin Oxide (ITO) Film Prepared by Ion Assisted Deposition (IAD)," *Proc. Soc. Vac. Coaters* **42**, 246-249 (1999).
397. R. Blacker, D. Bohling, M. Coda, M. Kolosey and N. Marechal: "Development of Intrinsically Conductive Antireflection Coatings for the Ophthalmic Industry," *Proc. Soc. Vac. Coaters* **43**, 212-216 (2000).
398. M. Gilo, R. Dahan, and N. Croitoru: "Transparent indium tin oxide films prepared by ion-assisted deposition with a single layer overcoat," *Opt. Eng.* **38**, 953-957 (1999).
399. I. Lubezky, O. Marcovitch, Z. Klein, and H. Zipin: "Activated reactive evaporation of a transparent conductive coating for the IR region," *Thin Solid Films* **148**, 83-92 (1987).
400. F. O. Adurodija, H. Izumi, T. Ishihara, H. Yoshioka, and M. Motoyama: "Influence of substrate temperature on the properties of indium oxide thin films," *J. Vac. Sci. Technol. A* **18**, 814-818 (2000).
401. I. Hamberg and C. G. Granqvist: "Optical properties of transparent and heat-reflecting indium tin oxide films: The role of ionized impurity scattering," *Appl. Phys. Lettr.* **44**, 721-723 (1984).
402. R. H. Wentorf, Jr.: *J. Chem. Phys.* **26**, 956 (1957).
403. M. Kagamida, H. Kanda, M. Akaishi, A. Nukui, T. Osawa, and S. Yamaoka: "Crystal growth of cubic boron nitride using $L1_3BN_2$ solvent under high temperature and pressure," *J. Crystal Growth* **94**, 261-269 (1989).
404. C. S. Yoo, J. Akella, H. Cynn, and M. Nicol: "Direct reactions of boron and nitrogen at high pressures and temperatures," *Phys. Rev. B* **56**, 140-146 (1997).
405. O. Mishima: "Growth and polar properties of cubic boron nitride," *Applications of Diamond Films and Related Materials*, 647-651 (Elsevier Science Publishers, Amsterdam, 1991).
406. J. Musil, P. Zeman, H. Hruby, and P. H. Mayrhofer: "ZrN/Cu nanocomposite film - a novel superhard material," *Surf. Coat. Technol.* **120-121**, 179-183 (1999).
407. Y. Yamada-Takamuru, O. Tsuda, H. Ichinose, and T. Yoshida: "Atomic-scale structure at the nucleation site of cubic boron nitride deposited from the vapor phase," *Phys. Rev. B* **59**, 10351-10355 (1999).
408. P. B. Mirkarimi, K. F. McCarthy, and D. L. Medlin: "Review of advances in cubic boron nitride film synthesis," *Mater. Sci. & Eng.* **R21**, 47-100 (1997).
409. M. Ben el Mekki, M. A. Djouadi, V. Mortet, E. Guiot, G. Nouet, and N. Mestres: "Synthesis and characterization of c-BN films prepared by ion beam assisted deposition and triode sputtering," *Thin Solid Films* **355-356**, 89-95 (1999).
410. D. Litvinov, R. Clarke, C. A. Taylor II, and D. Barlett: "Real-time strain monitoring in thin film growth: cubic boron nitride on Si (100)," *Mater. Sci. & Eng.* **B56**, (1999).
411. W. Otano-Rivera, L. J. Pilione, J. A. Zapien, and R. Messier: "Cubic boron nitride thin film deposition by unbalanced magnetron sputtering and dc pulsed substrate biasing," *J. Vac. Sci. Technol. A* **16**, 1331-1335 (1998).

412. W.-L. Zhou, Y. Ikuhara, and T. Suzuki: "Orientational relationship between cubic boron nitride and hexagonal boron nitride in a thin film synthesized by ion plating," *Appl. Phys. Lettr.* **67**, 3551-3553 (1995).
413. Z. Song, F. Zhang, Y. Guo, and G. Chen: "Textured growth of cubic boron nitride film on nickel substrates," *Appl. Phys. Lettr.* **65**, 2669-2671 (1994).
414. S. J. Harris, A. M. Weiner, G. L. Doll, and W.-J. Meng: "Selective chemical etching of hexagonal boron nitride compared to cubic boron nitride," *J. Mater. Res.* **12**, 412-415 (1997).
415. P. B. Mirkarimi, K. F. McCarthy, G. F. Cardinale, D. L. Medlin, D. K. Ottesen, and H. A. Johnsen: "Substrate effects in cubic boron nitride film formation," *J. Vac. Sci. Technol. A* **14**, 251-255 (1996).
416. E. Franke, M. Schubert, J. A. Woollam, J.-D. Hecht, G. Wagner, H. Neuman, and F. Bigl: "In-situ ellipsometry growth characterization of dual ion beam deposited boron nitride thin films," *J. Appl. Phys.* **87**, 2593-2599 (2000).
417. Y. Lifshitz, S. R. Kasi, J. W. Rabalais, and W. Eckstein: "Subplantation model for film growth from hyperthermal species," *Phys. Rev. B* **41**, 10468-10480 (1990).
418. H. Feldermann, R. Merk, H. Hofsass, C. Ronning, and T. Zheleva: "Room temperature growth of cubic boron nitride," *Appl. Phys, Lettr.* **74**, 1552-1554 (1999).
419. H. Hofsass, H. Feldermann, M. Sebastian, and C. Ronning: "Thresholds for the phase formation of cubic boron nitride thin films," *Phys. Rev. B* **55**, 13230-13233 (1997).
420. W. Otano-Rivera, L. J. Pilione, and R. Messier: "Pressure dependence of the negative bias voltage for stabilization of cubic boron nitride thin films deposited by sputtering," *Appl. Phys, Lettr.* **72**, 2523-2525 (1998).
421. M. P. Johansson, I. Ivanov, L. Hultman, E. P. Munger, and A. Schultze: "Low-temperature deposition of cubic BN:C films by unbalanced direct current magnetron sputtering of a B_4C target," *J. Vac. Sci. Technol. A* **14**, 3100-3107 (1996).
422. W. Donner, H. Dosch, S. Ulrich, H. Ehrhardt, and D. Abernathy: "Strain relaxation of boron nitride thin films on silicon," *Appl. Phys, Lettr.* **73**, 777-779 (1998).
423. I.-H. Kim, K.-S. Kim, S.-H. Kim, and S.-R.Lee: "Synthesis of cubic boron nitride films using a helicon wave plasma and reduction of compressive stress," *Thin Solid Films* **290-291**, 120-125 (1996).
424. H.-G. Boyen, P. Widmayer, D. Schwertberger, N. Deyneka, and P. Ziemann: "Sequential ion-induced stress relaxation and growth: A way to prepare stress-relieved thick films of cubic boron nitride," *Appl. Phys, Lettr.* **76**, 709-711 (2000).
425. K. F. Chan, C. W. Ong, and C. L. Choy: "Nanoscratch characterization of dual-ion-beam deposited C-doped boron nitride films," *J. Vac. Sci. Technol. A* **17**, 3351-3357 (1999).
426. C. Ronning, A. D. Banks, B. L. McCarson, R. Schlesser, Z. Sitar, B. L. Ward, and R. J. Nemanich: "Structural and electronic properties of boron nitride thin films containing silicon," *J. Appl. Phys.* **84**, 5046-5051 (1998).
427. D. Litvinov, C. A. Taylor II, and R. Clarke: "Semiconducting cubic boron nitride," *Diamond & Related Materials* **7**, 360-364 (1998).
428. C. Chardonnet, V. Bernard, C. Daussy, A. Gicquel, and E. Anger: "Polarization properties of thin films of diamond," *Appl. Opt.* **35**, 6692-6697 (1996).
429. A. Shabalin, M. Amann, M. Kishinevsky, and C. Quinn: "Industrial Ion Sources and Their Application for DLC Coating," *Proc. Soc. Vac. Coaters* **42**, 338-341 (1999).

430. A. Shabalin, M. Amann, M. Kishinevsky, C. Quinn and N. Capps: "Characterization of the Low-Voltage, High-Current Single-Cell Ion Source," *Proc. Soc. Vac. Coaters* **43**, 283-286 (2000).
431. A. Hieke, K. Bewilogua, K. Taube, I. Bialuch and K. Weigel: "Efficient Deposition Technique for Diamond-Like Carbon Coatings," *Proc. Soc. Vac. Coaters* **43**, 301-304 (2000).
432. L. Schaefer, J. Gaebler, S. Mulcahy, J. Brand, A. Hieke and R. Wittorf: "Tribological Applications of Amorphous Carbon and Crystalline Diamond Coatings," *Proc. Soc. Vac. Coaters* **43**, 311-315 (2000).
433. A. Bubenzer, B. Dischler, G. Brandt, and P. Koidl: "rf-plasma deposited amorphous hydrogenated hard carbon thin films: Preparation, properties, and applications," *J. Appl. Phys.* **54**, 4590-4595 (1983).
434. A. Bubenzer, B. Dischler, G. Brandt, and P. Koidl: "Role of hard carbon in the field of infrared coating materials," *Opt. Eng.* **23**, 153-156 (1984).
435. P. Koidl, A. Bubenzer, and B. Dischler: "Hard carbon coatings for IR-optical components," *SPIE* **381**, 177-182 (1983).
436. *Diamond Films and Coatings*, ed. R. F. Davis, (Noyes Pubs., Park Ridge, NJ, 1992).
437. P. J. Martin, S. W. Filipczuk, R. P. Netterfield, J. S. Field, D. F. Whitnall, and D. R. McKenzie: " Structure and hardness of diamond-like carbon films prepared by arc evaporation," *J. Mater. Sci. Lettr.* **7**, 410-412 (1988).
438. J. Robertson: "The deposition mechanism of diamond-like a-C and a-C:H," *Diamond & Related Materials* **3**, 361-368 (1994).
439. L. Martinu: "Plasma Deposition and Testing of Hard Coatings on Plastics," R. d'Agostino et al. (eds.) *Plasma Processing of Polymers*, 247-272 (Kluwer Academic Publishers, Netherlands, 1997).
440. S. Scaglione, F. Sarto, R. Pepe, A. Rizzo, and M. Alvisi: "Optical and mechanical properties of diamond like carbon produced by dual ion beam sputtering technique," *SPIE* **3738**, 58-65 (1999).
441. M. Adamik, I. Tomov, U. Kaiser, S. Laux, C. Schmidt, W. Richter, G. Sáfrán, and P. B. Barna: "Structure evolution of stratified NdF_3 optical thin films," *SPIE* **3133**, 123-131 (1997).
442. R. Thielsch, A. Duparré, U. Schulz, N. Kaiser, M. Mertin, and H. Bauer: "Properties of SiO_2 and Al_2O_3 films for use in UV-optical coatings," *SPIE* **3133**, 183-193 (1997).
443. J.-O. Carlsson: "Chapter 7, Chemical Vapor Deposition", *Handbook of Deposition Technologies for Films and Coatings,* R. F. Bunshah, ed. (Noyes Publications, Park Ridge, NJ 1994).
444. A. Sherman: "Chapter 8, Plasma-Enhanced Chemical Vapor Deposition", *Handbook of Deposition Technologies for Films and Coatings,* R. F. Bunshah, ed. (Noyes Publications, Park Ridge, NJ 1994).
445. J. E. Greene: "Chapter 13, Nucleation, Film Growth, and Microstructural Evolution," *Handbook of Deposition Technologies for Films and Coatings,* R. F. Bunshah, ed. (Noyes Publications, Park Ridge, NJ 1994).
446. Advanced Energy Industries, Inc., 1625 Sharp Point Drive, Fort Collins, CO 80525.
447. H. S. Niederwald, N. Kaiser, U. W. Schallenberg, A. Duparré, D. Ristau, M. Kennedy: "IAD of oxide coatings at low temperature: a comparison of processes based on different ion sources," *SPIE* **3133**, 205-213 (1997).

448. G. Doubinina, B. Poirier, E. Strouse and J. Cerino: "Non-Radioactive IR Anti-Reflective Coating," *Proc. Soc. Vac. Coaters* **41**, 310-312 (1998).
449. D. B. Fuller and D. S. Fisher: "Multilayer PECVD Antireflective Coatings for the Infrared," *Proc. Soc. Vac. Coaters* **38**, 168-171 (1995).
450. C. R. Seward, C. S. Pickles, R. Marrah, and J. E. Field: "Rain erosion data on window and dome materials," *SPIE* **1760**, 280-290 (1992).
451. W. F. Adler and D. J. Mihora: " Infrared-transmitting window survivability in hydrometeor environments," *SPIE* **1760**, 291-302 (1992).
452. W. F. Adler, J. W. Flavin, J. P. Richards, and P. L. Boland: "Multiple simulated waterdrop impact damage in zinc selenide at supersonic velocities," *SPIE* **1760**, 303-315 (1992).
453. W. Hasan and S. H. Propst: "Recent developments in highly durable protective-antireflection coatings for Ge and ZnS substrates," *SPIE* **2253**, 228-235 (1994).
454. H. Blackwell, E. M. Waddell, D. R. Gibson, and P. McDermott: "Broadband IR transparent rain and sand erosion protective coating for the AV8-B and GR-7 Harrier forward looking infra-red germanium window," *SPIE* **2776**, 144-158 (1996).
455. A. Zöller, R. Glötzelmann, K. Matl, and D. Cushing: "Temperature-stable bandpass filters deposited with plasma ion-assisted deposition," *Appl. Opt.* **35**, 5609-5612 (1996).
456. A. Zöller, R. Glötzelmann, H. Hagedorn, W. Klug, and K. Matl: "Plasma ion assisted deposition: a powerful technology for the production of optical coatings," *SPIE* **3133**, 196-203 (1997).
457. J.-C. Hsu and C.-C. Lee: "Single- and dual-ion-beam sputter deposition of titanium oxide films," *Appl. Opt.* **37**, 1171-1176 (1998).
458. R. Faber, K. Zhang, and A. Zoeller: "Design and manufacturing of narrow-band interference filters," *SPIE* **4094**, 58-64 (2000).
459. C. K. Carniglia, J. H. Apfel, G. B. Carrier, and D. Milam: "TEM Investigation of Effects of a Barrier Layer on Damage to 1.064μ AR Coatings," *NBS Spec. Pub. 541*, 218-221 (1978).
460. R. D. Mathis Co., P.O. Box 92916, Long Beach, CA 90809.
461. PS1500 Plasma/Ion Source, DynaVac, 110 Industrial Park Rd., Hingham, MA 02043.
462. D. E. Morton, "The effects of pumping speed on the operation of a cold cathode ion source," *Vacuum Technology & Coating*, **2**, June 2001.
463. R. R. Willey, "Achieving Stable Results with Titanium Dioxide," *Society of Vacuum Coaters Technical Conference Proceedings*, **39**, 207-210 (1996).
464. S. R. Schmidt and R. G. Launsby, Understanding Industrial Designed Experiments, Sec. 3.8, Air Academy Press, Colorado Springs, 1994.
465. DOE KISS, Ver. 97 for Windows, Air Academy Associates (and Digital Computations, Inc.), Colorado Springs (1997).
466. L. Martinu and D. Poitras: "Plasma deposition of optical films and coatings: A review," *J. Vac. Sci. Technol. A*, **18**, 2619-2645 (2000).

6

Process Development

6.1. INTRODUCTION

Before starting a coating development, it is well to review all of the requirements and options from the top down. Can you choose the substrate material or is it predetermined? This can change the situation considerably with respect to process variables such as temperature and the results such as adhesion. Can you choose the coating materials? This is usually the case, but not always. Do you have a choice of chambers and processes available to you? The choices made at the beginning, if there are any to make, can make a significant difference in the effort required and the success of the rest of the coating project. Doing the "homework" first is usually a good approach.

The development of a new optical coating process or the refinement of an existing one can sometimes turn into a "hairy" problem. Hamel[1] touched on this in his paper on "Big Foot ...." which deals with process optimization. The process variables might include: base pressure, substrate material, substrate preparation, substrate temperature, soak time, reactive gas partial pressure, gas mixture, deposition material, deposition rate, material evaporation pattern, E-gun crucible rotation speed and/or beam sweep pattern, E-gun high voltage and filament current, chamber cleaning, planet or calotte rotation speed, uniformity mask shape and position, etc., *just to name a few!* If we include an ion source, we can add: gas flow rate, chamber pumping speed, drive volts, drive current, filament current, at least five degrees of freedom of source positioning and aiming, and other variables. It is a wonder that we ever get a process to work well and reproducibly!

The common practice has been to attempt to hold all of the would-be variables constant except the one under investigation and find the best setting for that one variable. If we were to attempt to optimize a 20 variable process as described above by only two tests per variable with the assumption that the result was linear in all the variables, it would take 2×20 or 40 experiments just to adjust the variables to the best of the two settings tested for each. However, this would be generally a very naive approach since it is *very unlikely* that the optimum *combination* would be found this way. We would need to perform 2^{20} or about 1,048,576 experiments to find the best result for one output such as hardness, density, or index of refraction by the combination of variables in the tested range of each variable if they were all truly linear. This is the kind of Herculean task that many of us have unwittingly undertaken over the years. However, there is new hope for a way make a mole hill out of this mountain by reducing the number of experiments from a million to the order of 50, *and* to impart better knowledge of the process to the experimenter. The techniques come from the discipline called "Design of Experiments" (DOE). Schmidt and Launsby[2] have made these tools usable, understandable, and interesting to even those of us who have had an aversion to statistics.

One problem with the approach of "one variable at a time" is that it requires many test runs if there are many parameters to be optimized. However, the second and worse problem is that this approach may not find the optimum result for the combined parameters which have been varied. Figure 6.1 illustrates this effect. If variable B were first optimized with respect to the desired result Y and then the optimization of A were started from that point, it would appear that the optimum point in A and B was at point X. However, the real optimum is at point Z. With the proper statistical sampling techniques of DOE, it is possible to much more closely locate the true optimum at point Z by only 5 sample points. For example, these might be data at points in the four corners of Fig. 6.1 and at the center point.

The case of Fig. 6.1 is for only two variables. Typically there are three or more critical parameters to be optimized. When there are three variables, it is still possible to show the distribution of sampling points graphically. Figure 6.2 shows the positioning of sampling points for one of the preferred DOE configurations known as the Box-Wilson or Central Composite Design (CCD)[2]. This is very efficient and flexible for second order modeling. Such a design also has rotatability so that the predicted response can be estimated with equal variance regardless of the direction from the center of the design. Another frequently used design is the Box-Behnken Design shown in Fig. 6.3. These are potentially more efficient than CCDs for three factors (variables) and three levels. They allow estimation of linear and quadratic effects and all 2-way interactions. When the number of factors is greater than 4, the CCD would be more efficient. In both of

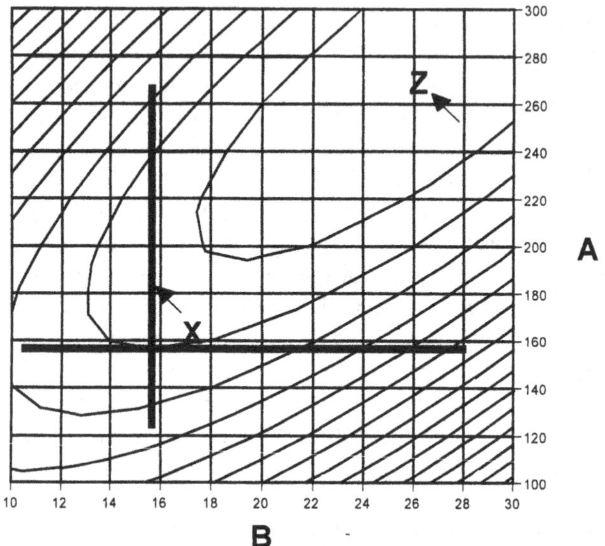

Fig. 6.1. Contour plot of result Y as a function of variables A and B.

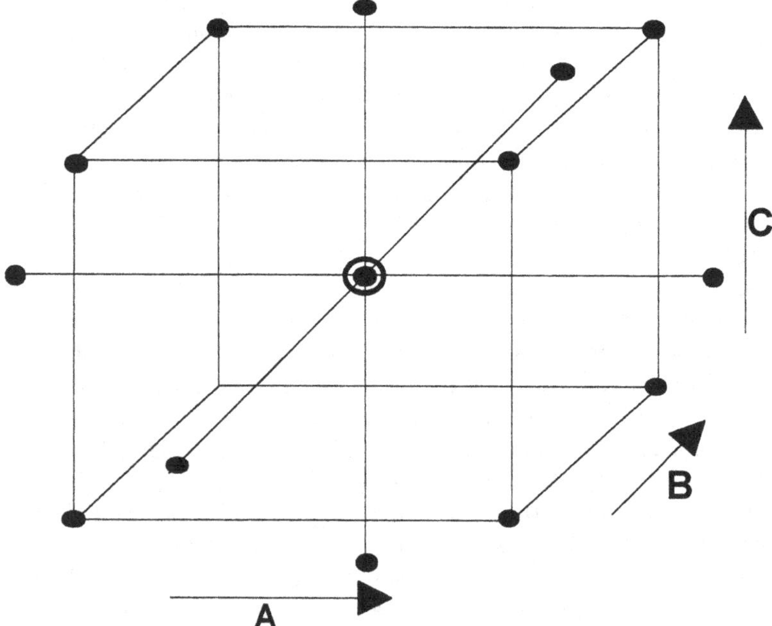

Fig. 6.2. CCD sampling scheme for 3 variables. Dots are sample points.

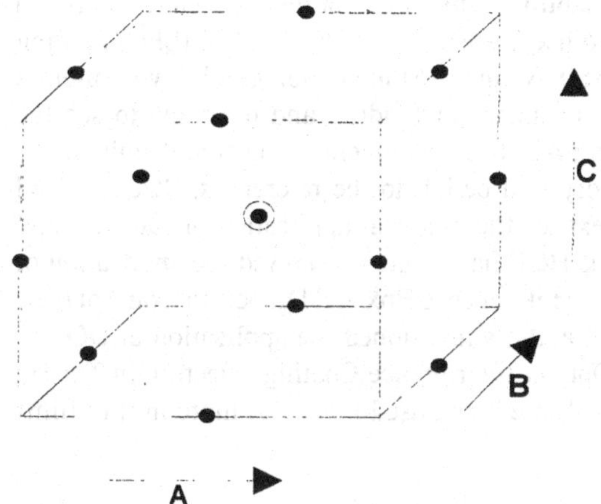

Fig. 6.3. Box-Behnken Design sample point scheme.

these designs, the central or axial point is sampled 3 or more times to measure the repeatability of the data and allow the estimation of the standard deviation.

Experiments are then conducted at the conditions of each of the sample points and the results are recorded. Note also that more than one type of result can usually be recorded for each data point, such as absorption, index, hardness, humidity shift, etc. The results are then processed in the DOE software[9] to fit (least squares) the data to a model for linear and quadratic effects and 2-way interactions. The model of the results can then be displayed in 2- and 3-dimensional graphics to aid visualizations of the process behavior. With the aid of these graphics, it is usually possible to find the values of each variable which will give the optimum process results.

These tools have come out of the practical application of statistical science in the past seven decades starting with Sir R. A. Fisher at the Rothamsted Laboratory in the 1920s. Early applications were to agricultural experiments and then to cotton and woolen industries. The more modern applications to the automotive industry are associated with the well-known names of Deming, Juran, Ishikawa, and Taguchi. Taguchi developed his approach to DOE in the 1950s and has done a great deal to bring the techniques to the attention of those who can benefit from them. American statisticians may not entirely embrace all of Taguchi's approaches and details, but owe a great deal to his work for bringing the field into more common use.

DOE provides a methodology to gain the maximum amount of process knowledge with a minimum number of experiments. This in turn implies a minimum cost to gain a maximum control of a process. This then should lead to reduced cost and increased repeatability. This is part of the same cultural growth in production which has come to use Statistical Process Control (SPC) and put forth the concepts of Motorola's Six Sigma (6σ) methodology. We will attempt here to give the reader enough of an understanding and overview to see the benefits of DOE and apply it to process development. The fine details of the derivations and background theory will be left to the references. Kiemele and Schmidt[3] have an appropriate text for the practical application of statistics and SPC. Knoll and Theil[4] as well as Heil[5] have recently showed the application of DOE to coating processes and make reference to Box and Draper[6] for some of their methodology. Several additional authors mentioned the application of DOE to their work at the International Optical Interference Coatings meeting in Tucson in June of 1995. The tools of DOE are being used more and more in thin films because of their benefits.

6.2. DESIGN OF EXPERIMENTS METHODOLOGY

The DOE methodology supplies a checklist or guideline for the orderly and efficient development or refinement of a process. Most of us have developed successful processes without the benefits of DOE, but some of us are now convinced that we could have done the same job better and easier with the tools of DOE.

Any new endeavor will benefit by starting with a clear statement of the needs and objectives. This then leads to the questions of how to proceed toward those goals and measure the results. A helpful checklist for process improvement is: Process Flow (PF), Cause & Effect (CE), Controlled-Noise-Experiment (CNX), and Standard Operating Procedure (SOP). We will briefly describe each of these in turn.

6.2.1. Process Flow Diagram

Preparing a Process Flow (PF) diagram or chart for a new or existing process often leads to new insights with respect to the process and its possible improvements. It certainly makes it much easier to communicate about the process with others. We have held the position for decades that software (processes) should have detailed flow charts before any code is written. In software or hardware, the PF is well worth the effort for its benefits in communicating, debugging, and refinement.

Figure 6.4 illustrates a simple PF for the aluminization of a mirror. Each step in the process might be further divided into more detailed processes as appropriate. The development of the PF often suggests areas of concern that need

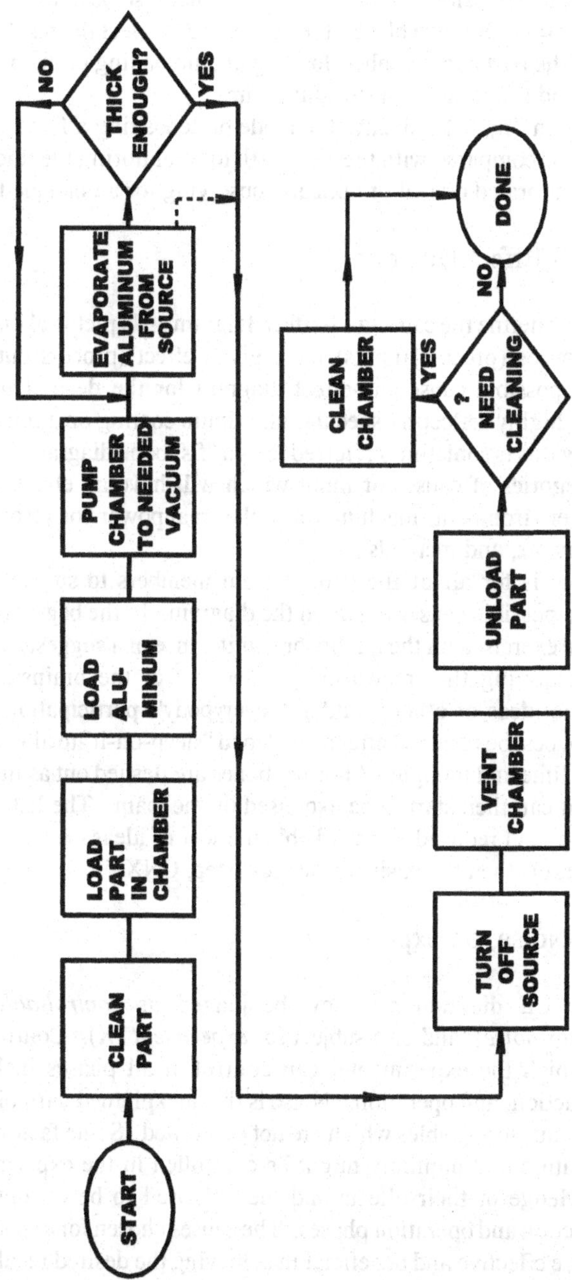

Fig. 6.4. A simple Process Flow Diagram (PF) for the aluminization of a mirror. The dotted line is if no thickness monitoring is used in the production process.

attention and brings to light things that might have otherwise been overlooked. As seen in Fig. 6.4, essentially each block of the process can affect the result of the process: cleaning of the part and chamber, loading and unloading of the part and material, pressure, and evaporation of the aluminum.

The dotted line in Fig. 6.4 indicates the mode of depositing a fixed weight charge of aluminum as compared with the solid path for monitoring the thickness of the coating by a calibrated crystal or optically observing for an opaque film.

6.2.2. Cause-and-Effect Diagram

Schmidt and Launsby[2] define the cause-and-effect diagram as "a pictorial diagram showing possible causes (process inputs) for a given effect (process output)." Figure 6.5 shows a possible cause-and-effect diagram for the desired process output of Fig. 6.4, a highly reflecting specular aluminum coating on a mirror. It is probably clear why this is sometimes referred to as a "fishbone diagram." There are usually six categories of causes or input which will have an effect on the output. These are: environment, machine (or tools), manpower (or resources), materials, measurements, and methods.

A good practice is for all of the project team members to sit down and brainstorm all of the possible causes to put on the diagram. In the beginning, no judgement should be exercised on the quality or significance of a suggested input. This is to avoid inhibiting the freewheeling discovery of the brainstorming process, build on the ideas of others, and get everybody's participation. The brainstorming might best be resumed after a break or a "sleep-on-it" to allow more ideas to evolve with time and thought. After the bones are fleshed out as much as practical, judgement can then start to be exercised by the team. The list can be critiqued, cleaned up, and reduced to a workable number of ideas.

The CE diagram is then the basis for the next step, CNX.

6.2.3. Control, Noise, or Experiment

The causes in the CE diagram can now be judged as *controllable* (C), uncontrollable (N for *noise*), and/or a subject for *experiment* (X). Controllable factors are those which the experimenter can control at all phases including experimental, production, and operation. Noise is the unexplained variability in a response, typically due to variables which are not controlled. Some factors such as ambient temperature and humidity might be controlled in the experimental stage to gain knowledge of their effects and then allowed to be uncontrolled (noise) in the production and operation phases. The causes chosen for experiment are those judged to be effective and beneficial in achieving the desired results and the factors to be optimized through designed experiments.

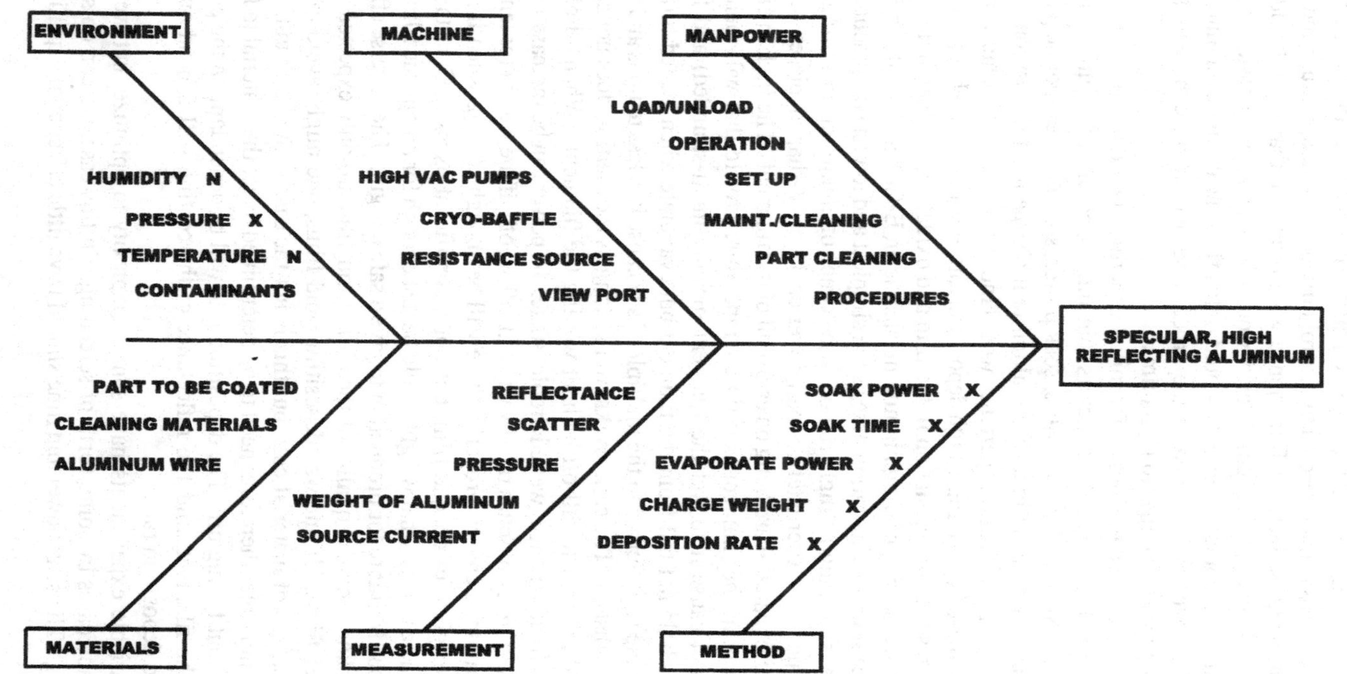

Fig. 6.5. A possible cause-and-effect diagram for the desired process output of Fig. 6.4.

On the CE diagram of Fig. 6.5, we have indicated by an "N" that the aluminizing process is not expected to control humidity and temperature and they will therefore be noise factors. Experience would indicate that they should not contribute much noise. The pumpdown would be to a given pressure before starting so that the ambient humidity would only affect the pumpdown time. No heat or cooling would be added to the process, so one would expect that the variations due to temperature to be minimal.

We have indicated the factors for initial experimentation by an "X" on Fig. 6.5. The rest will be controlled factors, but unmarked on the figure to avoid clutter. The six X-factors are: pressure, soak power, soak time, evaporate power, deposition rate, and the weight of the aluminum charge for a deposition. The literature (Hass, et al.[7,8]) and experience would indicate that the reflectance of the coating is increased by having a high deposition rate and a low chamber pressure. This minimizes the contamination in the film and oxidation. Although there are other processes for evaporating aluminum such as an E-Gun and sputtering, we use resistance source evaporation here as the simplest and lowest equipment cost. The practice is to hang a single rod or wire "cane" of aluminum in a tungsten heater filament. Alternately, small inverted U or V-shaped pieces of the aluminum wire material are hung on each coil of the filament. The trick in either case is to heat up the filament in a time-temperature profile that will melt the aluminum and cause it to wet the filament (usually multi-stranded) and form droplets which do not fall from the filament but evaporate when heated further. The single cane is preferred to the multiple V's because it is less tedious and time consuming to install. The cane works well as long as the melting does not cause sections of aluminum to fall off without wetting the filament. When all of the aluminum is in this melted, wetted droplet stage, the power can be increased such that the aluminum evaporates rapidly without falling off. If the weight (or length) of the aluminum is chosen properly, there will be enough coating when it is all evaporated to give an opaque film, but not a significant excess. Our starting choice would be to find what weight would be just enough for an opaque film and then use about 1/3 more for production to give a safe margin. The adverse effects of extra thickness on the quality of the aluminum film are only expected with much thicker films. If one was depositing gold films, one might avoid excess thickness because of material cost, but this is not the case with aluminum. The soak power and time, therefore, need to be determined to get the material to form droplets without falling off. The deposition rate will be dependant on the power applied to the filament, and the reflectance of the coating will depend on the pressure and deposition rate.

Six variable experiments have some complexity to optimize, but we can simplify many cases by some form of screening. In this case, we can do some prescreening. The soak power and time should have little effect on the quality of the film, but do need to be determined in advance. These can be set aside as a first and simple two variable experiment based on the desired result of "melted aluminum droplets which haven't fallen from the tungsten filament." Since the deposition rate depends on the "evaporate" power, we can simply observe the deposition rate while concentrating on our goal of the film quality, but use the evaporation power (current) as the variable. This leaves three variables to be the subject of the second designed experiment to optimize the specular, high reflecting aluminum. These variables are pressure, charge weight, and evaporation current.

6.2.4. Standard Operating Procedures

Any factors selected to be controlled should have Standard Operating Procedures (SOPs) written so that there will be a minimum variability in these factors from operator to operator and time to time. The objective is to eliminate noise in these variables so that they are truly controlled and their residual variations do not make a measurable contribution to the results. The SOPs will insure that the C's are held constant and that the process flow is complied with.

6.3. DESIGN OF THE EXPERIMENTS: EXAMPLES

The goal of the first experiment is to find an appropriate soak power and soak time to premelt the aluminum. This experiment is seeking a result which is not quantitative but has only three anticipated results (from experience). We expect: 1) that the aluminum won't melt at too low a current, 2) it will melt, wet the filament, and form droplets as desired, or 3) that it melts quickly and falls off the filament at too high a current. Normally, a multi-stranded filament is used for more surface area and better wetting characteristics. The results may also be nonlinear with time and current. As a result, this problem is not well suited to the common applications of DOE which depend on a smooth function of result with variables and quantifiable results. However, we will try a non-statistical, informal investigation for this part of the development of the aluminizing process.

Initially, we have no idea of what the "ballpark" value for soak current might be, but we might estimate that acceptable soak times could be between 2 and 20 seconds. A first test would be to load a filament with aluminum and *slowly* raise the current until the aluminum just melts, wets, and forms droplets on each loop of the filament. This recorded value would approximate the current setting required if the soak time were very long, and it could constitute the *lower bound* on the current to be applied. At this same time, we would then raise the current in perhaps 5% increments of this lower bound current to observe the current level at which the evaporation rate becomes significant (coats a test piece to opacity in about a minute). This might represent an *upper bound* on the current for soak because significant evaporation would occur during the soak phase if the soak time was too long. We now have invested one test run to gain an estimate of the two bounds on soak current. Next we might do an experiment with a fresh load of aluminum where we set the current to the average between the upper and lower bounds and time how long it takes to melt all of the aluminum. On the basis of the second experiment, we might use that average for the soak current and perhaps 1.2× the time found for the soak time. We have therefore found safe working values by just two experiments.

We now need to determine the values to use for pressure, charge weight, and evaporate current. Again, we can separate and simplify this from a three-variable problem to a two-variable problem. The opacity of the film will be complete if there is enough charge of aluminum, it will only significantly affect the reflectance if there is too little to produce an opaque film (or much too much thickness which might start to cause scattering). We might easily find out how much was enough weight by a single experiment. A large cane of aluminum and the filament should be weighed before installation. A test glass would be positioned at about the normal substrate distance through which we can observe the filament. The filament would then be soaked and slow evaporation started. By observing the filament through the substrate, the heating would be stopped when the filament could no longer be seen. The filament with residual aluminum would then be weighed to find how much weight had been evaporated to give the opaque coating. A reasonable charge weight might be 20– 40% more than this value to give some margin for error.

This then leaves only the pressure and deposition rate (current) to be determined for best reflectance. It is known from the literature[7,8] that low pressure and high rate usually give the highest reflectance with aluminum. We might best determine how much tolerance our process would have to variations of rate and pressure. The pumping system will have some ultimate low pressure limit which it can reach after a long pumping period. The high pressure limit might be the result of how quickly we want to start the process after closing the chamber. The high rate limit might be determined by how much current can be put to the filament without decreasing its life unduly. The low rate limit might be set by an acceptable process time for the evaporation.

Let us say that our chamber will pump to 1×10^{-4} torr in 30 seconds and 1×10^{-6} torr in 3000 seconds (50 minutes). We chose the former for our high pressure setting and 1×10^{-5} torr for our low pressure setting. Let us further imagine that a representative filament will burn out quickly at 100 amps and that in our earlier experiments we found the aluminum evaporated slowly at 50 amps. We chose 60 and 90 amps for our low and high current settings.

6.3.1. A Central Composite Design for Aluminizing

The Central Composite Design (CCD) or Box-Wilson Design is described in detail in Schmidt and Launsby[2] Sec. 3.9. It has an advantage of being very efficient in terms of the number of experiments and the information that it provides. In the case of the present aluminizing process, we have boiled it down to needing experiments to optimize only two variables (chamber pressure and deposition current). If the result (reflectance of the film) were only a straight line function of the two variables, the reflectance versus the variables might look like Fig. 6.6. Here it would just be a matter of doing four tests near the upper and lower limits of each of the two variables to find which corner of Fig. 6.6 gave the maximum reflectance. If, in another case, we wanted some specific reflectance between the maximum and minimum at these four points, we can find a line of values in the surface of Fig. 6.6 which would satisfy this reflectance. DOE gives the tools to quickly and easily get these values of the variables. The first four rows (runs) in Table 6.1 would represent the tests of these four combinations of the two variables.

If we just performed these four runs, we would have done a "full factorial" design and we would know everything there is to know about the "flat" (rectilinear) plane shown in Fig. 6.6 within the limits tested. However, if the surface is not a flat plane, but has some quadratic curvature, we would not know that without further runs. The next step would be to do a few runs at the mean

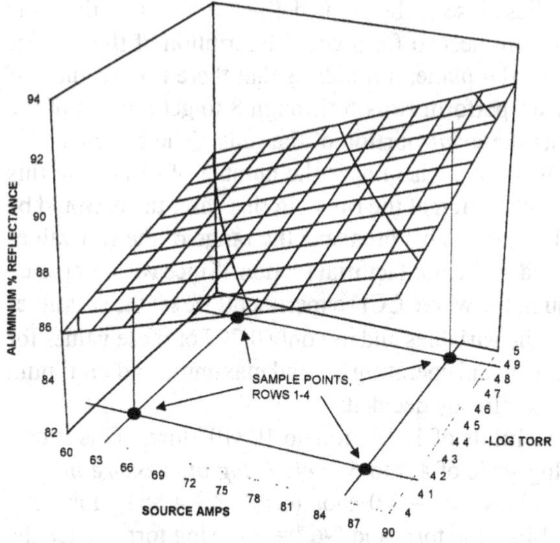

Fig. 6.6. A surface plot resulting from only a four (4) run, full factorial experiment as found in the first four rows of Table 6.1

Table 6.1 Design matrix and "experimental" values of percent reflectance for the Box-Wilson design (CCD) applied to pressure and current variations in the aluminization process. See page below for discussion of the − LOG TORR pressure scale.

ROW # & FACTOR		-LOG TORR	SOURCE AMPS	RESULT Y1
1	−,−	4.146	64.4	86.8
2	−,+	4.146	85.6	89.3
3	+,−	4.854	64.4	88.2
4	+,+	4.854	85.6	91.1
5	+α,0	5.0	75	90.0
6	−α,0	4.0	75	87.0
7	0,+α	4.5	90	90.5
8	0,−α	4.5	60	86.0
9	0,0	4.5	75	90.2
10	0,0	4.5	75	89.7
11	0,0	4.5	75	90.0

values of each of the two variables. Table 6.1 shows these as runs 9, 10, and 11 all at the same settings. The variations from one run to the next in these three can give an estimate of the variability in the measurements of the results (reflectance). If the mean of these new runs lies close to the plane defined in Fig. 6.6, then it is likely that no further test runs are needed for a good description of the process surface. If it does not lie in the flat plane, it indicates that there is curvature. If we then want to know more, we perform runs 5 through 8 to get the rest of the story. These are referred to as the *axial* portion of the runs. The extrema or α values are usually chosen to be $(n_F)^{1/4}$, where n_F is the number of rows (4 in this case) that are in the full factorial section of the runs. In this case, the α would be 1.414 times the full factorial values. This broadens the range of the run values from what would have been used in the full factorial section. Since we anticipated the need or interest to go through the whole CCD experiment, we set the α values to the maxima and minima of the variables and use only 0.707 of these values for the factorial section. This was to avoid operating beyond maximum and minimum experimental limits that we had already decided.

We have chosen pressure limits of 1×10^{-5} torr to 10×10^{-5} torr. This is best expressed in this case on a log scale of 4.0 to 5.0 *minus log of pressure in torr*. This gives the following run values: "α" = 5.0 − log torr, "+" = 4.854 − log torr, "0" = 4.5 − log torr, "− " = 4.146 − log torr, and "-α" = 4.0 − log torr. Since the pressure in the chamber will be continually decreasing, we will need to time the runs such that the evaporation starts when the indicated pressure is reached.

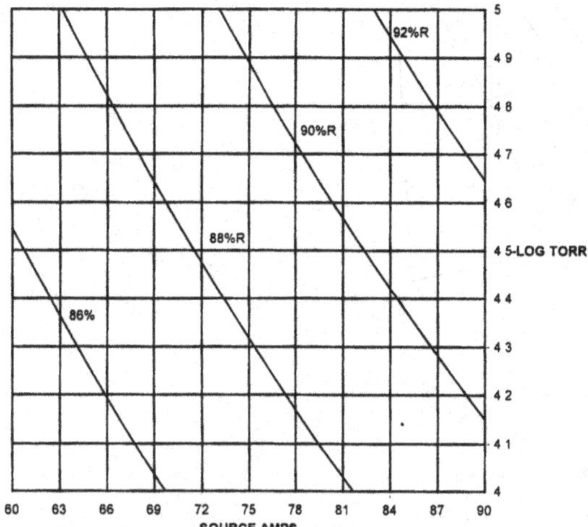

Fig. 6.7. A contour plot or top view of the surface shown in Fig. 6.6 for the four (4) run, full factorial experiment.

In the case of the evaporation current, we have chosen 60 to 90 amps as the range. This would give the following run values for the filament current: "α" = 90, "+" = 85.6, "0" = 75, "− " = 64.4, and "-α" = 60 amps.

Let us now imagine that we ran the 11 runs of this CCD design and obtained the resultant "percent reflectance" values shown in column Y1 of Table 6.1 for each run. These can be entered into a statistical data reduction program such as DOE KISS[9] to give all of the available information which can be extracted from these runs and produce figures like those shown here. Figure 6.7 is a contour plot or top view of the surface shown in Fig. 6.6 for the four run, full factorial experiment. The light area in the upper right is where the reflectance is predicted to be above 90%.

Figure 6.8 shows the surface plot when the full results of the CCD experiment in Table 6.1 is used. The data obviously point to quadratic behavior of the process. It can be seen from Fig. 6.8 and its contour plot in Fig. 6.9 that the region of reflectance above 90% is predicted to be of somewhat different shape and greater than that shown in the four run case. It would imply that, for highest reflectance, the best target settings for amps and pressure would be 85 amps and 4.75 − log torr or 1.78×10^{-5} torr. From a practical point of view, we might set the process start pressure requirement at "equal to or better than 2×10^{-5} torr" and 85 amps for the current. We can then use Figs. 6.8 and 6.9 to decide on reasonable

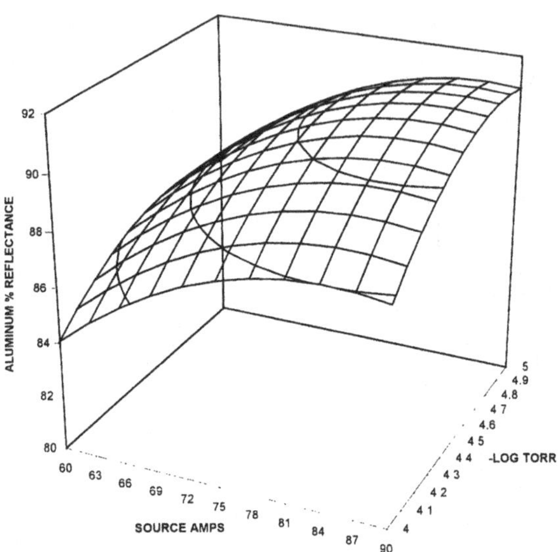

Fig. 6.8. The surface plot when the full results of the CCD experiment in Table 6.1 are used.

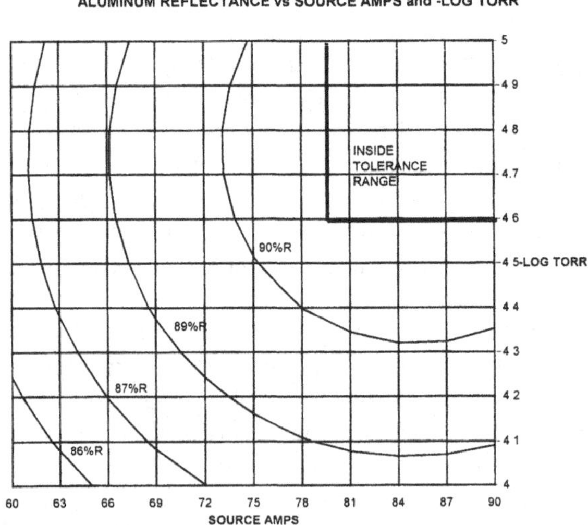

Fig. 6.9. A contour plot or top view of the surface shown in Fig. 6.8 showing the results of the full Box-Wilson CCD experiment.

tolerance bands for pressure and current to maintain reflectance above 90%. In this case we might choose $1\text{-}4\times10^{-5}$ torr (5-4.6 −LOG TORR) for a pressure tolerance band and 80 – 90 amps for evaporation current. However, please keep in mind that these numbers are fictitious and do not represent real data on aluminum evaporation; they are only to illustrate the DOE methodology.

6.3.2. A Box-Behnken Design for IAD Deposition of TiO_2

We will further illustrate the use of DOE with another fictitious example. Please note that this example is *not taken from real data*, it is only for illustration purposes. We will imagine that we want to optimize the deposition of titania (TiO_2) using ion-assisted deposition (IAD) with oxygen (O_2) and argon (Ar). The two major results desired are to have absorption (k) less than .001 and the spectral shift with humidity of less than 2 nm. There may be other results of secondary interest such as index of refraction and hardness, but adjusting for these could only be considered if the range of variables which satisfy the first two requirements leave some latitude to choose the best index and/or hardness results in that range.

We will further imagine that we have a conventional optical batch coater with heaters, electron beam gun (E-gun), ion gun, etc. What are the most important variables that could effect the desired results? From experience, we may know that some of these are: temperature, pressure, deposition rate, O_2/Ar mixture, ion current, ion voltage, etc. Our experience may allow us to bypass the screening experiments to determine which variables are important. The controls at our disposal are: deposition rate by E-gun power, ion gun current, temperature by heaters, and pressure and O_2/Ar mixture by mass flow controllers of the gasses to the ion gun. The flow is measured in Standard Cubic Centimeters per Minute (SCCM). Let us say that we know by experience that the temperature which would give the best results is the highest value that the equipment can provide. Therefore we will run the experiments at that temperature, eliminating it as a variable. Similarly, we know that the process runs most rapidly (and therefore economically) at the highest ion current that the source can provide. Again, we will run at that current and eliminate another variable. The process pressure and O_2/Ar mixture will be affected by the flow rates of both gasses, but the pressure will also be influenced by the deposition rate due to the gettering of the reactive TiO_2. As a result of all of the above considerations, we conclude that the three independent variables that we will optimize are: Rate of deposition (measured in Angstoms per second (A/S) by a crystal monitor), SCCM of Ar, and SCCM of O_2.

We next need to decide what would be a reasonable range over which to experiment with each of these variables. Due to our experience with the coating chamber and equipment to be used, we choose 20 to 60 SCCM for both gasses and 2 to 10 A/S for the range of the deposition rate.

We can now enter these choices in an appropriate DOE software program such as DOEKISS[9] to calculate the points to be sampled which satisfy the Box-Behnken Design (in this case). Figure 6.10 shows such a design sheet. We then need to choose which interactions of the variables will be included in the model to which the results will be fit. In this case, we will include all linear and quadratic interactions (as seen in Fig. 6.11). A, B, and C represent Rate, Ar, and O_2-SCCM. When the 15 experiments have been performed, we enter the data into the RESULTS column of Fig. 6.10 and then analyze the matrix with the software.

Figure 6.11 shows coefficients of the model (derived from the experiments) and all of the statistical detail such as the standard error of the least squares data fit. In the EXPER column, we can enter specific values for the three independent variables and use the software to compute the predicted value (Y) based on the model. We can see here that this point at the center of the parameter space sampled (6,40,40) does not meet our need for a k-value (x 1000) less than 1. There are many more details which can be gleaned from Fig. 6.11, but they are beyond the scope of this book.

The DOEKISS software facilitates the display of the results from the fit of the data to the model in a variety of graphs. Our first choice in a case such as this is usually to use a three dimensional plot of each type of result (such as "Humidity Shift" and "k × 1000") with respect to each of the independent variables taken two at a time. There would be three such plots per result if we were to view the "cube"

Column #	1	2	3		RESULTS
Row #	RATE-A/S	Ar-SCCM	O2-SCCM		k x 1000
1	2	20	40		1.25
2	2	60	40		0.75
3	10	20	40		2.5
4	10	60	40		2
5	2	40	20		4
6	2	40	60		0
7	10	40	20		9
8	10	40	60		0
9	6	20	20		7.5
10	6	20	60		0
11	6	60	20		5.5
12	6	60	60		0
13	6	40	40		1.5
14	6	40	40		1.3
15	6	40	40		1.6

Fig. 6.10. Design sheet for experimental points in Box-Behnken Design and k × 1000 results of those experiments.

FACTOR	COEF	P(2 TAIL)	TOL	LOW	HIGH	EXPER	ACTIVE
Constant	1.466667	0.002116					
RATE-A/S	0.9375	0.001747		2	10	6	X
Ar-SCCM	-0.375	0.05941		20	60	40	X
02-SCCM	-3.25	4.48E-06		20	60	40	X
AB	0	1					X
AC	-1.25	0.002272					X
BC	0.5	0.070593					X
AA	0.079167	0.74169					X
BB	0.079167	0.74169					X
CC	1.704167	0.000666					X
R Sq	0.991461						
Adj R Sq	0.976092						
Std Error	0.436559						
F	64.50843		PRED Y			1.4667	
Sig F	0.000123						

Fig. 6.11. Results of the multiple regression analysis of the data shown in Fig. 6.10.

from each of the three axes. Our general choice is to examine each of these at the plane containing the center point of the design. We see such a plot in Fig. 6.12 for the k-value as a function of O_2 flow rate and deposition rate. In Fig. 6.13 for the Humidity Shift as a function of both gas flow rates, it can be seen that the lowest shifts occur at the lowest gas flows (which result in the lowest chamber pressures). In Fig. 6.14 we see a similar plot for the k-value versus both gas flows. In this case, the lowest k-values are found at the highest gas flows. It can immediately be seen from Figs. 6.13 and 6.14 that the two results desired (of low humidity shift and low k-value) are in conflict, in that they "pull in opposite directions." The question to be resolved is whether there is a set of values within the range of variables where both results can be achieved.

We approach this question by using another graphic presentation, the Contour Plot. This is just a view from directly overhead of any of these plots, like a topographic map. In this case we choose to look down from the deposition rate axis to see the effects of the two gas flows on the k-value and the humidity shift. If there is an overlap between the regions of gas flow within some range of deposition rates that satisfy both of our objectives for these two results, we can solve the problem. Figure 6.15 is the contour plot of the humidity shift with gas flows at a Rate of 2 A/S. The white region meets the requirement of less than 2 nm shift. Figure 6.16 is the contour plot of the k-value with gas flows at a Rate of 2 A/S. The white region is where k meets the requirement to be less than .001. It can be seen that the two white areas of Figs. 6.15 and 6.16 only overlap at a very

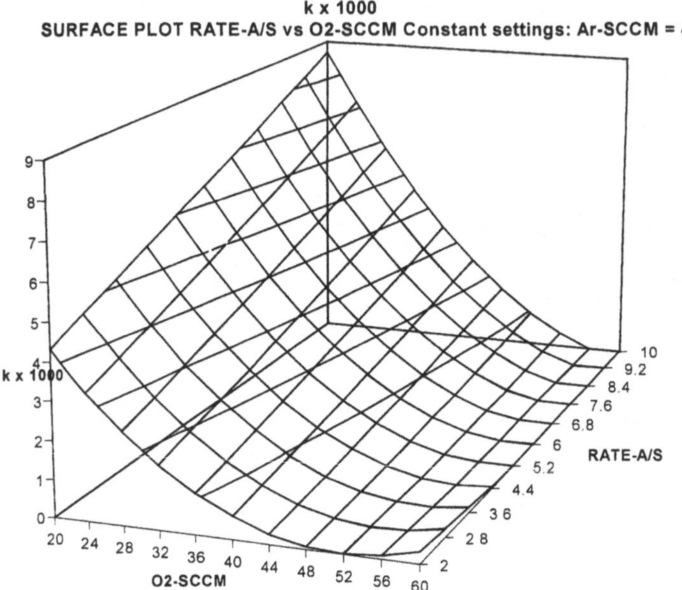

Fig. 6.12. Surface plot of k-values versus O_2 flow and deposition rate.

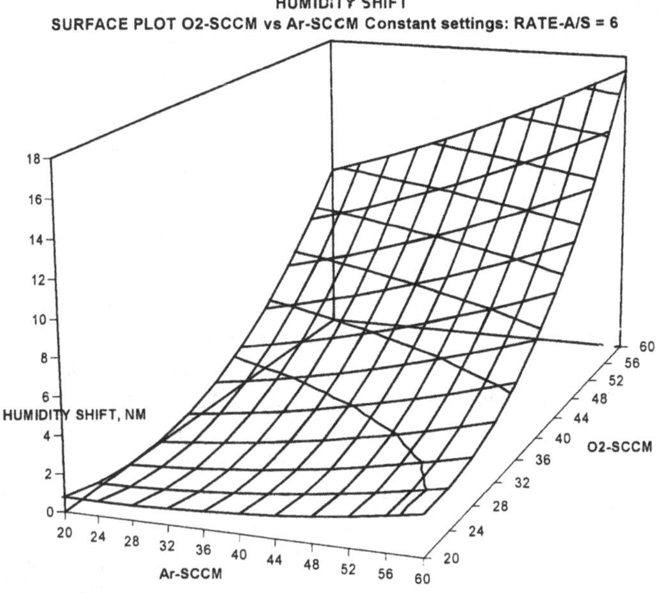

Fig. 6.13. Humidity shift versus both gas flows.

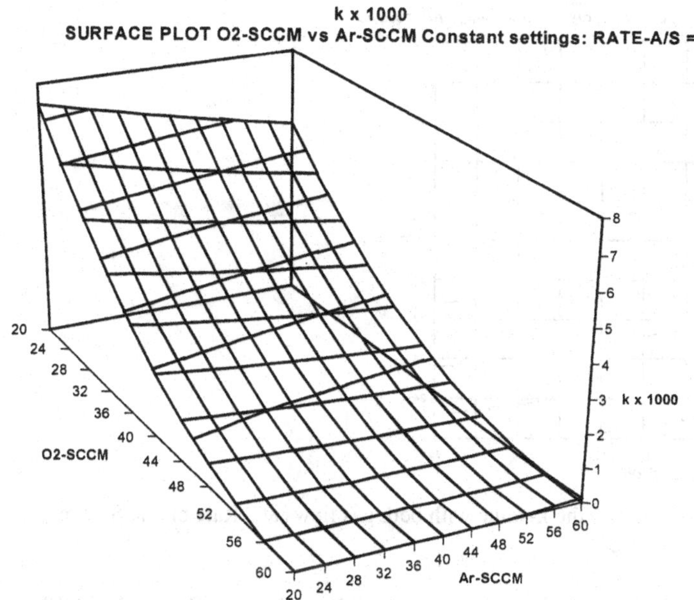

Fig. 6.14. K-values versus both gas flows.

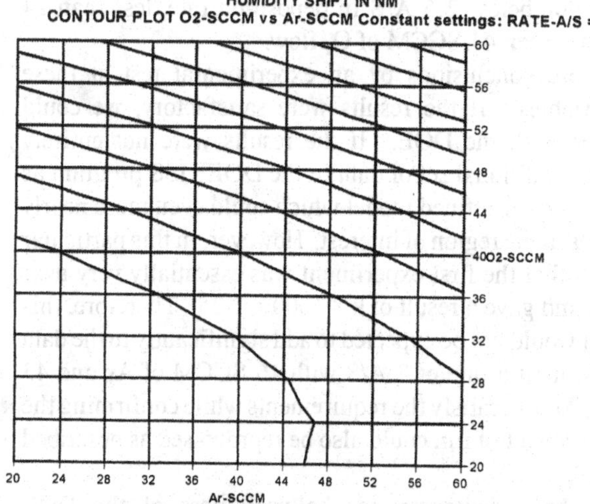

Fig. 6.15. A contour plot of the humidity shift with both gas flows at a Rate of 2 A/S, showing the acceptable region in white.

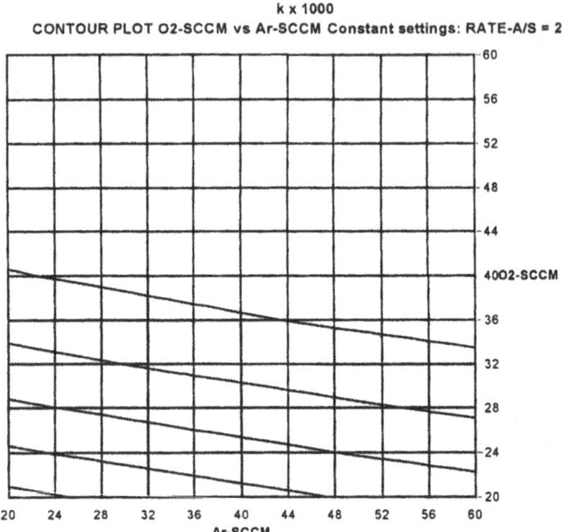

Fig. 6.16. A contour plot of the k-value with both gas flows at a Rate of 2 A/S, showing acceptable region in white.

small area in the vicinity of where the Ar flow is 20 and the O_2 flow is 41 SCCM. It was found from similar contour plots (not shown) of O_2 flow versus Rate at an Ar flow of 20 SCCM that there is no region of overlap at deposition rates greater than 2.5 A/S. We can conclude at this point that our result goals are only likely to be met in the variable region below 2.5 A/S in deposition rate, less than 24 SCCM of Ar flow, and greater then 41 SCCM of O_2 flow.

We might confirm these conclusions by an experimental test at these predicted values of the variables. If the results were satisfactory, we could consider that we were done with the DOE. If the results were not entirely satisfactory, we could use the additional set of data in the DOEKISS program as Historical Data to compute a further refined model which could even more nearly coincide with the experiments in the region of interest. However, in this particular case, we can see in Fig. 6.10 that the first experiment was essentially very near this variable point (2,20,40) and gave a result of $k \times 1000 = 1.25$. Therefore, this particular additional data set would not be expected to add significantly to the data base. It appears that a test run at a rate of 2 A/S with 16 SCCM of Ar and 43 SCCM of O_2 would be more likely to satisfy the requirements while confirming the predictions. The data from such a test run could also be reprocessed as described above.

In this example, we have demonstrated the usefulness of the DOE methodology in finding the characteristics of a process and the optimum parameters to achieve desired results.

6.4. SUMMARY

We have attempted here to give an introductory overview to DOE and its value in organizing and optimizing the process development task. There are an increasing number of papers appearing in which DOE methodology has been beneficially applied. The methodology is systematic and based on solid statistical concepts. The practical use of this tool or system is not dependent on deep understanding of the details of statistical mathematics. This like the fact that one can drive an automobile without being an experienced automotive engineer or mechanic. These tools are "user friendly". We have shown some of the available graphics which aid in process visualization, how these graphics might be used to gain insight, and how one might make process decisions. This methodology gleans the most information practical from a minimum number of tests or experiments. This has proven to be a great aid to efficient and successful process development. It is not expected that the reader would be fully grounded in DOE on the basis of what is presented here. However, by study of the references and/or (better yet) by participating in a short course by experts on the subject, the process developer should obtain significant benefit from the use of DOE.

6.5. REFERENCES

1. C. J. Hamel: "Big Foot--Little Foot: One is a Mystery; the Other Doesn't Have to Be!" *Proc. Soc. Vac. Coaters* **36**, 32-37 (1993).
2. S. R. Schmidt and R. G. Launsby, *Understanding Industrial Designed Experiments,* (Air Academy Press, Colorado Springs, 1994 Edition).
3. M. J. Kiemele and S. R. Schmidt, *Basic Statistics,* (Air Academy Press, Colorado Springs, 1993).
4. R. W. Knoll and J. A. Theil: "Effects of Process Parameters on PECVD Silicon-Oxide and Aluminum Oxide Barrier Films," *Proc. Soc. Vac. Coaters* **38**, (1995).
5. R. B. Heil: "Mechanical Properties of PECVD Silicon-Oxide Barrier Films on PET," *Proc. Soc. Vac. Coaters* **38**, (1995).
6. G. E. P. Box and N. R. Draper, *Empirical Model Building and Response Surfaces,* (John Wiley & Sons, New York, 1987).
7. G. Hass, W. R. Hunter, and R. Tousey: "Reflectance of Evaporated Aluminum in the Vacuum Ultraviolet," *J. Opt. Soc. Am.* **46**, 1009-1012 (1956).
8. G. Hass, W. R. Hunter, and R. Tousey: "Influence of Purity, Substrate Temperature, and Aging Conditions on the Extreme Ultraviolet Reflectance of Evaporated Aluminum," *J. Opt. Soc. Am.* **47**, 1070-1073 (1957).
9. ***DOEKISS*** Software, Digital Computations, Inc. and Air Academy Associates, LLC, 1155 Kelly Johnson Blvd., Suite 105, Colorado Springs, CO 80920, USA (1997).

7

Monitoring and Control of Thin Film Growth

7.1. INTRODUCTION

Let us put the monitoring and control requirements into a broad perspective. If we put down a QWOT stack of layers centered in the visible spectrum, it would require about 13,000 layers to make one millimeter of physical thickness! If we were trying to control each layer to about 1% of its nominal thickness, that would be about 1 millionth of a millimeter or one nanometer. The current thrust for "very small" detail in the WIDTH of semiconductor structures is now less than one micrometer. Therefore, we are working in the realm of two or three orders of magnitude smaller than "very small." The monitoring and control of what we are depositing to this type of tolerance is the subject of this chapter.

We will briefly take a long-range and somewhat philosophical view to define thin film monitoring for the purposes of this chapter. The purpose of thin film monitoring is to control the growth process such that the required properties of the thin film are achieved. Figure 7.1 is a simple flow chart of a typical thin film monitoring system. It consists of: defining the requirements, designing a plan of action to meet those requirements, the action of deposition of some thin film material, measurement of the properties of the deposited thin film, comparison of the results with the requirements, estimation of the next action needed, and feeding back this information to control the next action. The next action might be: to continue deposition, to stop because the goal has been achieved, or to stop because the goal is no longer achievable (failed process). In the case of most optical thin films, the requirements are in terms of reflectance and/or transmittance of a coated part at specific wavelengths. The action which attempts

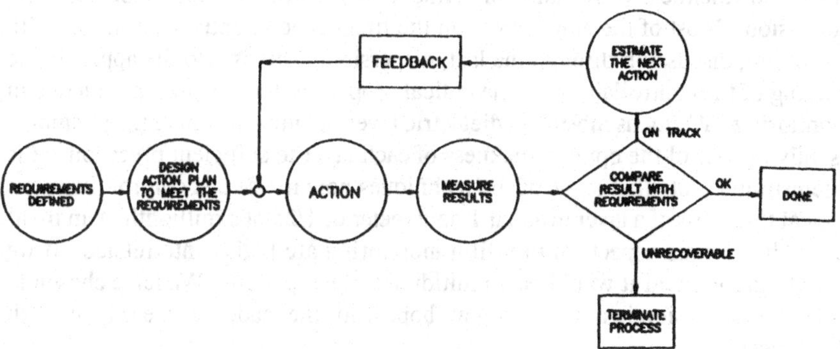

Fig. 7.1. Flow chart of major elements of a typical thin film monitoring system.

to meet these requirements is the application of some number of layers of material of appropriate index and absorptance with thicknesses that are designed to meet the requirements. It is necessary to be able to control these layer properties to within some tolerances determined by the requirements of the coating. In order to control the properties, it is necessary to measure the properties or something from which the properties can be derived. It is desirable to have some confidence at each step of the process that the requirements are still achievable and have not been made unattainable due to some uncompensatable error which has already occurred. Monitoring is a means to measure and know the status at a given point with sufficient accuracy to be able to exercise the necessary control (if possible) to get where we want to be. Thoeni[1] also has an interesting discussion of process control in optical thin film fabrication.

The purpose of this chapter is to discuss monitoring techniques which might be used to most efficiently allow requirements to be met. This ultimately relates to the yield of a given process and the fixed and operating costs of the equipment. We will not address these economic factors in detail here, but the economic motive underlies most of what we discuss. If it were not for economics, there could be a tendency to use the most precise and sophisticated process possible for even the simplest task. Another extreme is to use a process with inadequate control, and thereby have a yield which is so small or negligible that the product is prohibitively expensive.

We generally restrict the discussion of this chapter to the optical properties of optical thickness (nd), index of refraction(n), extinction coefficient (k), and dispersion. Most of the emphasis is on the first of these, optical thickness, with decreasing discussion through the last, dispersion. This is also the approximate ranking of the controllability of the optical properties and the present practice in monitoring. We focus mostly on dielectric layer monitoring where the attempt is usually to control the optical thickness of each layer to sufficient precision. It is interesting to note that some of the techniques now in use can control the mean optical thickness of a layer to about 1 nanometer or 1/25 of a millionth of an inch!

The various aspects of thin film monitoring are highly interrelated. If we were to graph them it would be a multidimensional picture. We have chosen to order our discussion in a way which we hope leads the reader most easily through the subject.

7.2. EFFECTS OF ERRORS

Let us look first at what would be the effect of less than perfect optical properties of layers on the end results of an optical thin film coating. As mentioned above, the major and most controllable factor is ordinarily the optical thickness of each layer. The next most critical is usually the index of refraction, over which we may or may not have some control in the process. Thoeni[1] demonstrated that, if one is concerned primarily with the results at a single wavelength as in a narrow bandpass filter (NBP), the optical thickness (nd) control is most important and index of refraction (n) is least important to control. Figures 7.2 and 7.3, after Thoeni, illustrate that the effects of errors in index of refraction are insignificant (when optical thickness is well controlled), but the effects of optical thickness errors are dramatic even if the index is constant. Thoeni further showed that, conversely, the index stability (if not controllability) is most critical to the broad band results on coatings such as beam-splitters. Figures 7.4 and 7.5 show that optical thickness control is much less important to the beamsplitter result than is the stability of the index of refraction. Thoeni concluded that NBP filters needed tight control of optical thickness and that partially transmitting mirrors, polarizers, etc. needed tight control of refractive index. He indicated that edge filters, broad band mirrors, and laser mirrors were in an intermediate condition where both n and nd have major effects. As we shall see in Section 7.4, it is possible to monitor optical thickness near the edge of an edge filter such that nd is all that is important for the result in the vicinity of the cutoff edge. Our conclusion is consistent with Thoeni's, but we would state it in this way: in the region of the monitoring wavelength, the control of the optical thickness (nd) at a specific monitoring wavelength has the major effect on the wavelength result,

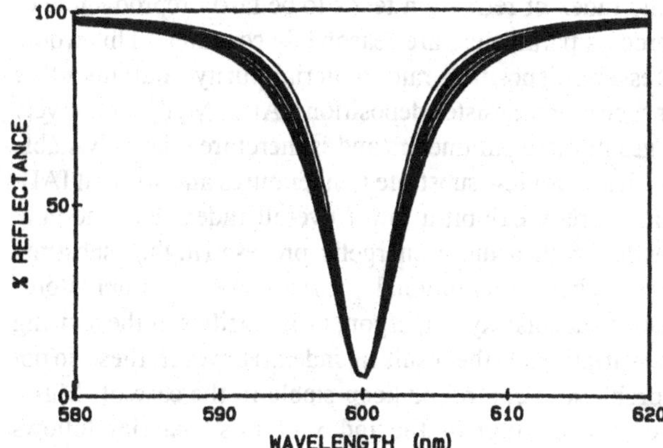

Fig. 7.2. Effects on a narrow bandpass filter of 1% RMS errors in index of refraction while there are no errors in optical thickness.

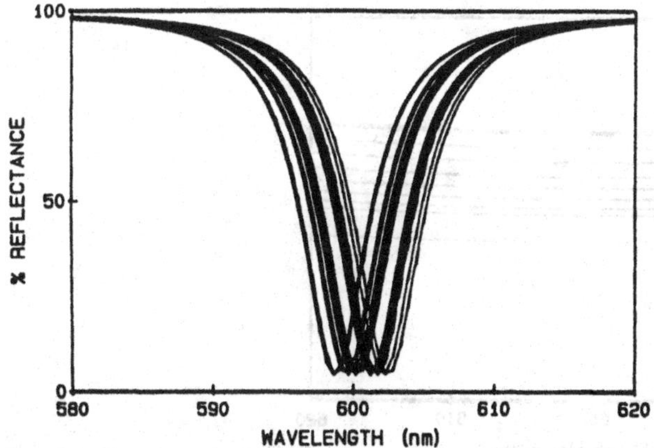

Fig. 7.3. Effects on a narrow bandpass filter of 1% RMS errors in optical thickness while there are no errors in index of refraction.

and the control (or stability) of the index of refraction (n) has a major effect on the photometric result in reflectance and/or transmittance.

The "control" of index of refraction deserves some discussion. With a material such as MgF_2, the index of refraction tends to be fairly reproducible or stable if the deposition process parameters are reasonably constant. This would include: temperature, pressure, deposition rate, material purity, and any other energizing process factor such as ion-assisted deposition (IAD). MgF_2 is relatively insensitive to small changes in these parameters and is therefore relatively stable in index. It is well known that, with low substrate temperatures and without IAD, the layers are porous and thereby exhibit a lower overall index than the bulk material or a film deposited with a more energetic process (higher substrate temperature and/or IAD). The unintentional variations of other atoms incorporated into the process such as oxygen, argon, or impurities in the starting material could cause some variations in the resulting index. However, these do not tend to be a significant problem to control or keep stable in the case of MgF_2. Therefore, the control or stabilization of the index of this material follows naturally by using reasonable precautions with purity, cleanliness, vacuum integrity, and temperature control. Because the physical/environmental properties of this, and other materials like it, are most satisfactory when in the most dense form practical, it is not usual to try to vary the index as a "controllable" parameter.

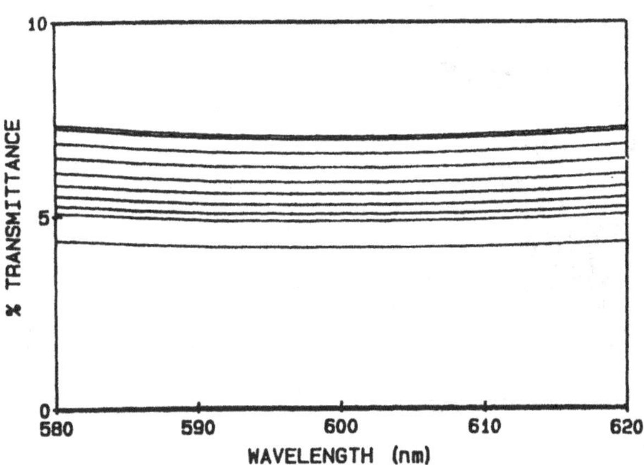

Fig. 7.4. Effects on a beamsplitter or partial reflector of 1% RMS errors in index of refraction while there are no errors in optical thickness.

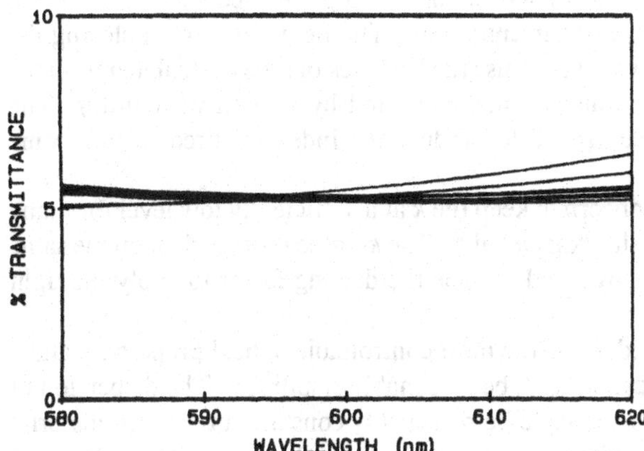

Fig. 7.5. Effects on a beamsplitter of 1% RMS errors in optical thickness while there are no errors in index of refraction.

The control of the index of materials such as titanium dioxide (TiO_2) is a different situation from that of MgF_2. This is the result of the sensitivity of this class of materials to the oxygen content or stoichiometry of the deposited material. This in turn is sensitive to the rate, temperature, oxygen partial pressure (O_2pp), and activation factors such as IAD and Reactive Ion Plating (RIP). There tend to be significant variations in both the refractive index (n) and extinction coefficient (k) which can be produced and are actually used in production[2-5]. The range of indices available with TiO_2 for which the films are environmentally robust seems to be greater than MgF_2. This implies that one would at least want a stable and reproducible process for these materials so that the index is known and under control. The interaction of rate and O_2pp have the most effect and sensitivity in non-IAD/RIP processes. The current practice in much of the industry seems to be to control the rate and O_2pp to constant levels independently. Usually the system is pumped to a low starting pressure, and then oxygen is admitted to the chamber to bring the total pressure up to the desired O_2pp. If the underlying chamber pressure without the oxygen varies due to leaks, outgassing, etc., the actual O_2pp will vary to make up the difference in total pressure. This can cause variations in the *n* and *k* of the resulting films. It is possible, but not yet common, with today's technology to control the O_2pp directly with a Residual Gas Analyzer (RGA) independent of the chamber's background pressure. It would also be possible to regulate the deposition rate to match the actual O_2pp or vice versa, but to the author's knowledge, this is not currently done. The next extreme in closing the control loop more tightly would be to use the feedback of index calculated from the actual reflectance and/or transmittance measured by an optical monitor. The rate/O_2pp could thus be controlled to produce the index required, within some modest range.

It is often of more concern to keep the *k* at a sufficiently low level for many applications of TiO_2 than the exact *n* value. The *k*-value is dependent on the same parameters as discussed above and may be the driving factor to apply the tight control loop mentioned.

As we have discussed, *nd* is the most controllable optical property, *n* and *k* are less controllable, but can at least be reasonably stabilized. The dispersion of optical thin film materials seems to be reasonably constant for a given material even if the index varies somewhat[5]. At present, the only way which we know to control dispersion is by the choice of materials, and not by any controllable deposition process.

We have discussed the effect of errors which are principally *nd, n,* and *k*. The *n* and *k* values are mostly controlled by maintaining a stable process by stable process parameters of pressure, rate, temperature, etc. The most common and important errors to control by monitoring are therefore the errors in optical thickness, *nd*.

7.3. WAYS TO MONITOR

As mentioned in the introduction, monitoring consists of definition, design, action, measurement, comparison, estimation, and feedback to control the ongoing actions. Figure 7.6 shows a typical physical vapor deposition (PVD) coating chamber with most of the sensors that could be used to control the thin films produced. Although we will concentrate on PVD by batch processes in "box" coaters or bell jars, much of the discussion is also applicable to sputtering, chemical vapor deposition (CVD), dip coating, and continuous (non-batch) processes. There are a broad range of ways in which the control the growth of optical thin films can be accomplished. We will address these from the simplest first to the progressively more sophisticated.

7.3.1. Measured Charge

An example of a very primitive but potentially economical monitoring system is the simple aluminization of a mirror. A measured amount of aluminum wire by length (or weight, if one adds more sophistication) is added to the evaporator.

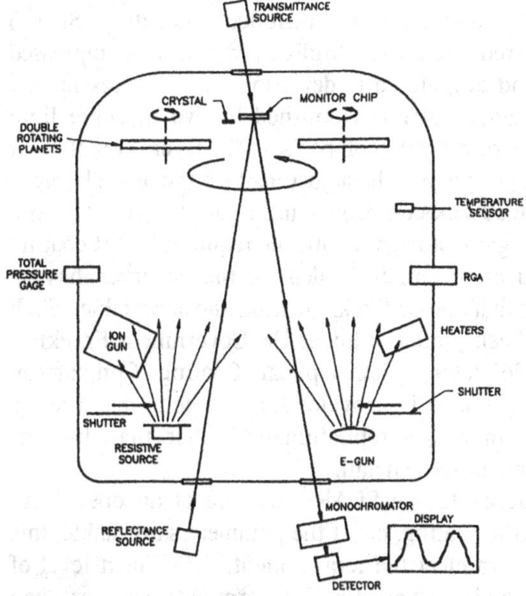

Fig. 7.6. Typical physical vapor deposition thin film coating chamber showing most of the sensors and components which might affect monitoring.

This charge is evaporated to depletion (the action). The mirror is held up to a light to see if it is opaque or nearly so (the measurement). If not (the comparison), the next evaporation will use a larger charge (the estimation) until the desired result is achieved (feedback-control). If the mirror layer were thicker than required, this might not be detected unless smaller charges were tried in subsequent evaporations. This example illustrates at least two things. First, the cost of the monitoring is minimal, but perhaps adequate. Second, the control occurs only after the process has either succeeded or failed to meet the requirement. This is a fairly open loop or loosely coupled process with a long time constant between the action and the correction of any error in the action.

One could enhance the above process by using the most common optical monitor, the eye of the operator. He can watch the filament of the source evaporating the aluminum through the substrate and bell jar or window in the chamber. As the aluminum is deposited, the mirror will become opaque. When it becomes opaque or nearly so, the process can be stopped. Because of the logarithmic response of the eye, when the operator estimates that the transmittance has dropped to 1% of its original uncoated level, it is probably less than 0.01% and the mirror has near maximal reflectance. This is usually necessary and sufficient for a good mirror and no excess aluminum has been deposited.

A similar process applied to a single layer antireflection coating (SLAR) might be as follows. A measured amount of MgF_2 by weight or compressed volume is loaded into a boat and evaporated to depletion onto a representative substrate (action). The reflectance color is examined by eye in white light (measurement). If it is yellowish or reddish (comparison) the layer is too thin, or bluish is too thick. The ideal is magenta. The adjustment for the next charge is made (estimation), and this (feedback-control) is used for the next test run (action). This loop is followed again and again until the required SLAR coating is achieved. A little sophistication can be added by the observer having comparison samples of coatings that are too thick, too thin, and acceptable. Such comparison sets have, in fact, been produced under US Government Frankfort Arsenal drawing number D7680600 ("Set, Optical Coating Comparison Standards") for just such use. The next improvement might be to measure the reflectance versus wavelength in a spectrophotometer rather than by eye. Macleod[3] discusses this technique in some detail.

Any of the above approaches to the SLAR are more of an open loop, cut-and-try, process to develop the coating, but if the parameters are stable, this may be a valid and simple way to meet the requirements. The next level of sophistication, which was common in past practice, is to terminate the layer when visual observation through a window in the vacuum chamber of the reflection of a white light from the part being coated shows the required color has been reached. This was done by turning off the source power or placing a shutter over the source at the appropriate moment. Mary Banning[51] gave one of the first descriptions of this visual optical monitoring in 1947. We will touch on this more in Section 7.3.4.2 under broadband optical monitors.

The measured charge means of controlling the optical thickness is seen to be simple and non-capital-intensive. The number of layers which can be done this way in a single run (pumpdown) is limited to the number of sources in the chamber. It might be practical to do a protected aluminum mirror with about a half wavelength optical thickness layer of SiO on top of the aluminum by measured charge. However, any more complex requirements than this would tend to favor using something other than the measured charge technique.

7.3.2. Time/Rate Monitoring

If a coating process can have a sufficiently constant rate of deposition of optical thickness, it is possible to control film growth and layer termination by time alone. This is the case in certain processes currently used such as sputtering and diamond-like-hard-carbon. The process development is similar to the measured charge technique, but it is the time that is measured in this case. Once the optical thickness as a function of time is well calibrated for each of the materials to be used, the thickness is controlled by deposition time. In this case, multilayers are possible up to any number for which the thickness errors of the layers are adequate. This is because the "charge" for this process is generally enough material for many layers and the layer termination is by time and not charge depletion.

7.3.3. Crystal Monitoring

The crystal monitor in common use on optical coating systems is a piezoelectric crystal that has a resonant frequency which changes as more mass condenses upon it from the growth of the thin film. This is sometimes referred to as a quartz microbalance and it has an electro-mechanical resonance which is highly sensitive and easily processed by the appropriate electronic circuitry. The readings typically have a precision of 0.1 nanometer. The crystal monitor is calibrated to optical thickness as in the measured charge technique by depositing a known crystal reading of the material to be calibrated by measuring the spectrophotometric result. The test layers are typically several quarterwaves of optical thickness (QWOTs) and the test pieces are measured by a spectrophotometer. Once the calibration factors are known for the materials to be used, the optical thicknesses of a layer system can be converted to the crystal thicknesses required. Many modern coating chambers are outfitted with automated process control systems using the crystal monitor as the layer thickness control system. This has proved to be quite practical for various types of production coatings such as SLARs, multilayer ARs, mirrors, some types of beamsplitters, and edge filters with modest requirements.

The crystal monitor is an indirect measure of the optical thickness. It is basically sensing mass change versus time. The mass is related to the amount of material deposited, but tells nothing about the density of the material, its index, or its optical thickness. Only by calibration can these be reasonably related to the crystal reading. The crystal monitor is quite precise, but not always very accurate. The change of the frequency of the crystal is not just a function of the new mass deposited, but also the previous mass deposited, temperature, stress, etc. The stability of the calibrations over a period of time tends to be a problem in cases where the tolerances on the layer thicknesses are tight. Another limitation is that the crystal will fail when the deposited material gets too thick or stressed, of if it delaminates. Still another problem is that the crystal cannot be where the workpiece is, and it is therefore an indirect measure in space of the required result on the workpiece.

There is a problem with monitoring a very thin layer by an optical monitor. A layer which is much less than a QWOT at the monitoring wavelength usually cannot be terminated optically with any accuracy because the photometric change generated by the layer may be too small to sense accurately. In cases such as this, the crystal (even with its limitations) can give a more accurate termination once it is calibrated to the optical thickness of a full QWOT or more. This is usually a consideration on layers thinner than 100 nanometers.

The crystal has various advantages also. It is precise, and we shall discuss ways to couple this advantage to the optical monitor in Section 1.3.3. Crystal monitors are well developed and proven, simple, compact, economical, and easy to operate. The feedback time constant is measured in milliseconds if not microseconds. One major advantage of the crystal monitor is for deposition rate control. Because it is precise and fast, the rate measurement can be used in "real time" to control the power to sources and thereby their rates. At the present time, the author believes that every sophisticated system should have a crystal monitor as the rate control sensor and for controlling very thin layers.

Our experience indicates that the random thickness errors of the crystal monitoring systems used in production are on the order of 4%. This is quite adequate for many optical coating applications. In fact, it is probably adequate for 90% of the volume optical coating production done in the world today. The remainder of this chapter will address the ways to monitor the other 10% of the coatings that cause 90% of the challenges in the industry today.

7.3.4. Optical Thickness Monitors

The monitoring of optical thickness of each layer in a thin film stack controls *nd* which is more closely related to the end performance of the layer system than the methods which we have discussed above. Figure 7.6 shows the elements of a typical optical monitor configuration. The workpiece or a monitor chip is mounted in the coating chamber and illuminated either by the reflectance source below or the transmittance source above. The reflected or transmitted light is passed through a NBP filter or a monochromator to select a specific monitoring wavelength. The selected light falls on a detector and the resulting signal is processed and displayed for the operator. Polster[52] was one of the first to describe this technique.

7.3.4.1. Single Wavelength Monitoring

The most common form of optical monitoring is the use of a single wavelength. Figure 7.7 shows what would be seen on a chart recorder trace of the reflectance of a crown glass substrate (n=1.52) as MgF_2 is deposited. The wavelength selected for monitoring a SLAR might be 510nm. The trace is nearly sinusoidal, and passes through the first turning point at A (one QWOT) where the reflectance is a minimum. This is the point to terminate the layer if a SLAR is required.

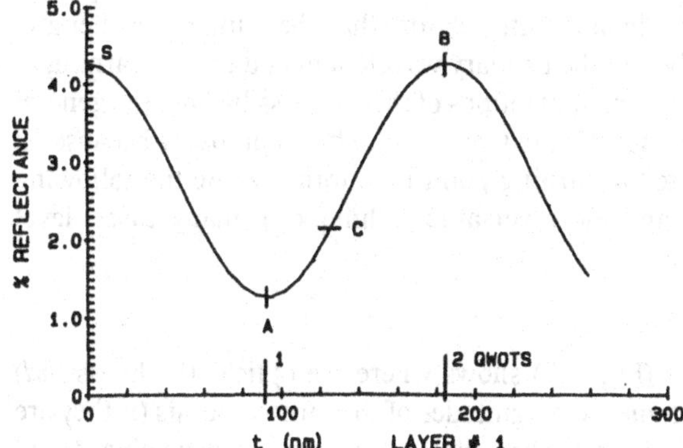

Fig. 7.7. Single wavelength monitor chart recorder trace of the reflectance versus thickness of magnesium fluoride being deposited on a crown glass substrate. Points A and B are the first and second turning points at one and two quarter waves of optical thickness. Point C is something more than one QWOT of thickness and at a predictable photometric level above the first turning point.

We will digress to point out to the neophyte that the common practice in layer deposition is to heat a source under a shutter which blocks the material from coating the substrate while stable evaporation conditions are being established. The shutter is then opened and the layer deposited on the substrate until the control terminates the layer deposition by closing the shutter. The source power may be shut off or kept at a lower hold or "soak" level to minimize heat up time for a subsequent layer.

Turning Point Monitoring

Referring again to Fig. 7.7, if the deposition were not stopped at point A but continued to the second turning point B, we would have deposited two QWOTs or one half wave. By terminating layers at any of these turning points, we can monitor a specific number of QWOTs at a specific wavelength. The turning points are simple and easy to recognize under ideal conditions, and this has been one of the most common techniques used in optical monitoring. Gibson et al.[6] also elaborated on 'turning point' monitoring. As we will discuss in Sect. 7.4 and 7.6, the effects of noise in the signal and other factors contribute errors to turning point monitoring that may not be acceptable in some cases. One approach to overcoming this problem is to monitor at two wavelengths simultaneously. If a second wavelength is chosen which is a little shorter than the primary wavelength, it will pass a turning point before the primary wavelength and give a warning of the imminent event. Various implementations of this are possible, but the general two-wavelength technique is not used extensively. This is probably because its usefulness is primarily limited to turning point monitoring while the following section describes an easier and more versatile technique in many cases, level monitoring.

Level Monitoring

The shape of the chart trace (Fig. 7.7) shows where the optical thickness (*nd*) QWOT points are. The photometric magnitudes of the turning points (if they are accurately known) can be used to find the results of the actual index values (*n*) of the layer and the effects of previous layers. An alternative is to chose a monitoring wavelength where the required cut or termination point of a layer is past a turning point such as point C. This new cut point is most often determined by its photometric level which can be precalculated from the design data and the assumed index of the layer. If the index of refraction of the layer and the photometrics of the monitor are well known and stable, the layer can be terminated at the calculated photometric level. However, at the present, it is rare that either of these conditions is sufficiently true, and even more rare that both are true. A more common and reasonable way around this problem is to precalculate the percentage change in reflectance from A to C as compared to the change from the start, S to A. Even if the actual signal change S to A is somewhat more or less than predicted, the percentage change A to C should remain nearly correct. We could say that S to A calibrates the scale for A to C and compensates for errors in both index and photometrics. The author uses this level monitoring technique extensively and with satisfactory results. In Sect. 7.6, we will discuss the sensitivity and accuracy advantages of the technique further.

Constant Level Monitoring

A thin film stack composed of identical pairs of high and low index layers is a common design for edge filters and selected line blocking filters. These structures lend themselves to a special case of monitoring which we call *constant level monitoring* (CLM) and has been described by Macleod[7] and Zhao[8]. At an appropriate wavelength for such a design, the optical monitor trace might look like Fig. 7.8. Here we have 1/2 QWOT first layer followed by QWOTs of high and low index layers. The first layer starts at S and terminates at a photometric level A. The second layer starts at A and terminates at the same photometric level when it reaches B. The third layer goes from B to C and stops at the same level. This continues through all of the pairs until the last 1/2-QWOT layer (typically) which

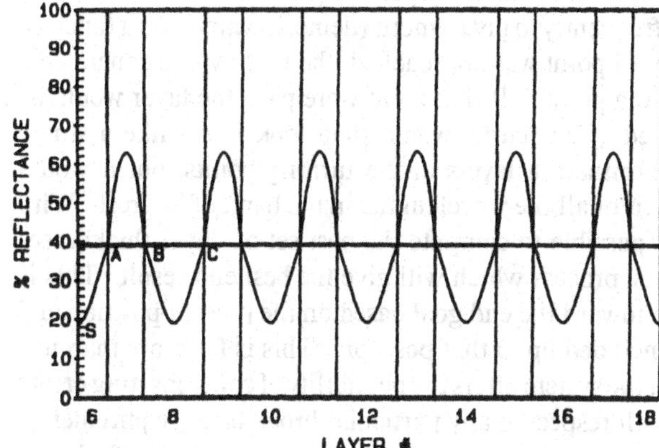

Fig. 7.8. Monitor trace of a stack of repetitive pairs of layers on a precoated monitor chip where all layers are terminated at the same photometric level (constant level monitoring) which is near the maximum sensitivity point for change in reflectance with thickness. Cut points for layers 6, 7, and 8 are at points A, B, and C respectively.

would terminate at the first turning point it reaches. The technique is very simple in principle and easy for the operator. It does depend on a reasonably stable photometric signal, but this requirement is true for all optical monitoring. This scheme has the advantages of level monitoring, described above, and error compensation and sensitivity as we will describe in Sects. 7.4 and 7.6.

7.3.4.2. Broad Band Optical Monitors

There have been several implementations of broad band optical monitors (BBOM) reported in the literature such as the work of Vidal et al.[9,10,14]. This has become technically practical with the advent of the high speed linear array where 1000 or more spectral elements can be sampled and processed at video rates. The application to practical broad band monitoring also depends on the use of high speed computers to process the data and control the depositions. Powell et al.[48] also describe their design and construction of a BBOM. Sullivan and Dobrowolski[49,50] provided a theoretical analysis and then experimental results using a BBOM to advantage.

Vidal et al.[9,10] typically compared the actual spectrum at a given point in the deposition with the expected/desired spectrum over the band. The difference between the present spectrum and the desired spectrum at the end of the layer (cut point) was computed very frequently to give a merit (demerit) value. As a function of film thickness when the cut point was approached, the merit value generated a curve approaching a "turning point." If that point were past, the layer would be getting thicker than desired. This curve would then look much like a single wavelength monitor for terminating layers at the turning points, but it would represent the weighted merit of all the wavelengths in the band of interest. With enough computation, it is possible to compute the best set of layer thicknesses from the present point in the process which will give the best end result. This is a continual reoptimization toward the end goal based on the present position and knowledge of the errors included up to that position. This is far more than just broad band monitoring, but the system offers the possibility of compensating errors in the most general way with respect to any particular broad band requirement.

It would appear that the greatest benefit of such a monitor would be for broad band coatings such as beamsplitters, color correction filters, and broadband AR coatings. It is hard to visualize how a BBOM would benefit the production of a DWDM filter which is the ultimate narrow-band task. Vidal and Pelletier[14] pointed out the potential to calculate the actual index, dispersion, and homogeneity of a layer. As we will discuss in Sect. 7.4, it is usually possible to design a single wavelength monitoring process which will automatically compensate for reasonable errors at and near the monitoring wavelength. In the case of the BBOM, since the monitoring wavelength is the whole band, it should be possible to extend this compensation to the broad band rather than just the small band around a single wavelength.

A possible interesting subset would be the broad band monitor where the single wavelength was selected electronically so that the monitor had no moving parts in the scanner. The broad spectral display would also serve as an *in situ* measure of the coating result before and after exposure to the external environment (air).

Note that the human eye is a broad band monitor and was the first optical monitor used for AR coatings.[51]

7.3.4.3. Ellipsometric Monitors

A detailed discussion of ellipsometry and its application to thin film monitoring is beyond the scope of this book. The field seems to be evolving rapidly in the research and development environment, and a few production applications. It certainly bears watching for its possible future benefit. Many laboratories now report the use of ellipsometers for determining optical properties of films. Some use variable angle spectrometric ellipsometry (VASE) as provided by Woollam.[35] A few have started using *in situ* ellipsometric monitoring of growing films.

As we already have discussed, the reflectance of a substrate or coating changes with polarization as a function of angle of incidence, and is only independent of polarization at normal incidence. Therefore, by measuring the polarization effects at non-normal incidence, it is possible to gain more information than with only the reflectance intensity. If this were further done over a broad spectral band, it should be possible to gain a great deal of information about the coating.

Ellipsometry has been available for about a century, but the understanding of its details had been mostly academic until recent decades. The theories and simplifying description and analysis techniques were evolving in the 1950s and since by notables such as Hans Mueller, R. Clark Jones, and others. The advent in recent times of inexpensive and fast computing power has made ellipsometry much faster and more practical (because of the computations required).

It is possible to know the n and k of a substrate and the films deposited on it plus the thickness of the films by using ellipsometry. For example, Martin and Netterfield[11] used it as a research tool to find the optimum ion densities to use in IAD to produce the maximum index of refraction of various materials. Sainty et al.[45] used *in situ* ellipsometry to study the evolution of rugate films. Various labs have applied the techniques to optical thin film monitoring, but to our knowledge it is not extensively used in the industry at this time. In a private communication with Bill Southwell and Bill Gunning of Rockwell International, they mentioned that it was a good tool for monitoring films which were too thin for standard monitors. This may be a good niche for the application of ellipsometry in our industry. The index values may be hard or even impossible to determine in very thin layers, however. The surface roughness of thick layers may also degrade the results of the measurement because ellipsometry depends on a smooth and non-scattering surface. The calibration of the system may also be tedious, and alignment and window birefringence need to be carefully handled.

Azzam et al.[40] analyzed reflection and transmission ellipsometry in some detail. Azzam et al.[42,43] also describe the details of a four detector photopolarimeter (FDP) with no moving parts. Masetti et al.[44] describe the use of the FDP as an *in situ* monitor. They mention that the main drawback was that it had to be precisely aligned but that appropriate procedures were developed to overcome the problem. For the interested reader, there are extensive reviews of the technology by Aspnes[32] and by Rivory[33]. Aspnes commented on the relative advantages and disadvantages of reflectometry and ellipsometry: "Ellipsometry is unquestionably the more powerful for a number of reasons. First, two parameters instead of one are independently determined in any single-measurement operation. Consequently, both real and imaginary parts of the complex dielectric function e of a homogeneous material can be obtained directly on a wavelength-by-wavelength basis without having to resort to multiple measurements or to

Kramers-Kronig analysis. Two independent parameters also place tighter constraints on models representing more complicated, e.g., laminar, microstructures. While two independent parameters R_s and R_p are also available in nonnormal-incidence reflectance measurements, these parameters can be obtained separately only after additional adjustments of system components. Second, ellipsometric measurements are relatively insensitive to intensity fluctuations of the source, temperature drifts of electronic components, and macroscopic roughness. Macroscopic roughness causes light loss by scattering the incident radiation out of the field of the instrument, which can be a serious problem in reflectometry but not in ellipsometry, for which absolute intensity measurements are not required. Third, accurate reflectometric measurements are difficult, in general requiring double-beam methods. In contrast, ellipsometry is intrinsically a double-beam method in which one polarization component serves as amplitude and phase reference for the other. Finally, ρ explicitly contains phase information that makes ellipsometry generally more sensitive to surface conditions. Insensitivity to surface conditions is often considered to be an advantage of reflectometry, but this is not correct if the objective is to obtain accurate values of the intrinsic dielectric responses of bulk materials. Because small differences in R can result in large differences in ϵ, reflectances in general must be measured more accurately. Surface artifacts affect ellipsometric results at the measurement level at which they can be identified and often corrected on the spot, whereas they affect reflectometric results at the data-reduction level at which it may be too late to do anything about them."

Vedam and Kim[41] gave an example of the use of ellipsometry to determine index as a function of depth and dispersion. Schubert et al.[46] used spectroscopic ellipsometry to perform detailed investigations of phase and microstructure of mixed-phase samples such as BN thin films.

Examples of the application of ellipsometry as thin film monitors (*in situ*) can be found in Netterfield et al.[34,36], Struempfel et al.[37], Vergöhl et al.[38,39], and Heitz et al.[47].

7.3.5. Trade-offs in Monitoring

If one is making a business or part of a business out of producing an optical coating, the life cycle cost of a given product will depend on the monitoring scheme chosen. This in turn would usually lead to which method to choose. If one has no existing equipment or process available, then it is necessary to consider the initial or capital cost of the equipment needed, the cost to develop the process for production, the direct production costs over the expected life of the product, and the overhead costs associated with the process.

The aluminized mirror example given in Sect. 7.3.1 is one of the simplest cases. A bell jar system with one resistive source might be adequate as capital equipment. The monitor would be the operator's eye looking at the opacity of the aluminum through the substrate at the glowing filament as it is deposited. The development of the process might consist of a few test runs to see how much aluminum by length to put in as a measured charge. Direct production costs would include the aluminum, the power to pump the vacuum chamber and evaporate the material, the operator labor, and substrate cleaning labor. The overhead cost would include rent for the space, cost of capital, maintenance, etc. This is about as simple as an optical coating process can be.

The SLAR process would add the capital cost of substrate heaters for a more durable coating. A light source (preferably a white fluorescent) would also need to be positioned for the operator to view the reflection in the rear of a substrate as the optical monitor. A few test runs might be required for the operator to find power settings for the resistive source which would give a workable rate which allows him to cut when the right color is reached. The production cost would be increased by the power for the heaters and the extra pumping time at higher temperature to get to the required pressure. The overhead costs would be greater in proportion to the greater process time.

As we go on to the multilayer AR, such as a four layer coating typical in the industry, the costs start to grow. An electron beam gun is usually required, and one might even have two. The eyeball is no longer as practical as a monitor, and at least a crystal monitor is usually employed. A spectrophotometer and some coating design software are needed to measure, develop, and adjust the results. The process time may not be a lot longer than the SLAR except by the time needed to deposit three more layers. The direct production costs are not dramatically greater than the SLAR, but the overhead of the cost of the capital is significantly greater in this case.

Having crossed this new threshold of capital investment, the addition of an optical monitor with filters, or a monochromator for more sophistication, opens the door to more than 95% of the coatings produced for optical instruments today. The ability to test coating durability in such environmental conditions as temperature, humidity, and salt fog will add further to the capital or outside testing costs. More complex coatings can take dozens of test runs to develop. They can require many hours to deposit. The yield of complex processes may be relatively low because there are so many more things which can go wrong, so many chances for errors.

This brings one to the art/science of selecting a monitoring/control system which is adequate in accuracy and reliability to give the best ratio of yield to cost per part. In this case, the cost must include the amortization share of the capital cost of the monitor. Another way of putting the task is to minimize the total cost to produce the parts. This is not easy to predict exactly, even when "infinitely" large numbers of identical parts are to be processed, but it is possible to come close enough to make a viable business decision on a "scientific" basis. However, with a wide mix of processes and parts to be coated in a given chamber, the decisions are more based on art than science. The chamber and equipment must be adequate for the most demanding process to be done with it, and lesser processes will not benefit as much, but will bear a slightly higher-than-necessary cost because of it. In today's instrumental optics coating industry, the author's choice is an optical monitor with a monochromator and a crystal monitor used for both rate control and thin (or special) layer termination. However, if large quantities of SLARs were to be coated, it might be worthwhile to look carefully at the total life-cycle-costs of using very simple chambers or bell jars and eyeball monitoring rather than tying up the expensive capital of a fully equipped box coater.

7.4. ERROR COMPENSATION AND DEGREE OF CONTROL

We have spoken above of the various ways to monitor and control the growth of optical thin films. We will confine our attention hereafter to optical and crystal monitoring because they are most common in the precision optical coating industry. Both crystal and optical monitors will have some errors in terminating layers accurately to the optical thicknesses desired. Our concern is to minimize any undesirable effects of these errors as much as possible. The application of certain simple principles to the monitoring scheme can make it possible for some of the errors to be compensated for by adjusting the thicknesses of later layers. This section will deal with the subject of the possibilities and limitations of error compensation.

Since the crystal monitor does not sense optical thickness directly, it is not usually possible for a later layer to compensate for errors in an earlier layer if only a crystal monitor is used. As discussed in Sect. 7.3.3, we find the crystal errors to be on the order of 4%. If a given design will have an unacceptable yield with this magnitude of random errors, it will be necessary to use something with smaller errors and/or some way to compensate for the errors. This generally implies an optical monitor in these cases. We will concentrate on the error compensation possible with optical monitoring now and add discussion about the combination of optical and crystal monitoring in Sect. 7.5.3. We now address an area which may apply to only 10% of the production work done today, but which may represent 90% of the effort expended in the development of new and complex coatings.

7.4.1. Narrow Bandpass Filter Monitoring

Narrow bandpass (NBP) interference filters have been made successfully for decades with amazingly simple equipment and monitoring systems. As one examines the possible effects shown in Fig. 7.9 of random errors of only 4% of a QWOT RMS on a single cavity NBP, it seems at first amazing that these can even be produced with today's technology and equipment. The mystery fades however as one learns that there is a compensation for the errors occurring in the optical monitoring scheme that has been used from the earliest days of NBP fabrication.

Figure 7.10 shows the typical optical monitor trace for the deposition of a single cavity NBP. This represents monitoring at the central wavelength of the required band by the light transmitted through the substrate being coated. The design is HLHLHL HH LHLHLH, where H is a QWOT of index 2.35 and L is 1.46 on a 1.52 substrate. As the first layer (H) is deposited, the transmittance drops to point A. The next QWOT of L raises the transmittance to point B, etc. This continues to the symmetric center point of the thin film stack in the HH spacer layer for this Fabry-Perot NBP filter. The transmittance then increases symmetrically as the later layers are deposited. The last layer restores the transmittance of the ideal filter to the starting value which the substrate had (at the monitoring wavelength).

Let us imagine that the first layer was terminated in error at A' instead of at A. The second layer would still be cut as close to the turning point B as possible. This means that the second layer will be shorter than the design by just enough to

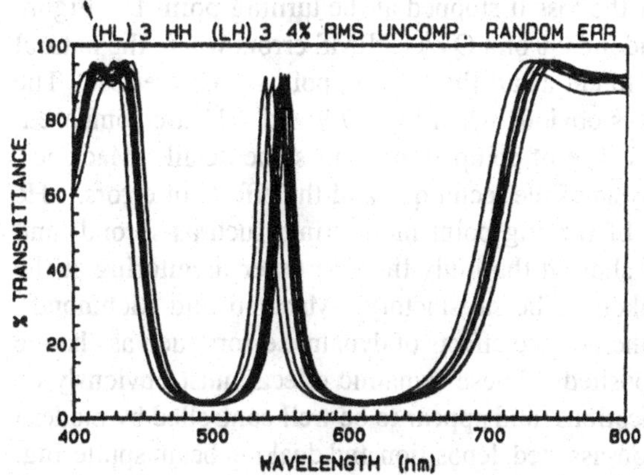

Fig. 7.9. Effects on a NBP filter of 4% of a QWOT RMS errors where there is no error compensation.

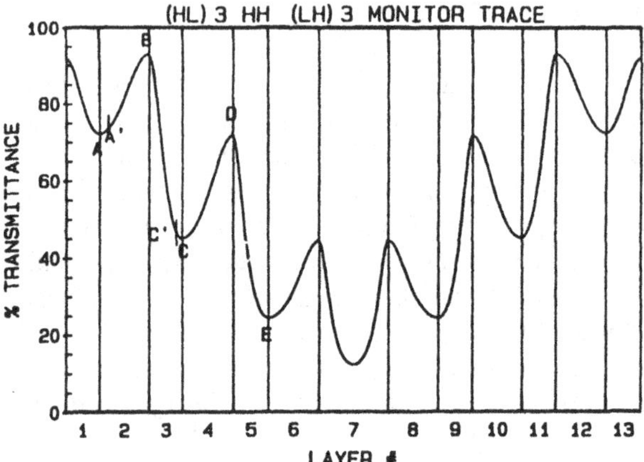

Fig. 7.10. Monitor trace for the deposition of a 13 layer NBP filter. The ideal layer terminations are at A, B, C, D, E, etc., while errors are shown at A' and C' which are partially compensated by terminating the next layer at the next turning point.

make up for the error in the first layer and produce two full QWOTs for the pair of layers. Similarly, if the third layer stopped short at C' instead of C, the fourth layer would just make up for the loss if stopped at the turning point D. Figure 7.11 shows the effects of random 4% of a QWOT RMS errors where the natural compensation of attempting to cut at all the turning points was in effect. The benefit of this compensation is obvious when Figs. 7.9 and 7.11 are compared. Bousquet et al.[12] discuss this type of compensation in some detail. Macleod[54] performed an extensive analysis of the technique and the effects of errors. He examined several variations of turning point monitoring, such as second- and third-order monitoring, and showed that only the first order monitoring at the passband wavelength was likely to be satisfactory. Macleod and Richmond[55] carried this work further to include the effects of dynamic errors such as change of index after a layer is deposited. These dynamic effects must obviously be minimized for DWDM applications, and appear to be well controlled by modern energetic processes such as ion-assisted deposition and dual ion beam sputtering. Regalando and Garcia-Llamas[56] also provided a general mathematical method for finding the relative stabilities of various types of coatings.

The keys to the effectiveness of this compensation are that the monitoring wavelength is right where we are most interested in the performance, the monitor is looking through the actual substrate to be coated, and the monitor sees all of the effects of all of the layers deposited from start to finish. If any of these elements are missing, this degree of error compensation cannot be expected. We will address each of these factors in more detail.

It can be seen in Fig. 7.11 that the performance of the error-compensated NBP away from the monitored wavelength departs somewhat from the unaberrated design. In a typical NBP design, the side bands where the filter starts to transmit again are blocked by edge filters on the long- and short-wavelength sides. These edges will have to be somewhat closer to the central peak of the NBP than indicated by the ideal design in order to allow for the effects of the errors, even though they are well compensated at the central wavelength. The phenomenon which we observe here will also be seen in the next subsection on compensation in edge filter monitoring. We could describe it as the same as holding an active garden snake in the middle. The position where we hold the snake is well controlled (error compensated), but the ends can move about extensively (due to errors which have been compensated at the holding [monitoring] point). Therefore, with single wavelength monitoring, it is usually possible to design the monitoring scheme to control well at the monitoring wavelength, but as one gets

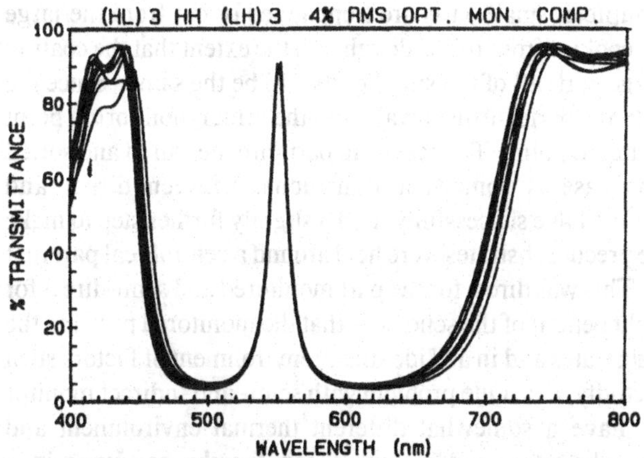

Fig. 7.11. Effects on a NBP filter of 4% of a QWOT errors where the compensation occurs by terminating each layer at the next turning point as closely as possible following an error.

further from the monitoring wavelength, the control and compensation tend to get worse and worse. This is why it is important to monitor as nearly as practical the wavelength which is most critical in the performance of the thin film. Zhao[8] pointed out these concepts with respect to edge filters which we will discuss in Sect. 7.4.3.

Let us digress briefly to define and discuss direct, semi-direct, and indirect optical monitoring. What we have described above is direct monitoring, where the transmittance of the actual workpiece is monitored. Semi-direct monitoring would be where the monitor is looking only at a monitor chip, while the deliverable parts are gyrating about the monitor in a calotte or planets. In this case, the error compensation would be good only to the extent that the ratio of the material deposited on the monitor chip to that on each area of each workpiece was identical for all the materials being used and during the entire deposition. This can only be approximated in common practice. The angular distribution of sources are not always the same. Aluminum droplets evaporating on a wire may be evaporating with almost the same intensity in all directions. However, magnesium fluoride melted in a round indentation in a "boat" may have a cosine fall-off of intensity with the angle from the surface normal to the pool of MgF_2. Materials subject to "tunnelling" in an electron beam gun, can emit a sometimes narrower, sometimes wider pattern of material as we discussed in Sec. 5.3.1.1. These factors mean that the monitor chip will receive a different ratio of material than the workpieces from material to material and from time to time, when semi-direct monitoring is used.

A common practice in the production of small filters is close to direct, but not quite. A large substrate is rotated about its center and monitored near that point. When the coating is complete, small parts are treppanned or cut from the large substrate like biscuits or cookies from rolled dough. To the extent that the coating was uniform over the large part, all of the small parts will be the same. Since the distribution is not perfect, the parts further away from the center monitoring point will differ from the monitored point. The resulting parts are measured and sorted for acceptability. This case is somewhat transitional between direct and semi-direct monitoring. We have successfully used a slightly further step to make a mini-calotte where the precut substrates were held around a central real part that was monitored directly. This was direct for the part monitored and semi-direct for the other parts. One slight benefit of the scheme is that the monitored part and the rest were all identical substrates and in an "identical" environment of factors such as thermal mass, conductivity, etc. One problem with semi- and indirect monitor chips is that they may have a somewhat different thermal environment and thermal mass than the substrates, and materials can thereby condense in a somewhat different thin film layer on these than on the other substrates.

The opportunity for any benefit of the "natural" error compensation described is lost by the use of *indirect optical monitoring*. Indirect monitoring is when only one or a few layers out of many are monitored on a chip which is then replaced by a new monitor chip. For example, a ten-layer coating could be monitored with each layer on one of ten separate chips, or two layers on each of five chips, etc. Any errors on one chip have no way to be sensed and compensated for on the next chip. As a result of this fact, the author has mostly gotten away from multi-chip monitoring where possible. We prefer to have all of the layers monitored on the same piece so that the maximum compensation for errors is possible, at least at the monitoring wavelength.

7.4.2. DWDM Filter Monitoring

With the increased interest in communications filters for DWDM, we have previously reported[57,58] on error sensitivity, monitoring, and compensation effects of such filters. That material is reviewed and summarized here. The three-cavity DWDM filter design shown in Fig. 1.26 would have 114 layers and an ideal monitor trace as shown in Fig. 7.12. The first cavity (37 layers) is shown in Fig. 7.13 and the most critical monitoring area near the spacer layer (number 19) is seen on an expanded scale in Fig. 7.14. It can be seen that the monitor signal has dropped two orders of magnitude from the starting layer to the spacer layer.

7.4.2.1. Signal to Noise in Monitoring

If the noise in the monitor signal were 1% peak-to-peak (p-p), the monitoring curve for the first layer would look like Fig. 7.15. Note, however, that this noisy monitor curve was terminated at 5% beyond the actual QWOT desired, but it would be very difficult for an operator to have sensed that overshoot. If one were to have this signal to noise ratio (SNR = 100 at the start) and reach the layers around 12 to 22, the noise would be as large as the signal and probably impossible for an operator to decide when to terminate a layer. If we were to improve the SNR by two orders of magnitude (SNR = 10,000), the monitor signal at layer 20 would look like Fig. 7.16 with a 5% overshoot. The situation is the worst at layer 20 where it would be difficult to decide from observation whether the 0% or 5% overshoot trace were the "perfect" QWOT. Therefore, as we will discuss below, the effective SNR must be quite high in order to expect to terminate layers within even a few percent of the desired QWOTs.

7.4.2.2. Special Layers in NBP Monitoring

The coupler layers, numbers 38 and 76 in this case, represent a special problem. Figure 7.17 shows the ideal monitoring curve for the first coupler layer (38) with

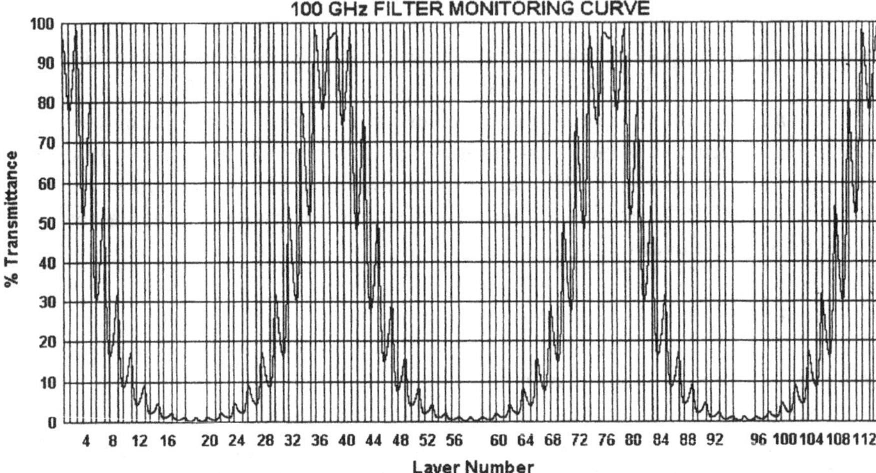

Fig. 7.12. Ideal monitor trace of the three-cavity DWDM filter shown in Fig. 1.26.

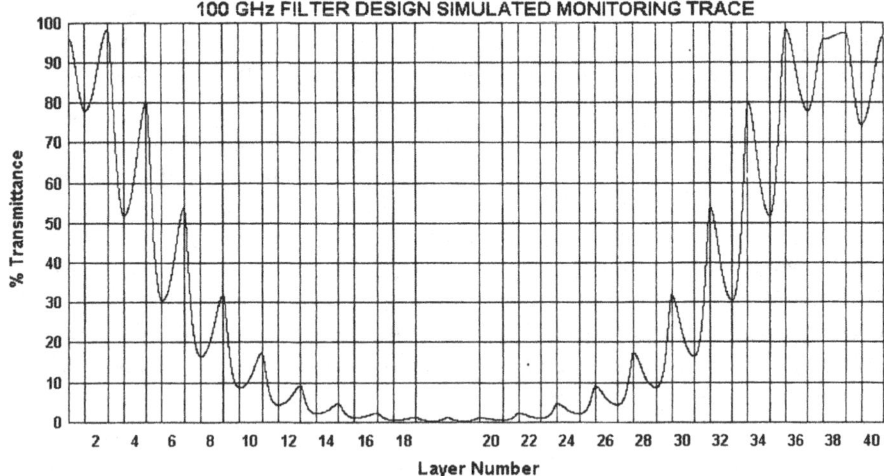

Fig. 7.13. Ideal monitor trace of the first cavity (37 layers) of the 114-layer three-cavity DWDM filter shown in Fig. 1.26.

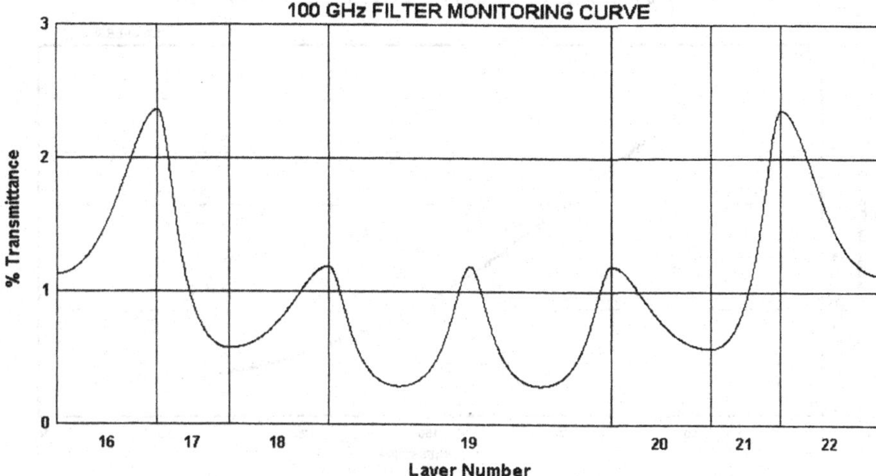

Fig. 7.14. Ideal monitor trace of the most critical monitoring area near the first spacer layer (19) of the DWDM filter shown in Fig. 1.26. Note 0-3% transmittance scale!

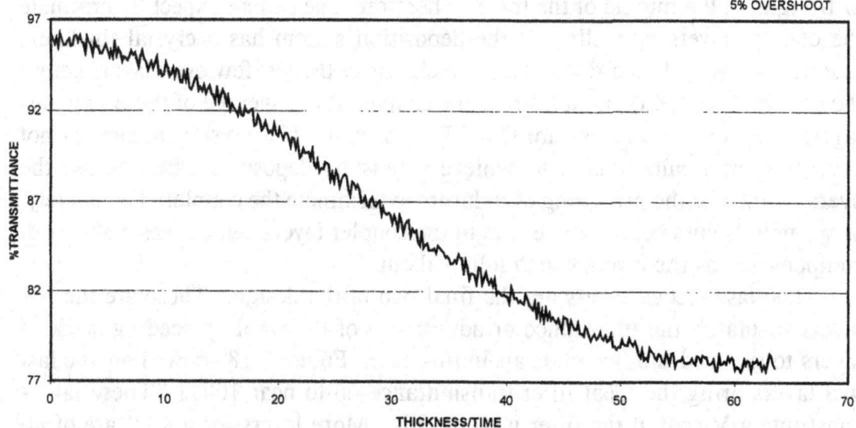

Fig. 7.15. Peak to peak noise of 1% in the monitoring signal of the first layer of the DWDM filter of the above figures.

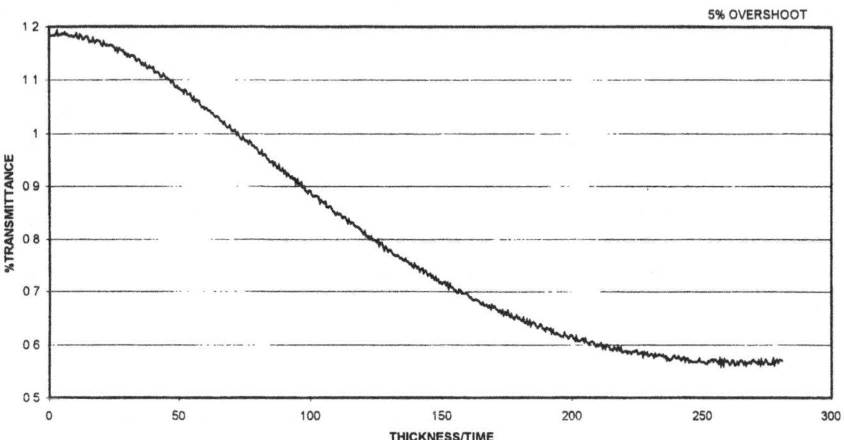

Fig. 7.16. Peak to peak noise of 0.01% in the monitoring signal of the most critical 20th layer of the DWDM filter of the above figures.

its adjacent layers and the second coupler (76) would appear symmetric to that. This would be difficult enough in this ideal form, but it turns out that any errors in the preceding layers would change the shape of this monitor trace. It can be shown that the build-up of errors in prior layers can cause the coupler monitoring trace to reverse or take on many other random looking shapes, including humps or troughs in the middle of the trace. Therefore, one cannot expect to terminate the coupler layers optically! If the deposition system has a crystal thickness monitor, we would record the crystal thickness of the last few even layers before the coupler (38, even) and terminate the coupler at the average of these previous layers. All of these layers are QWOT by design. If a crystal monitor is not available, we would attempt to achieve a constant deposition rate and use the average time for the preceding even layers to terminate the coupler. Fortunately, as we will discuss below, any errors in the coupler layers can be reasonably well compensated by the layers which follow them.

The last special layers are the final two of the design. These are the AR layers to match the impedance or admittance of the whole preceding stack of layers to the medium, which is air in this case. Figure 7.18 shows how the last two layers bring the ideal filter transmittance up to near 100%. These layers constitute a V-coat at the filter wavelength. More layers for an AR are of no advantage in this case. These two AR layers would be different if the filter was to be cemented to a glass surface.

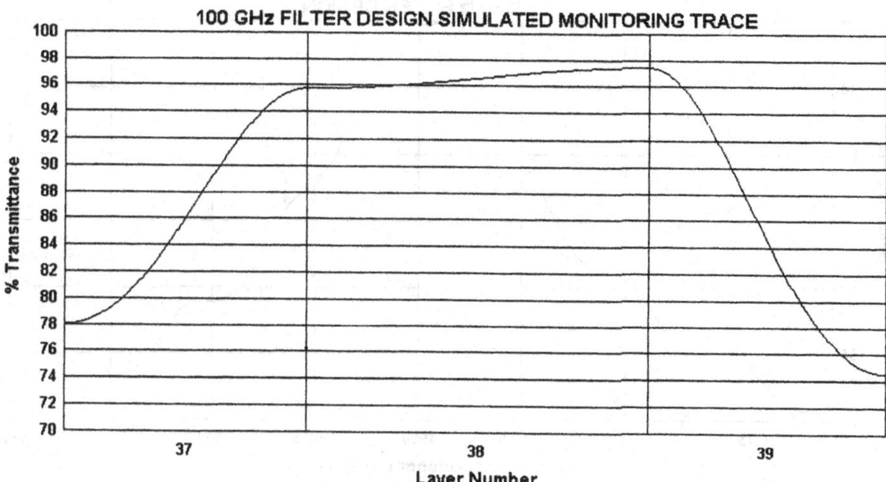

Fig. 7.17. Ideal monitoring curve for the first coupler layer (38) with its adjacent layers.

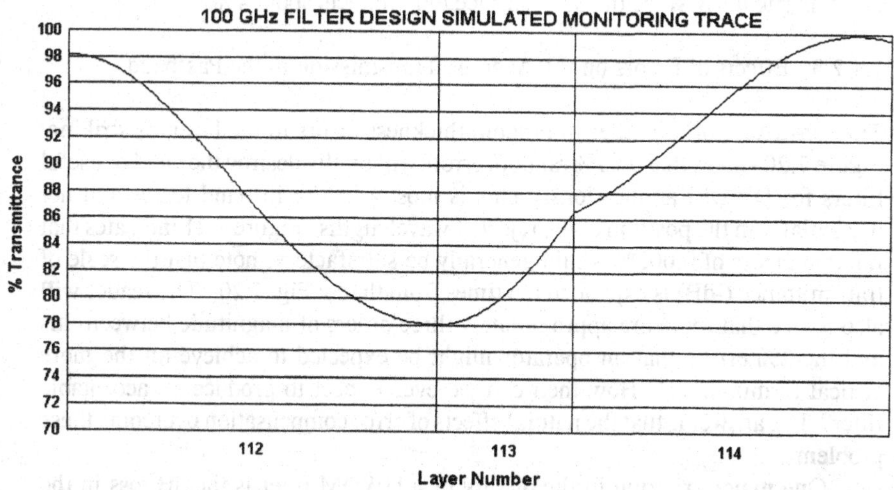

Fig. 7.18. Last layer of the periodic structure (112) plus the last two AR layers (113 and 114) of the DWDM filter of the above figures.

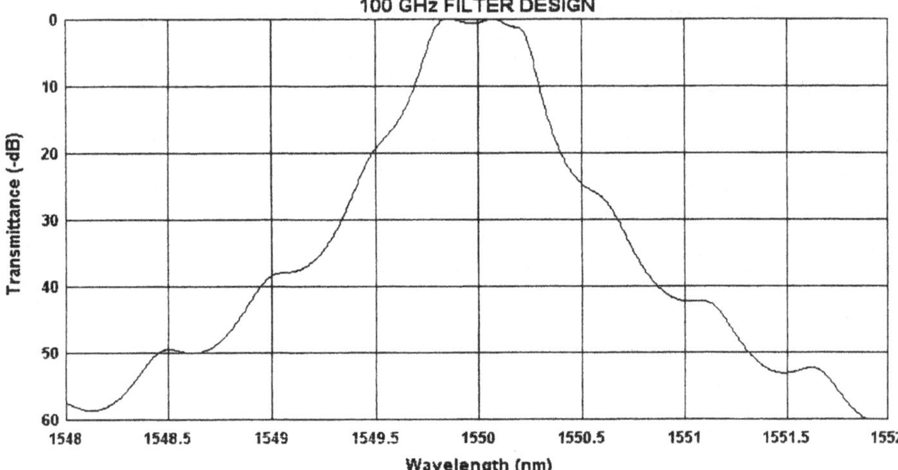

Fig. 7.19. Ripple effects which can occur if the back side of the substrate of the DWDM filter of the figures above does not have the proper AR coating.

Figure 7.19 shows the ripple effects which could occur if the filter were AR coated for air and then cemented instead (or vice versa, coated for cementing and left in air). The magnitude of this effect is at first counterintuitive because it is so large, particularly since the back surface only reflects about 4%!

7.4.2.3. Effects of Errors on the Average Transmission in the Passband

The sensitivity of this filter to random thickness errors in the layers is striking. Figure 7.20 shows that 0.01% random errors will totally destroy the yield of useful filters for DWDM applications. This is mostly in the in-band losses and not particularly in the position of the rejected wavelengths. Figure 7.21 indicates that random errors of 0.002% would generally be satisfactory; note that the scale of transmittance (-dB) is expanded ten times from that of Fig. 7.20. The reader will also notice that there are approximately three orders of magnitude between this and the 2% errors that an operator might be expected to achieve on the more critical terminations! How then can we ever expect to produce an acceptable filter? The answer is that the natural effects of error compensation overcome these problems.

One major criterion in the quality of a DWDM filter is the dB loss in the transmission (T) passband. The ripple is also specified, but we will concentrate now on the average transmission (T_{ave}) loss in the passband. We applied a 2% of

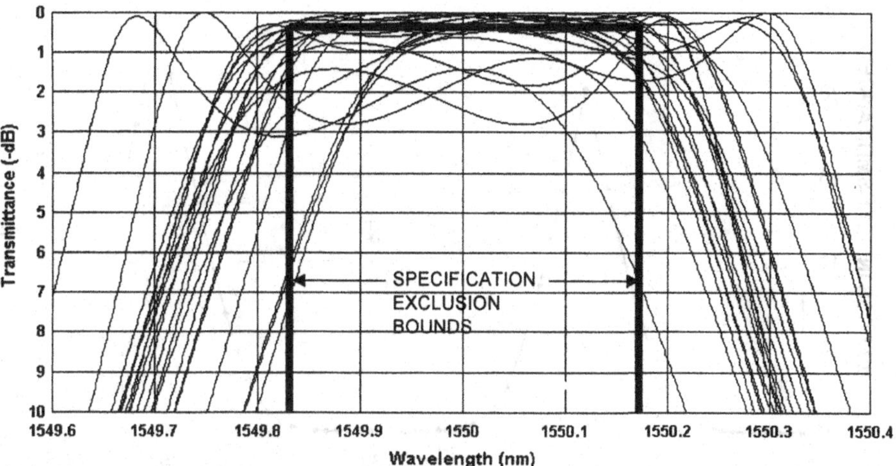

Fig. 7.20. The usefulness of a DWDM filter as in the figures above would be totally destroyed by 0.01% uncompensated random errors as illustrated here.

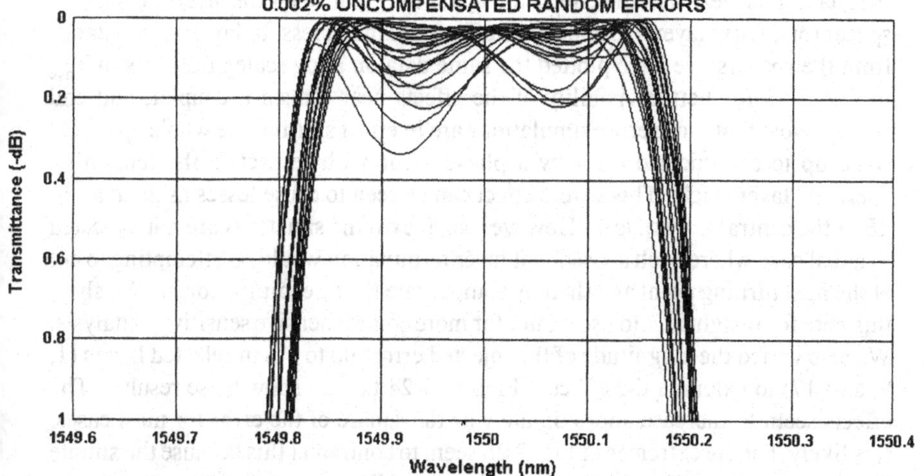

Fig. 7.21. The 0.002% uncompensated random errors as illustrated here might provide barely acceptable performance. Note transmittance scale difference from Fig. 7.20!

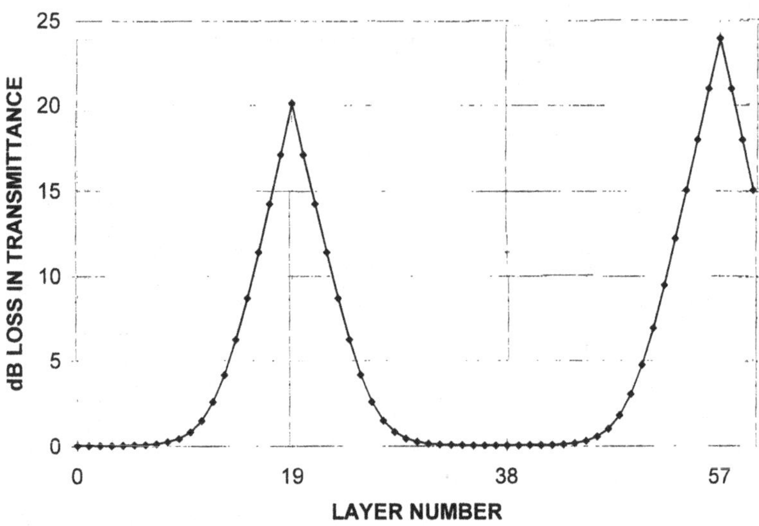

Fig. 7.22. Average transmittance loss in dB over the passband for 2% of a QWOT errors in each individual layer as a function of which layer.

a QWOT error to each layer in turn and computed the average T in the passband with this error. Figure 7.22 plots this dB loss for each layer up to a few more layers than the first half of this 3-avity filter. The second half is symmetric to the first half. It is clear that the effects are by far the greatest for the layers nearest the spacer (or cavity) layers. Because the effects are much less for layers at a distance from the spacers, we have plotted the same data on a log scale of dB loss in T_{ave} in Fig. 7.23 for better visibility of the effects near the start, coupler, and end layers. Note that these error simulations are in effect shifting the whole "perfect" stack up to the injected error by a phase error with respect to the remaining "perfect" layer stack. This severe effect can be seen to cause losses as great as 24 dB at the central (2nd) spacer. However, such extreme situations are not expected in a real case where each subsequent layer termination would be attempting to cut at the next turning point and thereby compensate the foregoing error(s). We show this here for insight and to use it later for more comprehensive sensitivity analysis. We also varied the magnitude of the injected errors up to 5% in selected layers (1, 9, and 17) to examine the effects. Figures 7.24 to 7.26 show these results. The effects seem to increase approximately as the square of the error for most cases. It is likely that the extremes of Fig. 7.26 seem to contradict this because the simple T_{ave} criterion is not the whole story in that case. We suspect that the total relative merit would in fact decrease approximately as the square of the error.

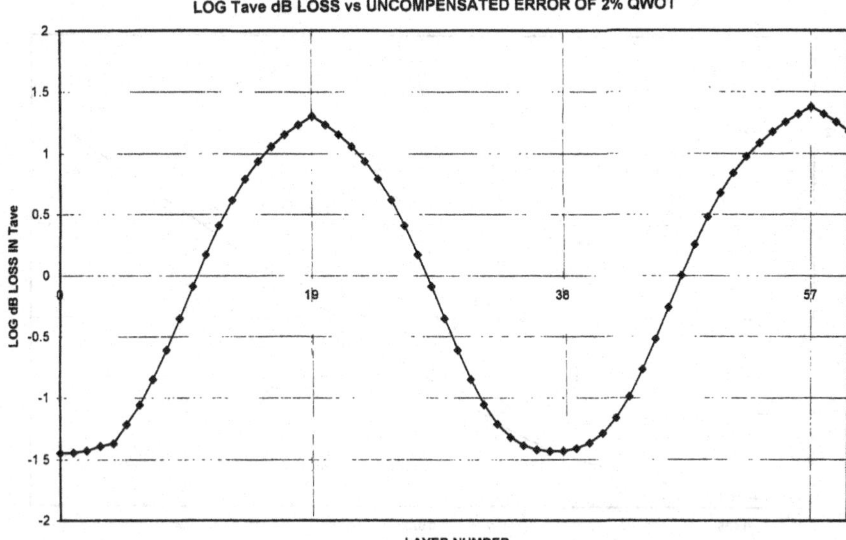

Fig. 7.23. Average transmittance loss in dB over the passband for 2% of a QWOT errors in each individual layer as a function of which layer, shown on a logarithmic scale.

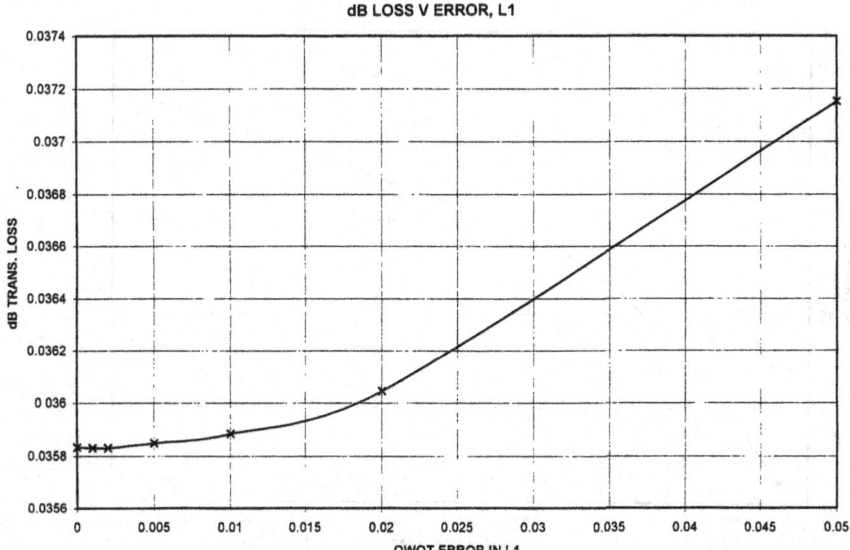

Fig. 7.24. Average dB transmittance loss in the passband versus up to 5% of a QWOT errors in layer #1.

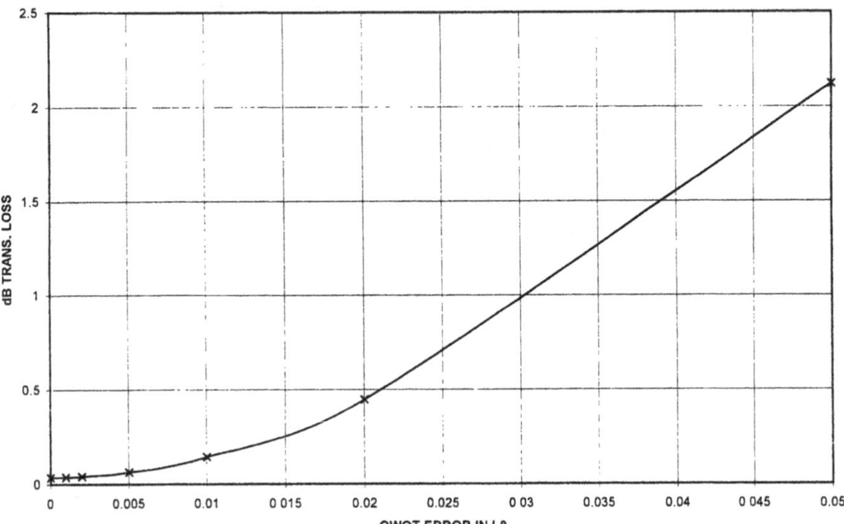

Fig. 7.25. Average dB transmittance loss in the passband versus up to 5% of a QWOT errors in layer #9.

dB LOSS V ERROR, L17

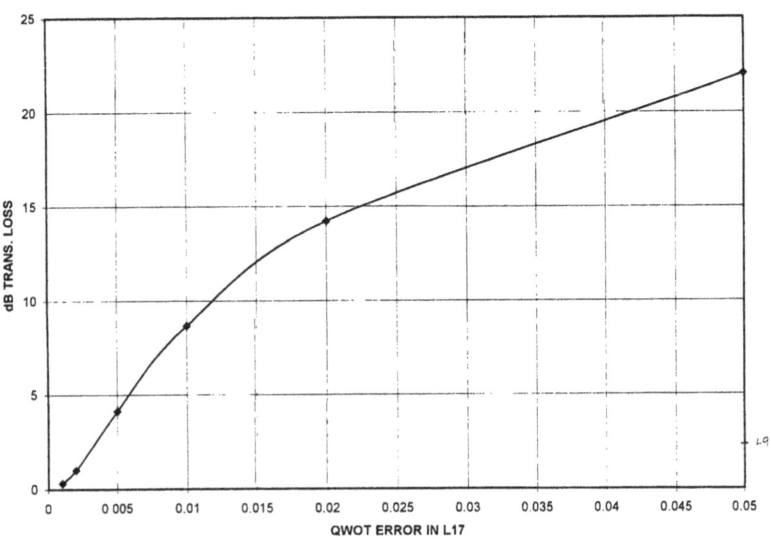

Fig. 7.26. Average dB transmittance loss in the passband versus up to 5% of a QWOT errors in layer #17.

7.4.2.4. Sensitivity of Turning Points in Monitoring

We devised a scheme to calculate the change in reflection with change in thickness of a layer at the turning points. Figure 7.27 illustrates the principle. This is the circle diagram view of layer 5, 6, and 7 as seen in the monitor view of Fig. 7.13. Each layer has a maximum and minimum reflectance as it crosses the real axis, the turning points. These define two points on the reflectance amplitude circle diagram which are on a diameter of the circular locus of that layer. The center of that diameter is the center of a circle for that layer which is not collocated with the center of the diagram defined by the intersection of the real and imaginary axes. An angle (shown by arrows in the figure) can be computed from these data which represents any given phase (QWOT) error from the turning point on the real axis. The change of reflectance from such an error point to the ideal turning point can be computed for these data. From this, we can calculate the expected error in ideal layer thickness as a function in the amount of reflectance change needed to sense a turning point. Figure 7.28 plots this turning point error sensitivity for the first 45 layers of this filter design. This can be seen to be intuitively consistent with Fig. 7.14 where the cavity layer (19) termination changes appear more dramatic and thereby more easy to detect than the next layer (20) termination, which has the worst sensitivity.

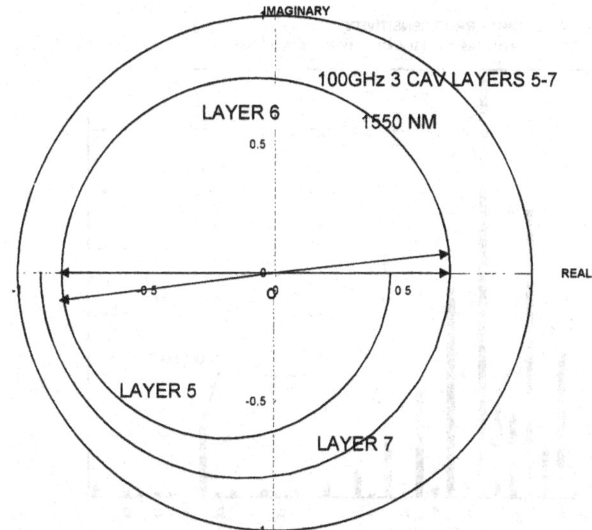

Fig. 7.27. Circle diagram view of layers 5, 6, and 7 as seen in the monitor trace view of Fig. 7.13 used to calculate the changes in reflectance with change in layer thickness.

7.4.2.5. Total Error Sensitivity of the Average Transmission in the Passband

The effect of errors on the T_{ave} as a function of layer position is estimated by the product of the data from Figs. 7.22 and 7.28. This is plotted in Fig. 7.29 and show how critical the layers near the cavities can be. The coupler layer effects are almost invisible on this scale. Therefore, we have replotted this data on a log scale of error sensitivity in Fig. 7.30. Here we can see that the coupler is more sensitive than its surrounding layers because of its small change in reflectance as we showed in Fig. 7.17.

7.4.2.6. Error Compensation in the Monitoring

As pointed out in Fig. 7.10 and subsequent comments, when an error occurs in this type of monitoring, attempting to cut each subsequent layer at its turning point will be the best possible compensation for the effects of the error. When we generated the data used in Fig. 7.22, we then also reoptimized the thickness of the layer after the layer with the error. In every case, the T_{ave} was restored to the ideal value and that layer was terminated at the turning point! This shows that even an error at the most sensitive (worst) layer can be essentially totally corrected for by the next turning point. The phase of all of the reflections except the layer with the error will be in perfect relationship. We have performed other simulations where

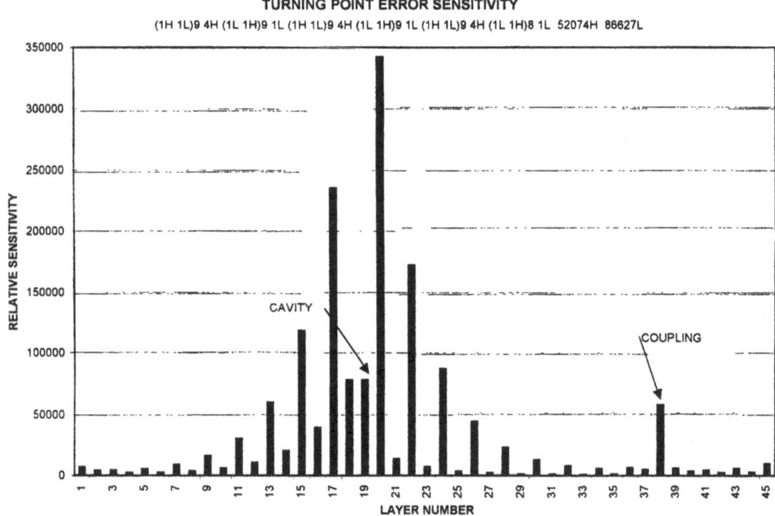

Fig. 7.28. Turning point error sensitivity over the first 45 layers based on the reciprocal of the change in reflectance with given QWOT error in thickness.

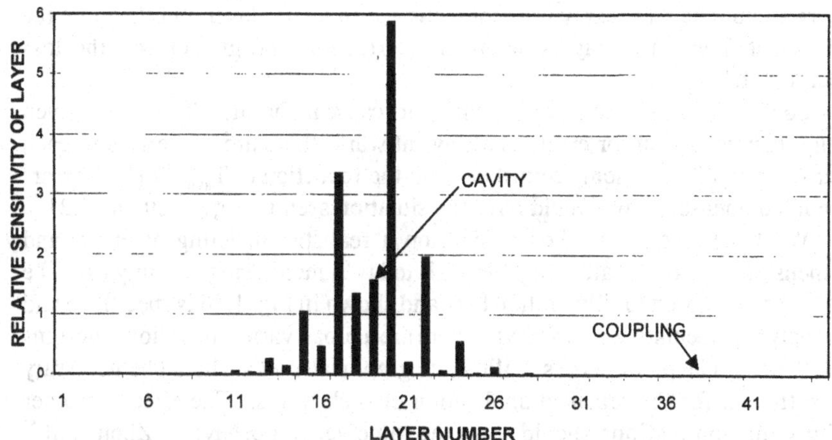

Fig. 7.29. The product of the data in Figs. 7.22 and 7.28 to give a measure of the relative turning point error sensitivity for the first 45 layers of the DWDM filter design.

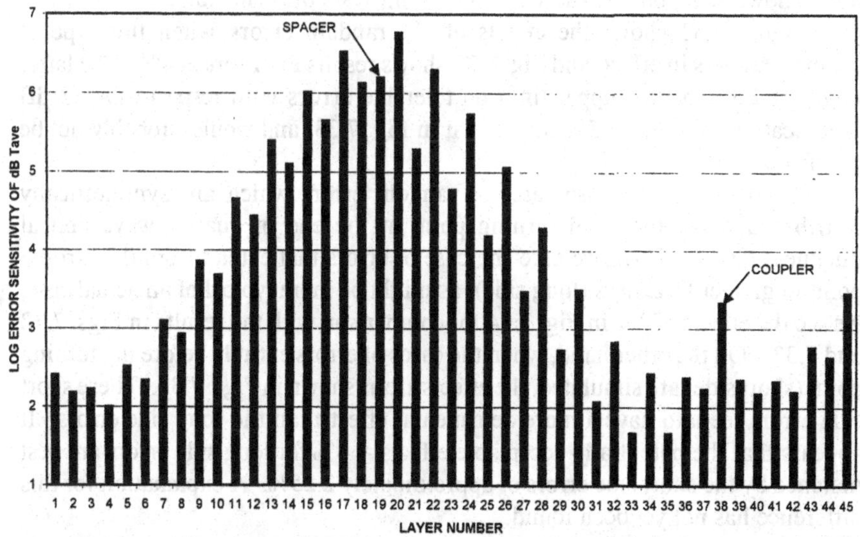

Fig. 7.30. The product of the data in Figs. 7.22 and 7.28 as shown in Fig. 7.29 but plotted on a logarithmic scale.

the error was not corrected until a subsequent layer. It was generally found that the error could be reasonably well compensated in a later layer, but the more the number of intervening layers before the correction, the greater was the loss generated in T_{ave}.

So then, what are we to say? That errors cause no harm? Clearly, no! Even though compensation for errors is always at work, there are real errors in each layer which will make some contribution to the reduction of T_{ave}. If those errors were uncompensated, we would have the situation seen in Figs. 7.20 and 7.21.

We have recently worked to develop a realistic modeling of errors and compensation as it actually happens with today's monitoring techniques. The single design chosen for illustration here and shown in Fig. 1.26 is specific to and relatively representative of DWDM. There are a broad variety of optional designs for DWDM. The point here is to illustrate graphically what should be necessary and sufficient for the practical application of such filters. The effects in other designs and applications should show the same general behavior. Zhou et al.[53] briefly reported work of this type with limited conclusions. The effects of random errors of optical thickness (QWOTs) in layer termination have been simulated and shown graphically here with the natural error compensation properties of the commonly used monitoring technique. Simulations are shown at various magnitudes of layer optical thickness error to allow the estimation of what level might be tolerable for given applications. Many more cases were examined than those shown here, but we show enough to illustrate the behavior.

Figure 7.31 shows the effects of 1% random errors when this type of compensation is in effect, and Fig. 7.32 shows results for errors of 4%. The latter might be a reasonable upper limit on tolerable errors with respect to a 0.3 dB specification. Errors of 5% are shown in Fig. 7.33, and would probably not be satisfactory.

All of the above cases are for random errors which are symmetrically distributed about the ideal turning point at the end of quarter wave optical thickness layers. When we take the case of errors that extend from the turning point to greater thickness (long side), as might be more typical of an actual case, we see the effects of 3% in Fig. 7.34 for comparison with the results in Figs. 7.32 and 7.33. On the other hand, when the cases of errors entirely before the turning point (short side) are simulated, the effects are as shown in Figs. 7.35. These short side errors seem to have a more detrimental effect than the long side errors. It appears that the potentially acceptable effects at 3% for long side errors are best matched by the short side errors of approximately 2.25%. A explanation for this difference has not yet been found.

In the work above, the errors were simulated as having a random distribution as a percentage of a QWOT imposed at each TP in sequence. In this extended work, we refine the model to simulate what should be a still more realistic

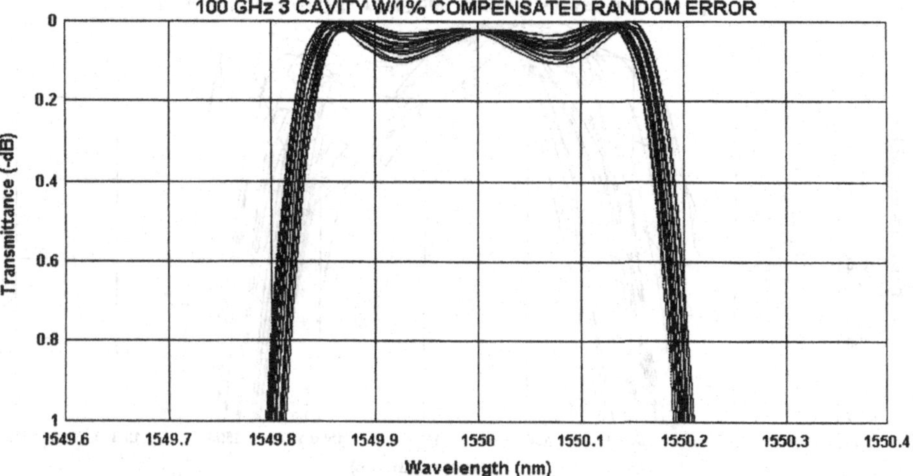

Fig. 7.31. Effects of 1% QWOT simulated random errors centered about the turning points when terminating the layers (monitoring) in a mode which provides compensation.

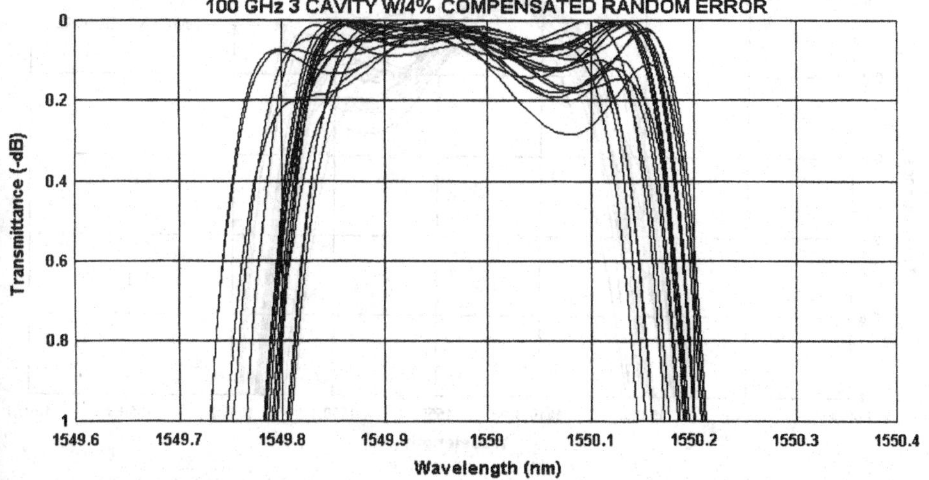

Fig. 7.32. Effects of errors of 4% QWOT with compensation as in Fig. 7.31.

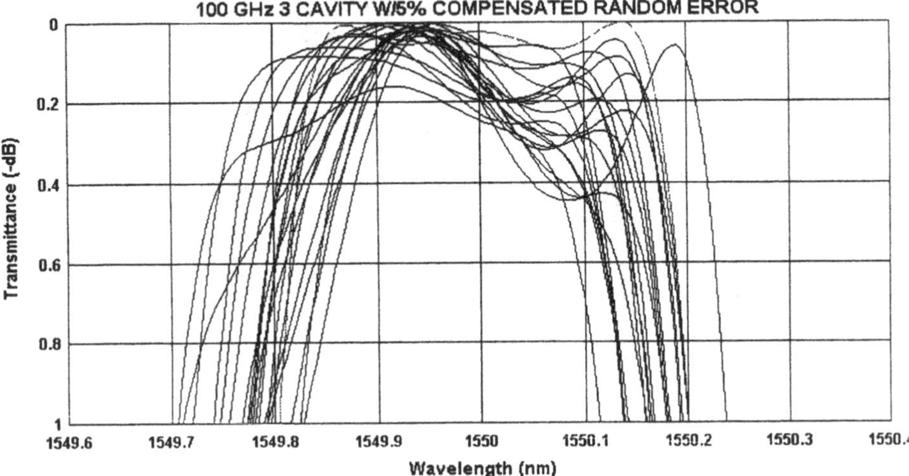

Fig. 7.33. Errors of 5% QWOT as shown in Figs. 7.31 and 7.32.

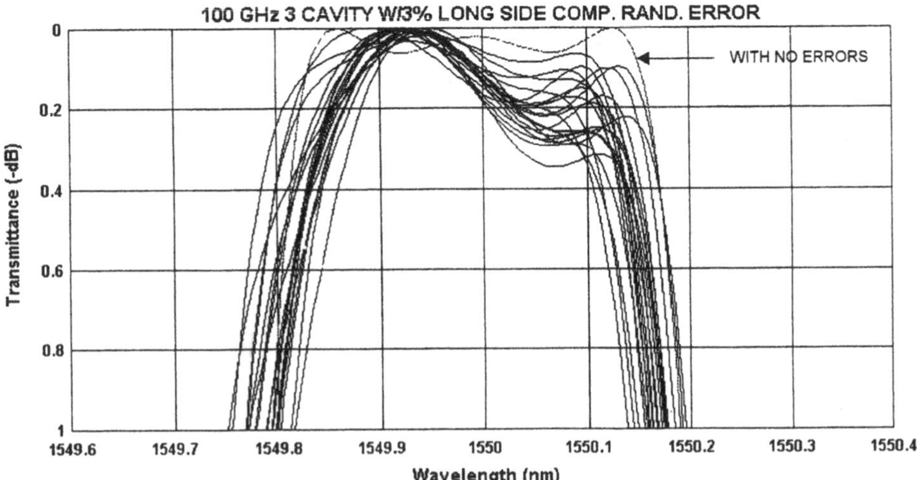

Fig. 7.34. Effects of 3% QWOT random compensated errors that extend only from the turning point to greater thickness, for comparison with the results in Figs. 7.32 and 7.33.

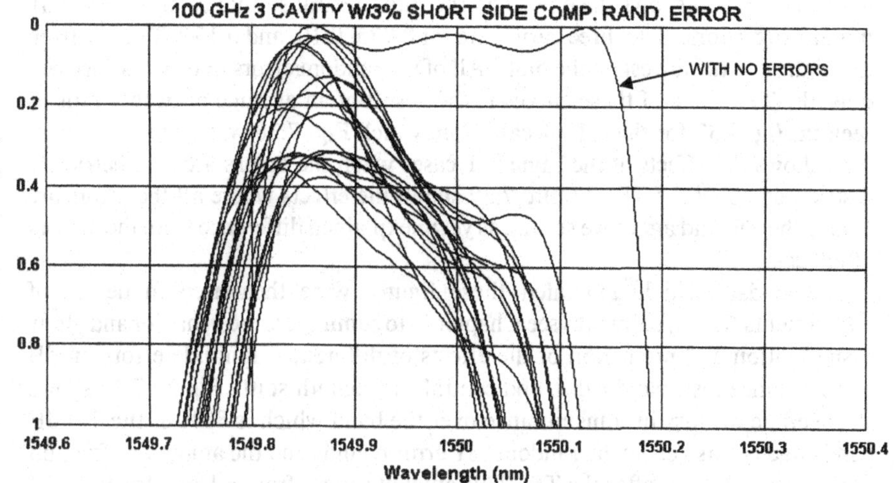

Fig. 7.35. Effects of 3% QWOT random compensated errors that occur entirely before the turning point, for comparison with the results in Figs. 7.34.

representation of actual practice. In attempting to terminate a layer at the TP, it is necessary to sense the changes in the *transmittance* of the monitoring signal at the TP. The errors that occur with respect to the monitoring signal at the TP are more likely to be measured "vertically" in errors of percent transmittance (%T) than "horizontally" in percent of a QWOT. The %T is a quadratic function of the layer thickness about a TP. As seen above, the change in %T from one TP to the next varies greatly from layer to layer. The change in %T is the smallest, and therefore most sensitive, at the layers nearest the spacer layers. We here use a random %T error in a defined range for each layer in the sequence, convert that to an equivalent QWOT error based on the sensitivity of each layer, and apply that error as each new TP is found.

The three types of cases previously simulated were also used here: centered errors about the TP, all of the errors after, and all before the TP. Similar behavior was found to the previous investigation, but the range of "probably satisfactory" results was found to be less than 0.15%T random errors as compared to 2-4% of a QWOT errors in the earlier work. In that case with errors applied as a percent of a QWOT, the magnitude of the errors were uniform throughout all of the layers, which is not likely to be the case due to the differing sensitivities of the layers. In the present case, the %T errors are uniform (randomly) throughout all of the layers, but the impact on thickness errors is greatest near the spacer layers where the sensitivity to %T errors is the greatest.

Figure 7.36 shows the effects on the filter dB transmittance of random %T monitoring errors. The three curves are for 0.075, 0.10, and 0.15 %T errors from the highest to the lowest on the plot. All of the random errors in this first case are *after* the TP. Each of these curves is the result of an average of twenty runs as seen in Fig. 7.37 for the 0.10% case from which Fig. 7.36 was derived. Figure 7.38 shows the effects at the same %T cases when the errors are symmetrically distributed about the TP. Figure 7.39 shows the effects where all the errors are *before* the TP, and again we see the as yet unexplained difference from the "errors after" case.

The data which generated these figures were then used in design of experiments (DOE) software (see Chapter 6) to summarize the behavior and allow visualization and prediction of the effects of different magnitude errors on dB transmittance loss, bandwidth, and central wavelength shift. Figure 7.40 shows the average dB loss in transmittance over the band which is 80% of the 1.0 dB band edge points versus the random %T error bounds and the amount or fraction of the error which is after the TP. This amount ranges from − 1 to 1 for 100% of the error before to 100% after the TP in each of these three (3) figures. As might be expected, the loss increases with error and is somewhat greater as the errors are before the TP. Figure 7.41 shows the average bandwidth in nanometers at the 0.5 dB points. Here the bandwith above 0.5 dB narrows rapidly as the errors become greater than 0.10%T. Figure 7.42 shows the percent shift in the center wavelength

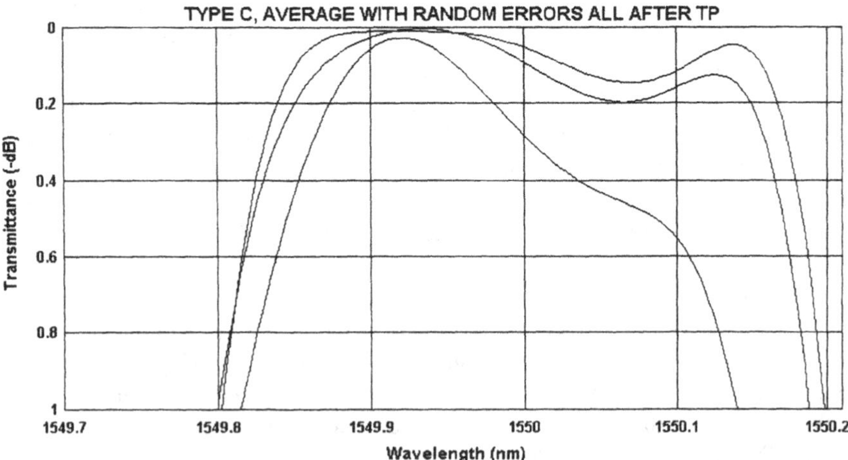

Fig. 7.36. Effects on the filter dB transmittance of random %T monitoring errors where the three curves are for 0.075, 0.10, and 0.15 %T errors from the highest to the lowest on the plot. All of the random errors in this case are ***after*** the TP. Each of these curves is the result of an average of twenty runs.

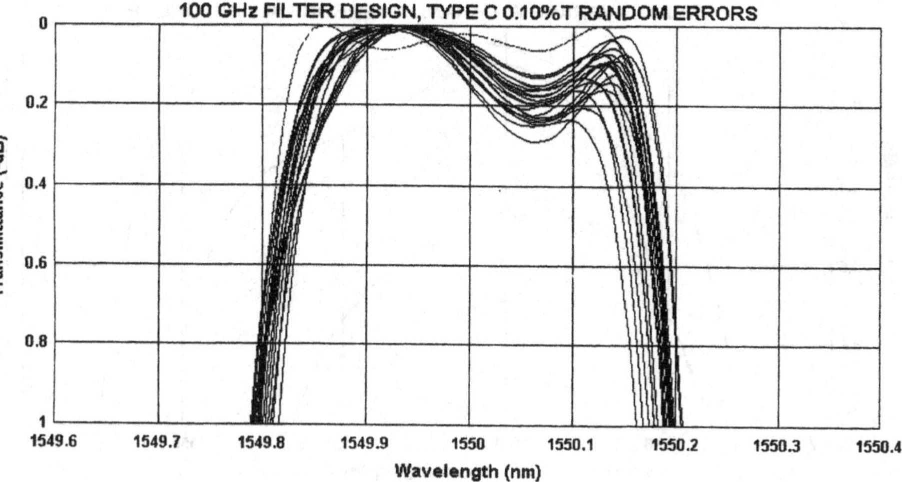

Fig. 7.37. Results of twenty runs for the 0.10% case from which Fig. 1 was derived.

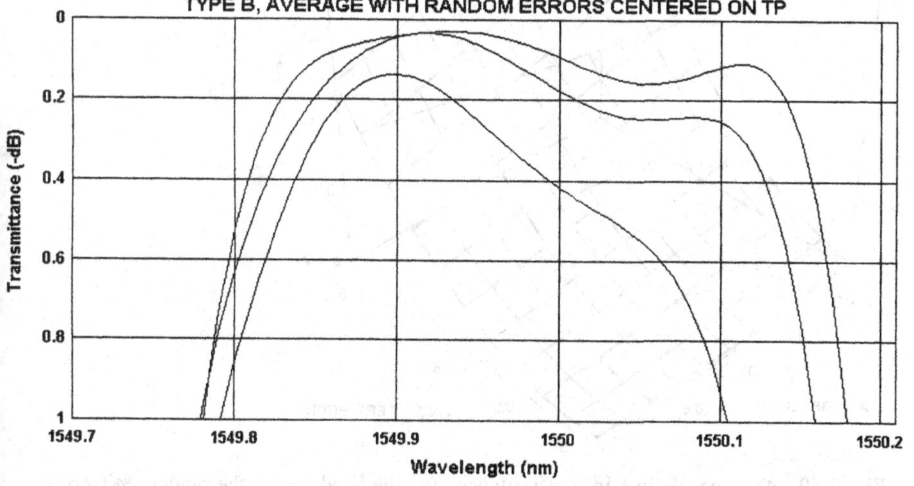

Fig. 7.38. Effects at the same %T cases as Fig. 1 when the errors are symmetrically ***centered*** about the TP.

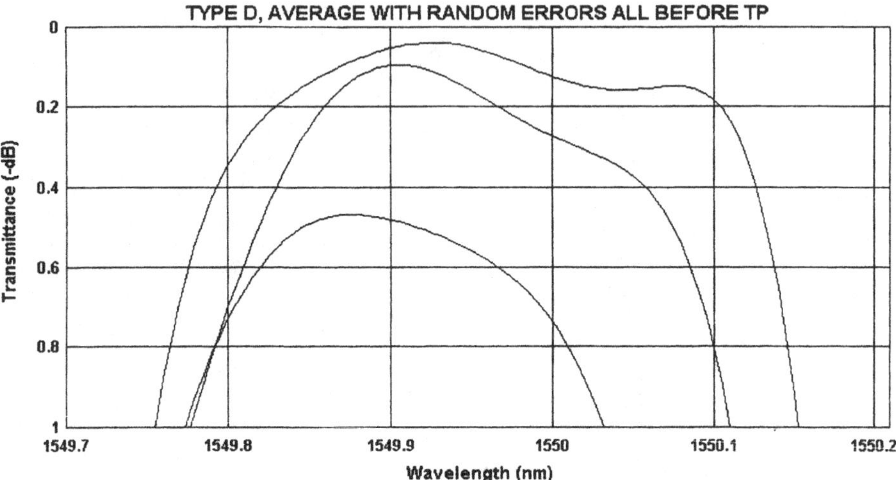

Fig. 7.39. Effects where all the errors are ***before*** the TP at the same %T cases as Fig. 1.

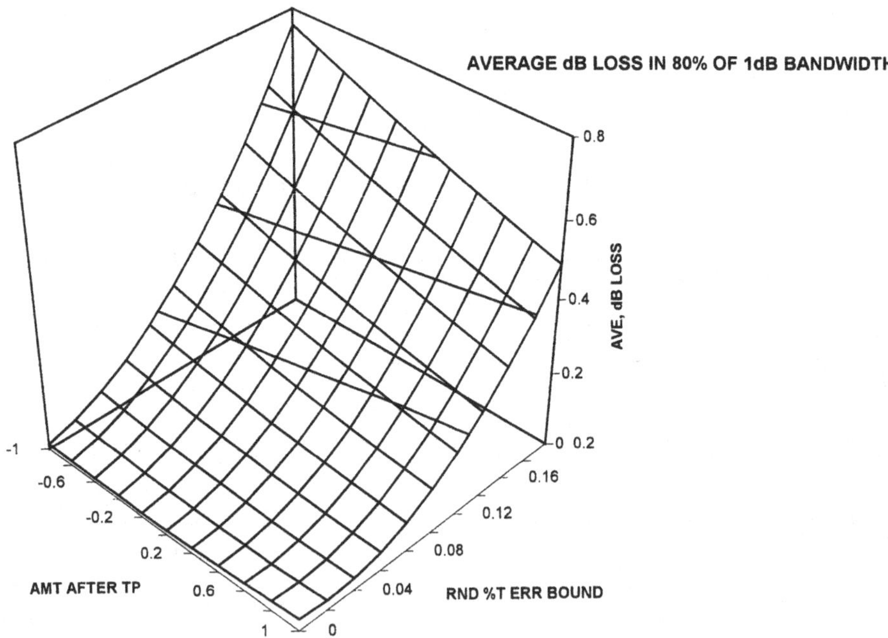

Fig. 7.40. Average dB loss in transmittance over the band versus the random %T error bounds and the fraction of the error which is after the TP. This fraction ranges from – 1 to 1 for 100% of the error before to 100% after the TP.

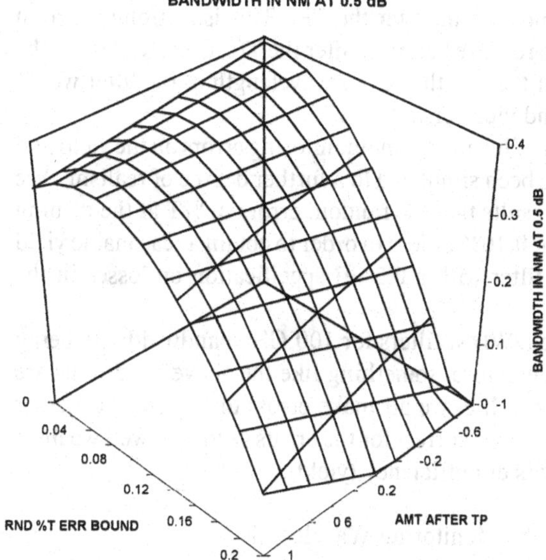

Fig. 7.41. Average bandwidth in nanometers at the 0.5 dB points versus the random %T error bounds and the fraction of the error which is after the TP.

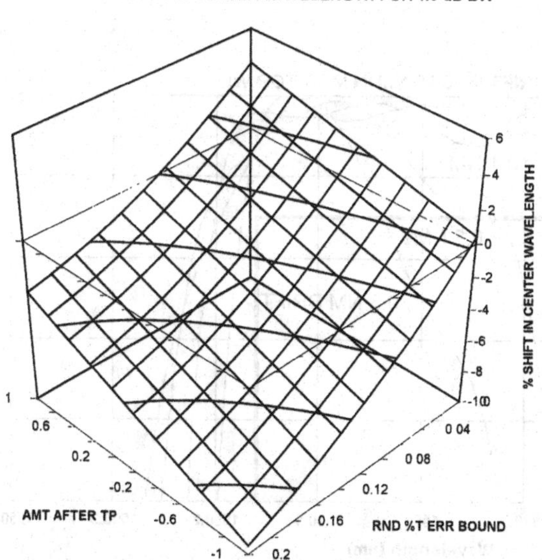

Fig. 7.42. Percent shift in the center wavelength of the passband versus % T error and fraction of the error which is after the TP.

of the passband versus error and amount after the TP. This is a strong function of both error and whether the errors are before of after the TP. Clearly, if all of the errors were on the long side of the TP, the center wavelength of the filter would be greater than the nominal and vice versa.

With the above, the effect of turning point monitoring errors on the yield and performance of NBP filters has been simulated to a further degree of realism. We conclude in examining these results that the random errors in %T at the turning points need to be on the order of 0.10% or less in order to obtain a reasonable yield for such a 100 GHz DWDM filter with a 0.3 dB specification on losses in the passband.

We know that acceptable DWDM filters for 100 GHz bandwidth are being produced in the year 2001. Therefore, something like the above conditions are apparently being satisfied. As in life, we all make errors to a greater or lesser extent; it is a question of how well we correct for the errors as to how well we meet our goals and how satisfactory is our efficiency (yield).

7.4.2.7. Error Due to Drift in the Monitoring Wavelength

At least one of the popular coating systems used for DWDM filter fabrication today uses a laser monitoring source such as might be used in the final communications application of the filters. The stability of the monitoring

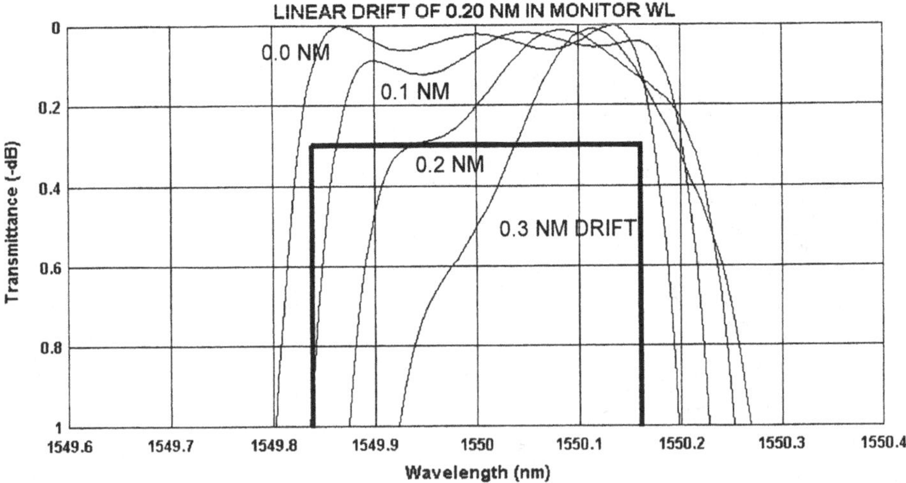

Fig. 7.43. Effect of linear monitor wavelength drifts of 0.0, 0.1, 0.2, and 0.3 nm over the time of the monitoring.

wavelength over the deposition time of the whole filter is critical. If the wavelength of the monitor drifts during the deposition, the layer thicknesses will drift with it even though the compensation effects for errors will still be functional. We have simulated linear drifts of 0.1, 0.2, and 0.3 nm over the time of the monitoring. The results are shown in Fig. 7.43 in comparison with the ideal filter. The filter is nominally 0.4 nm wide at 1.0 dB, and it can be seen that any drift more than about 0.1 nm will cause problems with the usual specifications of <0.3 dB ripple and losses. The implication is that the drift of the monitor wavelength should be almost an order of magnitude less than the filter bandwidth.

In Sections 7.4.6 and 7.5.4, we shall touch on the effects of the bandwidth of the monitor wavelength and a suggestion for computer-aided monitoring of DWDM filters.

7.4.3. Error Compensation in Edge Filters

Macleod[7] and later Zhao[8] showed how the constant level monitoring described in Sect. 7.3.4.1 can be used to control a repeating layer pair stack at any desired wavelength. They showed how the compensation effect at the monitoring wavelength is similar to that which occurs in the NBP filter. We reproduce figures from Zhao[8] in Figs. 7.44 and 7.45. These show the effect described above where the control is best at the monitoring wavelength and degrades with increasing distance from that wavelength. We believe this to be a key point in optical monitoring which was clarified by Zhao. Willey[13] expanded on the ways to enhance the sensitivity and reduce errors which we will expand upon in Sect. 7.6.

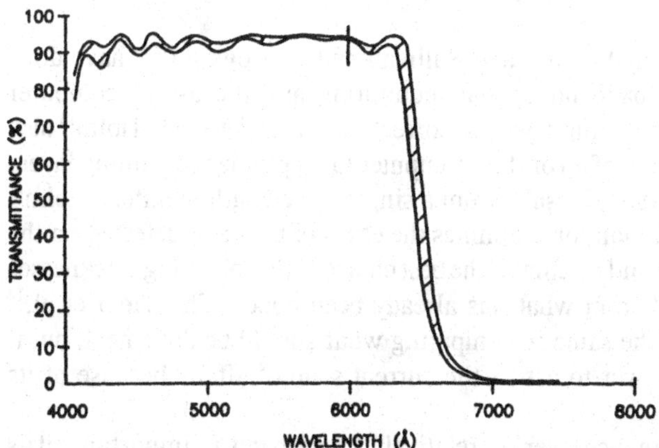

Fig. 7.44. Effects on an edge filter of the error compensation of constant level monitoring at 595 nm (5950 Å) as illustrated by Zhao.[8]

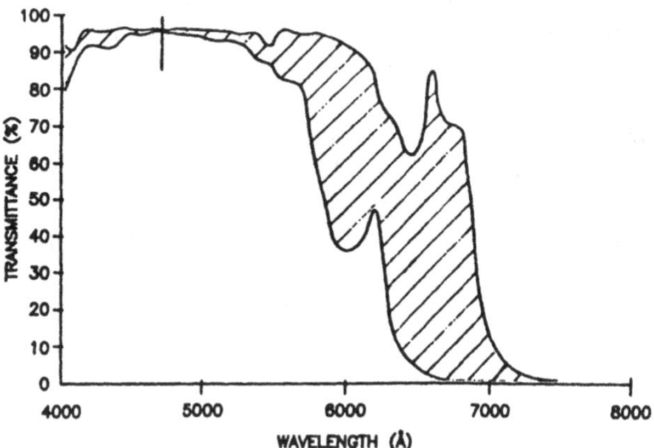

Fig. 7.45. Effects on an edge filter of the error compensation of constant level monitoring at 465 nm (4650 Å) as illustrated by Zhao.[8]

Zhao also pointed out that it was further possible for the operator to "drive" or control the monitor cuts in constant level monitoring to further make "in process corrections" for the errors detected. We elaborate on what this and other "driving" techniques by the operator could contribute to the result in Sect. 7.6.

7.4.4. Broad Band Monitoring Compensation

As mentioned above, Vidal et al.[14] and Sullivan and Dobrowoski[49,50] have done extensive work with broadband optical monitoring and the use of computer reoptimization of the remaining layers to correct for errors detected. Holm[15] also reported on the application of an on-line computer to reoptimize remaining layers on the basis of the analysis of results from a single wavelength monitor. Holm's subset of the broadband monitor examines the effect of the error detected on the broadband performance and calculates the best choice of the following thicknesses to make the best result from what has already been done. The Vidal et al.[14] approach is essentially the same in computing what should be done next, but it potentially has more power to assess the current state of affairs because of its broadband monitor.

Where the performance over a relatively broad band is important, it is desirable to have some means to sense what is happening in the whole band (per Vidal et al.) or at least what is predicted to be the impact on the whole band of a more narrowly sensed error (Holm). If the controllability of the layers involved is sufficient, it may be unnecessary to utilize these more sophisticated monitors. Willey[16] described a sample case of how to monitor a very broad band AR to tight tolerance without these techniques, and illustrated the desirability of the broadband monitor and computer correction if it were available. We will review that work in Sect. 7.7.

We can say in summary about error compensation that it is appropriate to "nail down" the performance where it is most important by monitoring it at that wavelength. Without the aid of computer reoptimization, the NBP filter and a broad class of edge filters can be monitored at one wavelength and will have certain "natural" error compensation. This is why NBP filters have been possible even though the state of the technology might not have seemed to be accurate enough to give an acceptable result. If a broad band must be controlled, the impact of each detected error must be considered on the broad band, and subsequent layers adjusted on the basis of computer "compensation." The very broadband AR to be described in Sect. 7.7 is a slight apparent violation of this last statement. However, it is really a case of not doing the computer work on-line in real time, but off-line between runs.

7.4.5. Effects of Thin Film Wedge on the Monitor Chip

Optical monitoring of a thick stack of layers can suffer from another problem if the layers are monitored on a single stationary monitor chip. Figure 7.46 shows that the angle of an evaporation source to the normal of the monitor chip surface causes a wedge in the deposited material. This is due to the fact that one edge of the chip is closer to the source and receives more material in approximate proportion to the inverse square of the distance from the source to that point on the chip. If the monitoring beam has a significant width, such as rays A and B in Fig. 7.46, the two rays will see significantly different path lengths as the layer thickness and, therefore, wedge increases. This will cause the beams through the different parts of the wedge to have interference patterns which get out of phase with each other and reduce the amplitude of the modulation of the signal from the overall combination of rays. Figure 7.47 shows an approximate and exaggerated case of this wash-out of the signal with increased thickness due to wedge.

The problem with the wedge effect is further compounded by the fact that each evaporation source is usually in a different position with respect to the chip and, therefore, the wedges are overlaid at differing orientations and may have different wedge angles.

There are at least three ways to deal with the wedge problem. If we were direct monitoring a rotating part as in a single NBP filter, there would be no effect (due to the rotation). We have also heard of a monitor with a rotating monitor

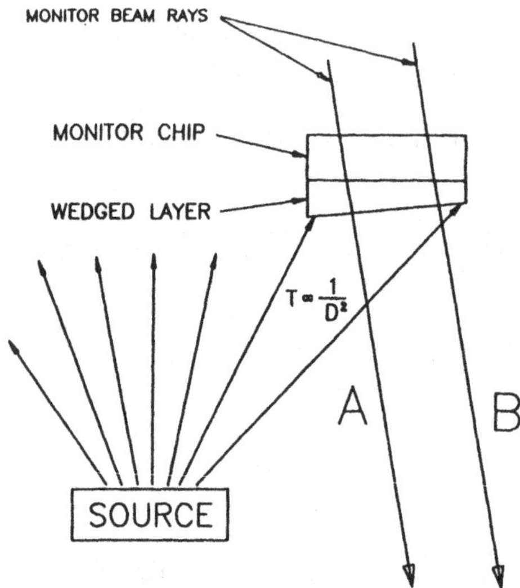

Fig. 7.46. Wedge of layer on optical monitor chip caused by the differing distances of parts of the chip from the source which causes the different rays (A and B) of the monitoring beam to pass through different thicknesses of the deposited layer.

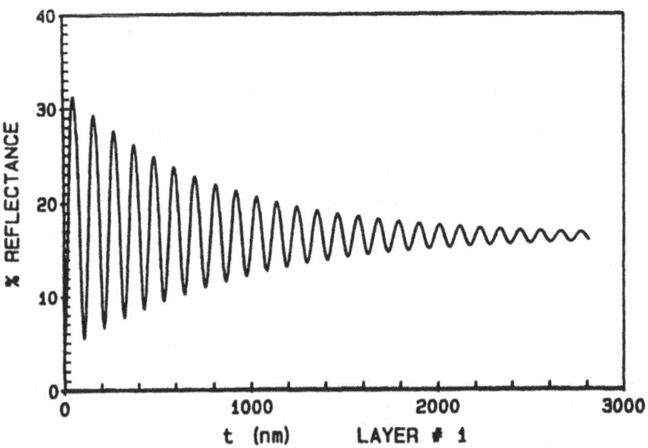

Fig. 7.47. Approximate (and exaggerated) effect on monitor signal of increasing wedge in the coating on a monitor chip with increasing thickness. This also illustrates the approximate effect of a monitoring bandwidth which is too wide so that the shortest and longest wavelengths generate interferences which get out of phase and wash out the contrast of the signal.

chip, but this can be quite a complication. A second solution would be to design the process such that the wash-out is not detrimental. A stack that is not too thick is one such case; a multilayer AR would be an example. Another such example is as described in Sect. 7.7 whereby reflection monitoring is at a wavelength where one or more of the materials has some absorption. Here the effect of the wedged layers underneath cannot be sensed through the absorption of the layers near the surface as they are deposited. A third solution is to use a small enough beam on the monitor chip so that the difference of the layer thicknesses over the beam is not significant. By reducing the beam size, however, the signal level is reduced in approximate proportion to the area and, thereby, the signal to noise ratio. There is therefore a tradeoff to be made in this third option to determine the best overall signal to noise ratio versus beam diameter on the chip. Positioning the sources and monitor chip to minimize the effect may be another option. We have seen systems where even resistive sources are all brought to the same position as would a multi-pocket E-gun before being evaporated, but this is seldom needed except in very special cases.

7.4.6. Error Due to Width of the Monitoring Passband

A similar effect to the wedge effect occurs if the bandwidth of a "single" wavelength monitor is too wide for the application. The wavelengths on either side of the band center produce interference fringes from the monitor chip that get more and more out of phase with layer thickness. The detector signal (highly exaggerated) versus thickness would again look something like Fig. 7.47. This kind of effect might show up in monitoring a very NBP filter if the monitoring bandwidth were too wide (see Sec. 7.4.2). The only solution that we are aware of to this problem is a narrower bandwidth monitoring beam which will reduce the signal level if all else is equal. Again, signal to noise will be the tradeoff.

The spectral bandwidth of a prism or grating monochromator (as has been typically used) is ultimately limited by diffraction effects if all else is perfect. The brightness across such a perfect slit image would vary as shown in Fig. 7.48. The bandwidth at half the power of the peak brightness would be also about half the width of the slit image between the first dark bands. The shape of this curve is a sinc^2(wavelength) where $\text{sinc}(x) = \sin(x)/x$. We have simulated the effects of various bandwidths on the monitoring signal using this diffraction limited image as the model. Figure 7.49 shows the monitor trace for layers 94 through 114 of the filter discussed in Sec. 7.4.2 with monitor bandwidths of 0.025 (very narrow), 0.1, 0.3, and 0.4 nm. The start of the signal "wash out" can particularly be seen for 0.4 nm. The effect of this on final filter performance would depend on whether it led to terminating layers too soon or too late with respect to the desired QWOT turning points. We found this result somewhat surprising in that it does not

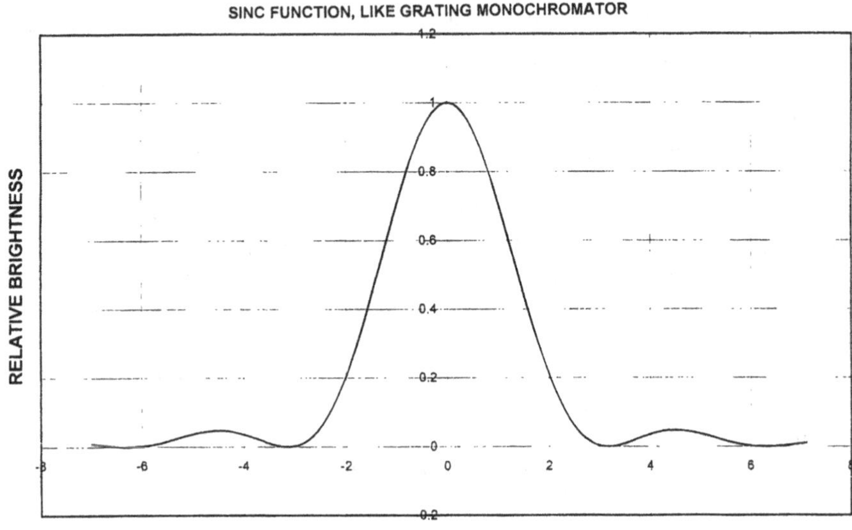

Fig. 7.48. Flux distribution across a perfect image of a very narrow slit due to diffraction effects.

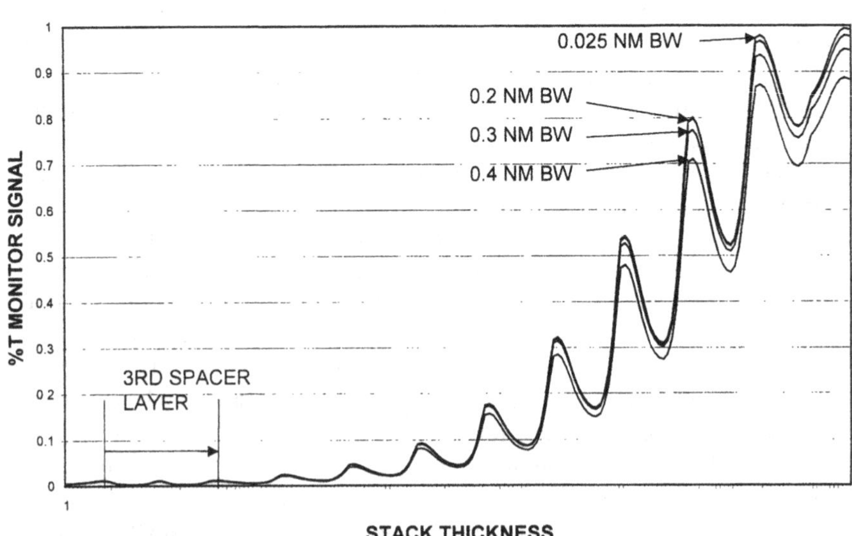

Fig. 7.49. Monitor trace for layers 94 through 114 of the filter discussed in Sec. 7.4.2 with monitor bandwidths of 0.025 (very narrow), 0.1, 0.3, and 0.4 nm.

appear that a monitor bandwidth as wide as the bandwidth of the filter being made would cause significant error! However, more than about 0.1 nm bandwidth in this case would distort the photometric results enough to cause other potential problems where photometric data was used to glean additional control in the monitoring.

This points to the fact that monitoring with a laser may not be absolutely necessary if a good monochromator is available. Note however that the wavelength stability must be better than 0.1 nm for this case as shown in Sec. 7.4.2.7. The other problem could be achieving the necessary SNR as discussed in Sec. 7.4.2.1. The laser has much greater brightness than any other source.

7.5. CALIBRATIONS AND VARIATIONS

As mentioned in Sect. 7.3.3, the frequency of a vibrating quartz crystal is not a direct indicator of the optical thickness of a layer deposited upon it. However, because the frequency varies almost rectilinearly with optical thickness, it is often practical to calibrate the crystal output with respect to optical thickness of a given material under given deposition parameters. The crystal readings can then be used as a practical surrogate for the control of the real parameter of interest, the optical thickness. When there is adequate stability and reproducibility for a given process requirement in the substitute measurement, simpler monitors can be substituted for more complex and expensive systems. For example, as mentioned earlier, it has been found that some sputtering processes can have a very stable and reproducible deposition rate when power settings and gas pressures are properly controlled. This means that the time that it takes to deposit a given optical thickness can be well calibrated, and then only the deposition time used to control the thickness of a single layer or even several layer coatings. This can save the cost of an optical monitor in many of the simpler cases.

Macleod[3] reviews many interesting sides of monitoring, including an extensive discussion of the use of only the operator's eye as an optical monitor. He shows how the apparent colors on reflection change with film thickness as both high and low index materials are deposited. As mentioned in Sect.7.3, this is a very adequate way to monitor SLAR coatings even today. We mentioned the sets of standards prepared by a US Government agency for different thickness SLAR coatings on glass so that the results can be accepted or rejected only by visual comparison. This calibration of the eyeball obviates the need for both an optical monitor and a spectrophotometer, if only SLARs are to be produced.

The simplest monochromator is a NBP interference filter at a specific wavelength. The wavelength cannot be changed over a broad wavelength range as a grating (or prism) monochromator, but it can be changed over a small range to shorter wavelengths by tilting the filter at an angle to the beam. The wavelength passed by the filter can be calibrated by measuring it in a spectrophotometer at the appropriate angles. The numerical aperture (F/number) which the filter passes will affect the effective bandwidth of the transmitted light. We have seen such filters used successfully for monitoring almost all types of coatings including NBPs. The major limitation over a grating monochromator is the lack of flexibility to select any desired wavelength, and secondarily, a less narrow potential bandwidth.

A grating monochromator used with an optical monitor can be easily calibrated for wavelength accuracy with a HeNe laser. However, it may be just as useful to deposit a specific number of quarter waves of a low or high index material using the optical monitor and then measure the monitor chip in a spectrophotometer to find the exact wavelength where the associated maximum or minimum occurs. This calibrates more than the monochromator, it includes the temperature, atmospheric, and geometric (tooling factor) effects, etc.

7.5.1. Tooling Factors

The term tooling factor is commonly used to describe the ratio between the coating as monitored on an optical monitor chip and the result of the coating deposited on the final workpiece. This ratio is generally not unity due to a variety of factors. If the optical monitoring is semi-direct, the geometry of the distances and exposure to the sources are different from the planets or calotte to the monitor chip. The angular distribution of material leaving a given source is significantly different from material to material. For example, silicon dioxide may have an angular "cloud" which decreases as the cosine to the third power of the angle to the surface normal (as we discussed in Chapter 5), while a pool of molten germanium may evaporate with a cosine to the first power distribution. Since the optical monitor is controlling the process in a vacuum and often at a temperature above ambient, there is often a change in the film properties at ambient temperature and after exposure to the humidity of the atmosphere. This has an effect even on direct monitored parts.

A typical calibration procedure for an optical monitor in a chamber with planets is to deposit several QWOTs of a given material as determined by the monitor. Witness pieces in the planets during the deposition are then measured in a spectrophotometer along with the monitor chip. The spectral shift in wavelength from what the monitor showed during deposition to witness chip after removal from the chamber and exposed to air is the tooling factor for that material under those conditions. Note that the shift in the monitor chip from time of deposition to spectrophotometer measurement is a calibration of the deposition temperature and vacuum to air at ambient temperature effects. The difference between the monitor and witness chips in air is a calibration of the geometrical effects of planet to monitor and the source "cloud" distributions. The most important calibration is usually the composite difference from what the monitor experienced during deposition to the final result on the workpiece. This is used to adjust what is seen on the monitor of the next run so that the deliverable parts meet the requirements.

For the convenience of utilizing available equipment, we have chosen to monitor infrared coatings for the 8 to 12 micrometer band at a wavelength of about one micrometer. This used a red sensitive photomultiplier and NBP interference filter. In this case, there was a great difference in the monitoring wavelength and the band which we need to control. The indices of the materials are different from the monitor wavelength to the band of interest due to dispersion. The optical thicknesses are therefore different from the simple ratio of wavelengths due to the inaccurate knowledge of the actual indices. These factors are calibrated by the same test runs mentioned above where tooling factors are found. The contribution of each individual factor is not usually determined. The sum of all of the effects is what is of interest. If all of these factors can be simply corrected for by a shift of monitoring wavelength, the process proceeds successfully. Most often this is the case, and one does not need to be concerned about how much each factor contributes. In production, it is the bottom line that counts. In a more research and development environment, doing the extra tests to separate out how much each factor contributes may be of interest and worthwhile.

7.5.2. Variations

One would like to be able to know that successive runs of a given process will be identical. This is not always the case to a satisfactory degree, and a process that worked well before may fail to meet the requirements when repeated. These variations in result are due to unintentional variations in process conditions or parameters. Some of the more common undesired variations are temperature, pressure, set-up, and "dirty" chambers.

We had an unpleasant experience where a good process failed for several runs until we noticed that the temperature probe had been repositioned after a chamber cleaning by about 10 cm from the normal position. The process included germanium layers at a temperature near the critical transition temperature described by Evangelisti et al.[17]. A small temperature change caused by the mispositioned sensor produced an unacceptable absorption or scattering in the germanium layer. When the sensor was properly repositioned, the process again produced good results. Fortunately, most processes are not this sensitive. However, zinc sulfide would be another material where temperature variations could cause trouble because its "sticking" coefficient is quite temperature-sensitive.

The residual pressure in a chamber during deposition can have a significant influence on some processes. An unwanted film on a surface before the first intentional layer can cause lack of adhesion, so a minimal residual pressure might be desirable. On the other hand, some materials under some conditions have less stress and are therefore more adherent when there is a somewhat higher residual pressure during deposition. The variation of index of refraction due to chamber pressure variations by leaks and outgassing was discussed in Sect. 7.3.

The geometric set-up of a chamber needs to be sufficiently constant from run to run to preserve the calibrated tooling factors. Planet and monitor height and "squareness" can affect this as well as source loading and temperature sensor placement (as mentioned above).

Even if all of the above-mentioned variations are sufficiently controlled, the effects of coating layers on the chamber walls from previous runs may be an ultimate limit on how many successive runs can be done to spec. The "dirty" chamber causes at least two variations. The infrared reflectance and emittance of the chamber walls or foiling in a freshly cleaned chamber is changed significantly as the process layers are deposited. This causes the radiative transfer changes and, therefore, substrate and monitor chip temperature changes. The layers on the walls after some runs also can radically change the outgassing of the chamber and, therefore, the residual gas in the chamber at deposition time.

Sufficient stability and reproducibility are needed to be able to achieve the required results from a process run after run. To achieve the required results in the first run, it is necessary to be able to either control the result directly, or at least calibrate out any indirectness. It might be said that reproducibility equates to precision, and by calibration against a known standard, accurate results can be achieved. The next subsection describes an interesting extension of these basic ideas to get the best from the optical and crystal monitor.

7.5.3. The Optical Monitor with Crystal Method of Schroedter

Schroedter[18] described a monitoring system which combines the best features of the optical monitor with the crystal monitor to get greater accuracy than is available from either monitor by itself. As we discussed in Sect. 7.3.3 on crystal monitoring, the crystal needs to be calibrated to a known optical thickness from an optical monitor reading or a spectrophotometer reading on the result of a test run. Schroedter did this in real time rather than after a test run. Figure 7.50 illustrates the procedure with a system that has both a crystal monitor and an optical monitor where the photometric level can be sampled and stored in the process computer. The photometric level is sampled and stored at equal intervals of crystal thickness readings. Variations in the deposition rate have little effect because the sampling is versus crystal thickness, not time. As a sufficient number

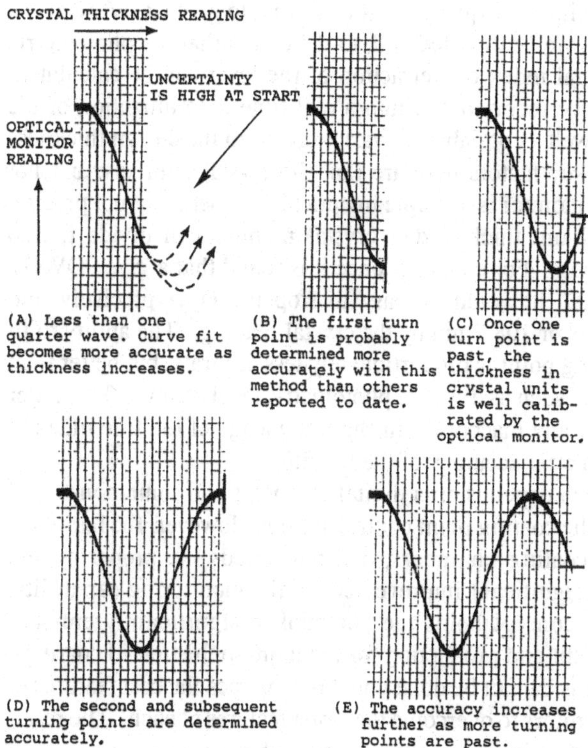

Fig. 7.50. Schroedter's[18] technique of recording photometric reading versus crystal reading and fitting results to expected shape to predict turning points and any general optical thickness as a function of crystal reading. This shows the constant updating of the calibration of the crystal readings by the optical monitor signal.

of points are gathered, it becomes possible to fit the data for the layer to a sinusoidal curve and predict where the turning point or one QWOT at the optical monitor wavelength will occur. It is further possible to fit the more exact form based on calculations of what the exact curve (which is not precisely sinusoidal) would be if the indices and previous layer were accurately known. The presence of noise in the signal adds proportionately to the uncertainty of the fit and turning point prediction; but this is still probably better than an operator can judge in the presence of noise. The more data are gathered, the more accurate is the prediction. As a turning point is passed, the accuracy of the calibration becomes potentially quite good. After two turning points are passed, the accuracy is probably as good as it will get because other instabilities may cause variations which make the earlier data points less relevant.

After two turning points, the quartz crystal reading to give a QWOT at the monitoring wavelength is well calibrated and, to the extent that the photometric scale is accurate, the actual index of refraction of the layer can be calculated. This information could be incorporated into a real time reoptimization of the design to compensate as well as possible for variations from the design thickness and index values. The key factors here are that the crystal is precise but not accurate, and the optical monitor is accurate in optical thickness, but not precise. A QWOT might be 100 nm thick and a crystal monitor can divide it into increments of 0.1 nm or less. For example, once it is found that a given QWOT is 85.6 nm on the crystal, it would be easy to stop the layer precisely and accurately at 1.27 QWOTs or 108.7 nm by the crystal reading. The ability to fit data before the first turning point is proportionately less accurate, but better than previously used techniques such as a simple photometric level change. This latter approach is influenced by any inaccuracy in the knowledge of the true index of refraction and the photometric accuracy of the monitor.

Schroedter's approach depends on a digital computer and some non-trivial programming as part of the coating plant control system. However, the hardware exists in many modern coating chambers, and only needs the proper system programming to make a significant improvement in the monitoring capability. The author believes this concept holds the potential to significantly enhance optical coating development and production for the more sophisticated coatings. There is no reason why this cannot be done now. It combines the best features of the optical and crystal monitors to overcome the worst features of each. The result of the combination of the two is greater than the sum of the parts.

7.5.4. Suggestion for Computer-Aided Monitoring

It appears that some attempts are being made to incorporate turning point prediction into DWDM fabrication, but we are not privy to whether they have the sophistication of the Schroedter technique. The case we have been closest to obviously did not, since it had no crystal monitor but assumed a constant deposition rate.

We would be inclined to have a crystal monitor for rate control and dealing with coupler layers (and in some cases the first of the last two AR layers). The data from this would then be used in a Schroedter-like technique.

For best results, one would need to predict the expected monitor shape of the layer being deposited. In order to do this, we need an estimate of the index of the layer and the reflectance at the start of the layer. If both of these were known exactly, the locus of the layer on a reflectance circle diagram (thereby transmittance) could be exactly predicted. The actual signal with its inherent noise could be continually fit to the predicted curve derived from this. At the point of best fit to the desired termination, the layer would be cut. If/since there are errors in the knowledge of the index of refraction of the layer and also the exact reflectance at its starting point, the software needs enough flexibility to handle this. It should also be able to handle the fact that the layer may have started before or after the previously intended turning point. The software also needs the ability as implied by Schroedter to predict a turning point enough in advance of its actual occurrence to get the termination process started in time to cut the layer at that point.

Another issue of value or necessity, depending on the photometric level stability of the laser (and detector system) with time, would be to monitor the level of the light into the monitor and divide the output by that level. This is in case there is any drift in that light level with time. This would be a first approximation of a true double beam spectrophotometer.

At the present time, the above approach would be the author's preference for reliable automatic monitoring of DWDM filters. This approach would seem highly advisable since we showed in Sec. 7.4.2.6 that 3% of a QWOT or a 0.10% T monitoring error is about all that can be tolerated.

7.6. SENSITIVITY AND STRATEGIES

It is necessary in most cases to minimize the optical thickness errors when controlling a deposition in order to achieve high yield with demanding requirements. If we confine our discussion to the use of single wavelength optical monitors which are sensing the change in reflectance (or transmittance) with optical thickness, the reflectance change with thickness change (dR/dT) is a most important factor. The change in reflectance is the basic signal information used to control optical thickness. The dR/dT is the sensitivity of the optical monitor to the optical thickness that we wish to control. This signal is superimposed on the noise in the monitor signal which degrades the certainty with which we know the reflectance at any point in the process. If the noise has been reduced to its practical limits, the only remaining possibility to improve signal to noise is to increase the signal to its practical limit. If the signal through the hardware of the monitor has been made as high as practical by the design, construction, and alignment of the monitor, then there remains the possibility to increase the signal change with optical thickness at the critical control points of the process. This sensitivity enhancement by proper choice of monitoring scheme or strategy is the subject of this section.

7.6.1. Sensitivity versus Layer Termination Point in Reflectance

We reported[13] on the general finding of how to determine where the rate of change of reflectance with thickness would be a maximum and minimum. Figure 7.51 shows a typical monitor trace of reflectance versus thickness as a HLHLHLHL stack of QWOTs is deposited. It can be seen that slopes on this curve (dR/dT) are steepest between turning points and even more so when those regions are in the vicinity of 30 to 40% reflectance. More extensive investigation showed the results plotted in Fig. 7.52. Here it was found that the maximum sensitivity occurs when a given layer passes through a reflectance amplitude (r) of about 0.6 or a reflectance intensity (R) of 36% and a phase of 90 degrees. This point turns out to be essentially independent of index, but the dR/dT is greater for higher indices in approximate proportion to the index minus one (n−1). This graph (Fig. 7.52) is our basic guide in the domain of the reflectance diagram to optimizing sensitivity of the monitoring scheme at the control or cut points of the process. The greatest sensitivity can be achieved if the cut point occurs at 0.6r on the imaginary axis of the reflectance diagram. The least sensitivity will be found as the reflectance crosses the real axis, which are the turning points or maximum and minimum points of reflection. The other region of minimum sensitivity is when the reflectance approaches 1.0. In this case, the reflectance is already quite high and additional thicknesses make only very small changes in reflectance. The axis

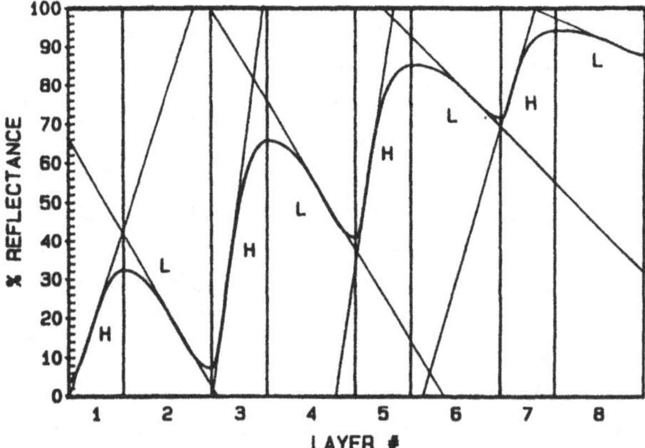

Fig. 7.51. Optical monitor trace of the reflectance versus thickness for the deposition of a (HL)4 stack. The steepest (most sensitive) points of dR/dT are nearly midway between the turning points and when the reflectance is nearest about 36% as in the third layer.

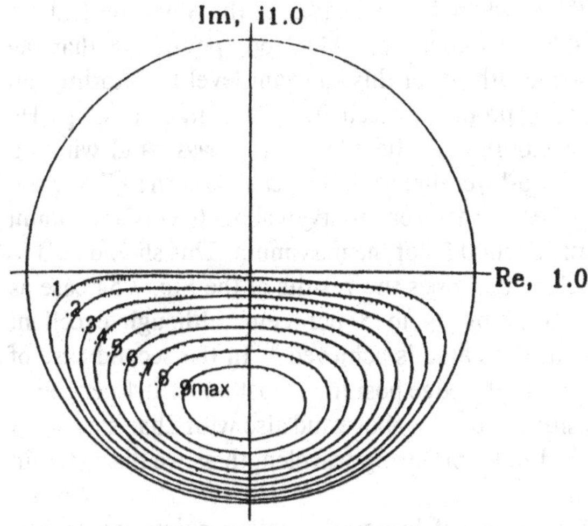

Fig. 7.52. Relative sensitivity of change in reflectance with change in thickness with respect to position on a reflectance diagram. The plot is symmetric above and below the real axis.

crossing turning points are actually to be preferred over the regions near 1.0 reflectance because the reflectance quickly moves toward and away from the axis so that the turning point can be sensed, but the reflectance stays close to 1.0 in the other case. Reference 12 discusses this derivation and application examples in some detail. The basic strategy is to select the monitoring wavelength and possibly a precoating on the monitor chip (which we will discuss shortly) that will cause the most critical cut points to occur as close as practical to the target point of 0.6r on the imaginary axis. This will maximize the sensitivity of the monitoring and thereby minimize the errors in optical thickness due to monitoring.

7.6.2. Sensitivity versus g-Value

We found that the application of the sensitivity concepts above leads to some interesting conclusions when applied to the periodic multilayer such as an edge filter[19]. Figure 7.53 shows the theoretical performance of a LWP filter from Macleod et al.[7]. This design is (H/2 L H/2)6 on a 1.52 index substrate where H is a QWOT of 2.35 and L is 1.35. The reflectance diagram for constant level monitoring of this design without a precoated monitor chip is seen in Fig. 7.54.

This results in a "g" value of 0.542 where "g" is the ratio of the wavelength at the QWOT to the wavelength under examination. Macleod points out that the position of the monitoring wavelength(g) for this constant level monitoring can be chosen at will by selecting an appropriate precoating of the monitor chip. He also gives an example which monitors at the edge of the pass band where g =.8086. It is necessary to use a two-layer precoat in this case as seen in Fig. 7.55. Note that the cut points in Fig. 7.55 are at a sensitivity of about 0.78 of maximum while those of Fig. 7.54 are only about 0.35 of the maximum. This should lead to smaller thickness errors. Macleod describes the results of the Fig. 7.54 case as follows: "that edge position varies by up to 5 per cent...although excellent performance around the monitoring value is achieved." In the second case of Fig.7.55 he says, "the worst error in the edge position is just under 0.4 per cent." This is consistent with our sensitivity concept above and also with Zhao's[8] concept of getting the best results around the monitoring wavelength as we discussed in Sect.7.4.

Figure 7.56 is a plot of the locus of layer termination points for various g-values on a reflectance diagram for the coating of Fig. 7.53 whose layers are of equal optical thickness. Guo et al.[20] shows results for other thickness ratios. We see that the range of g-values is from zero to about 0.825 in this case. It can be shown that there is a periodic repetition of these points for g-values greater than 1.0. Figure 7.57 is a plot of the sensitivity at the termination points of the layers versus the g-value of the monitoring wavelength. This is the result of combining

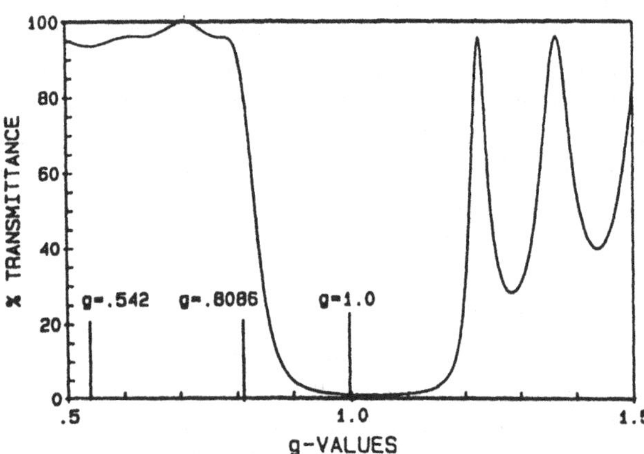

Fig. 7.53. Longwave pass filter design after Macleod[7] of (H/2 L H/2)6 where H is a QWOT of 2.35 index and L is 1.35 at the g = 1.0 wavelength. "g" is the ratio of the design (QWOT) wavelength to the wavelength under consideration.

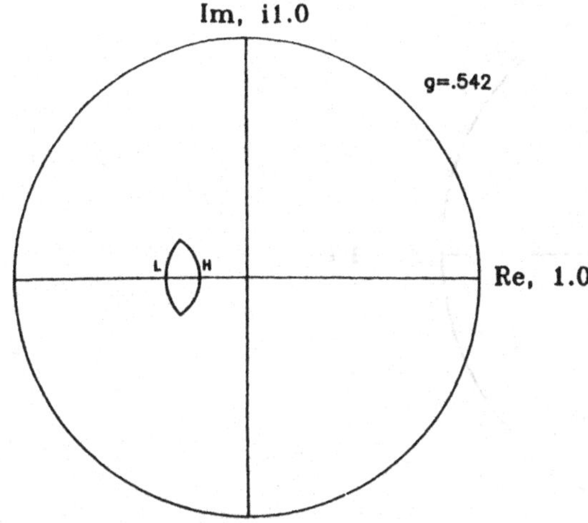

Fig. 7.54. Reflectance diagram of the locus of the monitored reflectance as the coating of Fig. 7.53 is deposited without any precoating on the monitor chip (g= .542).

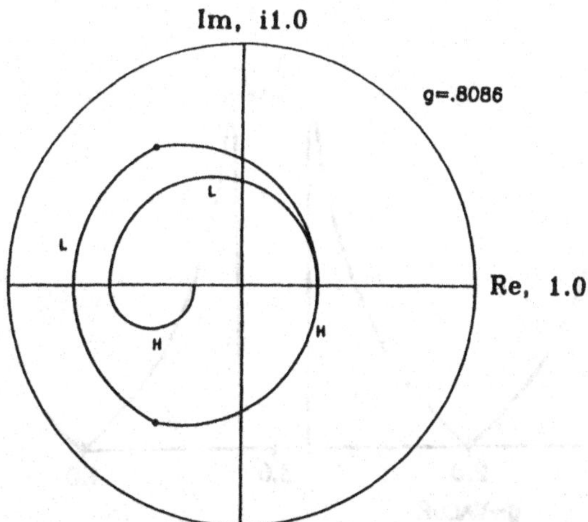

Fig. 7.55. Reflectance diagram of the locus of monitored reflectance as the coating of Fig. 7.53 is deposited with a two-layer precoating (HL at g = .8086) on the monitor chip.

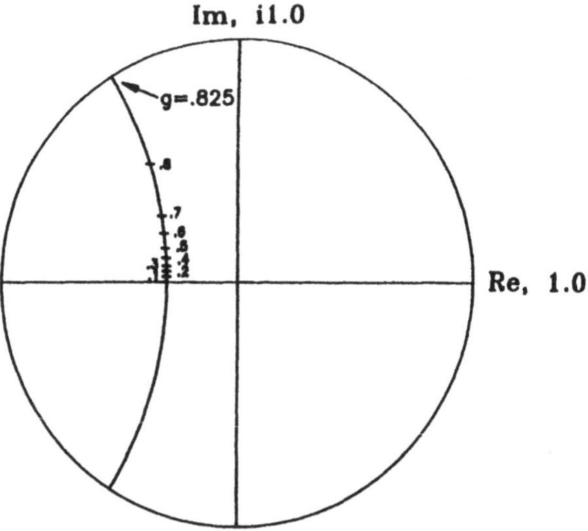

Fig. 7.56. Locus of layer termination points on a reflectance diagram for various g-values in the case of Fig. 7.53.

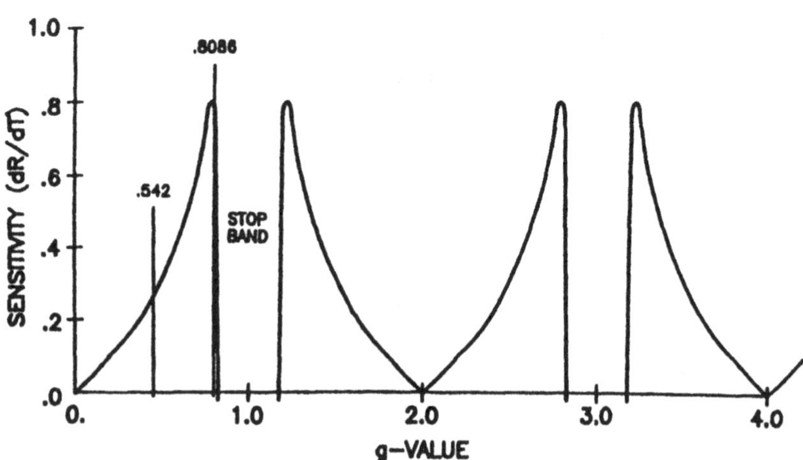

Fig. 7.57. Relative sensitivity at layer termination points versus g-value of monitoring wavelength in the case of Fig. 7.53.

the information contained in Figs. 7.52 and 7.56. We can see from Fig. 7.57 that the most sensitive wavelength with which to monitor is at either edge of the stop band. This is, in fact, where Macleod chose to monitor at a g-value of .8086. The least sensitive would be at the half wave points. The blocked band is excluded as a place to monitor because the reflectance rapidly approaches 1.0 where there is no change of reflectance with thickness.

7.6.3. Precoated Monitor Chips

Precoating refers to applying a coating to the monitor chip before the workpieces to be coated are put in the chamber. Thus the monitor chip has some coating which the final parts will not have. The purpose of precoating is to shift the starting point of the monitoring so that the cut points of the subsequent layers to be deposited are at more sensitive points than they would be without the precoating. Figure 7.55 is an example where the first high index QWOT layer of the precoat brings the reflectance of the substrate to a higher reflectance on the left and then the low index QWOT brings the reflectance to the point on the right where the process coating is to start. This starting point was chosen to give the g-value desired in this constant level monitoring case. Reference 12 discusses the application to non-periodic layer structures such as AR coatings. Precoating does require an extra coating run to produce the monitor chips, but many chips can be produced in that one run and used for as many production runs. The overhead cost of precoating is small for production which will involve many runs, but it could be high if only one or two runs are to be done. However, the result of sensitivity enhancement for even one production run may be worth the added effort.

The basic process of selecting a precoat which we use[13] is to decide where on a reflectance diagram the production coating should start to have the most critical cut points as close as practical to the target (0.6r) of maximum sensitivity at the chosen monitoring wavelength. We then use a reflectance diagram to design as simple a precoat as practical to get from the substrate reflectance to the desired starting point. Because of the compromises and adjustments to be made, the process is currently more a case of artful design than a simple one-time calculation. Zhao[8] mentions that it is sometimes possible in periodic layer structures to design the first antireflecting layers before the repetitive periods to serve the function of the precoating and also obviate the need for a precoat.

7.6.4. Eliminating the Precoated Chip

As discussed in earlier chapters, the AR coating is a series of layers to transition from the admittance of one medium to that of a coating stack or another medium. These generally cause a periodic oscillation of the admittance within an envelope going from the admittance of the medium to the extrema of the stack admittance at a given wavelength. In general, a three- to six-layer AR coating can be made to accomplish this in such a way that a chosen broad band can have low reflectance. We have found it practical in many cases to optimize such a coating so that it is the desired AR ***and*** brings the admittance to the required point to start the periodic stack with constant level monitoring.

It is critical that the admittance at the end of the AR/precoat at the monitoring wavelength for the stack have the correct amplitude **and phase.** We have had some success with this by optimizing the 3- to 6-layer part of the coating with a heavy weighting on the exact reflectance and phase (0 or 180°) at the monitoring wavelength and appropriate weighting on the AR band of interest. The stack monitoring wavelength has been previously chosen by the strategies described earlier in Sec. 7.6.

When well executed, such an AR/precoat has given very good AR properties and good control at the monitoring wavelength.

We will now describe the constrained optimization design procedure which we have used for this AR/precoat set of layers. The requirement can be stated as: the preliminary layers must move the reflectance magnitude and phase at the monitoring wavelength from that of the substrate to that of the start of the periodic stack needed for CLM. The constraints are that the magnitude and phase values at the end of the deposition of the preliminary layers satisfy these specific requirements. The constrained optimization can vary both the pre- and post-periodic matching layers (without varying the periodic stack itself) while attempting to meet the transmittance targets and simultaneously satisfy the constraints. When this is done, the resulting design has optimized transmittance ***and*** satisfies the required constant level optical monitoring conditions for the most reproducible results in production. Goldstein[28,29] has provided features in his software which allow this to be done automatically.

The underlying principle as described above was to use CLM as described by Macleod and Pelletier[7] and position that constant layer termination level for the least sensitivity to photometric and other layer termination errors. Willey and Machado showed[30] that the optimum layer termination levels were generally slightly above 50% reflectance. Such levels are not generally a natural consequence of an edge filter design which is intended to use CLM. A set of layers between the substrate and the periodic stack is needed which brings the cut level up to this >50% point and also provides the AR coating or matching layers needed for the pass band of interest. The preliminary layers must move the reflectance phase at the monitor wavelength from that of the substrate to the specific reflectance of the start of the periodic stack needed for CLM. The required phase at the end of the preliminary layers is usually either zero or pi. The constraints are that the magnitude and phase satisfy these specific values at the end of the deposition of the preliminary layers. Matching layers after the periodic stack also provide the interface to the final medium (usually air) in such a way as to minimize the reflection effects of that interface in the pass band. These post-stack layers can often be designed before the constrained pre-stack layers by unconstrained optimization over the band of interest. If a small number of post-stack layers are used which give a residual reflection that is significant with respect to the pre-stack layers, it may be beneficial to optimize the post-stack layers simultaneously with the pre-stack layers and constraints for the best results. On the other hand, if a large number (>4) of post-stack layers are used, that set of layers can be designed only once for a small reflectance value over the band and kept constant while the pre-stack layers are designed.

7.6.4.1 General Design Procedure

Figure 7.58 illustrates a long wavelength pass (LWP) filter with a blocking edge at 8568 cm^{-1} and a pass band from 4200 to 8200 cm^{-1}. This design is a periodic stack of (.5H 1L .5H)6 at 10000 cm^{-1} plus six (6) pre-stack and ten (10) post-stack matching layers, where the indices are L = 1.46, H = 2.2, and the substrate is 1.52. CLM is illustrated in Fig. 7.59 with the layer termination level of layers 9 through 20 slightly above 50% reflectance. Two negligibly thick layers (7 and 8) of high and low index have been inserted in the design to show the break point from the pre-stack layers to the periodic stack.

The first part of the requirement for a preliminary set of layers is illustrated in the reflectance circle diagram of Fig. 7.60. It can be stated as: the locus of the reflectance phase versus deposition thickness of the preliminary layers must move from the reflectance of the substrate to the specific reflectance of the start of the periodic stack needed for CLM at the monitoring wavelength. The monitoring frequency (cm^{-1}) and effective index of refraction at the start of the periodic stack have been determined by the techniques described previously[30] to be 8607 cm^{-1} and 0.42, respectively.

In this case with a substrate of index 1.52, the reflectance would start at $r = -0.20635 + i0$ ($R = 4.258\%$) and move clockwise with the deposition of successive matching layers to the point $r = 0.4084 + i0$ ($R = 16.68\%$) which has an effective index of refraction of $0.42 + i0$. The required phase at the end of the preliminary layers is usually either zero (as in Fig. 7.60) or pi. The reflectance magnitude and phase therefore must be constrained in the optimization to conform to these specific values at the end of the deposition of the preliminary layers. This makes it possible to have the constant layer termination level for the periodic layers 9 through 20 illustrated in Fig. 7.59. The periodic section, where CLM is used, might have many more layers than the 13 illustrated here (for simplicity), and the pre-stack layers might be more than six.

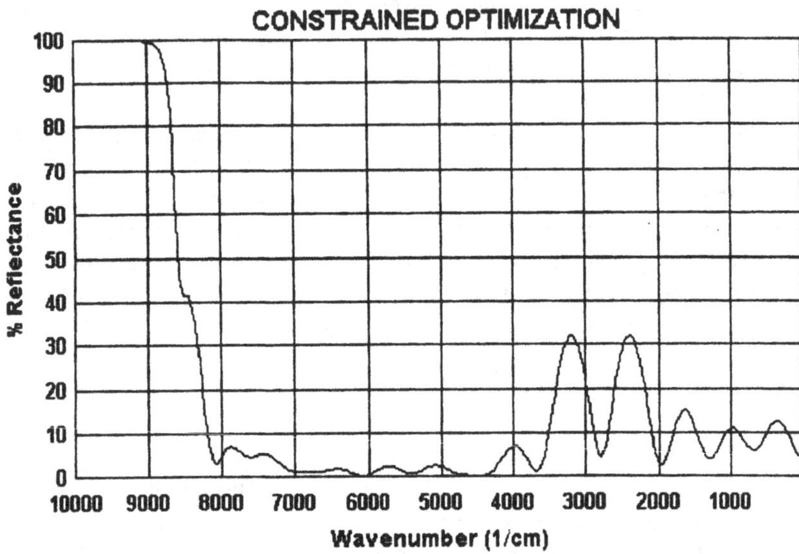

Fig. 7.58. LWP filter design with a periodic stack of (.5H 1L .5H)6 at 1000 cm^{-1} plus 6 pre-stack AR layers, where the indices are L = 1.46, H = 2.2, and the substrate is 1.52.

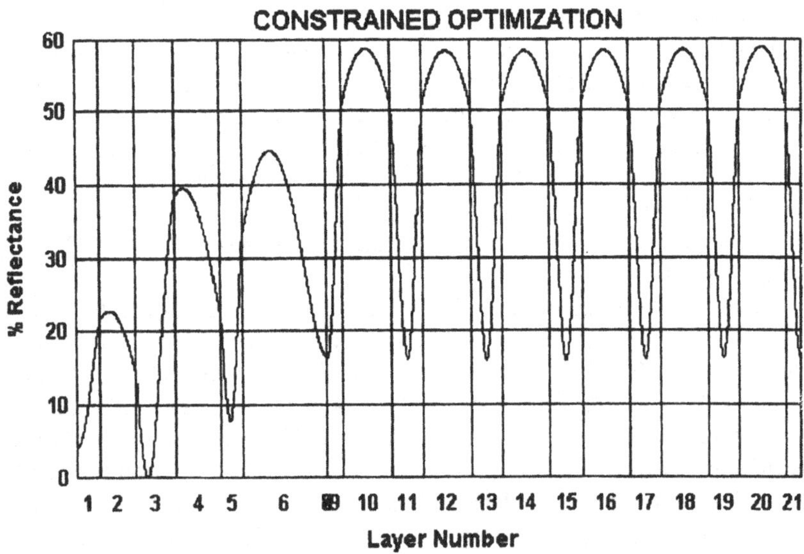

Fig. 7.59. Constant level monitoring at 8607 cm^{-1} of the design in Fig. 7.58. Two very thin layers (7 and 8) have been inserted to show the break point from pre-stack to stack.

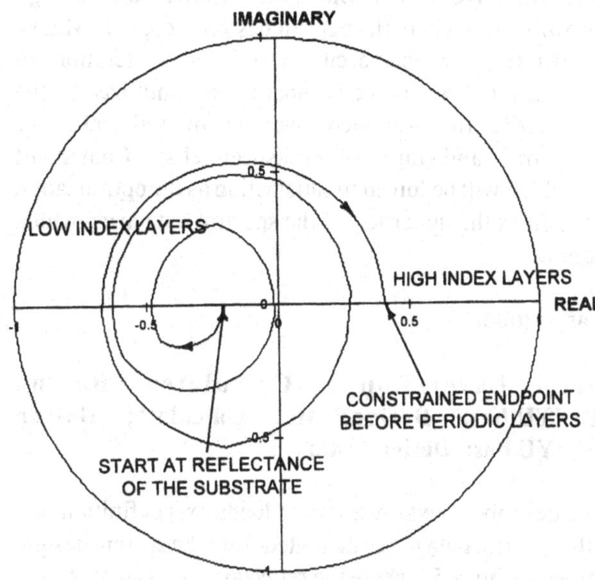

Fig. 7.60. Amplitude reflectance diagram of the first six layers of the design in Fig. 7.58 at 8607 cm^{-1}. Locus moves from substrate to start of the periodic stack for CLM.

7.6.4.2 Specific Design Procedure

An optical thin film optimization tool is needed which has the ability to impose constraints while optimizing. In the work reported here, we have used FilmStar[28] and the details reported refer to nomenclature used by that tool.

One of the goals is to design matching layers before and after the periodic stack which minimize the reflection in the pass band. This is an ordinary thin film design task which can be handled via: a starting design, performance targets, optimization variables, and an optimization routine. The extraordinary task is to also constrain the reflectance amplitude and phase to specific values at the end of the first set of matching layers. The constrained optimization does not vary layers 7 through 21. The free variables are layers 1-6 and 22-31. The targets were set to minimize the reflection from 4200 to 8200 cm^{-1}. We chose as a starting design: (.5H 1L .5H)15, where L = 1.46 and H = 2.3. After this "group design" was expanded to a "layers design," it was optimized (using the constraints) by varying the first 6 and last 10 layers.

It is necessary to use the Workbook spreadsheet capability of FilmStar to implement the necessary constraints. When optimizing from the Workbook, five named cells or ranges are key to the computational procedure which is executed.

These are: **Design, Macro, Objective, Constraint, and DataMarker**. **Design** indicates the section of the worksheet where the current design is copied. **Macro** defines a sequence of operations which occur during each iteration of optimization. **Objective** (merit function) is the number to be minimized by the optimization. This is ordinarily the weighted average of reflectance or transmittance over some spectral band (range of worksheet cells). **Constraint** defines multiple conditions which will be forced to satisfaction as the optimization progresses. **DataMarker** defines the upper left of the spectral data array which is updated during each iteration.

The Macro in this case was as follows:

AxesOpen CNSTR1; LayersCopy; Calculate; Basrun C:\winfilm\basic\DESPLITU.bas; DesignPaste; Calculate; Basrun C:\winfilm\basic\DESCOPYU.bas; DesignPaste;

This performs the sequence described next. **AxesOpen** loads axis definition file, **CNSTR1**, which defines the spectral data to be calculated from the current design. This might be the axes defined in Fig. 7.58 where the reflectance is evaluated from 30 to 10000 cm^{-1} in some specified increment. In the particular case described here, we chose an increment which caused the monitoring frequency (8607 cm^{-1}) to be one of the sample points and therefore a cell in the Workbook that can be addressed for constraint calculations. In this case, we were particularly interested in the reflectance phase at 8607 cm^{-1}. The next macro operation is to execute a **LayersCopy** command to copy the design to the defined **Design** place on the worksheet. Then a **Calculate** command is executed to evaluate the design at the spectral points defined by the axis file. These are automatically copied to the worksheet area defined by the **DataMarker**.

A basic program with the name **C:\winfilm\basic\DESPLITU.bas** is then executed. This program is as follows:

```
Sub Main
    WbCopy "$K$12:$K$42"
    WbPaste "$L$12:$L$42"
    WbCopy "$L$12:$L$17"
    WbPaste "$M$12:$M$17"
End Sub
```

This copies the whole design to the clipboard and then pastes it to an adjacent column on the worksheet for future use. It then copies to the clipboard the layers (6, in the case of Figs. 7.58 to 7.60) up to the point where the constraint is imposed on the reflectance phase, and then pastes it to another adjacent area. The next Macro command, **DesignPaste**, pastes the abbreviated design which is still on the clipboard onto the design area of the worksheet. The **Calculate** command is executed which evaluates this abbreviated design with respect to the axes definition and stores the results on the worksheet to the right of the previous evaluation of the whole design.

The **Objective**, which in this case was the average of the reflectance from 4200 to 8200 cm^{-1}, is calculated by the worksheet as the average over those cells where these reflectance values lie. A spectral weighting can be applied by making another column elsewhere on the worksheet which is the product of the calculated values times the spectral weighting function. The sum or average of this new column would then be the definition in the **Objective** cell.

The **Constraints** or violations thereof can now be evaluated by the worksheet from two cells of the second **Calculation** results with the abbreviated design. The cell with the reflected phase at 8607 cm^{-1} is compared with the target value for its constraint, and the cell with the reflected amplitude is compared with its constraint.

Before going on to the next iteration, it is necessary to restore the whole design. This is done by the macro commands:

Basrun C:\winfilm\basic\DESCOPYU.bas; DesignPaste;

This basic program is simply:

```
Sub Main
    WbCopy "$L$12:$L$42"
End Sub
```

Here, the whole design which was saved is now copied to the clipboard. The **DesignPaste** command puts it back in the correct position on the worksheet.

This completes the iteration and the cycle is repeated until the optimization procedure terminates itself or is terminated by the designer.

7.6.4.3 Results of the Procedure

The convergence of this design procedure has been generally satisfactory. However, it not as rapid as unconstrained optimization and appears to be more vulnerable to falling onto "local minima". We have found two potential pitfalls to avoid. One is that the number and indices of preliminary layers to be optimized must be sufficient to reach the reflectance and phase of the constraint point from the starting point. A single layer will almost never be enough and three may

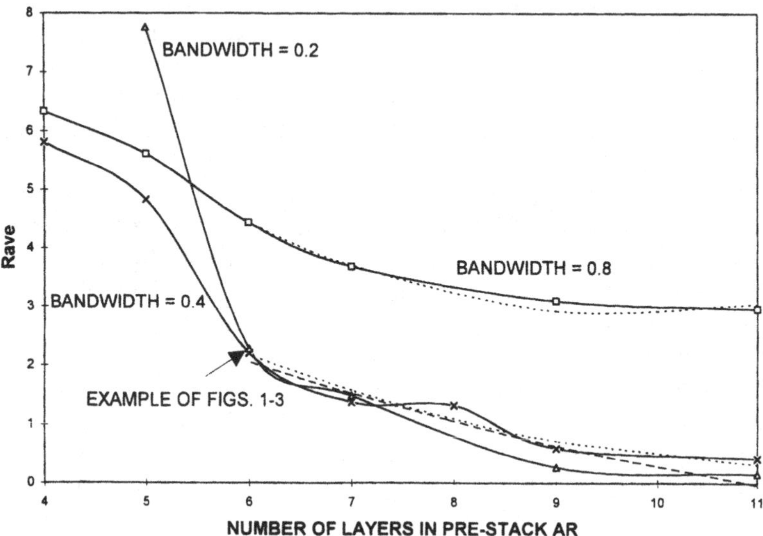

Fig. 7.61. Rave results of a systematic set of pre-stack layer designs at bandwidths of 0.2, 0.4, and 0.8. Solid lines for specific designs and dotted lines from equations fit to data.

occasionally be sufficient. Six or more layers are more commonly successful and appropriate when attempting to minimize the reflectance over a broad pass band. Figure 7.61 shows the Rave results of a systematic set of optimizations of pre-stack layer designs with 4 to 11 layers and at bandwidths of 0.2, 0.4, and 0.8. The bandwidth is defined as the range of wavenumbers in the pass band divided by the range from zero to the center of the block band (in this case 10000 cm^{-1}). The example in Figs. 7.58 to 7.60 has a bandwidth of 0.4. The AR behavior here has been found to be generally consistent with our earlier findings[31]. The second pitfall is the phase discontinuity found at the extremes of plus or minus pi when constraining the phase to pi. We overcame this by constraining the square of the phase to be equal to pi squared.

The resulting Raves from the above optimizations as a function of the two variables (total number of pre-stack (AR) layers from 6 to 11 and bandwidth from 0.2 to 0.8) were treated as Historical Data using design of experiments (DOE) methodology discussed in Chapter 6. The software provided the coefficients for equations to calculate any point on these surfaces that have been statistically fit to the data to the third order including interactions of the variables. The resulting equation is:

$$\mathbf{Rave = 7.6455 - 0.4641N - 16.423B - 2.127NB + 57.71B^2 - 35.2B^3 + 0.139N^2B} \quad (7.1)$$

Here B is the bandwidth and N is the number of pre-stack layers. This equation was used to generate the dotted lines in Fig. 7.61 to compare with the "experimental" data of the solid lines. In our recent work[31] on unconstrained optimizations of similar long wave pass filters (LWP), we found the following equation to be a good fit to the optimizations of that work:

$$\textbf{Rave (LWP)} = \mathbf{2.1678 - .7247N + 5.0606B + .0441N^2 - .0007N3} \qquad (7.2)$$

When these two equations are used to compare the number of extra layers required to satisfy the constraints and achieve similar Rave results, it is found that approximately 3 extra layers are required. If the bandwidth is greater than approximately 0.3, this number 3 varies by only a small fraction of a layer with bandwidth and number of layers. At bandwidths smaller than 0.3, six (6) layers may be needed for good constrained results; whereas only two might be adequate for an unconstrained AR.

We have previously[30] shown, graphically and with equations, how to find the proper monitoring wavelength and effective index needed to achieve stable control of edge filter deposition by constant termination level optical monitoring. We have shown here how to use that information to complete a design and monitoring scheme which is maximally robust and simultaneously controls a blocking edge position and pass band transmission. An additional three (3) layers are typically needed to satisfy the optimal monitoring constrained condition for CLM as compared to not using optimal CLM which might lead to far from optimal results.

7.6.5. Constant Level Monitoring Strategies

We will now examine the periodic edge filter described in Sect. 7.6.2 where we monitor at the most sensitive point. The ideal optical monitor signal (on a precoated monitor chip, or the AR/precoat described above) would look essentially like Fig. 7.8 if there were no errors. There are four strategies which the operator could use if errors can be detected as the monitoring proceeds. First, he could do nothing but cut each layer based on a crystal thickness indication; this would be basically uncompensated for errors. Second, he could make each layer cut when the photometric level reached the predetermined value as the technique implies. Third, he could always make the cut at a fixed percentage of the reflectance excursion between the last maximum and minimum. This might partially compensate for a drifting photometric level, or loss of modulation due to monitor chip coating wedge or too wide a monochromator bandwidth. Fourth, he could use the technique which we call the "quick fix" after the work of Zhao[8]. In this case, the layer after the one where an error is detected is adjusted to bring the magnitude of the swings as close as possible to the design values. Each of these four strategies has its advantages as we shall show.

If we use the uncompensated approach and have a 6% of a QWOT RMS error at the center of the blocking band in all layers, we will get spectral curves of the kind shown in Fig. 7.62. The error in the edge gets greater as the error is in a layer closer to the center of the stack. This scheme uses no optical monitor, only a crystal, so there is no level monitoring. As we shall see by comparison with the other strategies to follow, this approach gives the greatest variation in the edge position, but one of the least variation in the overall pass band (400-600 nm). This is also the easiest for the operator to carry out.

If we use the fixed level monitoring strategy (second case) and have a 12% of a QWOT error in the 11th layer, the monitor chart will look like Fig. 7.63. A five-layer precoated monitor chip makes the sixth layer of the coated part the 11th layer on the monitoring planning chart. Here the layer was too short, but the next layer was cut at the constant level after passing the next turning point. This produces spectral results as in Fig.7.64. This will be found to have next to the best stability of the edge wavelength. However, it has the worst performance at the wavelength of the pass band where the QWOTs of the blocking band become half waves. If only the edge of the blocking band and its vicinity are important, this might be the strategy of choice, but not if a broad pass band is required. This is the easiest of the optical monitoring strategies for the operator to carry out because no calculations are required between cuts.

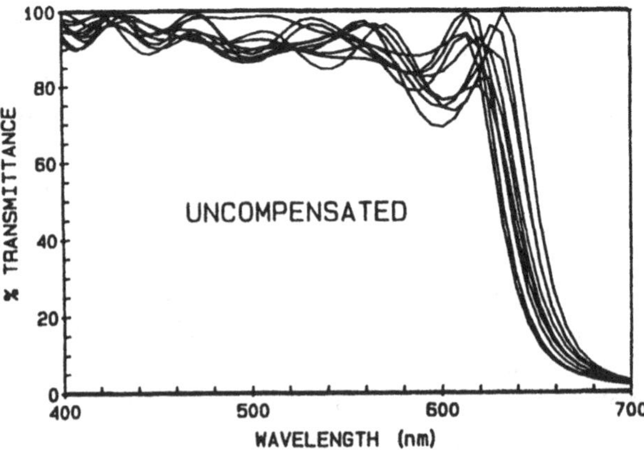

Fig. 7.62. Effect of uncompensated random errors of 6% of a QWOT RMS in the case of a design like that shown in Fig. 7.53.

In the third approach, the operator always makes the cut when the level reaches a fixed percentage of the excursion between the last maximum and minimum. The effect of an error in one layer is to cause small corrections in several subsequent layers as seen in Fig. 7.65. These corrections damp out eventually and a constant but offset level is restored. The spectral results of this "%-MAX/MIN" strategy are shown in Fig. 7.66. This has a medium performance at the edge and over the pass band. This requires the operator to make a new computation for each cut point as each new layer is deposited.

The optical monitor chart of the "quick-fix" strategy is illustrated in Fig. 7.67. An error is sensed in layer 11 where it was too thin. Point A is notably different from the previous minimum. The thickness of the next layer is adjusted to bring the pattern back to the design values in the 13th layer and thereafter by making the cut point at B. An interesting point is that the correction in layer 12 is just to stop the layer at the same level where the error occurred. This actually requires no calculation and is therefore easy for the operator. The spectral results in Fig. 7.68 show that the pass band is not quite as uniform as two of the other three approaches. However, this seems to be the best strategy by far for edge control. Constant optical level monitoring in one of these three forms appears to be the best strategy for edge type filters. The specific strategy can be chosen to best fit the application. The maximum sensitivity and error compensation with single wavelength monitoring can be achieved with this technique.

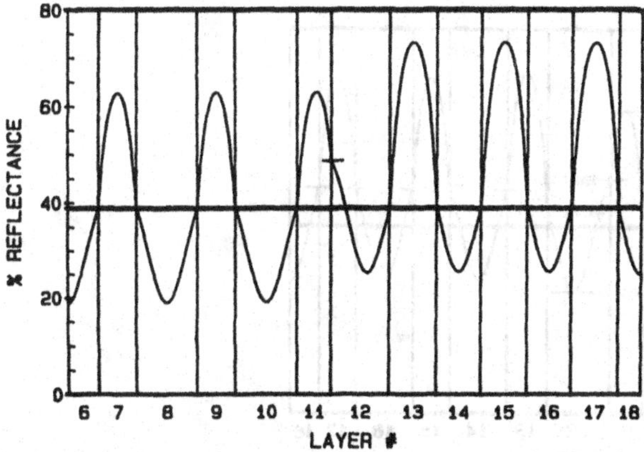

Fig. 7.63. Effect of a -12% of a QWOT error in layer 6 (layer 11 with precoat) on the monitor trace in the case of Fig. 7.53 monitored at g = .8086 when each subsequent termination is made at the predetermined constant photometric level.

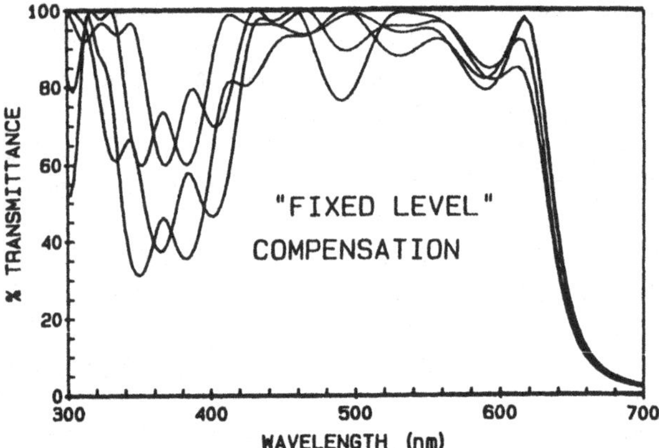

Fig. 7.64. Effect on the spectral performance of errors of ±12% in layers 6 and 7 in the case of Fig. 7.63.

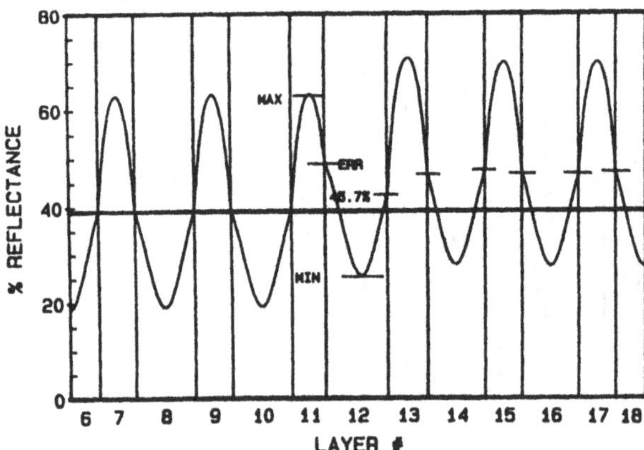

Fig. 7.65. Effect of the same error as in Fig. 7.63 when each subsequent termination is made at 45.7% up from the minimum of the last swing from maximum to minimum.

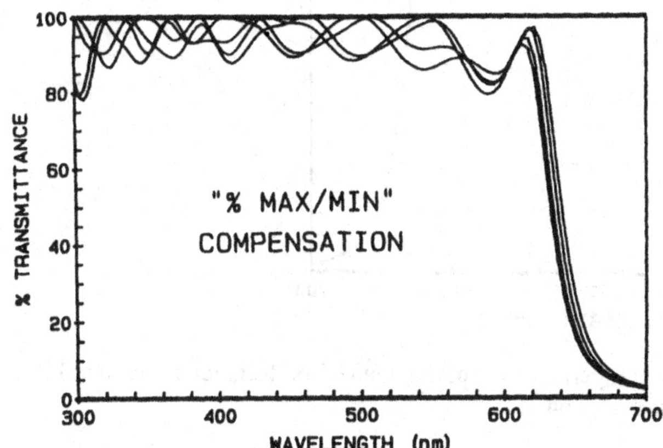

Fig. 7.66. Effect on the spectral performance of the "% Max/Min" technique to correct 12% errors in layers 6 and 7 as in Fig. 7.64.

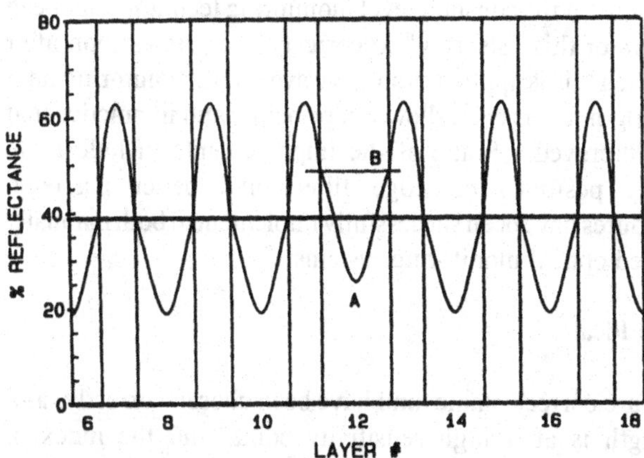

Fig. 7.67. Effect of the same error as in Fig. 7.63 where the next layer is terminated at the same level as the error per the "Quick Fix" technique.

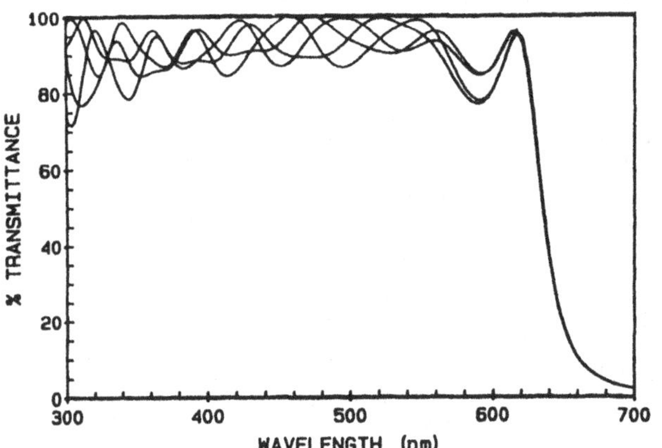

Fig. 7.68. Effect on the spectral performance of the "Quick Fix" technique to correct 12% errors in layers 6 and 7 as in Fig. 7.64.

7.6.6. Steering the Monitoring Signal Result

Zhao[8] briefly mentioned the possibility of the correction in real time of monitoring and process errors based on the observation of the actual turning values as compared to the expected values.

The control and stability of the constant level monitoring technique has been augmented by the addition of this "steering" to correct for errors as soon after detection as practical. When this is applied to single wavelength monitoring near the most sensitive wavelength, it has been shown in principle and in practice that the spectral results are improved. Some of the improvements include more repeatable spectral edge positions on edge filters and better passband transmittance. The procedures have been successfully implemented both manually and in computer-controlled optical monitoring systems.

7.6.6.1. Departures from Ideal

If the photometric levels are correct, stable, and have been chosen correctly, **and** the monitoring wavelength is at a high sensitivity point, **and** the index of refraction of the layers is constant, **and** the monitor spectral bandwidth is sufficiently narrow, etc., constant level monitoring works very well. **However**, in the real world, it is usually found that after many layers the process diverges or converges out of control. The probable causes of these instabilities are defects in the knowledge of the actual index of refraction of the layers being deposited and errors in the photometric scale of the monitor.

It is usually possible, through appropriate test coating runs, to characterize the index of refraction of the materials reasonably well. However, variations in rate, gas background pressure, temperature, etc., can cause index variations. If these occur, the maxima and minima or turning values will be at somewhat different levels than those predicted by the design. This will disturb the stability of the level monitoring.

It is probable that errors of the photometric scale of the monitor cause even greater instability. The absolute transmittance or reflectance is not usually reported correctly by an optical monitor due to factors such as coating buildup on some of the monitor optics within the chamber, and due to drifts in the source and detector circuitry. This latter factor could be eliminated by the use of a true double beam optical monitor, but the former factor might be more difficult to overcome.

There are few absolutes (at this time) in optical thin film production; most processes rely heavily on the calibration of most factors such as temperature, thickness, tooling factors, etc. In the case of the optical monitor, it would be highly advantageous to be able to calibrate the photometric level of the monitor just prior to the start of film deposition. If one is monitoring on a monitor chip, its reflectance or transmittance can be well characterized before loading it into the chamber. The monitor scale can then be calibrated by this reflectance or transmittance. If we monitor in reflectance with an uncoated crown glass monitor chip, the 4.2% reflectance is a somewhat poor calibration of where the real 100% scale point lies. In the case of transmittance monitoring, the 100% can be calibrated with much more confidence.

7.6.6.2. Steering Concept

From a knowledge of the indices of refraction of the materials and the design of the thin film which has the desired periodic structure for level monitoring, one can calculate what the reflectance or transmittance maxima, minima, and cut points versus optical thickness should be. As a real deposition proceeds, it is unlikely that the values will appear to be just as calculated. It is possible to "steer" the subsequent monitoring trace to the desired values by small adjustments in the cut points from the nominal values based on the error observed at the most recent turning point. The proportion of the percent adjustment (short or long) in the nominal cut point to the percent error at the turning point can be calculated at the design phase of the monitoring plan. For example, as illustrated in Fig. 7.69, if a maximum were 4% too low (layer 11), the correction would be to cut the layer at 3% higher reflectance (point X) than nominal. If a minimum were too high by 2% (layer 14), the layer should be terminated 4% higher (point Y) than nominal.

In principle, each correction has the power to bring the curve right back on track before the next turning point. This is because the correction is calculated to bring the cut point on the trace of admittance of the current layer to a point at the intersection of the admittance curve of the next layer which will bring it through the required turning value.

This procedure is similar to steering an automobile down a road. If the driver finds himself off of the center of his lane due to wind forces, inattention, etc., he makes the necessary corrections to bring the vehicle back on track. Our operators had learned to steer the monitoring process to stay on the road and not crash into the ditch. Arriving at the final destination of the monitoring will reasonably assure that the monitored wavelength will have the designed spectral properties. The properties of the results further away from the monitoring wavelength will be determined by how near the center of the track the process stayed during the whole trip.

7.6.6.3. Algorithm

We will now describe the algorithm which has been shown in principle and in practice to overcome these problems and give good spectral results in the filter products with which it has been used. This algorithm has been recently implemented on two automated optical coating machines with good results.

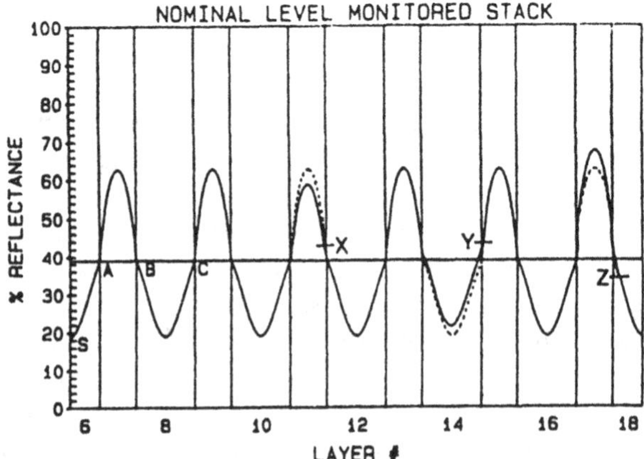

Fig. 7.69. Steering corrections (X, Y, and Z) for turning values that are too low (layer 11) or are too high (layers 14 and 17).

The algorithm has three sections which can be used or not as appropriate. The first section implements the basic "constant level" layer cutting scheme. The second section, which might be used instead of the first but not with it, makes it cut at a specified point which is a certain percentage or ratio of the distance between the last two extrema (max/min or min/max) of the monitoring curve. We will call the second section "general level" or "%MAX/MIN" monitoring. The third section is a correction factor to be applied to whichever of the first or second sections are used to account for the error sensed in the reflectance desired as the most recent turning point was past. This section has the power to correct for index variations, photometric errors, previous cut errors, and to some extent many of the other possible errors mentioned above. This third section is the steering correction factor.

The equation or algorithm for the photometric level at which to cut the layer is as follows:

$$
\begin{aligned}
C(I) &= J*(N*A(L)) && \text{"CONSTANT LEVEL CUT"} \\
&\quad + (1-J)*(A(I)-F*(A(I)-A(I-1))) && \text{"GENERAL LEVEL CUT"} \\
&\quad + K*(M*A(L)-A(I)) && \text{"STEERING CORRECTION"} \qquad (7.3)
\end{aligned}
$$

C(I) is the photometric reflectance (or transmittance) at which the Ith layer is to be terminated and I is the layer number.

F, I, J, K, L, M, and N are parameters entered to describe in advance the layer properties desired for each layer, I. The A's are the actual reflectances of the last extrema (max or min) of given layer numbers.

J is used to turn on either the "Constant" or "General" level cut sections. When J=1, the constant level section is used. When J=0, the general level section is used.

L allows us to refer to the photometric level reading of the extremum (A) of some previous layer number (L) as a photometric reference or calibration. For example, if the first layer was titania deposited on a fresh crown glass substrate, we might know from experience and other calibration that the maximum was expected to be 31% reflectance for A(L). If we then wanted to make constant level cuts at 15.5%, the value of N for each layer (I) would be set at .5 (and J=1 and L=1).

If we had a general level cut as in Fig. 7.69 at the eighth layer (I=8) where the cut is 45.4% up from that last minimum of the distance between the last minimum and the previous maximum. Then: I=8, J=0, F=.454. This leads to:

$$C(8) = (1) * (A(8) - .454 * (A(8) - A(7))).$$

When steering correction is to be used, K is not equal to zero ($K \neq 0$). The extremum of the current layer (as soon as it is past) is compared to the desired value M*A(L) which is a factor M times the reference extremum as previously described. In Fig. 7.69, L=1, A(L)=31, and therefore M=2.032 for the high layers and .613 for the low layers. The factor for the constant level is N=1.258. The difference from the desired values or error is multiplied by the K for layer I and added to the cut level determined in section one or two of the algorithm. The K for a given layer is determined by separate analysis by the coating designer. It may be positive or negative and is usually of magnitude from 0 to 3. In the example of Fig. 7.69 with silica and titania layers, it is −2.0 and.75 respectively.

7.6.6.4. More on Photometrics

If the photometric scale is significantly in error, it may be difficult for the operator or automatic control system to steer to the expected values. A more or less stable result may be possible, but the edge will be shifted from the expected point. This might be compensated for by an appropriate shift in the monitoring wavelength in the next run, if everything is otherwise reasonably reproducible. However, a more accurate photometric scale will produce more accurate results.

We discussed above the use of the reflectance of the uncoated monitor chip and also the peak of the first layer for photometric calibration. The uncoated chip is preferred because it is more reproducible even though it is less of the full scale to be calibrated. There are at least two other approaches, that we know of, which may be of use for further development of photometric accuracy. From the design process, we know where the nominal cut point should be as a percent of the swing between the last two turning points. We also know the ratio of the nominal cut point reflectance to the maxima reflectances and the ratio of the minima reflectances to the maxima. For a given design, these ratios can be derived as a function of monitoring wavelength and layer thicknesses for given indices of refraction. This information can be used to determine in real time what the photometric scale really is, **if** the indices can be correctly assumed. The details of these schemes are beyond the scope of this discussion, but our preliminary use of them has shown considerable promise.

7.6.6.5. Example Case

In most cases of an edge filter and a passband, as discussed in Sec. 7.6.4, we have been able to design an antireflection (AR) coating of 3 to 6 layers which also brings the monitoring reflectance to the required value for the constant level monitoring. If we liken the "steering" of the monitoring to driving down a highway, the first layers are like the "on-ramp" of the highway. This is illustrated in Fig. 7.70. The starting design for such an on-ramp can be done by designing

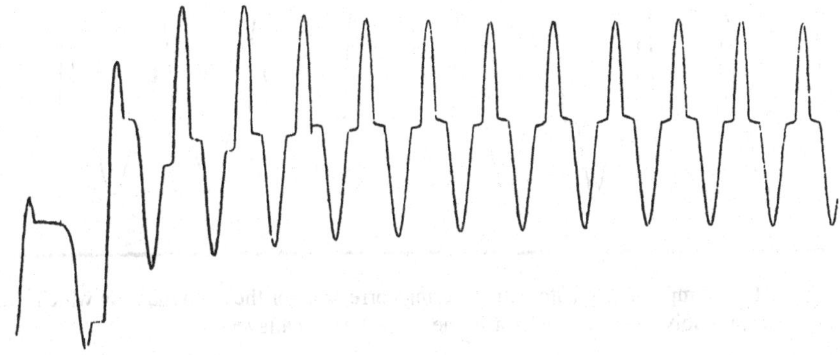

Fig. 7.70. Actual optical monitor chart trace from test run on the DynaVac 48" box coater system showing the "on-ramp" layers and constant level cuts with steering. Note the smooth transition of maxima and minima from start to constant levels.

an AR from the substrate to a fictitious media whose index is $n = (1-r)/(1+r)$, where r is the reflectance amplitude desired at the constant level cut points. The fictitious media is then replaced by the periodic stack and small adjustments are made as needed to the on-ramp layers to give the desired constant level monitoring for the periodic part of the structure. It can be noted as in Fig. 7.70 that the transition in reflectance versus thickness from the substrate to the highway portion is generally bounded by an envelope which is a smooth contour within which the oscillations from minima to maxima occur. This looks similar to and can be related to the quintic contours reported by Southwell et al.[21] and our own studies of the basic nature of AR coatings[22].

7.6.6.6. Lessons Learned

The steering techniques described were first applied in a Balzers BAK760 with an optical monitor by having the operators cut each layer at the appropriate level. This required the operator to observe the photometric level of the turning point for the layer and calculate (mentally or on a pocket calculator) the needed adjustment to the cut level. This was done with moderate success, but it can be tedious for the operator and divert his/her attention from other responsibilities during the process.

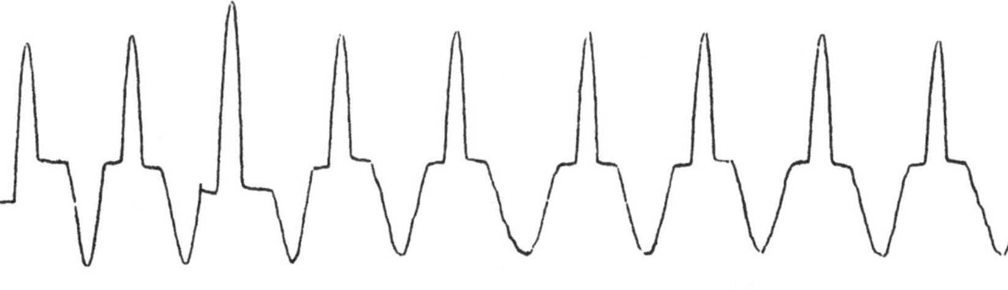

Fig. 7.71. Example of the automatic steering correction for the third maxima which was too high (probably due to a cut error in the second minima layer).

The algorithm was installed on two newly acquired chamber control systems: the Denton DVOCS system installed on a refurbished Leybold A1100 and the latest control system on a new DynaVac 48 inch box coater. The automated systems reduce the operator tedium and are more consistent in results (as might be expected). The repeatability of the automated systems allows better fine tuning of the parameters from test run results.

With reference to Fig. 7.69, we will point out some requirements and observations. Unless otherwise stated, these observation apply in both manual and automatic cases. It is necessary to determine how much further in reflectance each type of layer will coast after the cut command is issued and then incorporate this correction in anticipation of each cut point. For the given high and low index materials, the changes in cut point needed to get an errant turning point back on track must be figured at the monitoring design stage. This is what we have come to call the "gain" factor for the turning point errors of each material under the given design conditions. The target levels for the maxima, minima, and cut levels are all determined in advance from the knowledge of the indices, wavelength, and monitoring plan. These levels can be determined by the use of a thin film design program or other graphical methods. If the ratios of these three levels is in error, the proportions of the high and low index layer optical thicknesses will be in error and generate a "half-wave hole" in the transmittance which might not have been intended. This can be "tuned-out" in subsequent test runs by adjusting the cut level.

In the cases shown here, the monitoring is arranged such that the high index layers pass through a maxima. If the maximum is too high with respect to the target, the cut point will need to be lower than the nominal to get the next maximum to pass through the target level. This is the case illustrated in the third maximum of Fig. 7.71 which may have resulted from a spurious error in the cut of the previous layer. If a minimum is too high, the opposite is true, the cut will be higher than nominal. If the cut is made just right, the next layer after the cut will go exactly where it should (if the targets are correctly calculated from correct indices). The error, if any, in this next turning point can be used to tell if the gain of the most recent cut was too high or low. Changing the general cut level, as mentioned, will change to ratio of the high to low index layer optical thicknesses. It will also change the level of the maxima and minima. Lowering the cut level will raise the maxima and lower the minima (and vice versa). If the corrections become too large, the process can break down. For example, a turning point might be reached before the indicated cut level. This would not ordinarily be expected to happen in a well-behaved process.

The on-ramp layers in Fig. 7.70 use no steering, it comes into play at the 6th or 7th layer when the main highway begins. The extrema and cut points quickly settle in to relatively constant values if there are no perturbations as seen in Fig. 7.70. If the knowledge of the indices and photometric levels are correct, the steady state values will be as predicted. However, if there are differences from the expected index values, the values settle in to something different from nominal. We were surprised to discover that our centralized monitor chip was showing indices that were significantly lower than the parts being coated. We suspect that this was due to higher than average angle of incidence of the coating materials on the monitor chips as compared to the parts in the rotating calotte. This had to be taken into account. The resulting levels act like the stretching of a net to equalize the tensions between the target values, the actual values, and the force of the gain to correct the difference between them. This is also similar to driving down a highway with a strong crosswind. Constant pressure must be applied to the steering wheel to compensate for the force of the wind. Because of these and similar inaccuracies in available information, a reproducible edge wavelength may be achieved, but not at the correct wavelength. This can be corrected by adjusting the monitoring wavelength appropriately.

7.6.6.7. Results

Figure 7.72 shows the spectral scans of three successive production runs with the operators using the steering algorithm described here. It can be seen that the edge position was controlled to ±2.5 nm (0.4%) and that the passband in the region near the edge has relatively high transmittance. It should be noted that the substrate in each case is a heat-absorbing glass whose uncoated transmittance is also plotted in Fig. 7.72.

The use of the steering technique in either constant level or general level cut optical monitoring leads to better control and more reproducible results in the edge position and passband transmittance of edge filters of the periodic structure type. It can be employed by operator monitoring and/or automated optical coating systems.

Our experience points to the fact that automated systems have a higher likelihood of achieving good yields on complex and lengthy coatings of the type described. The steering algorithm is helpful if not critical in both manual and automated systems, but the automated systems remove the pressure on the operator and are more repeatable. The steering technique compensates for typical process variations and tends to stabilize the results. We believe that the approach described here offers the best control known to date for the edge position and adjacent passband of an edge filter.

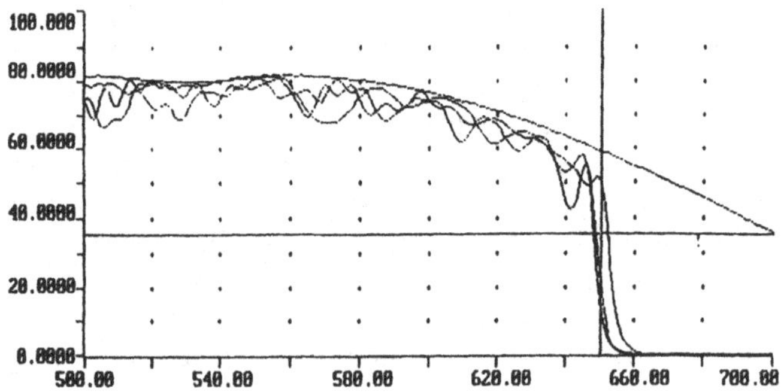

Fig. 7.72. Spectral results of three successive runs in the first implementation of the steering algorithm in an automated coater. Upper curve is the uncoated KG-3 glass.

7.6.7. Variation of Band-Edge Position with Monitoring Errors

We examined[30], by simulation, the effect of an error in the knowledge of the absolute value of the photometric termination level on the probable stability in the manufacture of the edge position of a blocked band. We developed empirical equations that allow the determination of the appropriate values of parameters associated with the optimum termination levels in order to minimize the effects of such errors. If the photometric level of transmittance and reflectance from the optical monitor of the part being coated is accurately known during the deposition of a periodic stack, then both the optical thickness and the index of refraction (and thereby the physical thickness) can be calculated. Because the photometric level is not usually as accurately known as might be desired, it is beneficial to attempt to minimize the sensitivity of the effects of errors in the absolute photometric level by the choice of the details of the monitoring strategy.

We have pointed out in Sec 7.6.1 above that the greatest sensitivity of change in reflectance or transmittance of the monitor signal with change in coating thickness occurs when the reflected intensity R equals 36%. We encountered disappointing results using this strategy to monitor an edge filter at a wavelength somewhat removed from the edge. The study reported here makes apparent the major cause of the problem. That is, the error in the knowledge of the absolute photometric level of the monitor signal has an increasingly more harmful effect as the monitor wavelength is farther from the band edge. As a consequence of this observation, we now choose to monitor as close to the edge as practical, and this implies layer termination levels slightly above 50% R as opposed to the 36% result of earlier studies. We have used simulation techniques, as described below, to reach these conclusions, and we developed equations to aid the design engineer in setting up such monitoring conditions.

The task of preparing for constant level monitoring is to find the effective index underlying the periodic layers that will yield the constant photometric level desired for layer termination and the monitor wavelength that is required. Alternately, the task of preparing for constant level monitoring at a specific monitor wavelength is to find the effective index and layer termination level that will give constant level monitoring at the wavelength required. Equations to aid in finding these values are given below.

This discussion will concern an edge filter as illustrated in Fig. 7.73. The monitoring curve represented in the form of Fig. 7.74 can be seen in the circle diagram Fig. 7.75 if the cut level were chosen at 53% R. If the design is of the form (0.5L 1H 0.5L)10 with the indices of H and L equal to 2.2 and 1.46 respectively as in Fig. 7.55, the reflectance needs to start at the point to the right of the origin in Fig. 7.75 where the locus intersects the real axis. This is the point where the reflectance amplitude r is equal to $0.588 + i0$, implying a phase of zero.

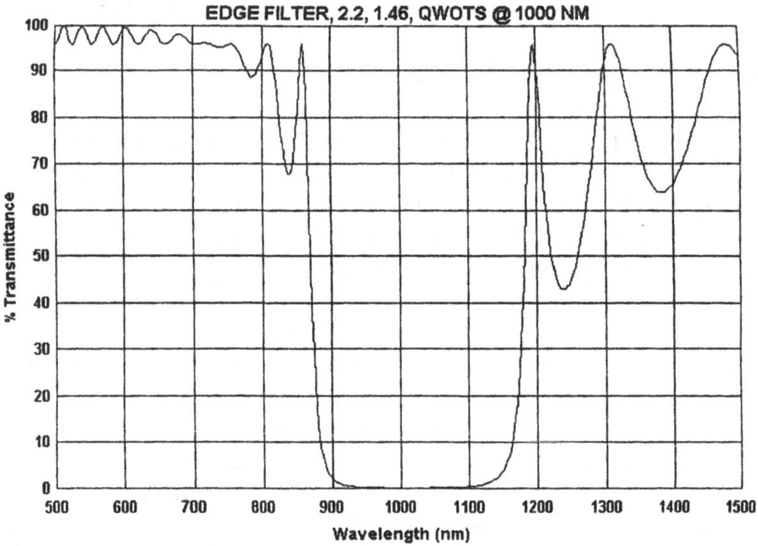

Fig. 7.73. Spectral transmittance of a short wavelength pass filter, (.5L 1H .5L)10 QWOTs at 1000 nm used for simulations of this work. H is index 2.2 and L is index 1.46.

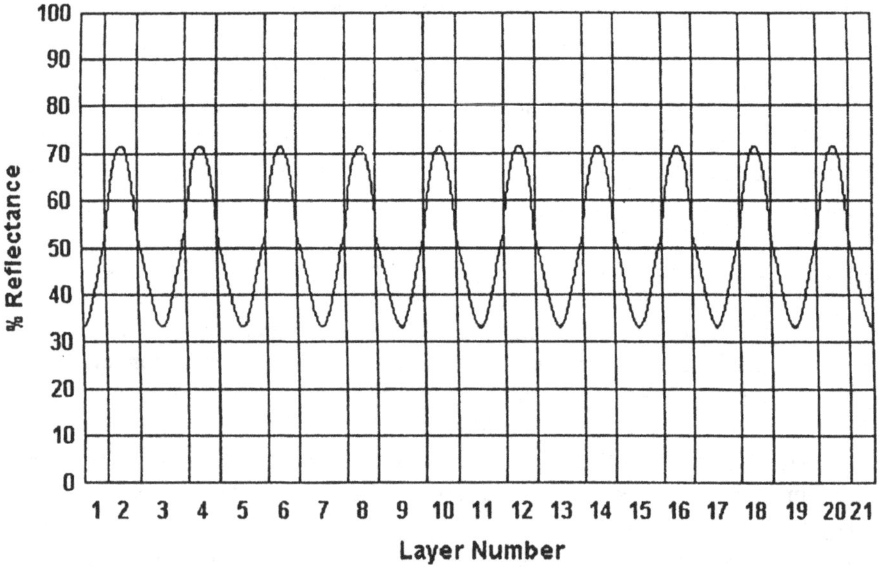

Fig. 7.74. Simulated monitoring trace where each of the periodic layers are terminated when the photometric level reaches 53%. This is monitored at 878.6 nm on an effective index of 0.2593.

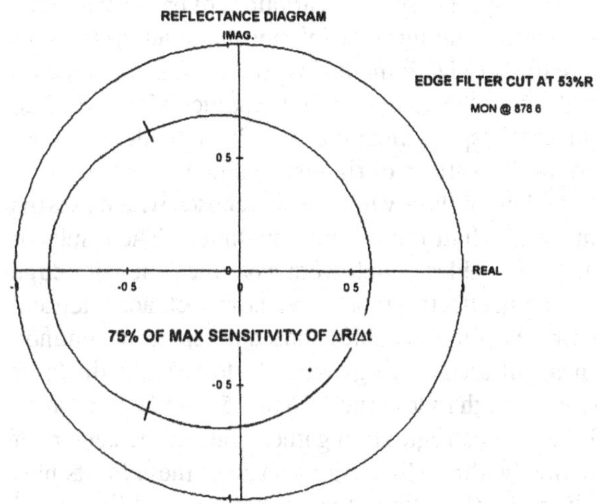

Fig. 7.75. The locus on a reflectance amplitude versus reflectance phase or "circle" diagram of the monitoring trace represented in Fig. 7.74 to achieve the desired result of Fig. 7.73.

When starting from this point, the first eighth-wave layer of low index (at the wavelength of the center of the stop band, not the wavelength of the edge) terminates at the marked point in the lower left quadrant. The next quarter-wave of high index terminates at the point marked in the upper left quadrant. Most optical monitors measure reflected (or transmitted) intensity R (or T). It is necessary to discuss both reflectance amplitude r and reflected intensity R = rr* in this type of investigation. The termination points discussed above are where R is 53% (and r is 0.728). The circle diagram often can give more insight with respect to some of these processes.

If the design were (0.5H 1L 0.5H)10 for a long wavelength pass (LWP) filter to be monitored near the edge between 1150 and 1200 nm, it would also be necessary to start where the locus cuts the real axis, but the effective index would be higher and the locus would begin with the high-index material. We confine our discussion here to the SWP filter cases, but the principles would be the same.

Because the substrate usually does not have an index equal to the required effective index, a preliminary coating of one or more layers must bring the reflectance amplitude point from the substrate to the desired start of the periodic structure as we have discussed above. In the case of a substrate of index 1.5, the reflectance would start at r = −.2 + i0 and move by successive layers to the point r = .588 + i0 which has an effective index of refraction of 0.2593.

The "experiments" used in this study were simulations derived by calculating the properties of the designs with respect to specific variations of parameters. The principal tools used to gain the understanding of the relationships and sensitivities in the work reported here were those of DOE methodology. With these standard tools, we are able to empirically find the equations and graphics of the required working relationships to adequate approximations for the intended purpose without recourse to the laborious derivations of rigorous mathematics.

The Box-Behnken type of DOE for three variables was chosen which uses the factors of Layer Termination Level, High Index, and Low Index. The results or responses as a function of these variables which were examined include: edge wavelength, monitoring wavelength, effective index, and the reflectance at top and bottom turning points. The ranges of the variables chosen to represent common ranges used for visible and near infrared coatings were: 16 to 56% for the layer termination level, 1.8 to 2.6 for the high index, and 1.36 to 1.56 for the low index. The Box-Behnken design for such a configuration gathers data at the centers of the twelve (12) edges that define the three-dimensional cube of these limits plus the center of the cube. Statistically, this allows the estimation of all linear and quadratic effects and all two-way interactions. Our preliminary studies showed that higher order effects were not of significance to our goals. The DOE methodology uses the ranges given to establish the values of the three variables at three levels for the thirteen (13) "experiments". A Box-Wilson or Central Composite Design might also have been used, but the Box-Behnken was chosen because it can model all of the quadratic and linear interactions, and the extremes of the experiments bound the region of interest in a cube.

The thirteen (13) simulations were carried out using thin film design software[28] to empirically find the effective index and monitor wavelength which would give constant level monitoring at the specific values of 16, 36, and 56% for the index combinations in the ranges given above. This also gave the maximum and minimum inflection or turning point reflected intensity levels as a result. The edge wavelengths were calculated directly from the design for the indices given. The DOE software was then used to performed a regression analysis to find the best fit of the data to the range of surfaces that can be modeled by the Box-Behnken design. Where coefficients were found to be statistically insignificant, they were dropped, and the analysis was rerun with the reduced configuration. The resulting coefficients were then used to generate surface plots of the results as a function of the variables, and equations were generated for the calculation of any point on the surfaces. As will be shown below, the most stable monitoring results are predicted to be with a layer termination level in the region of 53% reflectance when the materials are of index 1.46 and 2.3 (such as might be the case with SiO_2 and TiO_2). Equation 7.4 shows the variation of the necessary effective index with the high and low indices when the termination level is to be at 53%.

$$N_F = -0.07882 + 0.01522H + 0.2088L \tag{7.4}$$

Here, L and H represent the low and high indices of refraction, and K is the termination level in percent reflected intensity (%R). Equation 7.5 is for the general case at any termination level in the ranges used.

$$N_F = -0.624 - 0.00086K + 0.113H + 1.07L - 0.00184KH - 0.01625KL - 0.00021K^2 \tag{7.5}$$

Similarly, Eqn. 7.6 shows how the monitor wavelength varies at a 53% termination level with low and high index.

$$\lambda_{MON} = 1112.4 - 337.9H + 180L + 51.202H^2 \tag{7.6}$$

Equation 7.7 is for the general case at any termination level.

$$\lambda_{MON} = 1207.8 + 1.189K - 449.7H + 180L + 2.109KH - 0.0564K^2 + 51.202H^2 \tag{7.7}$$

The principal concern of this work is to have stable results from coating runs wherein the filter edge of interest has a reproducible offset from the monitor wavelength in the presence of possible errors in the absolute photometric level of the cut. We have therefore focused on the change in the difference between the monitor and edge wavelengths ($\Delta\lambda$) with the photometric cut level. It was somewhat surprising to find that the $\Delta\lambda$ is not a function of the low index of refraction, but only the termination level and the high index. All of the following figures have used L set to 1.46.

Figure 7.76 shows a view of $\Delta\lambda$ as a function of termination level and high index. It is apparent that there is a valley or trough where the change in $\Delta\lambda$ with termination level is zero. Figure 7.77 is an overhead or contour view with a line drawn approximately along the bottom of this trough. At such points, the effect of an error in the absolute photometric level of a constant level monitoring termination should have the minimal effect. That is the goal of this work.

Equation 7.8 represents $\Delta\lambda$ as a function of K and H, since L was found not to be a factor.

$$\Delta\lambda = -109.03 - 1.00065K + 112.5H - 2.11KH + 0.0538K^2 \tag{7.8}$$

The derivative of this with respect to K is given in Eqn. 7.9.

$$\Delta\lambda/\Delta K = -1.00065 - 2.11H + 0.1076K \tag{7.9}$$

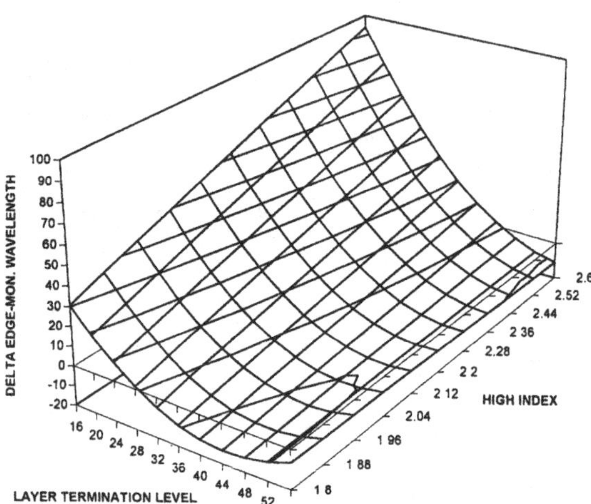

Fig. 7.76. View of $\Delta\lambda$, the difference between edge and monitoring wavelength, as a function of termination level, K and high index, H. Note the preferred area on lower right where the rate of change with both K and H approach zero.

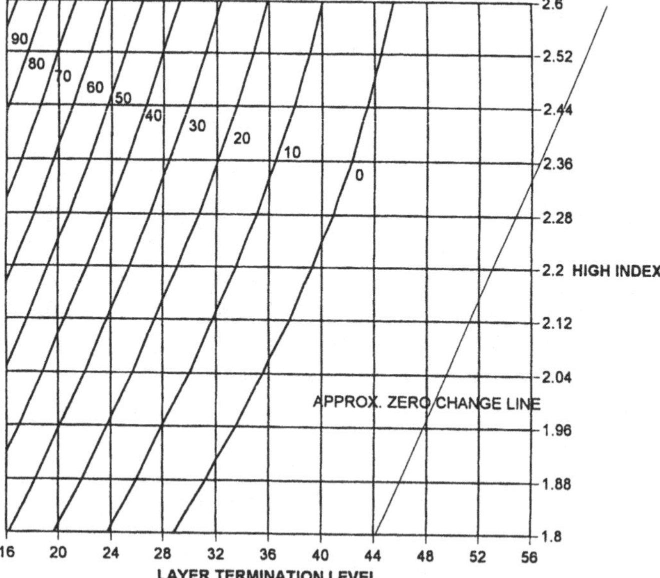

Fig. 7.77. Overhead or contour view of $\Delta\lambda$ as in Fig. 7.76 with a line drawn along the region of zero change of $\Delta\lambda$ with termination level.

When this $\Delta\lambda/\Delta K$ is set to zero, as in Eqn. 7.10, the minimum sensitivity to photometric error would occur.

$$K = 9.294 + 19.61H \tag{7.10}$$

This can be seen to be a linear function of H which would be 53% when H = 2.229. For the LWP case, the results are similar to everything discussed above and given in Eqn. 7.11.

$$K = -14.818 + 29.62H. \tag{7.11}$$

It can be found from these equations that the optimum termination point is the same (56.5%) for both the SWP and LWP filter when the high index is 2.4088.Figures 7.78 and 7.79 illustrate the effects on the edge position of a 1% photometric calibration error at a cut level of 16 (and 15) and 53% (and 52) respectively. These were generated empirically by finding the film characteristics when the termination was 1% different from that intended. When the variables of H = 2.2, L = 1.46, and K = 16 and 53 are inserted into Eqn. 7.9, we find the errors predicted are 0.06 and 3.9 nm respectively. This is in satisfactory agreement with Figs. 7.78 and 7.79.

Figures 7.80 and 7.81 show the predicted monitor trace and the reflectance circle diagrams for the 16% termination level case for comparison with those in Figs. 7.74 and 7.75 for the 53% case.

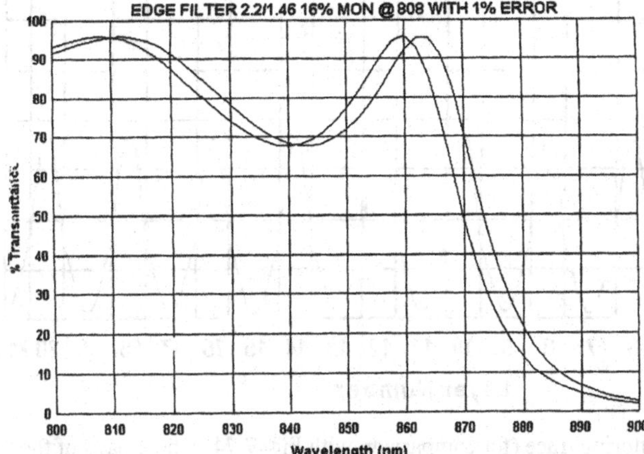

Fig. 7.78. Comparison of the edge position of the termination of each layer at 16% R with terminations having a photometric calibration error of 1% and a termination level of 15%. This is an amplified view of the same coating as in Fig. 7.73 and monitored at 808 nm.

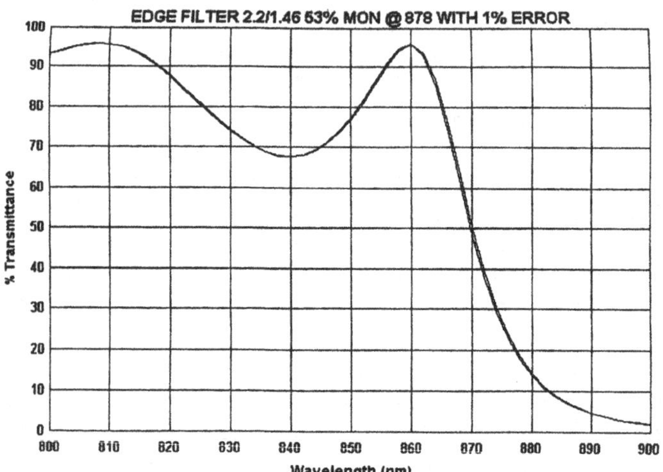

Fig. 7.79. Comparison of the edge position of the termination of each layer from the design of Fig. 7.73 at 53% R with terminations having a photometric calibration error of 1% and thereby a termination level of 52%. The monitoring wavelength is 878 nm.

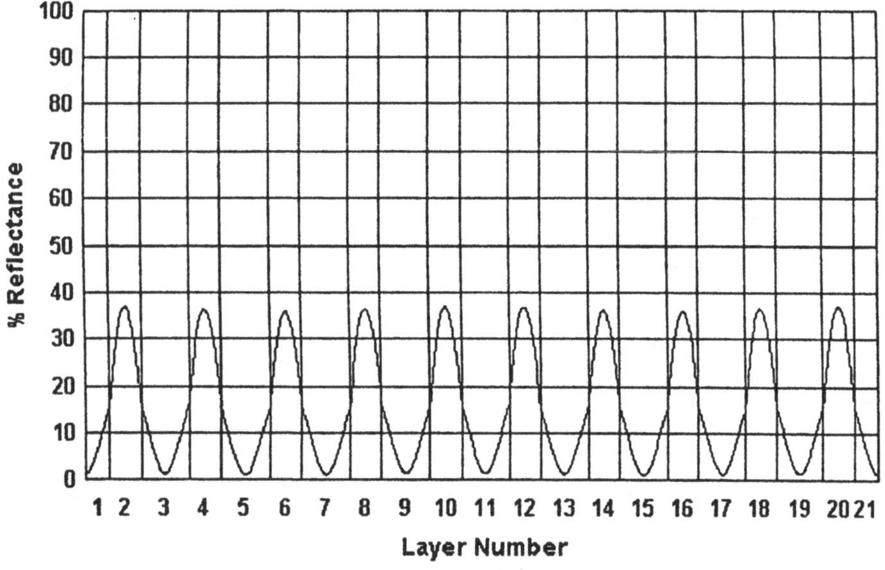

Fig. 7.80. Simulated monitoring trace (for comparison with Fig. 7.74) where each of the periodic layers are terminated when the photometric level reaches 16% and also with 1% error in photometric level at 15%. This is monitored at 808 nm on an effective index of 0.7826.

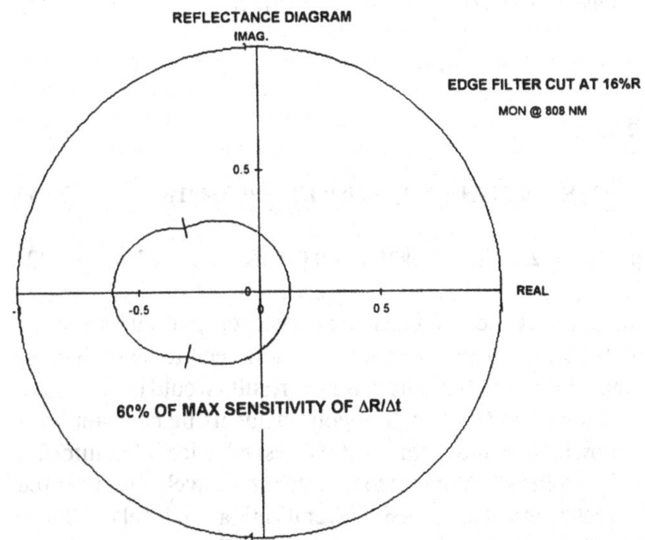

Fig. 7.81. The locus on a circle diagram of the monitoring trace represented in Fig. 7.80.

Some auxiliary results are shown in Eqns. 7.12 and 7.13, where the %R of the maxima ($\%R_T$) and minima turning points ($\%R_B$) in constant level monitoring as a function of termination level, high index, and low index.

$$\%R_T = -0.575 + 1.288K + 26.43H - 20.48L - 0.111KH + 0.283KL - 0.0074K^2 - 2.505H^2 \tag{7.12}$$

$$\%R_B = 12.64 + 1.122K - 2.603H - 8.25L - 0.565KL - 0.00822K^2 \tag{7.13}$$

If the monitoring is to be near the other edge of the block band for a LWP, the following equations are applicable as those above were for the SWP:

$$N_F = -2.175 + 0.1789H + 0.241L \quad \text{(@ 53\% TERMINATION LEVEL)} \tag{7.14}$$

$$N_F = -0.425 - 0.0159K + 0.852H + 0.241L - 0.0127KH - 0.000323K^2 \tag{7.15}$$

$$\lambda_{MON} = 1327.03 + 161.43H + 361.8L \quad \text{(@ 53\% TERMINATION LEVEL)} \tag{7.16}$$

$$\lambda_{MON} = 669 + 4.137K + 637.1H - 361.8L - 8.975KH + 0.1562K^2 \tag{7.17}$$

$$\Delta\lambda = -597.37 + 4.49K + 441.1H - 8.975KH + 0.1513K^2 \quad (7.18)$$

$$\Delta\lambda/\Delta K = +4.49 - 8.975H + 0.3026K \quad (7.19)$$

$$K = -14.818 + 29.62H \quad (7.20)$$

$$\%R_T = -5.34 + 1.233K - 6.044H + 13L + 0.075KH - 0.00491K^2 \quad (7.21)$$

$$\%R_B = 19.61 + 0.679K - 23.57H - 0.592KH + 0.0158K^2 + 8.026H^2 \quad (7.22)$$

We remind the reader that these studies were based on periodic stacks of equal quarter waves at 1000 nm. The wavelengths would need to be scaled for particular cases other than 1000 nm, but otherwise the results would be the same.

The root cause of some of our earlier disappointments using constant level monitoring have been shown. It now appears that the best practice in monitoring edge filters composed of periodic structures is to monitor at a wavelength near the edge of interest which gives a reflected intensity layer termination level of 50% to 55% (or 45% to 50% if monitored in transmittance). We have developed equations to select the most stable termination level (Eqns. 7.10 and 7.11) and to find the proper monitoring wavelength and effective index (Eqns. 7.4 to 7.7 and 7.14to 7.17) needed to achieve such monitoring.

7.6.8. Almost Achromatic Absentee Layers

A strategy which we have found useful on occasion is the use of what we call the almost achromatic absentee (AAA) layer[23]. A very thin layer of any index at the wavelength of spectral interest for the coating will have very little effect on the result in proportion to its thickness. This is commonly the case in the use of "glue" or adhesion enhancing layers. The index and often absorption of these layers is inconsequential because they are so thin.

We produced a large quantity of NBP filters in the 8 to 12 micrometer band using germanium and thorium fluoride. The monitoring hardware available had a long wavelength limit of about 1 micrometer. Germanium has a relatively large k-value (absorption) at 1 micrometer. By reflectance monitoring, we got monitor curves such as Fig. 7.82 where the thorium fluoride layers kept a relatively constant amplitude of modulation, but the modulation of the germanium layers rapidly decayed with thickness due to absorption. Some of the designs required layers of over 15 cycles of germanium at 1 micrometer, the modulation became highly obscured by noise at anything more than about 10 cycles. We can see what is happening to the monitor signal on the reflectance diagram in Fig. 7.83. As the germanium layer increases in thickness, the reflectance spirals inward on the point

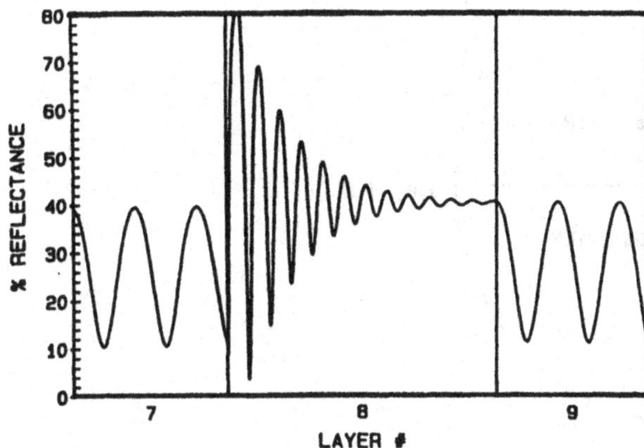

Fig. 7.82. Decay of the optical monitor signal in a layer which absorbs at the monitoring wavelength.

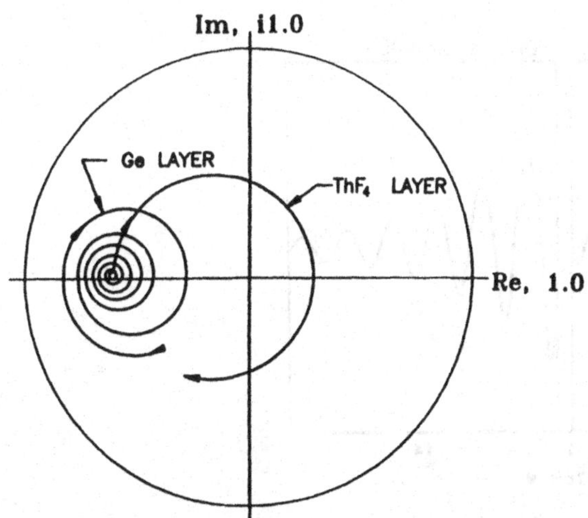

Fig. 7.83. Reflectance diagram of part of the monitoring trace from Fig. 7.82.

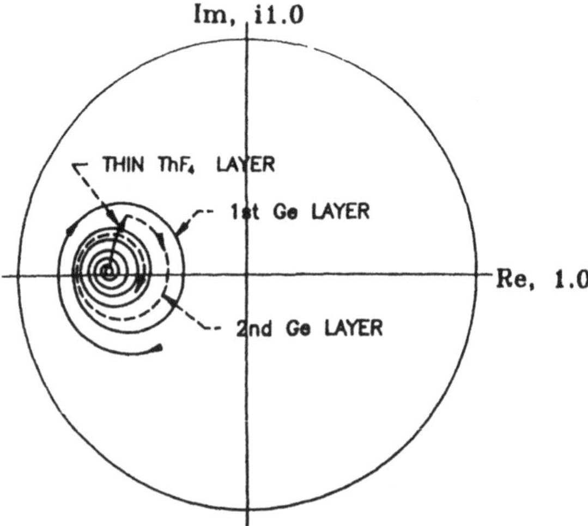

Fig. 7.84. Reflectance diagram showing the effect of a thin layer of low index injected in the middle of a long absorbing layer.

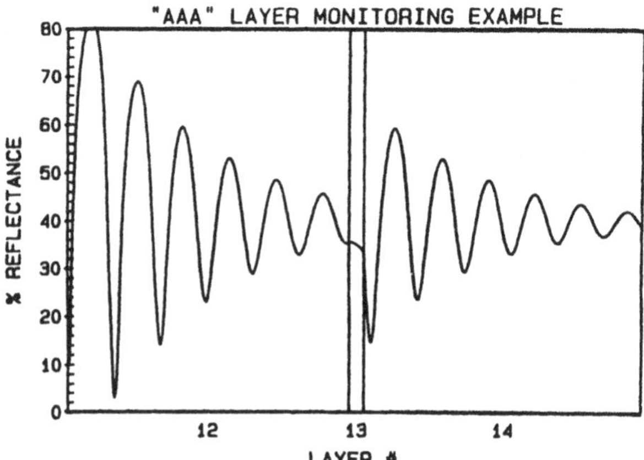

Fig. 7.85. Effect of the thin low index layer in Fig. 7.84 to "rejuvenate" the monitoring trace of Fig. 7.82.

which is the reflectance of a very thick layer of germanium. When the next low index layer is started, the reflectance locus travels a circular path from the start through a complete cycle at two QWOTs and then repeats itself because it does not absorb at this wavelength. Figure 7.84 shows that a small fraction of a QWOT of low index on top of the germanium will bring the reflectance out to where the germanium can be deposited and again produce adequate modulation. Figure 7.85 shows the effect on the monitoring signal. The modulation level is refreshed by this small injection of low index material. The effect on the NBP is quite small and can be adjusted for in the design. This AAA layer concept can thus be called upon to help when the modulation of the monitor signal "runs out of gas" due to absorption.

7.7. PRACTICAL CONSIDERATIONS

In this section, we will mention a few illustrative examples of the monitoring of thin films and then discuss some philosophical points and conjecture on the future of optical monitoring and automation.

7.7.1. A Narrow Bandpass Filter

At the end of the previous section, we mentioned the NBP filters for the 8 to 12 micrometer band which were monitored at 1 micrometer using the AAA technique. That family of filters successfully met requirements of bandwidth controlled to within 0.5% of the center wavelength and edges controlled to within 1.0% of the center wavelength. Many such parts were made in a single run where the center part was monitored directly in reflection and the others surrounded it in a single rotation calotte. This was a "pretreppanned" version of making a large single filter by direct monitoring and then treppanning out the smaller filters. The monitor used a NBP filter at 1064 nm instead of a monochromator. The operator made all of the layer cuts manually based on levels that were a given percentage of the change in reflectance from the preceding maximum and minimum in reflectance as recorded on a chart. A monochromator would probably have been required for transmittance monitoring because the bandwith of the 1064 nm filter would have washed out the signal as the stack of layers became thicker. The yield of the process was quite high. The success of these filters was an example of what can be done with relatively simple monitoring hardware.

7.7.2. A Special "Multichroic" Beamsplitter

Another practical example was a beamsplitter which had to be an AR coating at 40 degrees for the 8 to 11.5 micrometer band and a high reflector at 40 degrees from 700 to 950 nm and at 1064 nm and 1540 nm. The design selected consisted of 37 major layers plus glue layers. The substrate was germanium and the first eight layers were thorium fluoride and germanium which primarily functioned as the AR for the substrate. The rest of the layers used no germanium but zinc sulfide as the high index layer. This was because the *k* of the germanium would reduce the reflection in the region from 700 to 1540 nm. The layer thicknesses were optimized to give the required performance as shown in Fig. 7.86. The layers were not periodic. Semi-direct monitoring on a single germanium monitor chip by reflection was chosen. After some trials on the computer, the monitoring strategy selected was to monitor the first eight layers at 1064 nm and then change to 600 nm for the rest of the layers. The layers were cut by level monitoring as illustrated in Fig. 7.87. Here the scheme used was to make each level cut based on reaching a point which was a given percentage of the signal swing between the last maximum and minimum up or down from the last turning point. This made the monitoring less sensitive to slow drifts in the photometric level of the monitor signal, and it was unnecessary to know the actual photometric reflectance level.

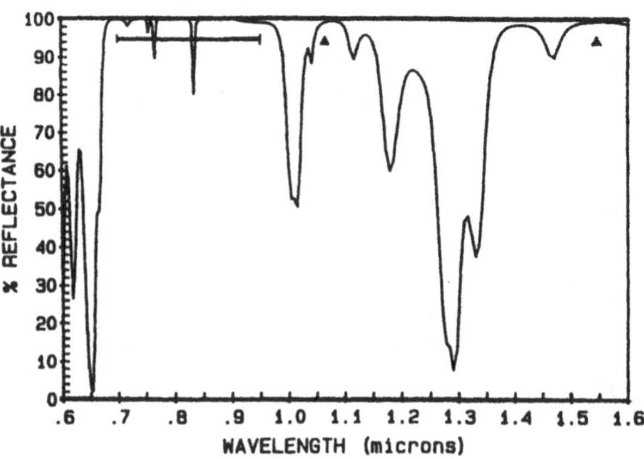

Fig. 7.86. Spectral performance of a "multichroic" beamsplitter which reflects (at 40 degrees) 700-950 nm, 1064 nm, and 1540 nm while transmitting the 8-11.5 μm band.

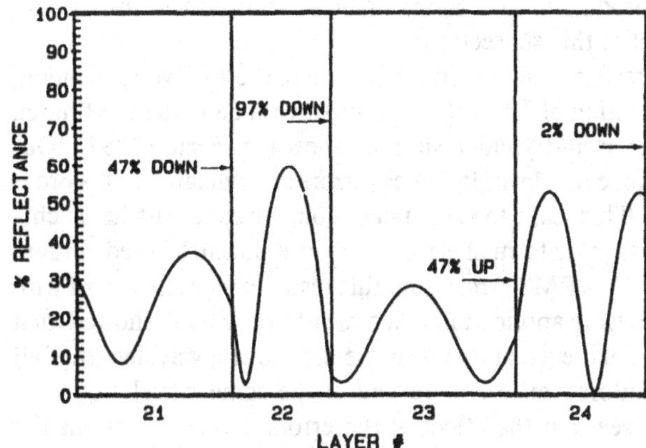

Fig. 7.87. Example of the optical monitor trace for level monitoring of some of the layers of the coating in Fig. 7.86 at 600 nm.

Once the tooling factors were determined by test runs, the monitoring showed sufficient control and reproducibility to give the required results. The most difficult part was to get the 1064 nm reflection peak at the right wavelength. This is an example of the general level monitoring technique.

7.7.3. A Very Broadband Antireflection Coating

In Chapter 1 and previous reports[24] we have described the underlying principles of very broadband antireflection coating design based on inhomogeneous layers. We further described the modification of such designs by approximating the inhomogeneous layers with a series of thin homogeneous layers. The number of these homogeneous layers was then reduced to a minimum which could satisfy certain spectral requirements. We will describe our experience in working to actually produce such a coating. This particular coating illustrates some of the current limitations of ordinary coating processes and the ability to control deposition over a large area.

We adapted a twelve-layer broadband design from the principles mentioned[25] and redesigned it to be optimum from 400 to 700 nm and at 1064 nm using SiO_2 and TiO_2. For reasons described below, we later changed the design to MgF_2 and TiO_2 and were able to eliminate one layer with no loss in performance. The resulting performance of the eleven-layer design is shown in Figure 7.88. The challenge of actually producing a coating which approaches the design performance is the subject of this subsection.

We simulated the effects of random errors in each of the 11 layers. Random errors with a standard deviation of 2.6% of a quarter wave at 584 nm (and limited to 5.2% maximum error) would yield results as shown in Figure 7.89. Our experience leads us to believe that layer thickness errors are typically of this order of magnitude in practice. This leads to some major concern as to whether such a coating is achievable unless some form of error compensation can be used. Figure 7.90 shows the effect of 0.65% RMS errors, and this result is more in accord with what is required for this coating application. We recall that Zhao[8] showed that error compensation can be achieved at and near the monitoring wavelength if all layers of a stack are monitored on one chip and at the same wavelength. In looking at Fig. 7.89, we see that the effect of the errors is most severe at the shortest wavelengths and not too severe at the 1064 nm band. This led us to choose the monitoring wavelength near 400 nm and on a single monitor chip. Figures 7.91 and 7.92 show the monitoring plan which evolved. We applied the previously discussed sensitivity principles and monitoring strategies, but we did not use precoated monitor chips to enhance sensitivity. We have not yet applied this additional step to the present case, but it might further improve the yield of the process. The design has several layers that tend to be too thin for easy optical

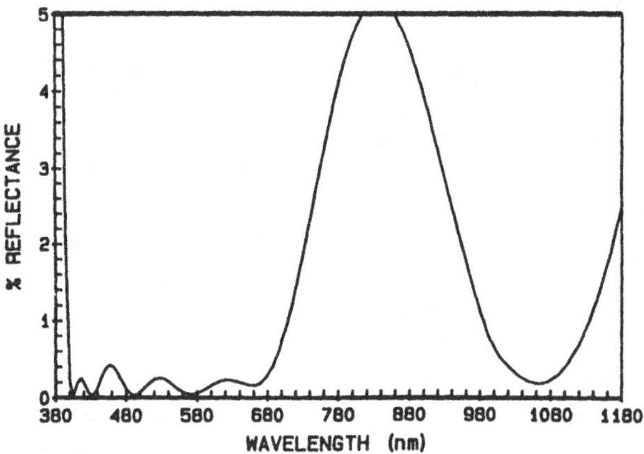

Fig. 7.88. Spectral performance of an 11-layer AR design for 400-700 nm and 1064 nm.

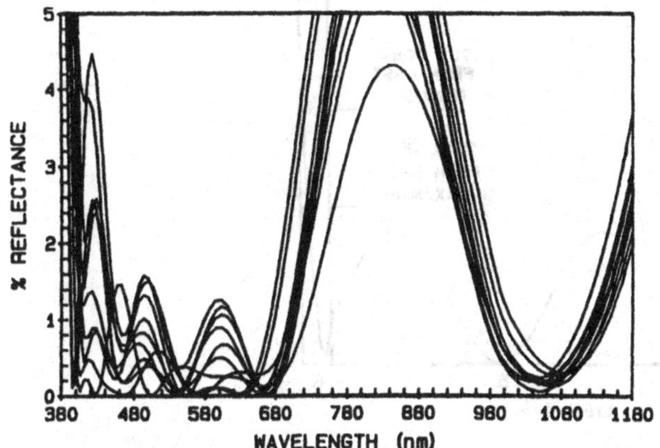

Fig. 7.89. Effect of 2.6% of a QWOT RMS random errors on the performance of the design of Fig. 7.88.

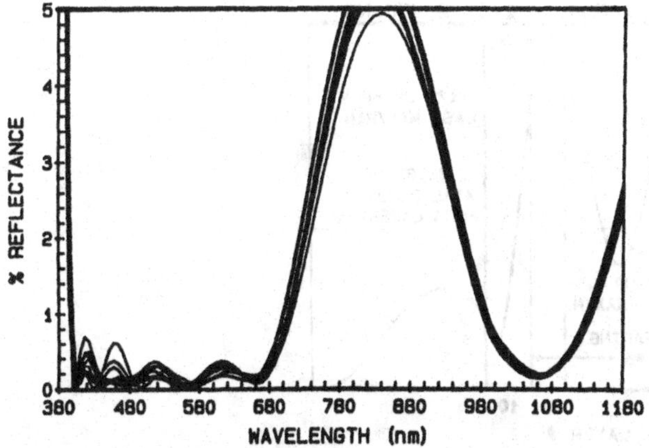

Fig. 7.90. Effect of 0.65% of a QWOT RMS random errors on the performance of the design of Fig. 7.88.

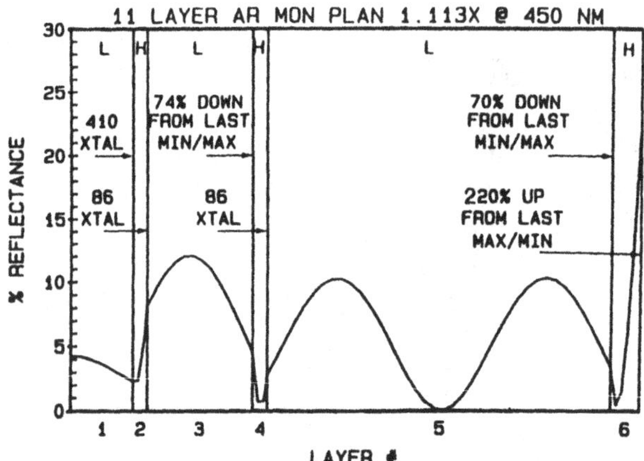

Fig. 7.91. Monitoring plan used to produce the design of Fig. 7.88 which combines optical and crystal monitoring in a manual adaptation of principles related to the work of Schroedter.[18]

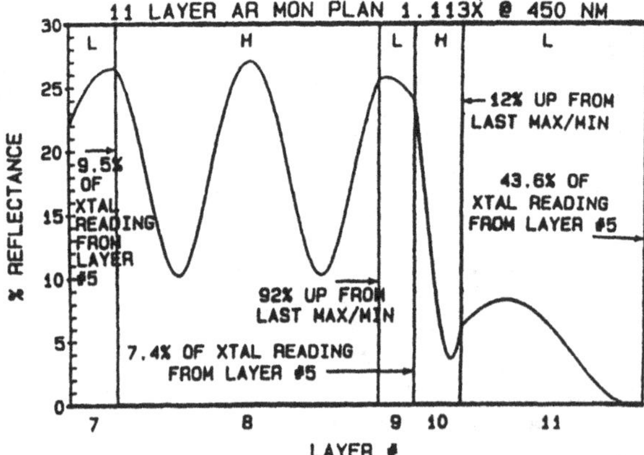

Fig. 7.92. Continuation of the plan in Fig. 7.91.

monitoring. This is common to many coatings of this type. Crystal monitoring is usually most appropriate in these cases, but there can be significant variability and errors if the crystal readings are not sufficiently calibrated. We usually favor optical monitoring because it controls the parameter most important to the result, the optical reflectance. Schroedter's[18] technique would have been an asset here, if we had the hardware and software to handle it. Therefore, we applied the basic concept of Schroedter in a manual mode. The first and second layers as seen in Fig. 7.91 are thin and not well suited to optical monitoring. The best calibration available in this case for the optical thickness of the high(H) and low(L) index materials versus the crystal readings is data from the most recent previous run under the same conditions. The crystal readings are calculated on the basis of the values of the readings from the previous run on layer 5 for L and layer 8 for H. Layer 3 is cut optically as 74% percent down from the previous peak, in terms of the reflectance change from the minimum of layer 2 to the maximum of layer 3. If the indices of the H and L layers are somewhat different from the design values, the absolute reflectance of the extrema will not be correct, but the relative reflectance will be nearly correct. Layer 4 is again a crystal cut based on the previous run calibration. Layer 5 is an optical 70% down from the last max/min. The crystal reading from layer 5 is a calibration of the L material to be used on later layers 7, 9, and 11 in the current run. Layer 6 could be cut either optically or by crystal, but we finally chose the optical cut as most sensitive and reliable in this case. Layer 8 is an optical cut and calibrates the crystal for H in future runs. From experiment, we found layer 10 to be best cut optically, although its thickness could be well controlled by the calibration from layer 8. We will come back to this issue below in the discussion of error compensation. Layer 11 could be cut either way, but we found the crystal somewhat preferable because the cut is too close to the turning point for good photometric sensitivity. An alternative for the system might be to monitor at a slightly shorter wavelength so that layer 11 would go enough beyond the turning point to get a good optical cut. This would have to be viewed with respect to its effects on the cuts of the other layers. This is the plan which was executed by the operator in producing the results described below.

The first attempts to produce this coating were with SiO_2 and TiO_2 from an electron beam gun. We encountered a major problem. The monitor chip could be made to have a very good result, but the witness chips in the planet would vary significantly from the monitor and (more importantly) from run to run. We attribute this to the difficulty of achieving a reproducible and constant angular material distribution from SiO_2 evaporated from a gun. We changed the design and process to TiO_2 from a gun and MgF_2 from a boat and found the results more reproducible from run to run.

The results of the first test run of the new design are shown in Figure 7.93. The actual spectral measurements from witnesses and monitor chips were inserted as design goals in a thin film design program. The thicknesses were optimized to fit the resulting reflectance spectrum to the measured values. The resulting thicknesses were compared to the design and used to adjust the crystal and optical monitor plan for the next run. We found, not surprisingly, a different tooling factor between the monitor and the witnesses in the planets for each material. As a result, the coating on the monitor chip needs to be something other than the "perfect AR" in order for the witnesses in the planets to have the coating desired. The results of the second test run based on these adjustments are shown in Fig. 7.94. The adjustment process was repeated. The results of the next run is the best of the curves seen in Fig. 7.95.

The results of two additional subsequent "production" runs with the same parameters as the third test run are plotted in Fig. 7.95. When we compare Fig. 7.95 with Fig. 7.90, we find similar results. This implies results that are nearly the same as those predicted for random layer errors of about 0.65% of a QWOT RMS. This is a pleasant surprise as compared to what might be expected based on Fig. 7.89. The measured data on the three results in Fig. 7.95 were fit as mentioned above to determine the thickness errors from the design. The RMS errors were 2.63, 2.95, and 4.40% respectively, and the worst errors in each run were 5.20, 7.60, and 10.84%. The effective monitoring wavelength was about 400 nm. It can be seen from Fig. 7.95 that a satisfactory result was achieved. The

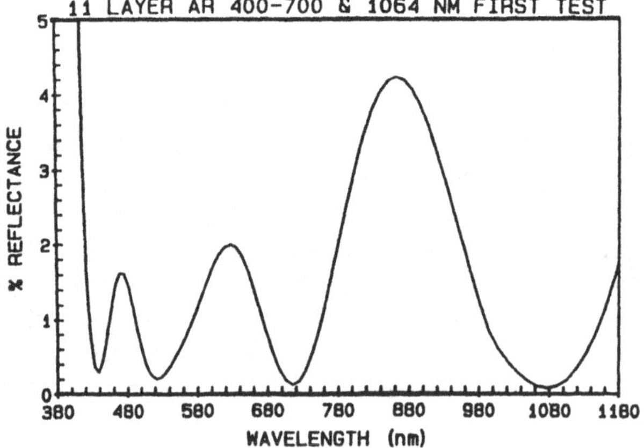

Fig. 7.93. Results of a first test run on the design of Fig. 7.88.

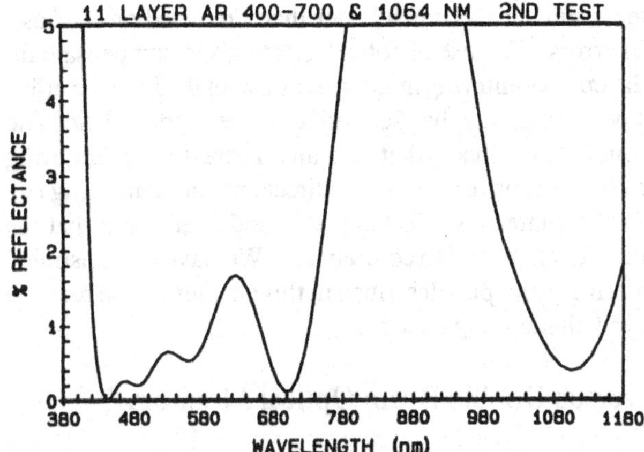

Fig. 7.94. Results of a second test run based on analyzing the results of the first run in Fig. 7.93.

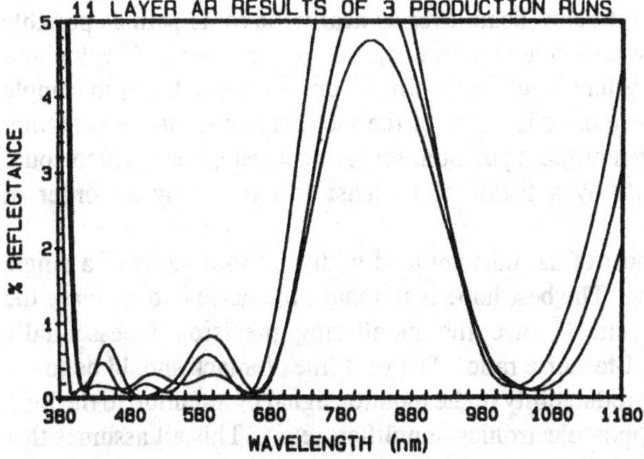

Fig. 7.95. Results of three production runs based on the adjustments from analyzing the second test run in Fig. 7.94.

1064 nm band was well controlled and reproducible with less than 0.25% reflectance in each case. The 400 to 700 nm range averaged less than 0.3% with a worst peak of 0.7%. Further refinement of the monitoring plan may also be possible to reduce the occurrence of reflection peaks above 0.5%.

We conclude that the scheme selected incorporates the benefits of error compensation because the results are so much better than predicted on the basis of random (uncorrelated) errors. The use of optical cuts tends to compensate for some errors in previous layers. Monitoring at the short wave end where the effect of random errors would be most severe, has seemed to be the correct choice and in agreement with the findings of Zhao[8] that the area nearest the monitoring wavelength is best controlled. We have seen that indirect optical monitoring can have sufficient control for the materials selected (TiO_2 and MgF_2), but that the SiO_2 process used was inadequate for this requirement. We have demonstrated that the design and monitoring principles described in this chapter can be reduced to practice to make real and challenging coatings.

7.7.4. Single Beam versus Double Beam Optical Monitors

Almost all of the optical monitors in use today are essentially single beam systems with all of the inherent problems of the single beam spectrophotometer which faded from the marketplace decades ago. The photometric level of reflectance or transmittance indicated by a single beam optical monitor will be in error and vary if any of the following factors change: light source brightness, detector response, electronic amplification, atmospheric and other absorptions in the optical path, etc. Double beam systems attempt to remove the effect of these changes by passing the sample beam and a reference beam over as nearly the same path as possible from a common source through common optics to a common detector and electronics. The ideal system would only differ from reference beam to sample beam by the transmittance or reflectance of the monitored part and its coating. This appears to be an area of great potential for future development which could reduce monitoring errors by a factor of at least two if not by an order of magnitude.

For the present, most of us must contend with the limitations of a single beam monitoring system. The best hope is to reduce the noise and increase the signal as much as practical since the monitoring precision is essentially proportional to the signal-to-noise ratio. The next line of attack should be to get as much stability and reproducibility in the monitor signal by attention to the light source, detector, and support electronics (amplifiers, etc.). This all assumes that the mechanical elements do not introduce problems. Some mechanical problems that we have seen are: vibration of the part being monitored, fluctuations due to tilts if the part is rotated, fixed monitor chip misalignments as the chamber pressure and temperature changes distort the structure, etc. A good system test which we have used is to turn on the monitor and chart recorder as the chamber door is closed, and see how constant the photometric level is as the chamber is pumped and heated, as the parts are rotated, and as the material sources are activated (before evaporation). None of these things should affect the signal or noise level. If we cannot depend on a stable photometric level which can be calibrated accurately, we cannot expect good results from monitoring schemes which depend on terminations of layers at specific levels of reflectance or transmittance. This may still be left for the future of true double beam systems. In today's common situation, it is most promising to make the single beam system as stable and noise-free as practical and use monitoring strategies which depend only on the shape of the monitor trace, not its absolute levels. By this we mean percentages of the photometric distances between turning points and the turning points themselves. We have already discussed extensively how to get the most sensitivity and least error using these techniques.

7.7.5. Automation versus Manual Monitoring

Some years ago we did a survey of the western world for commercially available automated "box" coaters[26]. Morton[27] had previously reported on the results of automating several box coaters at Texas Instruments. Morton's work actually started as early as 1976 and Leybold-Heraeus had started development of an automated optical monitor before that time. The thrust of these automated systems is to match or exceed the results of a skilled operator on a manually controlled system. This can then make it practical to accomplish more production through using more automated systems and fewer or lower skilled operators. Less experienced operators can achieve good results with the automated systems. As might be hoped, Morton reported that average throughput was increased by more than 20% and losses due to spectral and environmental problems were reduced by 50%. We believe this can be attributed to reducing operator errors and greater stability and reproducibility of all of the parameters controlled by the system.

However, manual control of optical coatings is not by any means a very inefficient process in general. The economic factors of each situation need to be examined. The skill and experience of the operator is very important, but the labor cost may not be as critical as the capital cost in many cases. The operator may be better able to deal with unpredictable variations if they exist. For example, the evaporation rate and angular distribution of SiO_2 and other materials can be difficult to control reproducibly and get a uniform deposition rate. This is a major difficulty to be overcome in most of today's systems, and is the cause of many of the shutdowns that we have seen in automated chambers.

We can use the machine tool industry as a historic guide because it is more mature than the optical coating equipment industry. Computer numerically controlled (CNC) machine tools such as lathes and milling machines have been developed to a high degree in the last three decades. They are typically controlling positions and feed rates of a limited number of axes. They are controlling less than a score of parameters. The CNC coating chamber, on the other hand, must control an order of magnitude more parameters, and these can be more unruly ones as mentioned above. The market for automated coaters is much smaller than for machine tools. As a result of these and other factors, CNC coaters are not as mature a product as machine tools and there are only a few choices in commercially available machines. It is rare, in fact, to find a CNC coater as an off-the-shelf item, they are usually only built on order. It appeared a few years ago that the additional cost beyond a manually monitored box coater for the automation to handle all normally required optically monitored processes was about 50%. As these systems mature, more and more aspects of even the "manual" systems will be automated and the differential cost should be less and less. We should soon be to the point where many more systems with all of the automation available will be economically justified by the payback in yield and efficiency. A word of caution may be in order, however. Our experience to date indicates that the reliability and maintainability of current systems may be most like the state of the microcomputers in the early 1980s. In that field, there has been a very significant maturing in the reliability of PCs, a great increase in capability, and a reduction in cost.

The development of the areas of automated source control and reliable optical monitoring seem most critical to the future of CNC coaters. The implementation of Schroedter's approach would seem to be a major improvement; a double beam system would further enhance the reproducibility. Although these things would help even a system that was not fully automated, full automation is a minor step beyond most of today's new box coaters as these improvements are added.

We think that Morton's summary in 1981 is still as true today as it was then. To quote Morton, "We feel that what we have done is only one more step in upgrading equipment used in the optical coating industry, and we challenge the equipment vendors to match and exceed what we have done here. We recognize that the expense for this kind of effort is considerable and that the market for optical coating equipment is limited. Nevertheless, the next levels of improvement in optical coatings probably will not come from the thin film designer. The future lies in improved control and innovative monitoring techniques."

7.8. SUMMARY

We have provided both an overview and a detailed discussion of most aspects of thin film growth monitoring as it relates to optical coatings. We showed that errors in index of refraction effect primarily the photometric properties of a coating while optical thickness errors affect the wavelength properties. The various systems of monitoring were discussed from the very simple measured charge to the more complex broad band optical monitors. It was seen that simple single layer ARs and mirrors might be successfully done with very simple equipment and monitoring while complex coatings require more complex equipment and techniques. Currently available monitoring techniques have been described and the relative merits of each for differing requirements were discussed.

It was demonstrated that error compensation mechanisms inherent in certain monitoring choices make it possible to achieve satisfactory results in the presence of surprisingly large errors. We have discussed strategies for taking advantage of this error compensation in several examples. It was seen that the control of the spectral results is greatest in the region surrounding the monitoring wavelength and less for wavelengths more remote from that point. This leads to the concept of choosing the monitoring wavelength which controls the most important and/or the most unruly part of the spectral result. We briefly discussed the impact of limiting factors such as monochromator bandwidth and wedge in the deposited layers on the monitor chip. We introduced the concept of Schroedter where the best features of both the optical and crystal monitors are combined to achieve better results than is possible with either monitor by itself.

The relative sensitivity of optical monitoring as a function of wavelength and photometric level has been shown. It was seen that the most sensitive point to terminate a layer is near 36% reflectance (R) and midway between a maximum and a minimum turning point. On a reflectance diagram, this is near 0.6 reflectance (r) and on the imaginary axis (phase equals 90 or 270 degrees). An example of using precoated monitor chips to apply this principle was given. It was further shown in Sec. 7.6.7 that the most stable place to monitor an edge filter is at the edge and where the reflectance is about 50%.

Three approaches to dealing with errors in constant level monitoring were compared. The "quick fix" technique was shown to give the best edge control while the "% max/min" gives slightly better pass band results. The "trick" of inserting a thin layer in the middle of an absorbing layer to restore modulation of the monitoring signal was shown.

Three coating examples were given to show how the various principles and techniques were applied in practice with equipment commonly in use today. The limitations of today's single beam optical monitors and optical coater automation were discussed and we conjectured on the possibilities for the future.

We have attempted to consolidate the findings and underlying principles of thin film monitoring which have evolved over decades by the efforts of those referenced and many more. It has been our intention to give the experienced reader a few new tools or viewpoints for his repertoire and to also provide an introduction and coherent overview for one relatively new to the field.

7.9. REFERENCES

1. W. P. Thoeni: "Deposition of Optical Coatings: Process Control and Automation," *Thin Solid Films* **88**, 385-397 (1982).
2. H. K. Pulker, G. Paesold, and E. Ritter: "Refractive indices of TiO_2 films produced by reactive evaporation of various titanium-oxygen phases," *Appl. Opt.* **15**, 2986-2991 (1976).
3. H. A. Macleod: "Monitoring of optical coatings," *Appl. Opt.* **20**, 82-89 (1981).
4. H. W. Lehmann and K. Frick: "Optimizing deposition parameters of electron beam evaporated TiO_2 films," *Appl. Opt.* **27**, 4920-4924 (1988).
5. J. M. Bennett, E. Pelletier, G. Albrand, J. P. Borgogno, B.Lazarides, C. K. Carniglia, R. A. Schmell, T. A. Allen, T.Tuttle-Hart, K. H. Guenther, and A. Saxer: "Comparison of the properties of titanium dioxide films prepared by various techniques," *Appl. Opt.* **28**, 3303-3317 (1989).
6. D. R. Gibson, P. H. Lissberger, I. Salter, and D. G. Sparks: "A high-precision adaptation of the 'turning point' method of monitoring the optical thickness of dielectric layers using microprocessors," *Opt. Acta* **29**, 221-234 (1982).
7. H. A. Macleod and E. Pelletier: "Error compensation mechanisms in some thin film monitoring systems," *Opt. Acta* **24**, 907 (1977).
8. F. Zhao: "Monitoring of periodic multilayers by the level method," *Appl. Opt.* **24**, 3339-3342 (1985).
9. B. Vidal, A. Fornier, and E. Pelletier: "Optical monitoring of nonquarterwave multilayer filters," *Appl. Opt.* **17**, 1038-1047 (1978).
10. B. Vidal, A. Fornier, E. Pelletier: "Wideband optical monitoring of nonquarterwave multilayer filters,"*Appl. Opt.*. **18**, 3851-3856 (1979).
11. P. J. Martin and R. P. Netterfield: "Optimization of Deposition Parameters in Ion-Assisted Deposition of Optical Thin Films," *Thin Solid Films* **199**, 351-358 (1991).
12. P. Bousquet, A. Fornier, R. Kowalczyk, E. Pelletier, and P.Roche: "Optical filters: monitoring process allowing the auto-correction of thickness errors," *Thin Solid Films* **13**, 285-290 (1972).
13. R. R. Willey: "Optical thickness monitoring sensitivity improvement using graphical methods," *Appl. Opt.* **26**, 729-737 (1987).
14. B. Vidal and E. Pelletier: "Nonquarterwave multilayer filters: optical monitoring with a minicomputer allowing corrections of thickness errors," *Appl. Opt.* **18**, 3857-3862 (1979).
15. C. Holm: "Optical thin film production with continuous reoptimization of layer thicknesses," *Appl. Opt.* **18**, 1978-1982 (1979).

16. R. R. Willey: "Realization of a Very Broad Band AR Coating," *Proc. Soc. Vac. Coaters* **33**, 232-236 (1990).
17. F. Evangelisti, M. Garozzo, and G. Conte: "Structure of vapor deposited Ge films as a function of substrate temperature," *J. Appl. Phys.* **53**, 7390 (1982).
18. C. Schroedter: "Evaporation monitoring system featuring software trigger points and on-line evaluation of refractive indices," *SPIE* **652**, 15-20 (1986).
19. R. R. Willey: "Sensitivity of Monitoring Strategies for Periodic Multilayers," *Proc. Soc. Vac. Coaters* **30**, 7-14 (1987).
20. P. Guo, S. Y. Zheng, C. N. Yen, and X. Ma: "Reflectance diagram-aided technique for optical coating design and monitoring," *Appl. Opt.* **28**, 2876-2885 (1989).
21. W. H. Southwell, R. L. Hall, and W. J. Gunning: "Using wavelets to design gradient-index interference coatings,"*SPIE* **2046**, 46-59 (1993).
22. R.R. Willey: "Basic Nature and Properties of Inhomogemeous Antireflection Coatings," *SPIE* **2046**, 69-77 (1993).
23. R. R. Willey: "Optical Monitoring Scheme for Narrow Bandpass Filters," Optical Society of America Thin Films Meeting, April 1988, Tucson.
24. R.R. Willey: "Rugate broadband antireflection coating design," in *Current Developments in Optical Engineering and Commercial Optics*, ed. by R. E. Fischer et al., *SPIE* **1168**, 224-228 (1989).
25. R. R. Willey: "Another Viewpoint on Antireflection Coating Design," in *Optical Systems for Space and Defense*, ed. by A.H. Lettington, *SPIE* **1191**, 181-185 (1989).
26. R. R. Willey: "Survey of the State of the Art of Automation of Optical Coaters," *Proc. Soc. Vac. Coaters* **29**, 262-272 (1986).
27. D. E. Morton: "Automated vacuum coater utilizing optical monitoring," (A) *JOSA* **71**, 1575 (1981).
28. F. T. Goldstein, *FilmStar*, FTG Software Associates, P.O. Box 579, Princeton, NJ 08542.
29. F. T. Goldstein, "Constrained optimization with user-specified functions," Paper WG1 at the Optical Interference Coatings Topical Meeting, June 7-12, 1998, Tucson, AR.
30. R. R. Willey and D. E. Machado, "The variation of band edge position with errors in monitoring layer termination level for long- and short-wave pass filters," *Appl. Opt.* **38**, 5447-5452 (1999).
31. R. R. Willey, "Estimating the Reflection Losses in the Passband of Edge Filters," *Proc. Soc. Vac. Coaters* **42**, 290-294 (1999).
32. D. E. Aspnes: "The Accurate Determination of Optical Properties by Ellipsometry," *Handbook of Optical Constants of Solids*, Ed. E. D. Palik, 89-112 (Academic Press, Inc., Orlando, 1985).
33. J. Rivory: "Ellipsomtric Measurments," *Thin Films for Optical Systems*, Ed. F. R. Flory, 299-328 (Marcel Dekker, Inc., New York, 1995).
34. R. P. Netterfield, P. J. Martin, W. G. Sainty, R. M. Duffy, and C. G. Pacey: "Characterization of growing thin films by *in situ* ellipsometry, spectral reflectance and transmittance measurements, and ion-scattering spectroscopy," *Rev. Sci. Instrum.* **56**, 1995-2003 (1985).
35. J. A. Woollam Company, Inc., 645 M Street, Suite 102 Lincoln, NE 68508 USA.

36. R. P. Netterfield, P. J. Martin, and Kinder: "Real-Time Monitoring of Optical Properties and Stress in Thin Films," *Proc. Soc. Vac. Coaters* **36**, 41-43 (1993).
37. J. Struempfel, C. Melde, E. Reinhold and J. Richter: "*In-Situ* Optical Measurements of Transmittance, and Reflectance by Ellipsometry on Glass, Strips and Webs in Large Area Coating Plants," *Proc. Soc. Vac. Coaters* **42**, 280-285 (1999).
38. M. Vergöhl, N. Malkomes, T. Matthée, G. Bräuer, U. Richter, F.-W. Nickolc, and J. Bruch: "*In-situ* Spectroscopic Ellipsometry and Plasma Control for Large-Scale Magnetron Sputter Deposition Processes," *Proc. Soc. Vac. Coaters* **43**, 11-16 (2000).
39. M. Vergöhl, B. Hunsche, N. Malkomes, T. Matthée, and B. Syszka: " Stabilization of high-deposition-rate reactive magnetron sputtering of oxides by *in situ* spectroscopic ellipsometry and plasma diagnostics," *J. Vac. Sci. Technol. A* **18**, 1709-1712 (2000).
40. R. M. A. Azzam, M. Elshazly-Zaghloul, and N. M. Bashara: Combined reflection and transmission thin-film ellipsometry: a unified linear analysis," *Appl. Opt.* **14**, 1652-1663 (1975).
41. K. Veedam and S. Y. Kim: "Simultaneous determination of refractive index, its dispersion and depth-profile of magnesium oxide thin film by spectroscopic ellipsometry," *Appl. Opt.* **28**, 2691-2694 (1989).
42. R. M. A. Azzam, I. M. Elminyawi, and A. M. El-Saba: "General analysis and optimization of the four-detector photopolarimeter," *J. Opt. Soc. Am. A* **5**, 681-689 (1988).
43. R. M. A. Azzam and K. A. Giardina: "Achieving a given reflection for unpolarized light by controlling the incidence angle and the thickness of a transparent thin film on an absorbing substrate: application to energy equipartition in the four-detector photopolarimeter," *Appl. Opt.* **31**, 935-942 (1992).
44. E. Masetti, M. Montecchi, R. Larciprete, and S. Cozzi: "*In situ* monitoring of film deposition with an ellipsometer based on a four-detector photopolarimeter," *Appl. Opt.* **35**, 5626-5629 (1996).
45. W. G. Sainty, W. D. McFall, D. R. McKenzie, and Y. Yin: "Time-dependent phenomena in plasma-assisted chemical vapor deposition of rugate optical films," *Appl. Opt.* **34**, 5659-5672 (1995).
46. M. Schubert, B. Reinländer, E. Franke, H. Neumann, T. E. Tiwald, J. A. Woollam, J. Hahn, and F. Richter: "Infrared optical properties of mixed-phase thin films studied by spectroscopic ellipsometry using boron nitride as an example," *Phys, Rev. B* **56**, 13306-13313 (1997).
47. T. Heitz, A. Hofrichter, P. Bulkin, and B. Drevillon: "Real time control of plasma deposited optical filters by multiwavelength ellipsometry," *J. Vac. Sci. Technol. A* **18**, 1303-1307 (2000).
48. I. Powell, J. C. M. Zwinkels, and A. R. Robertson: "Development of optical monitor for control of thin-film deposition," *Appl. Opt.* **25**, 3645-3652 (1986).
49. B. T. Sullivan and J. A. Dobrowolski: "Deposition error compensation for optical multilayer coatings. I. Theroretical description," *Appl. Opt.* **31**, 3821-3835 (1992).
50. B. T. Sullivan and J. A. Dobrowolski: "Deposition error compensation for optical multilayer coatings. II. Experimental results-sputtering system," *Appl. Opt.* **32**, 2351-2360 (1993).

51. M. Banning: "Practical methods of making and using multilayer filters," *J. Opt. Soc. Am.* **37**, 792-797 (1947).
52. H. D. Polster: "A symmetrical all-dielectric interference filter," *J. Opt. Soc. Am.* **42**, 21-24 (1952).
53. F. Q. Zhou, M. Zhou, and J. J. Pan, "optical coating computer simulation of narrow bandpass filters for dense wavelength division multiplexing," *Optical Interference Coatings Conference, Technical Digest Series* **9**, 223-225 (1998).
54. H. A. Macleod, "Turning value monitoring of narrow-band all-dielectric thin-film optical filters," *Optica Acta* **19**, 1-28 (1972).
55. H. A. Macleod and D. Richmond, The effects of errors on the optical monitoring of narrow-band all-dielectric thin film optical filters," *Optica Acta* **21**, 429-443 (1974).
56. L. E. Regalado and R. Garcia-Llamas, "Method for calculating optical coating error stabilities," *Appl. Opt.* **32**, 5677-5682 (1993).
57. R. R. Willey, "Estimating the Properties of DWDM Filters Before Designing and Their Error Sensitivity and Compensation Effects in Production," *Proc. Soc. Vac. Coaters* **44**, 262-266 (2001).
58. R. R. Willey, "Achieving Narrow Bandpass Filters Which Meet the Performance Required for DWDM," *Thin Solid Films* **398-399**, 1-9 (2001).

Appendix

Metallic and Semiconductor Material Graphs

A.1. INTRODUCTION

In Sec. 1.5.1, we discussed the design of coatings with absorbing materials. Some potentially useful graphical data is provided in this appendix for a selection of materials that are more commonly used for such designs. There are three types of plots provided: 1) Apfel Triangle Diagrams, 2) ***n*** and ***k*** of the opaque points plotted versus wavelength on the reflectance diagram format, and 3) reflectance diagrams at a given wavelength of the nature of the "spider" diagrams for dielectric materials. The latter can be used in the same way as spider diagrams for "back of the envelope" designs and/or better visualization and understanding of coatings using these materials.

Table 1 lists the materials in this appendix and the available graphs. The figures are numbered from A.1 through A.12 with a suffix of **T, N,** or **R** indicating Triangle, Index, or Reflectance Diagram plots, such as A.1.T.

Table 1. Materials shown in this appendix and their available graphs.

Fig. Number	Material	Triangle Diagram	Index of Refr.	Refl. Diagram
1	Al	X	X	X
2	Al_2O_3	-	X	-
3	Cr	X	X	2X
4	Ge	X	X	X
5	Au (gold)	X	X	X
6	InOx	-	X	-
7	ITO	-	X	-
8	Ni	X	X	X
9	Si	X	X	X
10	Ag(silver)	X	X	2X
11	Ta	X	X	-
12	$V_{1\ x}W_xO_2$	-	X	-

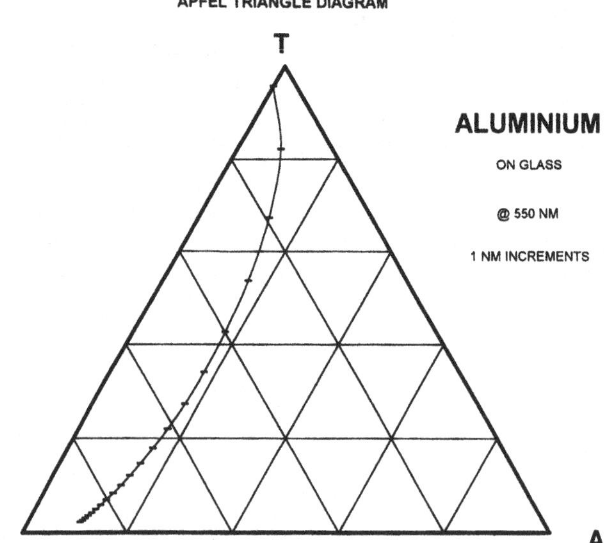

Fig. A.1.T. Triangle diagram for aluminum (Al) at 550 nm.

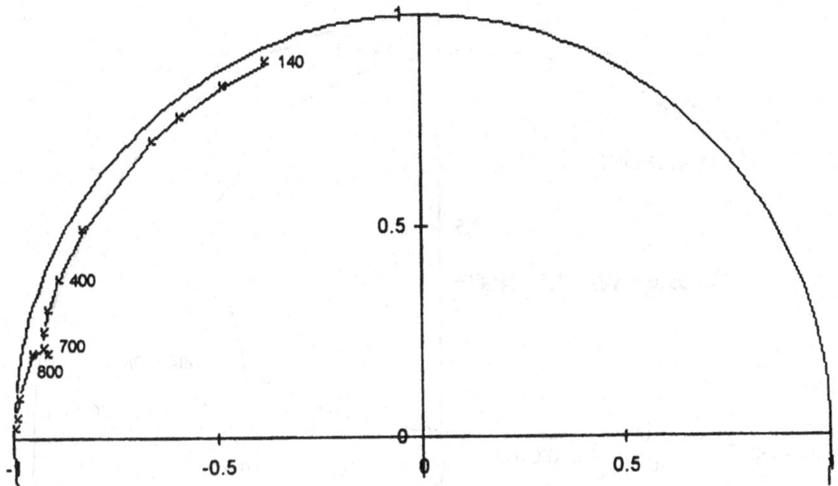

Fig. A.1.N. Index (n and k) of opaque point of aluminum (Al) versus wavelength.

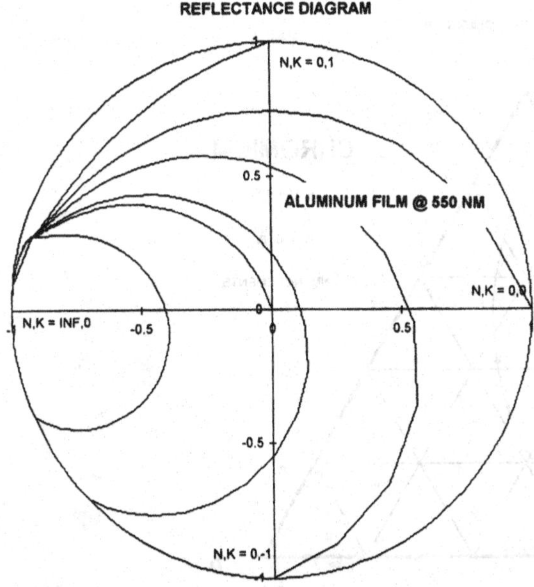

Fig. A.1.R. Reflectance diagram for aluminum (Al) at 550 nm.

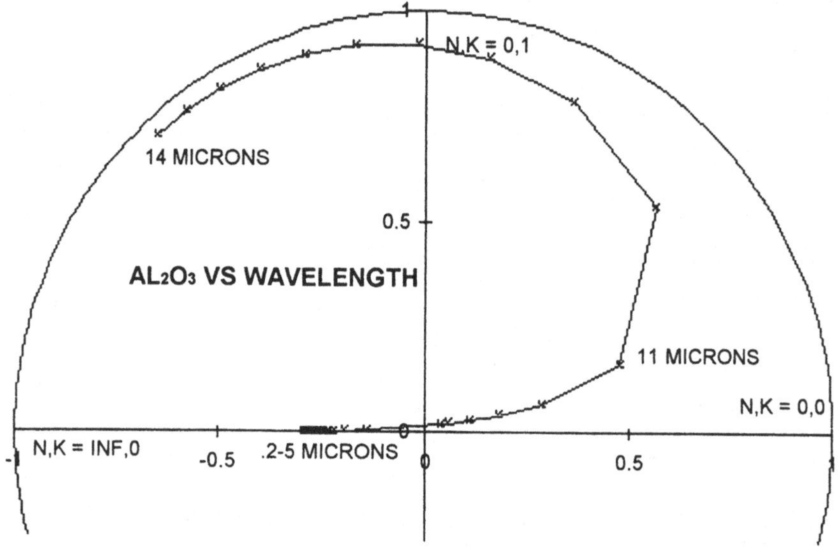

Fig. A.2.N. Index (n and k) of opaque point of alumina (Al_2O_3) versus wavelength.

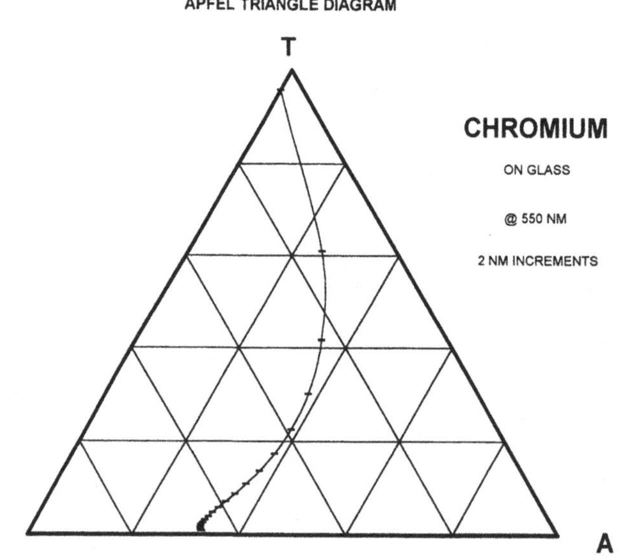

Fig. A.3.T. Triangle diagram for chromium (Cr).

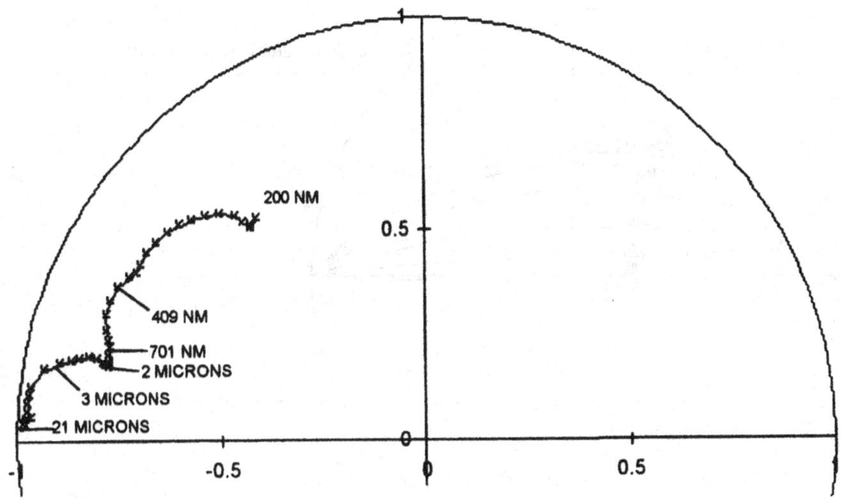

Fig. A.3.N. Index (n and k) of opaque point of chromium (Cr) versus wavelength.

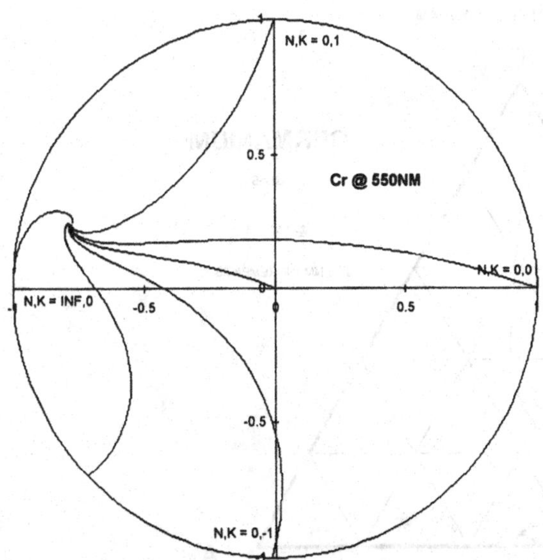

Fig. A.3.Ra. Reflectance diagram for chromium (Cr) at 550 nm.

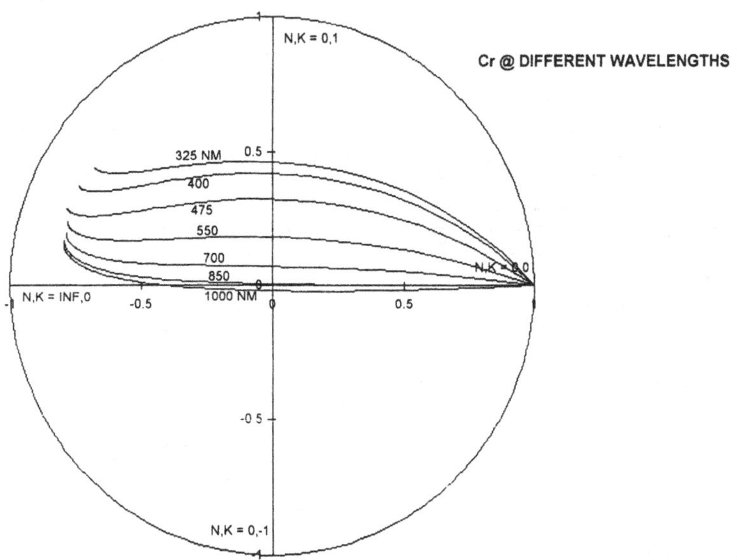

Fig. A.3.Rb. Reflectance diagram for chromium (Cr) at different wavelengths.

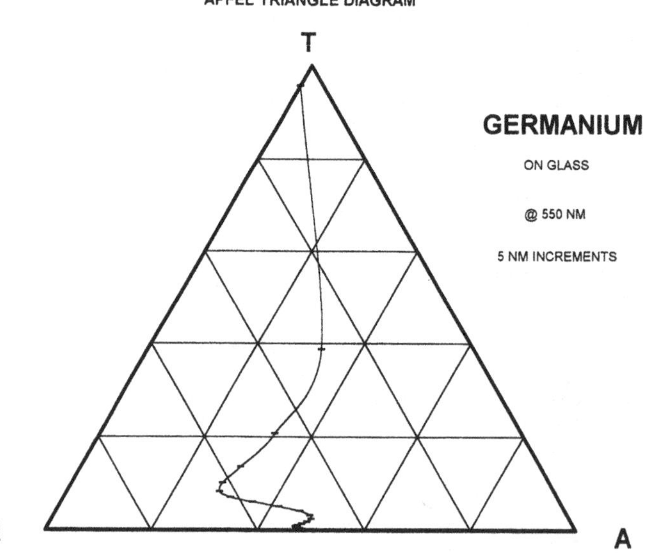

Fig. A.4.T. Triangle diagram for germanium (Ge) at 550 nm.

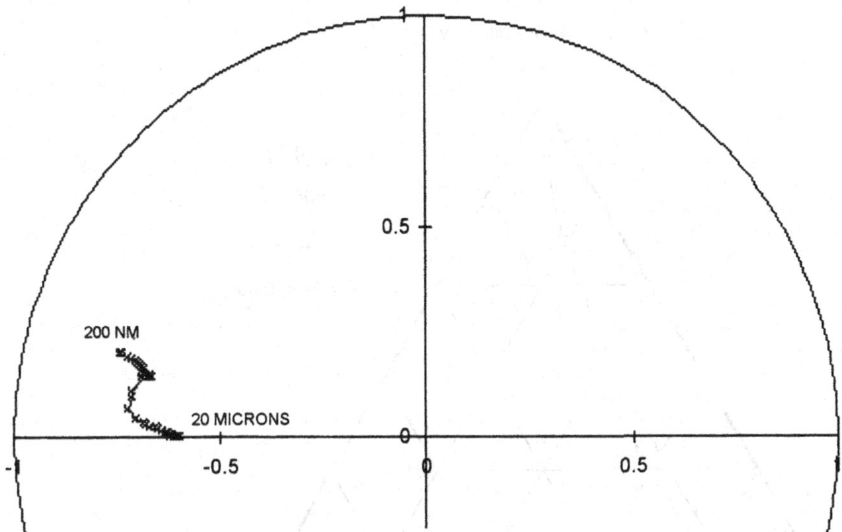

Fig. A.4.N. Index (n and k) of opaque point of germanium (Ge) versus wavelength.

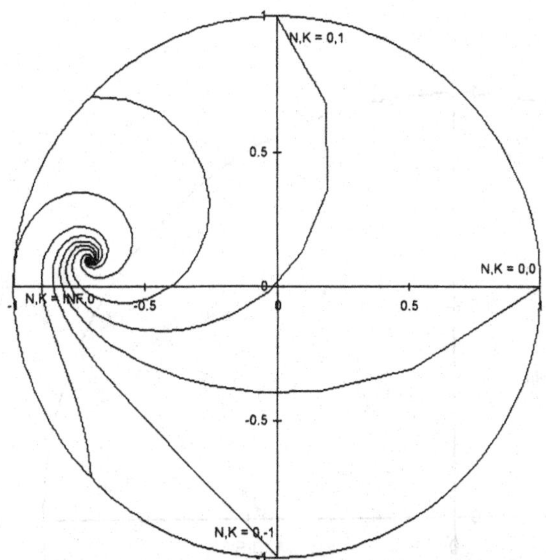

Fig. A.4.R. Reflectance diagram for germanium (Ge) at 550 nm.

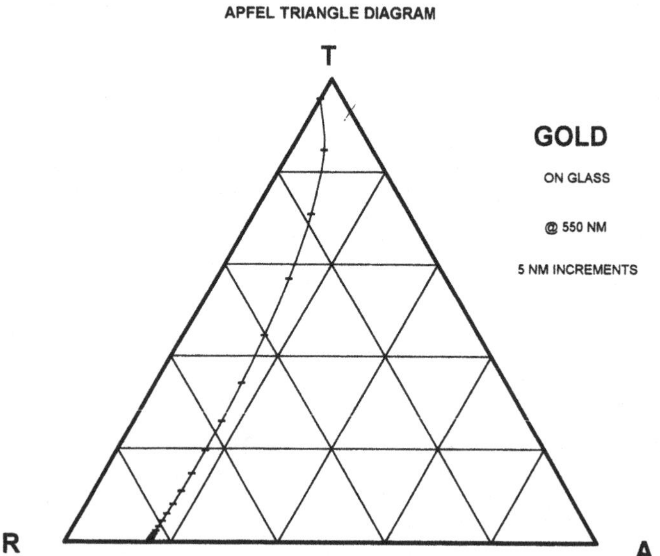

Fig. A.5.T. Triangle diagram for gold (Au) at 550 nm.

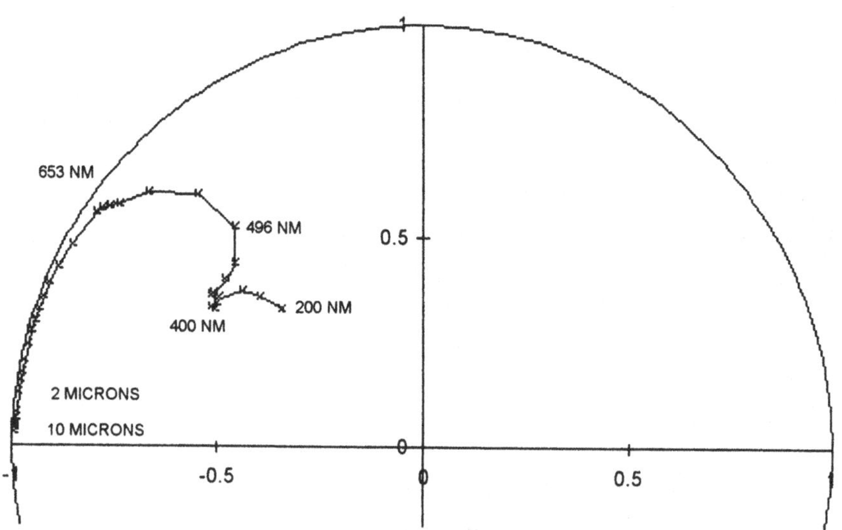

Fig. A.5.N. Index (n and k) of opaque point of gold (Au) versus wavelength.

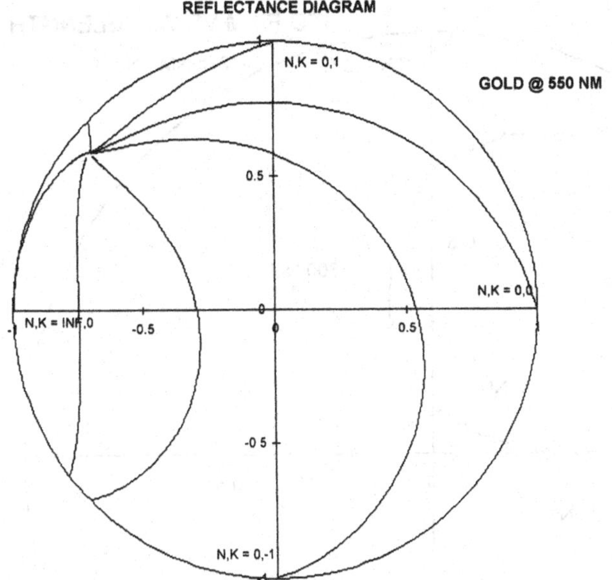

Fig. A.5.R. Reflectance diagram for gold (Au) at 550 nm.

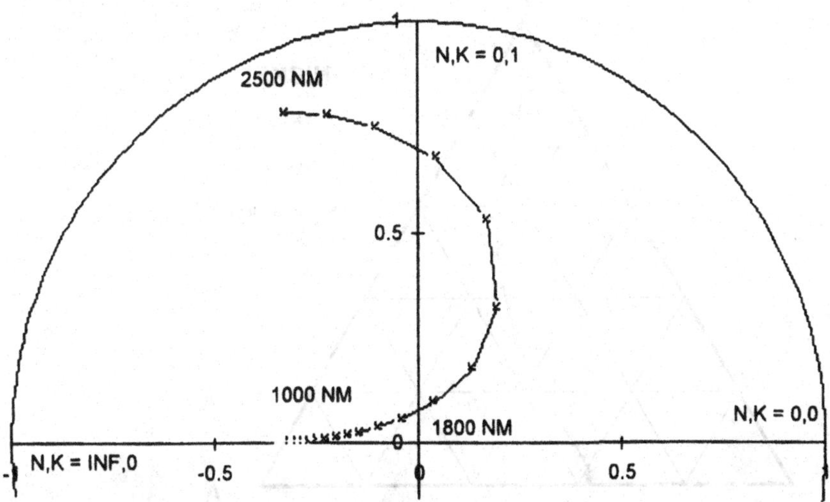

Fig. A.6.N. Index (n and k) of opaque point of indium oxide (InO_X) versus wavelength.

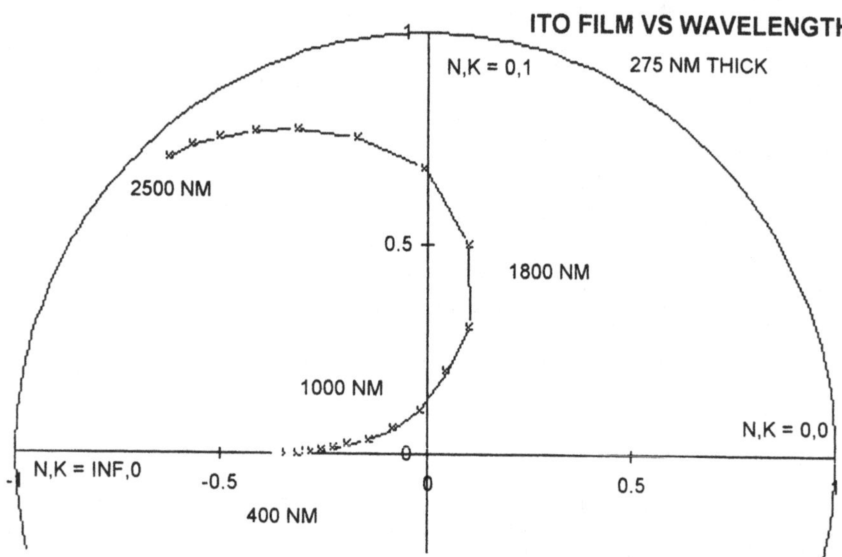

Fig. A.7.N. Index (n and k) of opaque point of indium tin oxide (ITO) versus wavelength.

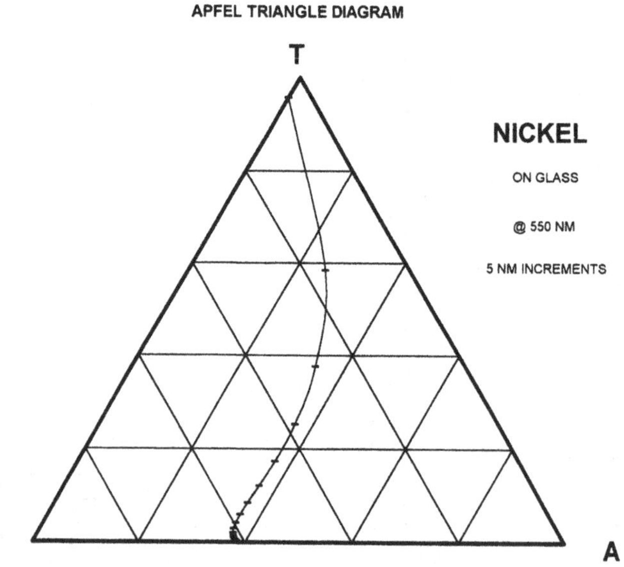

Fig. A.8.T. Triangle diagram for nickel (Ni) at 550 nm.

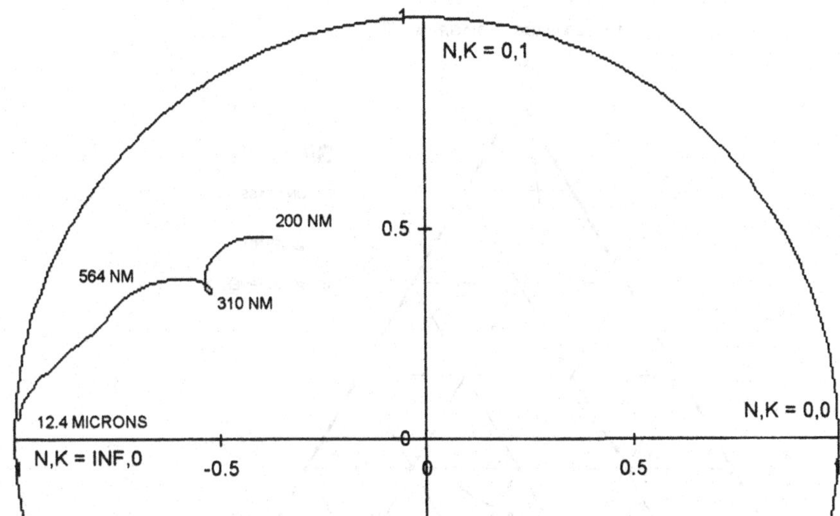

Fig. A.8.N. Index (n and k) of opaque point of nickel (Ni) versus wavelength.

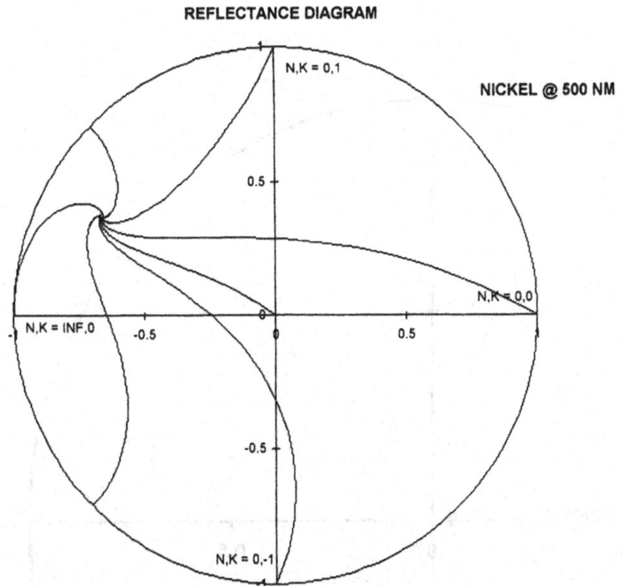

Fig. A.8.R. Reflectance diagram for nickel (Ni) at 550 nm.

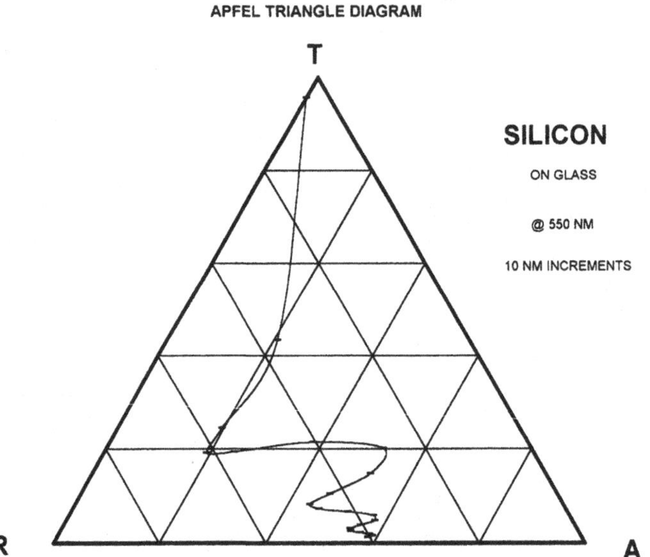

Fig. A.9.T. Triangle diagram for silicon (Si) at 550 nm.

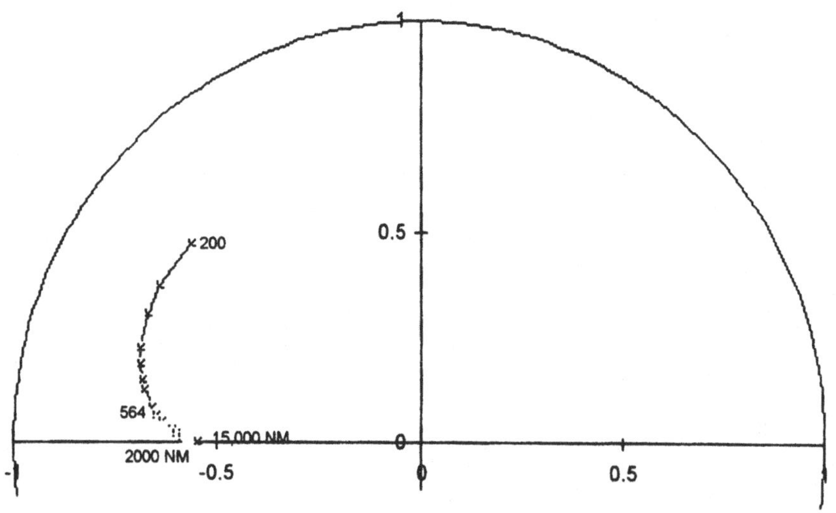

Fig. A.9.N. Index (*n* and *k*) of opaque point of silicon (Si) versus wavelength.

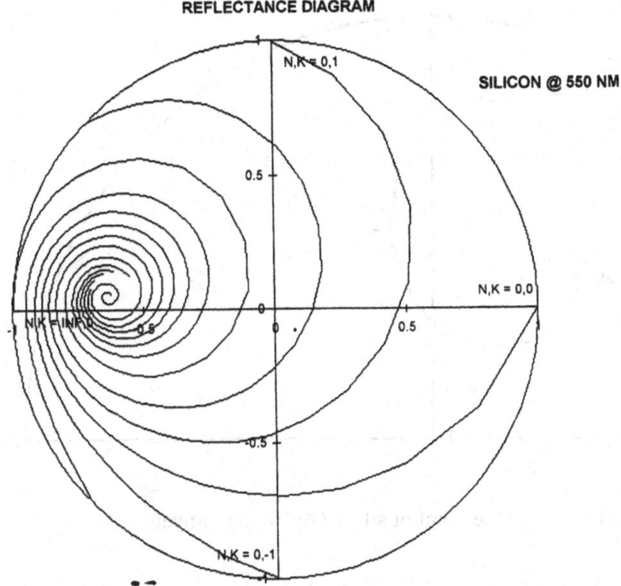

Fig. A.9.R. Reflectance diagram for silicon (Si) at 550 nm.

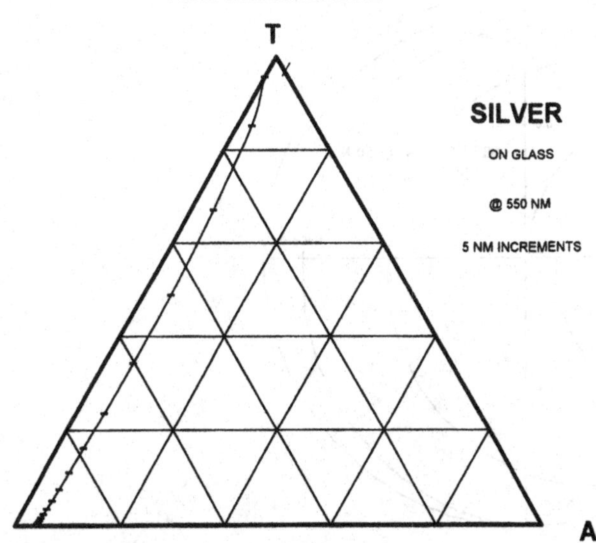

Fig. A.10.T. Triangle diagram for silver (Ag) at 550 nm.

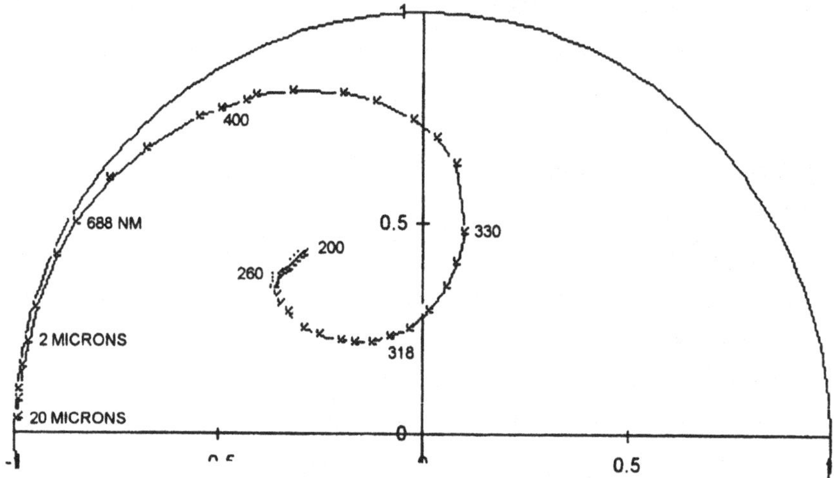

Fig. A.10.N. Index (n and k) of opaque point of silver (Ag) versus wavelength.

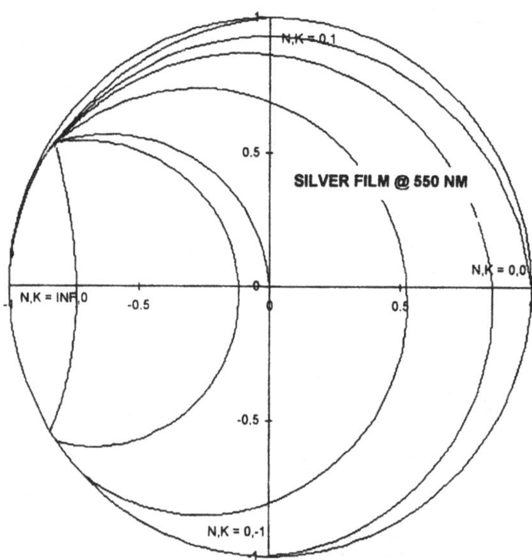

Fig. A.10.Ra. Reflectance diagram for silver (Ag) at 550 nm.

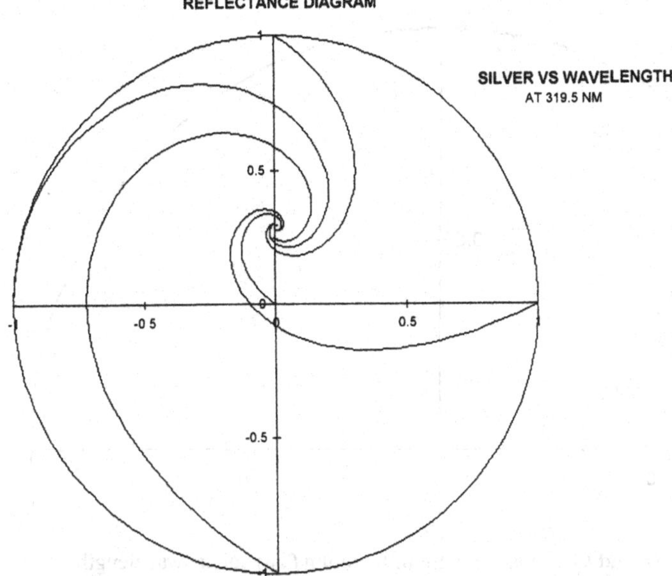

Fig. A.10.Rb. Reflectance diagram for silver (Ag) at 319.5 nm.

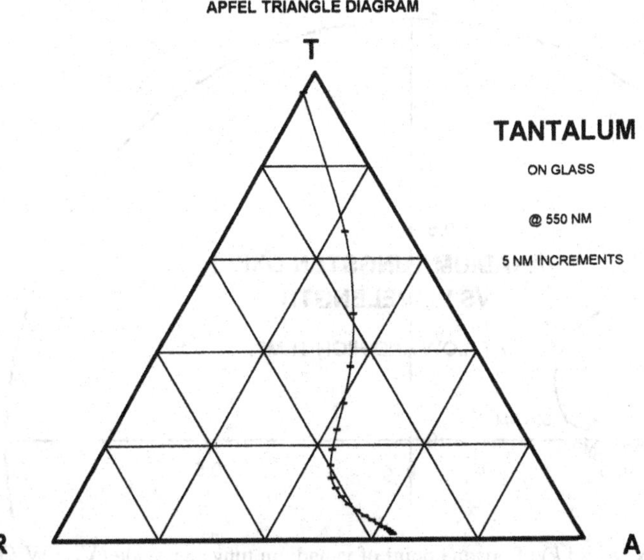

Fig. A.11.T. Triangle diagram for tantalum (Ta) at 550 nm.

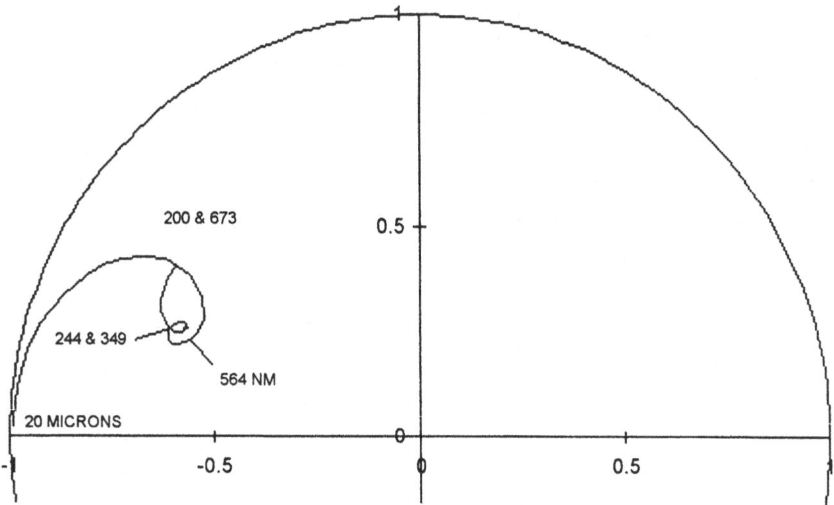

Fig. A.11.N. Index (n and k) of opaque point of tantalum (Ta) versus wavelength.

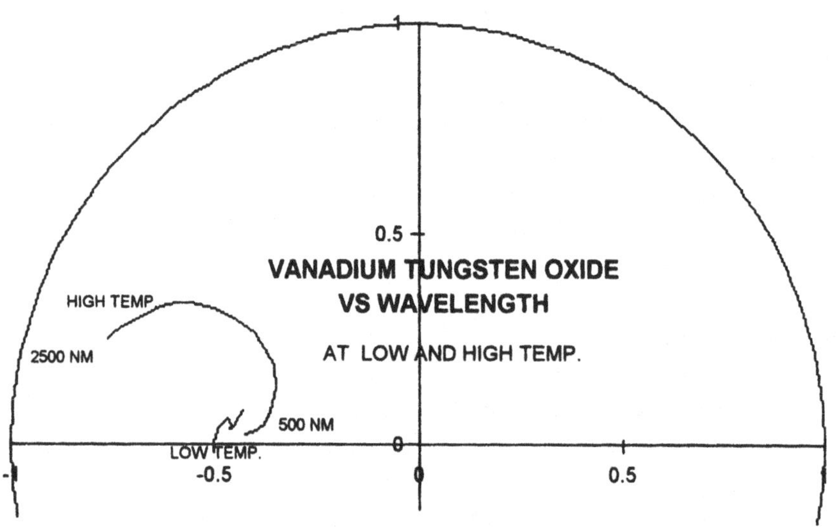

Fig. A.12.N. Index (n and k) of opaque point of vanadium tungsten oxide ($V_{1-x}W_xO_2$) versus wavelength at high and low temperatures (see page 47).

Author Index

Subject Index

Printed in the United States
by Baker & Taylor Publisher Services